Nutri-Cereals

The term "Nutri-Cereals" has been dedicated to ten cereals due to their unique nutritional benefits. *Nutri-Cereals: Nutraceutical and Techno-Functional Potential* covers these cereal grains, with each chapter focusing on nutrient composition and bioactive characterization followed by associated bio-functional properties and health benefits. Further, it covers techno-functionality of nutri-cereals including rheological properties, emulsification and foaming potential, gelation behavior, color profile and others which dictate the suitability of cereals in finished products.

Key Features:

- Covers diverse biological and functional features of nutri-cereals to dictate their potential as functional ingredients in value-added products.
- Discusses the nutraceutical potential of ten cereals: sorghum, pearl millet, finger millet, foxtail millet, barnyard millet, kodo millet, little millet, proso millet, black wheat and Amaranthus.
- Explains how these grains are ideal ingredients for gluten free food formulations with enhanced bio- and techno-functional characteristics.

Although many of the nutri-cereals have been known for thousands of years, due to their coarse nature and lack of processing they escaped the human diet. Now, thanks to their excellent agro-economic potential and numerous health benefits, they are once again recognized as functional ingredients. Recently, earmarked investment and funding have been observed for valorization of these crops and thus, this book will help academicians to strengthen future investigations.

Nutri-Cereals
Nutraceutical and Techno-Functional Potential

Edited by
Rajan Sharma,
Vikas Nanda
and Savita Sharma

CRC Press
Taylor & Francis Group
Boca Raton London New York

CRC Press is an imprint of the
Taylor & Francis Group, an **informa** business

Designed cover image: © Shutterstock

First edition published 2024
by CRC Press
2385 NW Executive Center Drive, Suite 320, Boca Raton FL 33431

and by CRC Press
4 Park Square, Milton Park, Abingdon, Oxon, OX14 4RN

CRC Press is an imprint of Taylor & Francis Group, LLC

ISBN: 978-1-032-13584-7 (hbk)
ISBN: 978-1-032-16984-2 (pbk)
ISBN: 978-1-003-25127-9 (ebk)

DOI: 10.1201/9781003251279

Typeset in Garamond
by Newgen Publishing UK

Contents

About the Editors

Rajan Sharma, PhD, is Teaching Assistant at Punjab Agricultural University, Ludhiana, India. Dr Sharma holds specialization in the domain of cereal technology. His doctoral dissertation focused on characterization, processing and value addition of Nutri-cereals. He obtained a B.Tech. (Hons) in Food Technology and M.E. Food Technology with merit from University gold medals. He has authored more than 25 peer-reviewed articles, 5 book chapters and 2 feature articles in reputed international and national journals. He is an active participant in international conferences, seminars and workshops. To his credit, he has won two prestigious awards – IFI Best Feature Article Award 2016, by the Association of Food Scientists and Technologists (India) and Outstanding Paper 2019 in Emerald Literati Awards.

Vikas Nanda, PhD, is Professor cum Head, Department of Food Engineering and Technology, Sant Longowal Institute of Engineering and Technology, Longowal, India. He is keenly involved in research and teaching activities over the last 20 years. His major areas of interest include the following: development of extrudates with nutraceutical properties; spray dried honey powder as a functional ingredient in value-added products; quantification of sugars from domestic honey variants; and extraction of bioactive compounds from food products using chromatographic techniques. He has attended and chaired several international conferences in India, Greece, France, Czech Republic, Argentina, Dubai and the Netherlands. He is co-chairperson of the International Honey Commission. He has supervised several graduate students and published more than 95 research papers and 30 book chapters in internationally known journals and books. He is also the recipient of several awards and distinctions dedicated to outstanding research work.

Savita Sharma, PhD, is Principal Food Technologist (Dough Rheology) cum Head of the Department of Food Science and Technology at Punjab Agricultural University, Ludhiana, India. She has contributed more than 32 years of research, teaching and extension activities related to cereal science and technology. She has been working to develop novel technologies for cereal products such as functional foods and convenience products in addition to extensive characterization and utilization of bioprocessed grains, including cereals and pulses. She has supervised more than 30 MSc and 13 PhD scholars. She has completed 8 research projects funded by national agencies. She has authored 6 books and more than 150 peer-reviewed articles in renowned journals. She has attended and chaired several technical sessions in international conferences, conventions and seminars and is the recipient of several prestigious awards and distinctions, including Fellow of National Academy of Dairy Science (India).

Contributors

Ranjana Acharya
Govind Ballabh Pant University of
  Agriculture and Technology
Pantnagar, Uttarakhand, India

Uday S. Annapure
Institute of Chemical Technology
Mumbai, Maharashtra, India
Institute of Chemical Technology,
  Marathwada campus
Jalna, India

V. Baskaran
CSIR-Central Food Technological Research
  Institute,
Mysuru, Karnataka, India

Shayani Bose
Govind Ballabh Pant University of
  Agriculture and Technology
Pantnagar, Uttarakhand, India

Anil Kumar Chauhan
Banaras Hindu University
Varanasi, India

B. N. Dar
Islamic University of Science and
  Technology
Awantipora, J&K, India

Rahel Suchintita Das
University College Cork
Cork, Ireland

Gunjana Deka
Mizoram University
Aizawl, India

Subhamoy Dhua
Tezpur University
Assam, India

Ankita Dobhal
Govind Ballabh Pant University of
  Agriculture and Technology
Pantnagar, Uttarakhand, India

Anuradha Dutta
Govind Ballabh Pant University of
  Agriculture and Technology
Pantnagar, Uttarakhand, India

Himjyoti Dutta
Mizoram University
Aizawl, India

Gunaseelan Eazhumalai
Institute of Chemical Technology
Mumbai, Maharashtra, India

P. S. Gaikwad
National Institute of Food Technology,
  Entrepreneurship and Management,
  Thanjavur (NIFTEM-T), MOFPI, GOI,
Thanjavur, Tamil Nadu, India

Jyoti Goyat
GD Goenka University
Gurgaon, India

Hardeep Singh Gujral
Guru Nanak Dev University
Amritsar, India

Arun Kumar Gupta
Tezpur University
Tezpur, Assam, India
Graphic Era (Deemed to be University)
Dehradun, Uttarakhand, India

Rachna Gupta
National Institute of Food
　Technology Entrepreneurship and
　Management
Kundli, Sonipat, Haryana, India

Arunima S. H.
Lovely Professional University
Punjab, India

Monjurul Hoque
Teagasc Food Research Centre
Dublin, Ireland
University College Dublin
Belfield, Dublin (Ireland)

Deepa Joshi
Govind Ballabh Pant University of
　Agriculture and Technology
Pantnagar, Uttarakhand, India

Swati Joshi
Punjab Agricultural University
Ludhiana, Punjab, India

Sandhya K.
National Institute of Food Technology,
　Entrepreneurship and Management,
　Thanjavur (NIFTEM-T), MOFPI, GOI,
Thanjavur, Tamil Nadu, India

A. Kalaiselvan
Tamil Nadu Agricultural University
Madurai – Tamil Nadu, India

Ranjitha Gracy T. Kalaivendan
Institute of Chemical Technology
Mumbai, Maharashtra, India

L. Karpagapandi
Tamil Nadu Agricultural University
Thirupathisaram – Kanyakumari District,
　Tamil Nadu, India

Sawinder Kaur
Lovely Professional University
Punjab, India

Bababode Adesegun Kehinde
Amity University
Jaipur, Rajasthan, India

Gopika S. Kumar
Lovely Professional University
Punjab, India

R. Prasanth Kumar
National Institute of Food Technology
　Entrepreneurship and Management
Kundli, Sonipat, Haryana, India

Rushda Malik
Govind Ballabh Pant University of
　Agriculture and Technology
Pantnagar, Uttarakhand, India

Murlidhar Meghwal
National Institute of Food Technology
　Entrepreneurship and Management
Kundli, Sonipat, Haryana, India

Karine Sayuri Lima Miki
Federal University of Fronteira Sul
Laranjeiras do Sul, PR, Brazil

Anusha Mishra
Institute of Chemical Technology
Mumbai, Maharashtra, India

Poonam Mishra
Tezpur University
Tezpur, Assam, India

Vikas Nanda
Sant Longowal Institute of Engineering
　and Technology
Longowal, Punjab, India

P. Prabhasankar
CSIR-Central Food Technological Research
Institute,
Mysore, India

Shafiya Rafiq
University of Kentucky
Lexington, KY

Rajeev Ranjan
Banaras Hindu University
Varanasi, India

Prasad Rasane
Lovely Professional University
Punjab, India

Muzamil A. Rather
Tezpur University
Tezpur, Assam, India

Nalla Bhanu Prakash Reddy
National Institute of Food Technology,
Entrepreneurship and Management,
Thanjavur (NIFTEM-T), MOFPI, GOI,
Thanjavur, Tamil Nadu, India

Rajan Sharma
Punjab Agricultural University
Ludhiana, Punjab, India

Shivani Sharma
CSIR-Central Food Technological Research
Institute,
Mysore, India

Himani Singh
National Institute of Food Technology
Entrepreneurship and Management
Kundli, Sonipat, Haryana, India

Jyoti Singh
Lovely Professional University
Punjab, India

Pratibha Singh
NIFTEM-T, Liaison Office Bathinda
Bathinda, Punjab, India

Shubhendra Singh
Banaras Hindu University
Varanasi, India

Shweta Suri
Govind Ballabh Pant University of
Agriculture and Technology
Pantnagar, Uttarakhand, India

T. Tamilselvan
CSIR-Central Food Technological Research
Institute
Mysore, India

Barbara Elisabeth Teixeira-Costa
Federal University of Amazonas
Manaus, Brazil

Mamta Thakur
ITM University
Gwalior, Madhya Pradesh, India

Rahul Thakur
National Institute of Technology
Rourkela, India

Nalini Trivedi
Govind Ballabh Pant University of
Agriculture and Technology
Pantnagar, Uttarakhand, India

B. K. Yadav
National Institute of Food Technology,
Entrepreneurship and Management,
Thanjavur
(NIFTEM-T), MOFPI, GOI,
Bathinda, Punjab, India

Preface

Nutri-cereals, a term defined by the Government of India in 2018, has been dedicated to ten cereal grains owing to their agrarian and nutritive excellence, and comprising sorghum, millets and two pseudocereals. They have been appreciated for being substantial crops sustaining the challenges of the present day, such as global warming, decreasing ground water level and food security, ascribing to their suitability to drylands and soils having poor fertility. They have been termed as "cereals of the future" due to their nutraceutical potential and environmental sustainability. Further, these grains possess better compositional and functional characteristics in comparison with conventional cereals such as wheat and rice, which make them ideal candidates for functional ingredients in value added food products. Researchers have indicated that these grains are rich sources of biologically active constituents, also known as phytochemicals, such as polyphenolic compounds, phytosterols, dietary fibre, minerals and vitamins which show nutraceutical potential through preventive mechanisms against several modern lifestyle health complications like cancers, neurodegenerative disorders and cardiovascular diseases. Moreover, biochemical and structural characterization of nutri-cereals suggests that they hold excellent techno-functional properties enhancing the utility of cereals in finished products.

The present book aims to popularize nutritional and agrarian potential of nutri-cereals around the globe. Many of the nutri-cereals have been known for more than 4,000 years, but due to their coarse nature and lack of processing, they escaped for a time from the mainstream human diet. However, because of their excellent agro-economic potential and numerous health benefits, they are now once again recognized as functional ingredients. Recently, earmarked investment and funding have been observed for valorization of these crops and, thus, the present book will help academicians to strengthen future investigations. This book presents phytochemical characterization, techno-functional potential and pharmacological impact of nutri-cereals. Further, processing, valorization and utilization of these grains is also discussed.

1

Nutri-Cereals
Niche to Mainstream

Swati Joshi and Rajan Sharma

Contents

1.1 Introduction

Millets previously denoted as "yesterday's coarse grain and present nutri-cereals" are considered as the future crops. Owing to their excellent nutritional profile, resistance towards most of the diseases and pests and adaptability towards harsh climatic conditions of the semi-arid and arid regions, nutri-cereals are making a wonderful comeback in the grain production and utilization sector in Africa and Asia. Nutri-cereals in India contribute around 40.62 percent to the global millet production.

Nutri-cereals provide energy about 320–370 kcal/100 g of consumption and have a higher proportion of dietary fiber and non-starchy polysaccharides in comparison to staple cereals like wheat and rice. High dietary fiber holds several health benefits like improving blood lipid profile, gastrointestinal health and blood glucose levels. With a low glycemic index

DOI: 10.1201/9781003251279-1

1

and gluten content, millets are excellent options for celiac and diabetic patients (Gowda et al. 2022). The millets are 3–5 times more nutritious than conventional cereals like rice, wheat and maize in terms of fiber, vitamins, proteins, health-promoting phytochemicals such as polyphenols, phytosterols, phytocyanins, phyto-oestrogens, lignins and minerals (phosphorus, manganese, calcium potassium, magnesium and iron) with an additional advantage of being gluten-free; hence, they are referred as "superfoods." The nutri-millets are therefore a viable remedy to overcome the rising instances of metabolic disorders and malnutrition. Millets are a good source of slowly digestible fiber, resistant starch and total dietary fiber, and therefore provide a sustained glucose release and satiety providing multiple benefits like reducing the incidence of diabetes, cardiovascular diseases and gastrointestinal problems and manage hyperglycemia. This makes millets a perfect food for the diabetic population and a potential modern diet.

Although nutri-cereals have a high nutritive profile, their utilization as food, declined at the time of the urbanization and green revolution that shifted attention towards major cereals like rice, wheat and towards mass relocation to urban areas. Post the green revolution era, high production of major cereals like rice, wheat, and so forth shifted the focus of scientists, industrialists and food-based industries as well as consumers towards development of fine cereal food and food products. Such food patterns in no time changed the mindset of consumers and, as food, nutri cereals became totally neglected. Further, drastic urbanization made the people switch to processed and convenient wheat-rice based value-added food products.

Despite India being first in nutri-cereals production globally and second in pulses and rice, it ranks second in malnutrition cases among children. India is home to more than one-third of the world's malnourished children (Gowda et al. 2022). Also, the consumption of hyperglycemic, low fibre foods have triggered the country to become a hub for an overweight and diabetic population. Recent researches regarding a rise in the instances of malnutrition and other metabolic disorders have again highlighted the need of nutri-cereals in food, slowly fueling their consumption. In the last few years, when the incidences of metabolic disorders like obesity, hypertension rose, irrigated land availability became low, and millets that were best suited for inter and mixed cropping and required less water and tolerated extreme temperatures offered food security and better land-use options to farmers as well as consumers (Oelke et al. 1990). In addition, all the global bodies are promoting millets' production with the thought that these might reduce carbon footprints related with agriculture along with ensuring food and nutritional security providing them the title of "miracle-crops" because of their multiple uses.

In order to mainstream the nutri-cereals into the diets of common people, a committee was constituted by the Indian Central Government for inclusion of millets in the Public Distribution System and to reach to all strata of the population. Owing to millets high nutritive value, stress resistance and excellent antidiabetic nature, the Government declared the food category as Major Millets, comprising Finger Millet (Ragi), Pearl Millet (Bajra), Sorghum (Jowar), Minor Millets, including Proso Millet (Cheena), Barnyard Millet (Sawa/Sanwa/ Jhangora), Foxtail Millet (Kangani/Kakun), Little Millet (Kutki), Kodo Millet (Kodo) and pseudo millets like Blackwheat (Kuttu) and Ameranthus (Chaulai) as "NutriCereals."

1.2 Impact of green revolution on nutri-cereals and their major comeback

The Consultative Group related with International Agriculture Research (CGIAR) reported that millets may play a crucial role in the developing nations, where nutrient and food security are the major issues (Sagar Maitra 2020). In the later years of green revolution, after facing declines in the popularity of nutri cereals as food, the demand of nutri-cereals rose again owing to the following reasons.

- The wheat–rice cropping system is known to be a water-guzzler, and agriculture specialists demanded suitable inter- or mixed cropping systems to rehabilitate water productivity and sustainable agriculture for the future. The irrigated lands in the Asian countries are overexploited causing a rise in the salinization, plateauing of yield, degradation of natural resources and depletion in the levels of ground water. The land replenishment to attain increased productivity is more required in rainfed lands than irrigated areas as they can handle various limitations like washed and poor soil. Therefore, farmers and smallholders are coping with the ill effects of climate change, global warming, malnutrition and poverty.
- Currently, agriculture is dealing with issues related to global warming and climate change. The major consequences of climate change are uncertainties in the rainfall, rise in the temperature, and a boost in greenhouse gas emissions, with the major portion contributed by carbon-dioxide. Nutri-cereals are eco-friendly crops and regarded as climate-smart since they can withstand stresses like higher temperatures and drought conditions, and generate minimal carbon footprints.
- Being C_4 plants, millets can utilize the rising environmental CO_2 and convert it into harmless biomass (Brahmachari et al. 2018). Millets fall under the category of C_4 cereals that intake more CO_2 from the atmosphere, have better water usage efficiency, demand low input, making them an eco-friendly crop that aids in mitigating climate change.
- Being an ecologically hardy crop with greater storability, millets easily overcome the uncertainties of monsoon that agriculture in India suffers and, hence, denoted as "famine reserves" (Passi and Jain 2014). A large number of farmers in the developing countries are now switching to millets cultivation due to their rising importance and demands with respect to dietary and nutraceutical uses.
- Intercropping has been reported as quite advantageous in several ways as it provides better resource utilization, combating the undesired growth of harmful biotic agents, ensuring soil health and better sustainability and production in the agriculture system (Maitra and Shankar 2019). Several crops are grown together on one land for better utilization of the soil nutrients, atmospheric CO2, sunlight and moisture. Intercropping also improves soil health and prevents surface run-off, nutrient loss and soil erosion from the upper fertile soil. It has been reported that intercropping facilitates soil fertility when nutri-cereals were intercropped with legumes and pulses and enabled growth of beneficial microflora in the soil. Foxtail millet have been reported to show outstanding results in the intercropping patterns (Maitra et al. 2020). When dealing with dryland conditions, intercropping provides a natural security against total crop failure and therefore supports sustainability of agriculture even if the major crops fail. Studies have denoted that intercropping of black gram or green gram with fox-tail millet in the ratio 3:3 is quite beneficial (AICRPSM 2017). In the state of Andhra Pradesh, Indian intercropping system with ground nut + foxtail millet (1:2), foxtail millet and pigeon pea (5:1) and cotton + foxtail millet (1:5) are quite common.
- Amaranth has the capacity to adapt and grow in harsh temperature and weather conditions (Olufolaji and Ojo et al. 2010). Studies have reported that it can be easily cultivated for leaves and grains in harsh seasons and climatic conditions where other crops can not survive (Grundya et al. 2020). Therefore, the development of value chains for utilization of amaranth grains from cultivation in the field to manufacture

of several value-added products, and exploiting its marvelous health benefits could be an excellent way to uplift the majority of rural agriculture, nurturing food-based industries as well as developing countries where malnutrition and deficiency disorders are glaring with their devastating health consequences.

- The demand for food grows proportionally with the rise in global population. At present about half of the world's total calories are accounted for from cereals – the majority wheat, rice and maize (Kumar et al. 2018). Studies conducted by Verge et al. (2007) reported that the overcultivation of crops that have high water requirements like cotton, sugarcane and rice, has caused a severe decline in the ground water levels. An increase of 0.009 percent in the distance between the ground water table and ground level, equivalent to a loss of 71,911 of ground water/hectare has been reported. Since, the world is already facing the consequences of overexploitation of lands, increasing dry lands and depletion of the ground water, a much less possibility of elevating the production staple grains is visible.
- An alarming note has been sounded by the National Rainfed Area Authority (NRAA) that, even if the current irrigation potential is utilized to its full potential, about 50 percent of the net sown area would continue to remain only rainfed. This forced the agriculturists and farmers to promote a shift to the alternative grains from conventional cereal staples. Cultivation of nutri-cereals can therefore be a solution since they can easily be sustained on shallow and low fertile soils and within the wide pH range of soil between acidic (4.5) to basic soils (8.0). Staple cereals like rice is highly sensitive to soils with high salinity; retarded growth and yield were observed on soil with salinity above 3dS/m. On the contrary, finger millet and pearl millet can grow up to a higher soil salinity of about 11–12 dS/m and have very low water requirements during the overall growth. The minimum requirement of rainfall for proso millet and pearl millet has been reported as low as 20 cm, contrasting against rice, which demands an average rain of around 120–140 cm. In addition, the majority of millets grow and mature in 2–2.5 months after sowing, barnyard millet having the minimal maturation time (45–70 days) among nutri-cereals, which is almost half rice maturation time (120–140 days), minimizing both land and water requirements (Kumar et al. 2018).

Along with above-mentioned points, the scientific interventions with regard to the development of new and high-yielding varieties, sequence information, molecular biomarkers and desirable mutations have led to the development high yielding varieties of nutri-cereals globally. Newly developed varieties are more resistant to disease and have better per hectare production in comparison to their native parent varieties. Pearl millet initially had approximately 140 species and subspecies and further development in the gene bank enhanced its count to 65,400. The International Crops Research Institute for the Semi-Arid Tropics (ICRISAT) owns 33 percent of the pearl millet gene bank accessions in the world, making it one of the largest primary global collection of pearl millet. The Bureau for Plant Genetic Resources, India, holds approximately 27 percent of the global finger millet gene accessions. The Institute of Crop Germplasm Resources (ICGR), China, maintains 56 percent of the world's accessions of foxtail millet (*Setaria italica*) (Kumar et al 2018). The improvement in varieties has also paved a way towards the development in the post-harvest processing and invention of millet-specific decorticators, threshers, polishers and destoners. In previous years labour-intensive conventional methods like pounding, winnowing, and so forth, were utilized for the decortication and milling of millets, limiting the production, sale and distribution of edible millets at large scale. Therefore, the recent developments in varietal improvement and post-harvest operations of millets have cleared the way for utilization of nutri-cereals in the production of food products and provided a sustainable solution to the existing agrarian problems. Also, their efficient utilization and can provide a milestone in attaining the end of malnutrition by 2030 as committed to by the United Nations.

1.3 Conventional utilization of nutri-cereals in the early decades

"A rice eater is always weightless like bird, a jowar eater is as strong as a wolf and a ragi eater will always be disease-free." History and the early dietary patterns reveal that the nutri-cereals, one of the ancient food crops that humans discovered have been under cultivation for thousands of years. Historically, nutri-cereals (millets) were the first known cereal used for food purposes because nutri-cereals, previously known as millets, have been among the most underutilized crops in the Asian and African countries. Evidence has been found of their consumption in the Middle Ages by Romans and Gauls in the form of porridges (Ambati and Sucharitha 2019). Only small- or home-scale utilization of nutri-cereals in the early years for food and animal feed have been observed.

With a rise in demand for cheap and nutrient dense animal-feed ingredients, the demand for millets also increased. Complete dependence on raw materials such as maize is not possible, since the human demand for such food products is also very high. Hence, there is an urgent need for alternative crops rich in energy and protein. Utilization of millets for feed and fodder purposes can reduce the burden on production of staple cereals. Using major and minor millets not only serves as an energy source in animal feeds but also fulfills the demand of protein recommended dietary allowance (RDA). Amaranth, owing to its protein and energy balance, has an immense potential to complement or substitute for any of the staple cereals (Maitri et al. 2020). The energy gained by amaranth grain consumption is similar to other cereals, but the protein content is two times higher than cereals. Amaranth grain leaves have been used globally as a silage crop and forage for animals, including chickens, cattle, rabbits and pigs (Peiretti 2018).

Kimata et al. (2000) reported that the foxtail millet had been utilized in several food preparations in the early centuries like boiled grain *(anna)*, unleavened bread *(roti)*, porridge (sankati) and thin gruel (ganji or peja) in the states of India.

Mageu (amahewu, mahewu or magou) is a traditional non-alcoholic fermented beverage widely consumed in Zimbabwe and South Africa. It is prepared by steaming any thin porridge and contains solid content around 8–10 percent. The lactic acid fermented product had a sour and refreshing taste (Taylor and Duodo 2011). *Malwa,* a Ugandan original is a fermented beverage derived from finger millet and is still being widely consumed at household level in the north-eastern and northern regions (Muyanja et al. 2010). The beverage is ready to consume when a sour taste develops and bubbling ceases (Muyanja et al. 2010). China is well known for its traditional sorghum derived distilled spirits called *baijiu* (Zheng and Han 2016). These spirits have alcohol around 38 percent (v/v) and often higher. *Hulu-mur* is a conventional Sudanese non-alcoholic drink prepared from thin flakes derived from malted and unmalted sorghum flour along with spices (Ibnouf 2012; Baidab et al. 2016). Coarse grains individually or in combination have been utilized to produce different types of beers. *Omuramba, kweete* and African *ajon* were the most common beers produced from sorghum, maize and millet and finger millet respectively (Mwesigye and Okurut 1995).

Nutri-cereal based thin and thick porridges had been used since years with solids content in the range of 10–20 percent (Taylor et al. 2013). A wide range of fermented and unfermented porridges are prepared from millets and sorghum. The fermentation improved the shelf-life of the porridge. Couscous is a granulated and steamed product made from cereal and millet flours. Sorghum rice analogous to boiled rice generally required a much longer cooking time and the texture of cooked grain is chewy and firmer than rice. Also, it has a nutty flavor (Taylor and Duodo 2011). Arguelles-Lopez et al. (2018) prepared an amaranth chia-based beverage utilizing extrusion followed by germination of amaranth grains and chia seeds.

Barahnaja was a conventional mixed cropping system incorporating millet crops along with cereals in the Himalayan region. Being a sustainable crop and with a promised harvest in case of erratic weather. Many of the nutri-cereals have been utilized for the same. The main Kharif crops were grown during April to October including foxtail millet, paddy, and barnyard millet in the down fields and other millets like finger millets in the upward fields. The Rabi crops, which include barley, wheat and oilseeds, are planted in the downward fields. *Barahnaja* offered sustainability since the nutrient requirement of one crop is fulfilled by the other (Gururani et al. 2021). The detailed utilization of millets in the early centuries have been presented in **Table 1.1**.

1.4 Nutritive value

Nutritional content is crucial for development, maintenance of body's metabolism, maximizing the genetic potential and wellness. The nutritive profile of millet is quite comparable to staple cereals and constitute an exceptional source of carbohydrates, protein, phytochemicals, dietary fiber and vitamins. Nutri-cereals provide energy lying between 320–370 Kcal/ 100 g. The nutritive value of nutri-cereals has been presented in **Table 1.2**.

The nutritional content of most popular nutri-cereals is listed below (Gowda et al. 2022):

- **Proso millet**: It contains higher amounts of dietary fiber and minerals as compared to staple cereals. Proso millet is an excellent source of iron (Fe), magnesium (Mg), calcium (Ca), zinc (Zn), potassium (K), phosphorus (P), vitamin B-complex, niacin and folic acid. It contains a majority of essential amino acids except lysine and has a better essential amino acid index than wheat. Food products based on proso millet exhibits quite lower glycemic index than maize or wheat-based products.
- **Pearl millet** Pearl millet has a low amount of carbohydrates in comparison to staple cereals, contains insoluble dietary fibers and amylose rich starch (20–22%) and therefore exhibits a lower glycemic index. It is a gluten-free grain with higher prolamins, making it an excellent option for people with gluten sensitivity. Pearl millet is a rich source of omega-3 fatty acids and essential fatty acids like eicosapentaenoic acid, docosahexaenoic acid and alpha-linolenic acid.
- **Kodo millet** It has higher concentration of vitamins (B6, B-complex, niacin, folic acid, Ca, Fe, K, Mg and Zn) and minerals. Owing to its higher digestibility, it is widely used in product formulation for infants and geriatrics.
- **Foxtail millet** It is nutritionally superior to rice and wheat due to its higher fiber, vitamins, minerals, resistant starch and lysine-methionine ratio highest than most of the cereals. It is also a rich source of stearic acids and linoleic and aids in retaining a good lipid profile.
- **Finger millet** It contains a high amount of slowly digestible starch as well as resistant starch, offering an additional advantage of being a low glycemic food. With 7 percent protein and high lysine, threonine and valine, the amino acid score is better than that of cereals.

Table 1.1 Conventional Utilization

Products	Nutri-cereal	Method of processing	Country/Region	
Non-alcoholic beverages				
Motoho	Sorghum	Lactic acid bacteria fermentation	Lesotho	Gadaga et al. 2013
Kunun-zaki	Sorghum, millets (and maize)	Lactic acid bacteria fermentation; cereal malt may be included; flavored with spices and sweetened with sugar	Northern Nigeria	Akoma et al. 2006
Malwa	Finger millet	Lactic acid bacteria fermentation of a composite of finger millet sourdough and finger millet malt flour	Uganda	Muyanja et al. 2010
Borde	Sorghum, millet, teff (and maize)	Lactic acid bacteria fermentation of a mixture of malted cereals	Ethiopia	Abegaz 2007
Alcoholic beverages				
Cloudy beers	Malted sorghum or sometimes pearl millet or fonio, sometimes plus adjunct, e.g., cassava flour	-	West Africa	Zheng & Han 2016
Spirits (*Baijiu, Kaoliang liquor*)	Sorghum	-	China/Taiwan	Zheng & Han 2016
Wine	Foxtail millet	Semi-opaque alcoholic beverage with a hazy appearance followed by microfiltration	Taiwan	Wang et al. 2006
Ikigage	Sorghum based beverage	*Saccharomyces cerevisiae* and *Issatkenkia orientalis* and *Lactobacillus fermentum* used as culture	Rwanda	Murty & Kumar 1995
Tortillas	sorghum	Grains were decorticated to remove outer bran layers, cooked and steeped prior tortilla preparation	-	Choto et al. 1985
Porridges				
Bogobe, Bosokwane	sorghum	Non-fermented	Botswana	Bolling et al.1981
Couscous	sorghum	Steamed and agglomerated to prepare porridge like product	-	Galiba et al. 1988
Ogi	sorghum	a traditional gruel or porridge prepared from fermented millet	West Africa	Blandino et al. 2003
Ben saalga	semi-solid millet-based gruel	soaking, milling, sieving, fermentation and settling followed by a cooking	Burkina Faso	Songré-Ouattara et al. 2008
Boiled Rice-Like Products	sorghum	The dehulled grain is cooked in water in the proportion of 1: 3.	India	Subramanian 1981
Parboiled millets		Rice many millets are reported to be parboiled in India	-	Kimata et al. 1999

(continued)

Table 1.1 (Continued)

Products	Nutri-cereal	Method of processing	Country/Region	
Mixed cropping *Barahnaja* system	Mandua (Eleusine coracana), Ramdana/ Chaulai (Amaranthus spp), Kuttu/Ugal (Fagopyrum esculentum and F. tataricum), Jowar/Jonyala (Sorghum bicolor)	All crops planted together on the same terraced fields and includes the combination of cereals, pulses and other creeper legumes, vegetables, and root crops	India	Gururani 2021
Baked and fried products				
Breads	Brachiaria ramosa	-	South India	Kimata et al. 2000
Bhakri Suhali Khichri Churma	Pearl millet and sorghum	-	India	Rekha 1997; Chaudhary 1993
Feed and Fodder				
Millet silage for lactating dairy cows	pearl millet	NDF (%) 66.9; ADF (%) 38.8 Soluble protein (%) 68.9 Neutral detergent insoluble protein (% of crude protein) 17.5 Acid detergent insoluble protein (% of crude protein) 8.0 In vitro dry matter disappearance (48 h), % 66.6 In vitro NDF disappearance (48 h), % of NDF 58.3 In situ DM disappearance (48 h), %62.3 In situ NDF disappearance (48 h), % of NDF 52.0	-	Amer & Mustafa 2010

Table 1.2 Nutritional Composition of Different Nutri-Cereals

Nutri-cereals	Major Millets		Minor Millets					Pseudo Millets	
	Finger millet	Pearl millet	Foxtail Millet	Proso Millet	Kodo Millet	Barnyard Millet	Little Millet	Black-wheat	Amaranthus
Macronutrients									
Carbohydrate	86–89	63.0–76	61–74.7	65.45–77	65.6–71.3	51.5–62.7	69.5–78.2	64.8	65.25
Protein (%)	7.5–10.4	8.5–19.3	11.2–12.5	10.45–15	6.0–11.0	11.5–12.6	10.4–13.7	12	13.56
Fat (%)	1.5–1.9	1.6–6.8	3.6–3.8	1.45–3.66	3.5–5.1	2.7–6.1	3.8–4.3	1.2	7.2
Ash (%)	2.9–3.4	2.5–3.6	3.0–3.4	2.0–4.0	3.0–4.5	4.5–5.2	3.1–3.7–3.6	-	2.88
Crude fibre (%)	3.7–4.0	1.4–10	4.0–7.0	2.0–8.0	8.5–11.8	13.9–14.8	4.0–8.0	12	6.7
Micronutrients									
Iron	377–695	50–110	208–386	423–550	1082–1413	301–381	457–515	45mg	7.61
Zinc	23–93	29–31	34–82	74–91	60–75	60–100	40–174	35mg	287
Magnesium	1.3–3.76	1.3–1.36	1.4–3.02	1.95–3.0	2.10–3.01	2.40–3.0	2.33–3.40	-	248
Manganese	55–160	11.5–16	11.6–36	21–43	47–90	36–41	26–33	-	3.3
Potassium	4.08–11.0	3.90–4.3	3.62–9.20	1.95–5.12	1.40–6.20	7.34–7.92	1.23–4.6	-	508
Calcium	0.90 3.20	0.26–0.45	0.20–0.35	0.15–0.20	0.22–0.36	0.20–0.28	0.17–0.24	-	159

Source: Kumari and Tzudir 2021; Espinosa-Ramırez 2019; Sharma et al. 2021

- **Little millet** It contains about 8.7 percent protein and is a wonderful source of lysine and sulphur, containing cysteine and methionine, which are deficient in staple cereals. Known to induce a lower glycemic response on consumption, is also a good source of Fe, P and niacin.

1.5 Phytochemical profile and antioxidative nature of nutri-cereals

Millets contain phytochemicals, for example, polyphenols, lignin, phytosterols and phyto-estrogens. These act as immune modulators, antioxidants and detoxifying agents, reducing degenerative metabolic illnesses such as type-2 diabetes, cancer and cardiovascular diseases. A study reported that nutri-cereals contain 50 different phenolic groups such as flavanols, flavones, ferulic acid and flavanones with excellent antioxidant capacity. The concentration of phytochemicals are presented in **Table 1.3**. Significant amounts of phenolic components in millets are found either in free form and in bounded form. The proso millet comprises of phytochemicals like chlorogenic acid, syringic acid, caffeic acid, p-coumaric and ferulic acid. It has been reported that 65 percent of the total phenolic compounds are present in the bound form (Gowda et al. 2022).

Phenolic acids, tannins, flavonoids, coumarins and alkyl resorcinol account for the wide range of phenolic compounds. They are responsible for the texture, flavor, taste, color and oxidative stability of the foods (Naczk and Shahidi 2004). The bran holds nutraceutical properties with two classes of phenolic acids: hydroxyl benzoic acid and hydroxycinnamic acid. The hydroxybenzoic acids include vanillic, p-hydroxy benzoic acid, protocatechuic acid and hydroxycinnamic acids include caffeic acid, coumaric, sinapic acid and ferulic acid. Flavonoids include flavanols, anthocyanins, flavonoids, flavones, and flavanones. The common anthocyanidins found in nutri-cereals are malvinidin, pelragonidin, cyanidin, petunidin, peonidin and delphinidin. The 3-deoxyanthocyanidin present in sorghum, a unique anthocyanidin group that is very stable at alkaline pH, provides sorghum with great food coloring properties. Barley contains cyanidin-3-glucosides, delphinidine and pelargonidin glycosides. Several flavones identified in ragi are orientin, vitexin, isoorientin, isovitexin, valanthin, saponarin and lucinenin-1-tricin (Subbarao and Murlikrishna 2002). Nutri-cereals contain condensed tannins that protect them from molds and deterioration (Waniska 2000).

1.6 Effect of antinutritional factors and processing treatments for their reduction/removal

Nutri-cereals are an excellent source of macronutrients as well as phytochemicals and bio-active compounds, which aid in combatting chronic health issues. Millets could become the best option to deal with deficiency and metabolic disorders if the impact of antinutrients is controlled in them. Hence, present need of the hour is to develop methods for reduction or removal of anti-nutrients and improve the nutrient bioavailability for its better utilization and production of new products. Phytate and tannins have been identified as a major antinutritional factor in finger millet that reduces the bioavailability of minerals like Ca and Mg and trace minerals (Zn, Fe, Cu and Mn) Mbithi-Mwikya et al. 2000. Phytates have been known to have a chelating action on iron, prohibiting the lipid peroxidation and formation of hydroxyl radicals. Guiragossian et al. 1978 concluded that the high content of tannin in sorghum reduced protein digestibility. Daiber and Taylor (1982) denoted that the high tannin/poly-phenol content

Table 1.3 Phytochemical Profile and Bioactive Composition

Nutri-cereal	Bioactive compounds, phytosterols and antioxidative profile	References
Pearl millet	α-Amylase (IU/gds): 79.8; β-Glucosidase (IU/gds) 9.9; Xylanase (IU/gds) 18.4; Total phenolic compounds (mg GAE/g dwb) 6.4; DPPH (% inhibition) 88.8; ABTS (% inhibition) 96.5; HFRSA (%)12.1; CUPRAC (mg AAE/g dwb) 30.3; TAC (mg AAE/g dwb) 5.4; RPA (mg QE/g dwb) 3.4; MCA (%)67.9; Ascorbic acid 0.32 mg/g dwb; Gallic acid 0.12 mg/g dwb; p-Coumaric acid 0.16 mg/g dwb	Salar et al.2017
Finger millet	gentisic acid 4.5 (µg/g); vanillic acid 20.0 (µg/g); gallic acid 3.91–30.0 (µg/g); syringic acid 10.0–60.0 (µg/g) salicylic 5.12–413.0 (µg/g); protocatechuic 119.8–405.0 (µg/g); p-hydroxyberzoic 6.3–370.0 (µg/g); caffeic 5.9–10.4(µg/g); sinapic 11.0–24.8 (µg/g); p-coumaric 1.81–41.1(µg/g); trans-cinnamic 35–100.0 (µg/g) trans-ferulic 41–405.0 (µg/g)	Udeh et al. 2018
Foxtail millet	Total phenolic content (mg of FAE/100g) 72.70 ± 2.29 Total Benzoics µg/g defatted meal 59.25 ± 6.65 Total Cinnamics µg/g defatted meal 326.37 ± 4.36	Hutabarat & Bowie 2022
Proso millet	DPPH (%): free 19.78–26.94; bound 31.15 ABTS values (mmol/g): free 0.11-0.58 bound 0.26-1.09 FRAP value (mmol/g): free 1.36-3.82 bound 0.57-0.89	Li et al. 2021
Kodo millet	Phenolic content (mg gallic acid equivalents (GAE)/100 g) – Free phenolics 16.46, Bound phenolics 38.08 Flavonoid content (mg/g plant extract in Rutin Equivalents). – Free flavonoids 19.25, Bound flavonoids 33.34	Sharma et al. 2017
Little millet	Total phenolic content (µmol ferulic acid eq/g) 23.95 - 24.05 p-coumaric acid (µg/g): soluble 1.12-4.12; bound 80.77-169 p-hydroxybenzoic acid (µg/g): soluble 2.84-11.2; bound 0.67-0.83 Total flavonoid content (µmol catechin eq/g): soluble 9.92-13.64 bound 0.79-1.45 Condensed tannin contents (CTC): 89.12- 107.56 mg catechin equivalents/100 g Total anthocyanin content (TAC): 44.69- 62.51 µg of cyaniding 3-O-glucoside equivalents/g Gallic acid: Soluble 1.53-7.20 (µg/g)	Pradeep et al. 2018
Amaranth	Squalene 469.96 mg/100 g; Campesterol 3.49 mg/100 g; Stigmastenol 20.85 mg/100 g; Δ7-Ergosterol 28.64 mg/100 g; Δ5-Avenasterol 0.61mg/100g; Δ7-Avenasterol 14.37 61mg/100g α-Tocopherol 1.91 (mg/100 g); β-Tocopherol 4.07 (mg/100 g) γ-Tocopherol 1.29 (mg/100 g); δ-Tocopherol 3.34 (mg/100 g)	Ogrodowska et al. 2014
Barnyard millet	TPC (mg GAE/g) 0.41; AOA (%DPPH) 1.68; (%) DPPHinhibition 32.90	Sharma et al. 2016
Kodo millet	Phytic acid (mg/g) 225, Polyphenols (mg/100g) 143 Tannin (%) 5.04; mg TAE/g 50.46; Antioxidant Activity (%) 80.56 mg AAE/100g 24.64	Sharma et al. 2017

that gave sorghum the status of "bird-proof" inhibited enzymatic reactions and microflora development essential for the brewing of sorghum beer. The effect of different treatments on antinutritional factors have been discussed in **Table 1.4**.

On sprouting, some minor changes in the *in vitro* protein digestibility and viscosity were observed after 96 hrs of sprouting in finger millet. A significant decline in the antinutritional factors, with insignificant change in tannins and phytates was reported. Activity of trypsin

Table 1.4 Antinutritional Factors and Methods for Their Reduction/Removal

Antinutritional factor	Found in	Native content	Treatment	Post-treatment	References
Phytic acid	Foxtail millet	7.46 (mg/g)	Soaking	6.34 (mg/g)	Sharma & Sharma 2022
			Germination	4.58 (mg/g)	
			Fermentation	5.22 (mg/g)	
Phytic acid	Finger	7.89 (mg/g; dry basis)	Soaking	6.72 (mg/g; dry basis)	Sharma & Gujral 2019
			Sprouting (48 hr)	4.84 (mg/g; dry basis)	
Phytic acid	Barnyard	9.14 (mg/g; dry basis)	Soaking	8.94 (mg/g; dry basis)	Sharma & Gujral 2019
			Sprouting	8.94 (mg/g; dry basis)	
Phytic acid	Kodo	8.30 (mg/g; dry basis)	Soaking	8.09 (mg/g; dry basis)	Sharma & Gujral 2019
			Sprouting	4.19 (mg/g; dry basis)	
Phytic acid	Little	8.34 (mg/g; dry basis)	Soaking	8.10 (mg/g; dry basis)	Sharma & Gujral 2019
			Sprouting	4.56 (mg/g; dry basis)	
Saponins	Foxtail millet	87.38 (mg/100 g)	Soaking	13.64 (mg/100 g)	Sharma & Sharma 2022
			Germination	113.23 (mg/100 g)	
			Fermentation	65.47 (mg/100 g)	
Oxalates	Barnyard	0.44 (g/100 g)	Germination	0.44 (g/100 g)	Kulla et al 2021
	Foxtail	0.56 (g/100 g)	Germination	0.37 (g/100 g)	Kulla et al 2021
	Little	0.44 (g/100 g)	Germination	0.37 (g/100 g)	Kulla et al 2021
Tannins	Barnyard	4.72 (%)	Germination	5.78 (%)	Kulla et al 2021
	Foxtail	4.94 (%)	Germination	6.27 (%)	Kulla et al 2021
	Little	6.14 (%)	Germination	9.1 (%)	Kulla et al 2021
	Foxtail millet	86.54 (mg TA/100 g)	Soaking	73.39 (mg TA/100 g)	Sharma & Sharma 2022
			Germination	53.54 (mg TA/100 g)	
			Fermentation	60.36 (mg TA/100 g)	

inhibitor declined 3 times when sprouted. Significant improvement *in vitro* protein digestibility was reported however 13.3 percent dry matter of the seeds was lost within 96 hrs of sprouting. Prolonged sprouting resulted in significant dry matter losses without any prominent nutritional benefits. It was finally concluded that sprouting time of finger millet meant for utilization in weaning foods should not exceed 48 hrs (Mbithi-Mwikya et al. 2000).

Foxtail millet contains tannins, saponins and phytates. When the grains were exposed to soaking, fermentation and germination, enzymic activation led to the significant variation in the functional properties due to hydrolytic action of activated enzymes (Sharma and Sharma 2022). Bioprocessing of millet flours were characterized by structural transformation due to macromolecular arrangement and reduction in the starch crystallinity. It was clearly indicated that bioprocessed flours with reduced antinutrients could serve as a potential ingredient in functional food products.

Barnyard, foxtail and little millet grains were germinated for 1 day, dried, milled to obtain flour. Germination caused a decline in the energy value, oxalates, and improved the content of tannins, crude fiber, total flavonoid and phenolic compounds. DPPH radical scavenging activity boosted in the case of germinated millet flours (Kulla et al. 2021). Since, germinated flours are an excellent source of antioxidants; they can be exploited for the development of functional and nutraceutical foods.

Decortication employed in millets significantly declined crude fiber, protein, oil, and ash contents. Also, it lowered the tannins (343.9–174 mg/ 100 g), calcium (129–115 mg/ 100 g), phytic acid (301.10–77.95 mg/100 g), phosphorous contents (509–380 mg/100 g) and iron (5.1–3.1 mg/100 g) (Babiker, Abdelseed, Hassan, & Adiamo 2018).

Soaking reduced phenolic compounds in finger and pearl millet from 161.64 to 128.97 GAE mg/100 g and 241.47 to 184.43 GAE mg/100 g respectively. This might be attributed to the activation of the polyphenol oxidase enzyme as a result of soaking. *In vitro* digestibility in pearl millet significantly increased when naturally fermented for 72 hrs due to the breakdown of starch to smaller oligosaccharides and enzymic hydrolysis that reduced the phytate content and resulted in better starch digestibility (Naveena & Bhaskarachary 2013). Shigihalli et al. (2021) soaked finger millet for different time interval at 175–200 °C and popped causing a reduction in the trypsin inhibitor activity, tannins and phytic acid (1 hr soaked) in popped samples.

1.7 Techno-functionality of the nutri-cereals

1.7.1 Functional properties

Water absorption capacity (WAC) of finger millet flours ranged from 0.93 to 1.23 mL/g with milky flour having the highest value (Wani et al. 2021). Lower WAC was found suitable for preparation of thinner gruels. High WAC values denotes loose starch granules whereas low values indicated the starch compactness. The details have been provided in **Table 1.5**.

The cold viscosity of the flour samples ranged from 5.00 to 6.00 cP and cooked viscosity lied between 57.67 to 306.7 cP (Ramashia et al. 2018). Barnyard flour had WAC (5.28 g/g), which increased to 5.49–5.77 g/g when irradiated with gamma rays (Wani et al. 2021). An increase in WAC of treated flours might be attributed to the formation of simpler breakdown products like disaccharides, dextrins and monosaccharides that are more hydrophilic as compared to their native starch (Luo et al. 2008). Meals and flours with high water absorption power can be mixed to gravies, soups and so forth for imparting consistency to them. Oil absorption capacity (OAC) indicating the ability of lipids to associate with proteins in foods, is often related with flavor, mouth feel and advancing the shelf life of food products (Bhat et al. 2008). OAC of untreated sample was 4.90 g/g. which increased to 5.02 to 5.15 g/g with treatment (Wani et al. 2021). Higher OAC of irradiated flours might be attributed to exposure of apolar peptides as a result of by denaturation caused by irradiation or due to the distortion of native starch structure to entrap oil (Abu et al. 2006).

The solubility and swelling power of the starch samples improved with the rise in temperature from 50 to 90 °C. This might be due to the damage caused to the granular starch structure due to disruption of H-bonds between starch when exposed to hydrothermal treatment. Additionally, the new H-bonds that formed between the exposed hydroxyl groups of amylose-amylopectin and water molecules might cause an increase in swelling properties of starch (Li et al. 2018).

With rise in the temperature from 50–90 °C, the swelling index increased from 2.18 to 9.66 g/g (Wani et al. 2021). Exposure to high temperature reduces affinity by lowering the forces that bind individual flour granules causing a high swelling index (Thanuja et al. 2011). On exposure to high radiation doses, higher damage to starch occurs causing loss in the ability of starch to associate with water, a lesser rise in swelling index (Wani et al. 2021).

The solubility index for irradiated millet flour samples was found to be higher. This might be attributed to the fragmentation of starch, breakdown of H bonds induced as a result of irradiation. This improved the polarity causing an increase in its solubility in a polar solvent.

Table 1.5 Techno-Functional Properties of Different Nutria Cereals

Parameters	Nutri-cereal	Inferences	Reference
Functional properties			
Oil absorption capacity, Water absorption capacity	Pearl millet	Oil absorption capacity 150%; Water absorption capacity 226%; Least gelation capacity 10%	Sade et al. 2009
Hydration Index, Swelling capacity, Swelling Index, Gel consistency, Flour Solubility	Pearl millet	Hydration Index: 0.36–0.43; Swelling capacity (µl/seed): 4–6 Swelling Index: 0.26–0.50; Gel consistency (mm) 53.55–65.20 Emulsification capacity(g/g):0.50–0.75 Flour Solubility (%): 9.93–13.53 Swelling Power (%):109–116 Cooking time (min): 21.52–27.55	Pawase et al. 2021
Bulk density, Dispersibility, Viscosity, cold paste	Finger millet	WAC (mL/g) 0.93–1.23; Bulk density (kg/m³) 0.89–0.93 Dispersibility (%) 84.73–92.03; Viscosity, cold paste (cP) 5.00–6.00 Viscosity, cooked paste (cP) 57.67–288.3	Ramashia et al. 2018
Syneresis, Foaming capacity, Light transmittance	Barnyard millet	Swelling power (g/g) 90(°C): 9.66 Water solubility index (g/g) 90(°C): 0.20 Syneresis (g/100g)- (48 h) 60.10 Light transmittance (%):48 h 0.91 Emulsifying capacity (%) 1.31 : Emulsion solubility (%) 100.0; Foaming capacity (%) 13.6	Wani et al. 2021
Cooking quality	Kodo millet	Gain in weight after cooking (%) 184.8 Cooked volume (ml 349.66 Gain in volume after cooking (%) 249.83±0.28 Cooking time (min) 14; Swelling power (g/g) 3.28 Solubility (%) 5.71; Solid loss (%) 14.02	Patil et al. 2020
Thermal and pasting properties			
DSC parameters	Prosomillet	Onset temperature 64.16; Peak temperature 68.45; Conclusion temperature 79.09; Gelatinization temperature range 14.93; Gelatinization temperature 35.75 cP, Breakdown 35.67 cP, Enthalpy of gelatinization 10.58	Li et al. 2018
RVA parameters	Pearl millet	Peak viscosity 71.42 cP, Trough 35.67 cP, Breakdown 35.75 cP, Final viscosity 91.50 cP, Setback 55.83 cP, Peak time 4.86 minutes	Sade et al. 2009

Structural and morphological attributes

SEM	Finger millet	Milky cream, brown and black flours showed that the loosened starch granules had various shapes which were mainly isolated, oval/ spherical or polygon	Ramashia et al. 2018
XRD	Proso millet	Degree of crystallinity (%): 38.87 The crystallinity degree was increased with the increase in retrogradation time and the recrystallized starch showed a relative crystallinity degree after 192 h of storage. The main character of B-type crystallinity is the distinct peak occurred at 17. This could be attributed to the melting of the amorphous starch as a result of the increase in the amylopectin fraction during storage	Li et al. 2018

Color values

Lab values and hue angle	Finger millet	L* values ranged from 19.23 ± 0.42 for black to 52.97 ± 1.76 for milky cream grain cultivars. The H° values ranged from 35.73° ± 1.06 to 68.63° ± 0.06, with the highest hue angle obtained from milky cream cultivar and the lowest hue angle from black cultivar for the grain cultivars.	Ramashia et al. 2018

In vitro nutrient digestibility

Starch digestibility	Finger millet	The FMF contains higher RS (30.17 g/100g) and low RDS (12.17 g/100g) as compared to WF (19.27 g/100g and 24.16 g/100g, respectively) IVSD of puffed FM (29.6–35.8 mg/g) was approximately four times higher than raw FM (7.2–9.9 mg/g) indicating gelatinization of starch during puffing	Sharma & Gujral 2019
Starch digestibility	Proso millet	Uncooked: RDS (%)14.89; SDS (%)13.81; RS (%)71.31; Cooked: RDS (%) 81.67; SDS (%) 10.91; RS (%)7.42	Xiao et al. 2021
Protein digestibility	Finger millet	IVPD of various minor millets ranging between 72.01 and 75.41% with FMF exhibiting 72.01% which is markedly lower than WF (80.84%). The high content of ANFs in FM plausibly interferes with the protein digestion resulting in low IVPD	Sharma & Gujral 2019
Protein digestibility	Foxtail millet	IVPD of untreated FMF was 68.65 g/100 g while bioprocessed flours exhibited IVPD ranging between 70.56 and 77.92 g/100 g	Sharma & Sharma 2022
Starch digestibility	Foxtail millet	IVSD of untreated FMF was lowest which substantially enhanced during biological processing and the most profound effect was caused by combination of germination and fermentation treatments	Sharma & Sharma 2022

A significant decrease in the syneresis of irradiated flours as compared to control sample was observed. At initial stages, syneresis in the flour gels ranged between 56.09 to 57.0 g/100 g and increased to 63.69–65.24 g/100 g after 120 h. This might be attributed to the formation of small breakdown products of starch fractions that hold water poorly, causing water exudation from flour gel with progression of storage (Wani et al. 2021).

With increase of the storage, light transmittance of flour gels showed a significant decline. Untreated barnyard flour gels displayed the light transmittance between 0.08–1.79 percent during low temperature storage. Initially, the high light transmittance (1.79%) declined with progression of storage period (0.08%). This might be due to retrogradation of starch gel, amylose reorganization into aggregates causing light to refract more and cause a decline in the light transmittance of resulting starchy gel (Wani et al. 2021).

Emulsion capacity of irradiated samples and native flour denoted a significant difference (Wani et al. 2021). Irradiation caused a loss in the native structure as a result of deamination in proteins. Denaturation of proteins improved the solubility and favored their adsorption at the interfacial area due to the reduction in interfacial energy and prevents separation of two phases, showed excellent emulsion capacity (Malik et al. 2017). Native barnyard flour represented 100 percent emulsion stability. The emulsion stability of treated flours was lower than untreated sample and ranged between 29.16–79.62 percent.

1.7.2 Structural and morphological attributes

Scanning electron micrographs of finger millet were studied to determine the microscopic structure of finger millet starch at a voltage of 5.000 kV. Loosened, isolated polygonal or spherical starch were observed and the smooth surface might be attributed to soaking and drying followed by milling into whole meal flour (Ramashia et al. 2018). Soaking technique enhanced the bioavailability of minerals Saleh et al. (2013). Milling showed a negative impact on nutritional composition, however, digestibility as well as bioaccessibility was found higher in treated grains (Saleh et al. 2013).

Li et al. (2018) presented irregular, polygonal and oval-shaped native starch granule with diameter between 3–10 μm. At lower pressures of 150 MPa and 300 MPa, the starch retained its native structure. whereas, granules began to lose their structure when exposed to higher pressures (450 MPa 600 MPa); the granules disintegrated into gel-like moities. The birefringent properties remained unaffected even after UHP treatment at 150 and 300 MPa.

The proso starch had a typical "A" type starch XRD pattern (Li et al. 2018). The UHP pressures (150 to 450 MPa) did not cause any change in the crystallinity of starch granules, but the crystallinity as well as the peak intensities declined significantly as the pressure increased. It can be observed that the endothermic peak shifted to greater temperatures, the peak turned weak and broad with the elevation in pressure. No characteristic endothermic melting peak or thermal transition was detected up to 600 MPa signifying complete gelatinization of proso millet (Li et al. 2018).

1.7.3 Pasting properties

The UHP-treatment caused a significant increase in final viscosity, trough viscosity, peak time and pasting temperature; whereas, the peak and breakdown viscosity declined on treatment. The pasting temperature and starch viscosity are altered by pressure treatment due to transformation in granular crystalline structure of the starch due to rapid water diffusion into starch, breaking the H bonds within the crystalline region and amorphous area under high pressure, and thereby resulting in restructuring of the crystal structure of starch and its gelatinization. Significant decrease in peak viscosity was reported as the level of irradiation increased up to 5 kGy, ranging between 1392.5–1441.0 cP (Wani et al. 2021). Loss of rigidity in starch granules due to fragmentation of starch chains might be the cause of decline in peak viscosity.

1.7.4 *In-vitro* nutrient digestibility

Millet proteins are known to have lower digestibility caused because of enzyme inhibition by antinutrients, protease inhibitors, tannins and limited enzyme exposure due to the hard seed coat . However, the *in-vitro* protein digestibility (IVPD) declined when soaked and dehulled prior to cooking (Pawar and Machewad 2006).

Studies conducted by Sharma & Gujral (2019) revealed finger millet flour showed lower glycaemic index (42.24) when compared to wheat flour (62.59). Additionally, finger millet contains higher resistant starch (30.17 g/100g) and lower rapid digestible starch (12.17 g/100g) in comparison to wheat flour. Being a rich source of arabinoxylan, dietary fiber, ANFs and phenolic compounds, finger millet flour restricts enzymatic starch hydrolysis. Additionally, the phenolic compounds and fats in millets gets directly adsorbed on the starch surface restricting its enzymatic breakdown and thereby a lower glycemic index. *In vitro* starch digestibility (IVSD) of puffed finger millet improved by four times when compared to native grains signifying starch gelatinization during puffing (Sharma & Gujral 2019)

1.8 Recent utilization

With changing food patterns, escalating population with prevalence of metabolic diseases and diet induced disorders, cancer, diabetes, celiac diseases and cardiovascular diseases, the food industry today, targets development of new convenience, functional and wholegrain foods. This has triggered the incorporation of underutilized nutricereals in food products providing a healthier alternative to staple food (Alaunyte et al. 2012). The details regarding trending utilization of millets have been provided in **Table 1.6**.

Rhados et al. (2022) developed gluten free bread by incorporating millet bran. Bran was sieved into 3 fractions with different composition. The results indicated that the fraction with average diameter of 500 μm retained highest nutrients (9% fat, 26% starch, 14% protein, 36% dietary fibre, and 3 mg GAE/g phenolics). Gluten-free bread with 10 percent of the bran had higher dietary fibre and total phenolic content, better crumb and volume softness, irrespective of the particle size. Biradar et al. (2021) prepared little millet flour-based cookies (20% wheat flour and 80%) with excellent overall acceptability till 90th day.

Rashwan et al. (2021) reported that the size of the sorghum starch is typically 16 μm on average and studied the potential of sorghum starch as a food and as a pharmaceutical ingredient. Higher gel consistency, gelatinization temperature synersis degree were observed in sorghum malt in comparison to other cereals. However, the β-amylase activity was lower in sorghum malt compared to other malts. Suri et al. (2020) optimized the variables for production of iron-rich, ready-to-eat (RTE) extruded snacks from barnyard millet, amla and defatted soy flour and rice flour. The optimized conditions, blend ratio (6:1), temperature (115 °C), and amla (12.25% of flour) produced a snack with high iron and protein content.

Majid and Priyadarshini (2020) studied the millet based bioactive peptides (BAPs) and defined peptides as protein fragments with molecular weight less than 10 kDa (Farrokhi, Whitelegge, and Brusslan 2008). They exist either naturally or may be derived from the precursor protein

Table 1.6 Mainstream Utilization of Nutri-Cereals

Utilization	Nutri-cereal	Method of development	Inferences	References
Extruded products				
Pasta	Foxtail Millet	Pasta was made with 70% foxtail millet and 30% tapioca incorporation	Cooking time of millet-based pasta was found significantly lower than control.	Dhas et al 2021
Chakli	Proso Millet	Chakli was standardized with 65 % proso millet flour and rice flour and studied for its storage.	Chakli can be stored up to 1 month in aluminium pouches. Chakli contains 50.26 % carbohydrates, 9.51 % protein, 30.47 % fat, 2.48 % ash, 7.28 % moisture, 2.37 % crude fiber, 40.10 mg/kg of iron, 260.30 mg/kg calcium and 20.10 mg/kg of zinc	Sarojani et al. 2021
Vermicelli	Barnyard Millet	Barnyard millet flour, rice flour, malt flour from whole green gram, carrot powder and xanthan gum were used to develop enriched vermicelli	The optimized product is nutritionally enriched having 3.81mg/100g of iron and 1039µg/100g of beta-carotene present in it.	Goel et al. 2021
Hot extruded	Little Millet (Kutki)	The blend of little millet, rice and maize were prepared and flour was allowed to equilibrate for 15 min.	Increasing proportion of little millet (0–20) % in the extrusion blend causes increase in ash content, fat content, protein content, fiber content, hardness and bulk density of the extruded products Expansion ratio, carbohydrate content and calorific value declined	Saini & Yadav 2018
Hot extruded	Amaranthus (Chaulai)	The process was optimized at extrusion temperature of 190 °C and 14% initial moisture	61.5% water solubility index, 101.2 µmol trolox/g DPPH-RSA, 364.2 µmol trolox/g AC-ABTS and total phenols 34.5 mg GAE/100 g.	Basilio-Atencio et al 2020
Bakery				
Gluten-free bread	Millets	Millet bran was sieved into three fractions with substantially different nutritional profile and baked for 18 min at 190°C with initial steaming in a deck oven.	Micronization increased the antioxidant activity as well as the content of soluble fibre. Bread where 10% of rice flour was replaced with millet bran was found optimum.	Mustac et al 2020
Cookies	little millet	Cookies were prepared from 20% wheat flour and 80% little millet flour	protein 8.40%, crude fat 29.10%, crude fiber 6.46%, carbohydrates 67.48% and iron 8.42 mg/ 100 g; stored up to 3 month in LDPE.	Biradar et al 2020
Biscuits	kodo millet	The quality biscuits were prepared from 50% maida and 50% kodo millet flour	The fresh biscuits had 4.07 % moisture, 10.05% protein, 26.62% fat, 4.30% crude fibre, 69.62% carbohydrates, 24.88 mg/100g calcium and 2.17mg/100g iron.	Mitkal et al. 2021
Cookies	Foxtail millet	The foxtail millet (20, 30 and 40%) and ginger powder (5, 10 and 15%) blended wheat flours	WAC increased with the increase in the foxtail millet proportion. The crude protein content (3.85–11.52%), Dietary fibre ranging from 2.43–8.78 g/100 g.	Marak et al. 2019
Cake	Finger Millet	Sugar, margarine, eggs, wheat flour, baking powder and salt were combined and baked.	Protein content of the flour increased with increase in millet flour addition from 9.36–13.74%. Supplementation of wheat flour with millet flour up to 40% in baking was found optimum	Maharajan et al. 2021

Product	Millet type	Process	Findings	Reference
Beverages				
Beverages	Sorghum and finger millet	Sorghum, finger-millet, sprouted soy flour and green-gram flour, dairy whitener and desiccated coconut were used.	The nutritional quality of beverage was higher in terms of iron and calcium with 90% antioxidant activity. Beverage shoed pseudo-plastic nature Significant increase in viscosity was also observed with the increase in temperature	Agrahar-Murugkar et al. 2020
Multigrain functional beverage	barnyard, foxtail and kodo millet	The basic processing steps included roasting, boiling and grinding with 100 mL distilled water and filtering.	Contained 5.72 g/100 g total dietary fibre, 47.69 mg ferulic acid equivalents (FAE)/100 mL; prebiotic activity of 1.56 and 45.07 of glycaemic index; categorized as functional low GI beverage	Arya & Shakya 2021
Flaked and popped				
Flaked snack bar	flakes of foxtail and proso millets	standardized flakes of foxtail and proso millets, jaggery, liquid glucose, skimmed milk powder, cocoa powder, dates, flax seeds, sesame seeds, groundnuts and soya granules.	compare to control bar (CNB) there was decrease in protein, fat, ash and energy whereas increase in crude fiber and carbohydrates	Sohan et al 2021
Ready-to-eat (RTE) breakfast cereal	pearl millet	The tempered grains were popped by high temperature and short time (HTST) treatment in a domestic grain popper	contain good amount of total dietary fiber The per serving of breakfast cereal was 22.8% for energy, 12.80 % for protein, 13% for fat, 34.5% for calcium, 20.5% for iron, and 55.75% of the RDA.	Kumari et al. 2019
Bioactive peptides				
Protein hydrolysates	Buckwheat	Hydrolysis through digestive enzymes, proteolytic microorganisms and fermentation	Protein hydrolysates show antiviral activity/HIV-1 reverse transcriptase. Globulin protein is the precursor. LQAFEPLR and EFLLAGN peptides have antidiabetic effect	Leung & Ng 2007
Protein hydrolysates	Finger Millet	Hydrolysis through digestive enzymes, proteolytic microorganisms and fermentation	Protein hydrolysates antibacterial effect/Pseudomonas aeruginosa (MTCC 424), and Salmonella entrica (MTCC 739 TSSLNM VRGGLTR and STTVGLGISMRSASVR peptide sequence have antioxidant	Bisht, Thapliyal, & Singh 2016 Agrawal, Joshi, & Gupta 2019
Peptide sequence	Pearl millet Total protein)	Hydrolysis through digestive enzymes, proteolytic microorganisms and fermentation	Peptide sequence SDRDLLGPNNQYLPK have antioxidant effect	Agrawal, Joshi, & Gupta 2019
Protein hydrolysates	Little millet	Hydrolysis through digestive enzymes, proteolytic microorganisms and fermentation	Lowers blood glucose and cholesterol levels; prevention of diabetes mellitus; antioxidant potential	Nithiyanantham et al. 2019
Edible Films				
Starch like films	Proso-Millet	Films were prepared from proso millet starch (native) and a starch–κ-carrageenan blend.	The water vapor permeability and solubility of films prepared from native starch (2.38 g/Pa·s·m² and 28%) were lower than those prepared using the κ-carrageenan blend	Bangar et al. 2021

by hydrolysis either proteolytic or digestive. BAPs have been known to cause certain beneficial biochemical impacts on the human body and health owing to its antihypertensive activity, anticancer/antiproliferative, anticancer/antiproliferative, antioxidant property and antimicrobial activity.

He et al. (2002) derived squalene from amaranth. Squalene derived from amaranth oil is a compound present in animal and vegetable cells, and is highly demanded because of its biological value. It has been reported that the amaranth plant is the richest plant source of squalene (Alvarez et al. 2010; Venskutonis and Kraujalis 2013). The compound is utilized in personal care products and cosmetics and shows a high antioxidant potential and eliminates free-radical induced oxidative changes in the skin (Huang et al. 2009).

Kafirin, a sorghum based prolamin protein has been reported to have hydrophobic properties essential for manufacturing of bioplastics (Belton et al. 2006). Similarly, coatings made from sorghum wax and medium chain triglyceride oil were found comparable to edible coatings developed from gelatin (Weller et al. 1998).

Wu et al. (2006) found that the pearl millet could be potentially utilized for the manufacturing of alcohol with an efficiency of 94.2 percent similar to maize and sorghum. Similarly, the research on the production of bioethanol from sweet sorghum has been reported (Schaffert 1995). Additionally, sorghum can be used for alcohol, sugar syrup, fodder, jaggery, bedding, fuel, roofing, chewing, fencing and paper. Also, sweet sorghum had been potentially utilized for paper production.

1.9 Conclusions

Millets can efficiently survive in harsh environmental conditions like drought, swampy grounds and even flooded areas. These nutri cereals possess abundant gluten free protein, has low glycemic index and are an excellent source of minerals (iron, magnesium, calcium, copper), antioxidants and B-vitamins. These unique properties make nutri-cereals a climate-change compliant and nutritious crop. These hardy crops on one hand can serve as an income source for farmers and on the other hand enhance the community health as a whole. Some limitations like less sensory acceptability and ample anti-nutritional factors in the millets can be combated by the processing interventions like, roasting, fermentation, cooking and germination. Additionally, the utilization of millets in commercially processed food will sensitize farmers to cultivate millets and will create better opportunities for the farmers.

References

Abegaz, K. (2007). Isolation, characterization and identification of lactic acid bacteria involved in traditional fermentation of *borde*, an Ethiopian cereal beverage. *African Journal of Biotechnology*, 6(12), 1469–1478.

Abu, J. O., Duodu, K. G., & Minnaar, A. (2006). Effect of γ-irradiation on some physicochemical and thermal properties of cowpea (Vigna unguiculata L. Walp) starch. *Food Chemistry*, 95(3), 386–393.

Agrahar-Murugkar, D., Bajpai-Dixit, P., & Kotwaliwale, N. (2020). Rheological, nutritional, functional and sensory properties of millets and sprouted legume based beverages. *Journal of Food Science and Technology*, 57(5), 1671–1679.

Agrawal, H., Joshi, R., & Gupta, M. (2019). Purification, identification and characterization of two novel antioxidant peptides from finger millet (Eleusine coracana) protein hydrolysate. *Food Research International*, 120, 697–707.

AICRPSM (2017). Annual progress report: 2016–17, AICRP on Small Millets, 2. Agronomy, Bengaluru, p. 10, available at: www.aicrpsm.res.in/Downloads/Reports/2: Agronomy-report.pdf (accessed on: 12 March 2020).

Akoma, O., Jiya, E. A., Akumka, D. D., & Mshelia, E. (2006). Influence of malting on the nutritional characteristics of kunun-zaki. *African Journal of Biotechnology*, 5(10), 996–1000.

Alaunyte, I., Stojceska, V., Plunkett, A., Ainsworth, P., & Derbyshire, E. (2012). Improving the quality of nutrient-rich Teff (Eragrostis tef) breads by combination of enzymes in straight dough and sourdough breadmaking. *Journal of Cereal Science*, 55(1), 22–30.

Alvarez-Jubete, L., Wijngaard, H., Arendt, E. K., & Gallagher, E. (2010). Polyphenol composition and in vitro antioxidant activity of amaranth, quinoa buckwheat and wheat as affected by sprouting and baking. *Food Chemistry*, 119(2), 770–778.

Ambati, K., & Sucharitha, K. V. (2019). Millets-review on nutritional profiles and health benefits. *Int J Recent Sci Res*, 10(7), 33943–33948.

Amer, S., & Mustafa, A. F. (2010). Effects of feeding pearl millet silage on milk production of lactating dairy cows. *Journal of Dairy Science*, 93(12), 5921–5925.

Argüelles-López, O. D., Reyes-Moreno, C., Gutiérrez-Dorado, R. R., Sánchez-Osuna, M. F., López-Cervantes, J., Cuevas-Rodríguez, E. O., ... & Perales-Sánchez, J. X. K. (2018). Functional beverages elaborated from amaranth and chia flours processed by germination and extrusion. *Biotecnia*, 20(3), 135–145.

Arya, S. S., & Shakya, N. K. (2021). High fiber, low glycaemic index (GI) prebiotic multigrain functional beverage from barnyard, foxtail and kodo millet. *LWT*, 135, 109991.

Babiker, E., Abdelseed, B., Hassan, H., & Adiamo, O. (2018). Effect of decortication methods on the chemical composition, antinutrients, Ca, P and Fe contents of two pearl millet cultivars during storage. *World Journal of Science, Technology and Sustainable Development*, 5(3), 278–286.

Baidab, S. F., Hamad, S. A., Ahmed, A. H. R., & Ahmed, I. A. M. (2016). Preparation of Hulu-mur flavored carbonated beverage based on Feterita sorghum (Sorghum bicolor) malt. *International Journal of Food Studies*, 5(2), 120–130.

Baligar, V. C., Anghinoni, I., Pitta, G. V. E., Dos Santos, H. L., Filho, E. C., & Schaffert, R. E. (1995). Aluminum effects on plant and nutrient uptake parameters of soil and solution grown sorghum genotypes. *Journal of Plant Nutrition*, 18(11), 2325–2338.

Basilio-Atencio, J., Condezo-Hoyos, L., & Repo-Carrasco-Valencia, R. (2020). Effect of extrusion cooking on the physical-chemical properties of whole kiwicha (Amaranthus caudatus L) flour variety centenario: Process optimization. *LWT*, 128, 109426.

Belton, P. S., Delgadillo, I., Halford, N. G., & Shewry, P. R. (2006). Kafirin structure and functionality. *Journal of Cereal Science*, 44(3), 272–286.

Biradar, S. D., Kotecha, P. M., Godase, S. N., & Chavan, U. D. (2021). Studies on nutritional quality of cookies prepared from wheat flour and little millet. *International Journal of Chemical studies*, 9(1), 1675–1680.

Bisht, A. N. J. A. L. I., Thapliyal, M. A. N. I. S. H. A., & Singh, A. J. E. E. T. (2016). Screening and isolation of antibacterial proteins/peptides from seeds of millets. *International Journal of Current Pharmaceutical Research*, 8(3), 96–99.

Blandino, A., Al-Aseeri, M. E., Pandiella, S. S., Cantero, D., & Webb, C. (2003). Cereal-based fermented foods and beverages. *Food Research International*, 36(6), 527–543.

Boling, M. B., & Eisener, N. (1981). Bogobe: Sorghum porridge of Botswana. *Sorghum Grain Quality*, 32(1), 32–35.

Brahmachari, K., Sarkar, S., Santra, D. K., & Maitra, S. (2019). Millet for food and nutritional security in drought prone and red laterite region of Eastern India. *International Journal of Plant & Soil Science*, 26(6), 1–7.

Chaudhary S. 1993. Preparation of nutritional evaluation of some traditional foods from pearl millet [MSc thesis]. Hisar, Haryana, India: CCS Haryana Agricultural Univ. 98 pp.

Choto, C. E., Morad, M. M., & Rooney, L. W. (1985). The quality of tortillas containing whole sorghum and pearled sorghum alone and blended with yellow maize. *Cereal Chemistry (USA)*, 62, 51–55.

Daiber, K. H., & Taylor, J. R. (1982). Effects of formaldehyde on protein extraction and quality of high-and low-tannin sorghum. *Journal of Agricultural and Food Chemistry*, 30(1), 70–72.

Dhas, T. A., Rao, V. A., Sharon, M. E. M., Manoharan, A. P., & Nithyalakshmi, V. (2021). Development of fibre enriched ready-to-cook pasta with foxtail millet and tapioca flour. *Journal of Pharmacy Research*, 1(3), 68–69.

Ding, R. X., Jia, Z. K., Han, Q. F., Ren, G. X., & Wang, J. P. (2006). Border effect and physiological characteristic responses of Foxtail Millet to different micro-catchment stripshapes in semi-arid region of south Ningxia. *Scientia Agricultura Sinica*, 39(3), 494–501.

Duodu, K. G. (2011). Effects of processing on antioxidant phenolics of cereal and legume grains. In *Advances in Cereal Science: Implications to Food Processing and Health Promotion* (pp. 31–54). American Chemical Society.

Farrokhi, N., Whitelegge, J. P., & Brusslan, J. A. (2008). Plant peptides and peptidomics. *Plant Biotechnology Journal*, 6(2), 105–134.

Gadaga, T. H., Lehohla, M., & Ntuli, V. (2021). Traditional fermented foods of Lesotho. *Journal of Microbiology, Biotechnology and Food Sciences*, 2021, 2387–2391.

Goel, K., Goomer, S., & Aggarwal, D. (2021). Formulation and optimization of value-added barnyard millet vermicelli using response surface methodology. *Asian Journal of Dairy and Food Research*, 40(1), 55–61.

Gowda, N. N., Siliveru, K., Prasad, P. V., Bhatt, Y., Netravati, B. P., & Gurikar, C. (2022). Modern Processing of Indian Millets: A Perspective on Changes in Nutritional Properties. *Foods*, 11(4), 499.

Grundy, M. M., Momanyi, D. K., Holland, C., Kawaka, F., Tan, S., Salim, M., ... & Owino, W. O. (2020). Effects of grain source and processing methods on the nutritional profile and digestibility of grain amaranth. *Journal of Functional Foods*, 72, 104065.

Guiragossian, V., Chibber, B. A., Van Scoyoc, S., Jambunathan, R., Mertz, E. T., & Axtell, J. D. (1978). Characteristics of proteins from normal, high lysine, and high tannin sorghums. *Journal of Agricultural and Food Chemistry*, 26(1), 219–223.

Gururani, K., Sood, S., Kumar, A., Joshi, D. C., Pandey, D., & Sharma, A. R. (2021). Mainstreaming Barahnaja cultivation for food and nutritional security in the Himalayan region. *Biodiversity and Conservation*, 30(3), 551–574.

Han, Y., Huang, B., Liu, S., Zou, N., Yang, J., Zhong, Z., ... & Pan, C. (2016). Residue levels of five grain-storage-use insecticides during the production process of sorghum distilled spirits. *Food Chemistry*, 206, 12–17.

He, H. P., Cai, Y., Sun, M., & Corke, H. (2002). Extraction and purification of squalene from Amaranthus grain. *Journal of Agricultural and Food Chemistry*, 50(2), 368–372.

Hutabarat, D. J. C., & Bowie, V. A. (2022, February). Bioactive compounds in foxtail millet (Setaria italica)-extraction, biochemical activity, and health functional: A Review. In *IOP Conference Series: Earth and Environmental Science* (Vol. 998, No. 1, p. 012060). IOP Publishing.

Ibnouf, F. O. (2012). The value of women's indigenous knowledge in food processing and preservation for achieving household food security in rural Sudan. *Journal of food Research*, 1(1), 238.

Karim, A. A., & Bhat, R. (2008). Gelatin alternatives for the food industry: Recent developments, challenges and prospects. *Trends in Food Science & Technology*, 19(12), 644–656.

Karkannavar, S. J., Shigihalli, S., Nayak, G., & Bharati, P. (2021). Physico-chemical and nutritional composition of proso millet varieties. *The Pharma Innovation Journal*, 10(1), 136–140.

Kiran, P., Mammen, D., & Mammen, D. (2012). Ecdysterone, Phospholipids and Phenolics in the seeds of Amaramthus. Hypochondriacus Linn. *Journal of Pharmacy Research*, ISSN 2278–2818.

Klmata, M., Ashok, E. G., & Seetharam, A. (2000). Domestication, cultivation and utilization of two small millets, Brachiaria ramosa and Setaria glauca (Poaceae), in south India. *Economic Botany*, 54(2), 217–227.

Kulla, S., Hymavathi, T. V., Kumari, B. A., Reddy, R. G., & Rani, C. V. D. (2021). Impact of germination on the nutritional, antioxidant and antinutrient characteristics of selected minor millet flours. *Annals of Phytomedicine*, 10(1), 178–184.

Kumar, A., Tomer, V., Kaur, A., Kumar, V., & Gupta, K. (2018). Millets: A solution to agrarian and nutritional challenges. *Agriculture & Food Security*, 7(1), 1–15.

Kumari, R., Singh, K., Singh, R., Bhatia, N., & Nain, M. S. (2019). Development of healthy ready-to-eat (RTE) breakfast cereal from popped pearl millet. *Indian Journal of Agricultural Sciences*, 89(5), 877–881.

Kumari, S., & Tzudir, L. (2021). Black Wheat: Next Big Thing in India's Agricultural Landscape. *Biotica Research Today*, 3(4), 240–242.

Leung, E. H., & Ng, T. B. (2007). A relatively stable antifungal peptide from buckwheat seeds with antiproliferative activity toward cancer cells. *Journal of Peptide Science: An official publication of the European Peptide Society*, 13(11), 762–767.

Li, W., Gao, J., Saleh, A. S., Tian, X., Wang, P., Jiang, H., & Zhang, G. (2018). The modifications in physico-chemical and functional properties of proso millet starch after Ultra-High pressure (UHP) process. *Starch-Stärke*, 70(5–6), 1700235.

Li, W., Wen, L., Chen, Z., Zhang, Z., Pang, X., Deng, Z., ... & Guo, Y. (2021). Study on metabolic variation in whole grains of four proso millet varieties reveals metabolites important for antioxidant properties and quality traits. *Food Chemistry*, 357, 129791.

Luo, Z., Fu, X., He, X., Luo, F., Gao, Q., & Yu, S. (2008). Effect of ultrasonic treatment on the physicochemical properties of maize starches differing in amylose content. *Starch-Stärke*, 60(11), 646–653.

Maharajan, T., Ceasar, S. A., & Ajeesh Krishna, T. P. (2022). Finger Millet (Eleusine coracana (L.) Gaertn): Nutritional Importance and Nutrient Transporters. *Critical Reviews in Plant Sciences*, 41(1), 1–31.

Maitra, S. (2020). Intercropping of small millets for agricultural sustainability in drylands: A review. *Crop Research (0970–4884)*, 55(3), 162–173.

Maitra, S., & Shankar, T. (2019). Agronomic management in little millet (Panicum sumatrense L.) for enhancement of productivity and sustainability. *International Journal of Bioresource Science*, 6(2), 91–96.

Majid, A., & Poornima Priyadarshini C. G. (2020). Millet derived bioactive peptides: A review on their functional properties and health benefits. *Critical Reviews in Food Science and Nutrition*, 60(19), 3342–3351.

Malik, M. A., Sharma, H. K., & Saini, C. S. (2017). High intensity ultrasound treatment of protein isolate extracted from dephenolized sunflower meal: Effect on physicochemical and functional properties. *Ultrasonics Sonochemistry*, 39, 511–519.

Marak, N. R., Malemnganbi, C. C., Marak, C. R., & Mishra, L. K. (2019). Functional and antioxidant properties of cookies incorporated with foxtail millet and ginger powder. *Journal of Food Science and Technology*, 56(11), 5087–5096.

Mbithi-Mwikya, S., Van Camp, J., Yiru, Y., & Huyghebaert, A. (2000). Nutrient and antinutrient changes in finger millet (Eleusine coracan) during sprouting. *LWT-Food Science and Technology*, 33(1), 9–14.

Mitkal, K. T., Kotecha, P. M., Godase, S. N., & UD Chavan (2021). Studies on nutritional quality of Kodo millet cookies. *IJCS*, 9(1), 1669–1674.

Moza, J., & Gujral, H. S. (2018). Mixolab, retrogradation and digestibility behavior of chapatti made from hulless barley flours. *Journal of Cereal Science*, 79, 383–389.

Murty, D. S., & Kumar, K. A. (1995). Traditional uses of sorghum and millets. *Sorghum and Millets: Chemistry and Technology*, 221, 153–177.

Muyanja, C., Birungi, S., Ahimbisibwe, M., Semanda, J., & Namugumya, B. S. (2010). Traditional processing, microbial and physicochemical changes during fermentation of malwa. *African Journal of Food, Agriculture, Nutrition and Development*, 10(10), 2125–2138.

Mwesigye, P. K., & Okurut, T. O. (1995). A survey of the production and consumption of traditional alcoholic beverages in Uganda. *Process Biochemistry*, 30(6), 497–501.

Naczk, M., & Shahidi, F. (2004). Extraction and analysis of phenolics in food. *Journal of Chromatography A*, 1054(1–2), 95–111.

Naveena, N., & Bhaskarachary, K. (2013). Effects of soaking and germination of total and individual polyphenols content in the commonly consumed millets and legumes in India. *International Journal of Food and Nutritional Sciences*, 2(3), 12.

Nithiyanantham, S., Kalaiselvi, P., Mahomoodally, M. F., Zengin, G., Abirami, A., & Srinivasan, G. (2019). Nutritional and functional roles of millets – A review. *Journal of Food Biochemistry*, 43(7), e12859.

Oelke, E. A., Oplingerm E. S., Putnam, D. H., Durgan, B. R., Doll, and J. D. Millets. (1990). In: *Alternative Field Crops Manual*. Univ. of Wisc-Ext Serv. Univ. of Minn. Ext Serv and Univ. of Minn. CAPAP.

Ogrodowska, D., Zadernowski, R., Czaplicki, S., Derewiaka, D., & Wronowska, B. (2014). Amaranth seeds and products–the source of bioactive compounds. *Polish Journal of Food and Nutrition Sciences*, 64(3), 165–170.

Olufolaji, A. O., Odeleye, F. O., & Ojo, O. D. (2010). Effect of soil moisture stress on the emergence, establishment and productivity of Amaranthus (Amaranthus cruentus L.). *Agriculture and Biology Journal of North America*, 1(6), 1169–1181.

Passi, S. J. and Jain, A. (2014). *Millets: The Nutrient Rich Counterparts of Wheat and Rice*. Government of India: Press Information Bureau, New Delhi.

Patil, R. B., Vijayalakshmi, K. G., & Vijayalakshmi, D. (2020). Physical, functional, nutritional, phytochemical and antioxidant properties of kodo millet (Paspalum scrobiculatum). *Journal of Pharmacognosy and Phytochemistry*, 9(5), 2390–2393.

Pawar, V. D., & Machewad, G. M. (2006). Processing of foxtail millet for improved nutrient availability. *Journal of Food Processing and Preservation*, 30(3), 269–279.

Pawase, P. A., Chavan, U. D., & Kotecha, P. M. (2021). Evaluation of functional properties of different pearl millet cultivars. *The Pharma Innovation Journal*, 10(6), 1234–1240.

Peiretti, P. G. (2018). Amaranth in animal nutrition: A review. *Livestock Research for Rural Development*, 30(5), 2018.

Pradeep, P. M., & Sreerama, Y. N. (2018). Phenolic antioxidants of foxtail and little millet cultivars and their inhibitory effects on α-amylase and α-glucosidase activities. *Food Chemistry*, 247, 46–55.

Punia Bangar, S., Nehra, M., Siroha, A. K., Petrů, M., Ilyas, R. A., Devi, U., & Devi, P. (2021). Development and characterization of physical modified pearl millet starch-based films. *Foods*, 10(7), 1609.

Pushparaj, F. S., & Urooj, A. (2017). Impact of household processing methods on the nutritional characteristics of pearl millet (Pennisetum typhoideum): A review. *MOJ Food Processing & Technology*, 4(1), 28–32.

Radoš, K., Čukelj Mustač, N., Varga, K., Drakula, S., Voučko, B., Ćurić, D., & Novotni, D. (2022). Development of High-Fibre and Low-FODMAP Crackers. *Foods*, 11(17), 2577.

Ramashia, S. E., Gwata, E. T., Meddows-Taylor, S., Anyasi, T. A., & Jideani, A. I. O. (2018). Some physical and functional properties of finger millet (Eleusine coracana) obtained in sub-Saharan Africa. *Food Research International, 104*, 110–118.

Rashwan, A. K., Yones, H. A., Karim, N., Taha, E. M., & Chen, W. (2021). Potential processing technologies for developing sorghum-based food products: An update and comprehensive review. *Trends in Food Science & Technology, 110*, 168–182.

Rekha (1997). Efficacy of processing techniques in the utilisation of pearl millet for value-added products. *M. Sc. Thesis.* CCSHAU. Hisar. India.

Sade, F. O. (2009). Proximate, antinutritional factors and functional properties of processed pearl millet (Pennisetum glaucum). *Journal of Food Technology, 7*(3), 92–97.

Saini, R., & Yadav, K. C. (2018). Development and quality evaluation of little millet (Panicum sumatrense) based extruded product. *Journal of Pharmacognosy and Phytochemistry, 7*(3), 3457–3463.

Salar, R. K., Purewal, S. S., & Sandhu, K. S. (2017). Fermented pearl millet (Pennisetum glaucum) with in vitro DNA damage protection activity, bioactive compounds and antioxidant potential. *Food Research International, 100*, 204–210.

Saleh, A. S., Zhang, Q., Chen, J., & Shen, Q. (2013). Millet grains: Nutritional quality, processing, and potential health benefits. *Comprehensive Reviews in Food Science and Food Safety, 12*(3), 281–295.

Sarojani, J. K., Hegde, S. C., Desai, S. R., & Naik, B. K.(2021). *Standardization and Nutrient Composition of the Proso Millet Chakli. Biological Forum – An International Journal (SI-AAEBSSD-2021)* 13(3b): 44–50.

Serna-Saldivar, S. O., & Espinosa-Ramírez, J. (2019). Grain structure and grain chemical composition. In *Sorghum and Millets* (pp. 85–129). AACC International Press.

Sharma, B., & Gujral, H. S. (2019). Influence of nutritional and antinutritional components on dough rheology and in vitro protein & starch digestibility of minor millets. *Food Chemistry, 299*, 125115.

Sharma, R., & Sharma, S. (2022). Anti-nutrient & bioactive profile, in vitro nutrient digestibility, techno-functionality, molecular and structural interactions of foxtail millet (Setaria italica L.) as influenced by biological processing techniques. *Food Chemistry, f368*, 130815.

Sharma, R., Sharma, S., Dar, B. N., & Singh, B. (2021). Millets as potential nutri-cereals: A review of nutrient composition, phytochemical profile and techno-functionality. *International Journal of Food Science & Technology, 56*(8), 3703–3718.

Sharma, S., Saxena, D. C., & Riar, C. S. (2016). Analysing the effect of germination on phenolics, dietary fibres, minerals and γ-amino butyric acid contents of barnyard millet (Echinochloa frumentaceae). *Food Bioscience, 13*, 60–68.

Sharma, S., Saxena, D. C., & Riar, C. S. (2017). Using combined optimization, GC–MS and analytical technique to analyze the germination effect on phenolics, dietary fibers, minerals and GABA contents of Kodo millet (Paspalum scrobiculatum). *Food Chemistry, 233*, 20–28.

Sharma, S., Sharma, N., Handa, S., & Pathania, S. (2017). Evaluation of health potential of nutritionally enriched Kodo millet (Eleusine coracana) grown in Himachal Pradesh, India. *Food Chemistry, 214*, 162–168.

Sohan, S. Z., Kumari, B. A., Suneetha, J., & Gayatri, B. (2021). *Formulation and Quality Evaluation of Millet Flaked Snack Bar. The Pharma Innovation Journal,* 10(11S): 1937–1942.

Songré-Ouattara, L. T., Mouquet-Rivier, C., Icard-Vernière, C., Humblot, C., Diawara, B., & Guyot, J. P. (2008). Enzyme activities of lactic acid bacteria from a pearl millet fermented gruel (ben-saalga) of functional interest in nutrition. *International Journal of Food Microbiology, 128*(2), 395–400.

Subba Rao, M. V. S. S. T., & Muralikrishna, G. (2002). Evaluation of the antioxidant properties of free and bound phenolic acids from native and malted finger millet (Ragi, Eleusine coracana Indaf-15). *Journal of Agricultural and Food Chemistry, 50*(4), 889–892.

Subramanian, V., Murty, D. S., Jambunathan, R., & House, L. R. (1981). Boiled Sorghum Characteristics and Their Relationship to Starch Properties. *Materials Science, 23*, 21–25.

Suri, S., Dutta, A., Shahi, N. C., Raghuvanshi, R. S., Singh, A., & Chopra, C. S. (2020). Numerical optimization of process parameters of ready-to-eat (RTE) iron rich extruded snacks for anemic population. *LWT, 134*, 110164.

Udeh, H. O., Duodu, K. G., & Jideani, A. I. (2018). Effect of malting period on physicochemical properties, minerals, and phytic acid of finger millet (Eleusine coracana) flour varieties. *Food Science & Nutrition, 6*(7), 1858–1869.

Venskutonis, P. R., & Kraujalis, P. (2013). Nutritional components of amaranth seeds and vegetables: A review on composition, properties, and uses. *Comprehensive Reviews in Food Science and Food Safety, 12*(4), 381–412.

Verge, X. P. C., De Kimpe, C., & Desjardins, R. L. (2007). Agricultural production, greenhouse gas emissions and mitigation potential. *Agricultural and Forest Meteorology, 142*(2–4), 255–269.

Vilakati, N., MacIntyre, U., Oelofse, A., & Taylor, J. R. (2015). Influence of micronization (infrared treatment) on the protein and functional quality of a ready-to-eat sorghum-cowpea African porridge for young child-feeding. *LWT-Food Science and Technology*, *63*(2), 1191–1198.

Wani, H. M., Sharma, P., Wani, I. A., Kothari, S. L., & Wani, A. A. (2021). Influence of γ-irradiation on antioxidant, thermal and rheological properties of native and irradiated whole grain millet flours. *International Journal of Food Science & Technology*, *56*(8), 3752–3762.

Waniska, R. D. (2000, May). Structure, phenolic compounds, and antifungal proteins of sorghum caryopses. In *Technical and Institutional Options for Sorghum Grain Mold Management: Proceedings of an international consultation* (Vol. 18, p. 19).

Weller, C. L., Gennadios, A., & Saraiva, R. A. (1998). Edible bilayer films from zein and grain sorghum wax or carnauba wax. *LWT–Food Science and Technology*, *31*(3), 279–285.

Wu, C. X., Wu, Q. H., Wang, C., & Wang, Z. (2011). A novel method for the determination of trace copper in cereals by dispersive liquid–liquid microextraction based on solidification of floating organic drop coupled with flame atomic absorption spectrometry. *Chinese Chemical Letters*, *22*(4), 473–476.

Wu, X., Wang, D., Bean, S. R., & Wilson, J. P. (2006). Ethanol production from pearl millet by using Saccharomyces cerevisiae. In *2006 ASAE Annual Meeting* (p. 1). American Society of Agricultural and Biological Engineers.

Xiao, Y., Zheng, M., Yang, S., Li, Z., Liu, M., Yang, X., … & Liu, J. (2021). Physicochemical properties and in vitro digestibility of proso millet starch after addition of Proanthocyanidins. *International Journal of Biological Macromolecules*, *168*, 784–791.

Xu, W., Zhang, F., Luo, Y., Ma, L., Kou, X., & Huang, K. (2009). Antioxidant activity of a water-soluble polysaccharide purified from Pteridium aquilinum. *Carbohydrate Research*, *344*(2), 217–222.

Yadav, S., Mishra, S., & Pradhan, R. C. (2021). Ultrasound-assisted hydration of finger millet (Eleusine Coracana) and its effects on starch isolates and antinutrients. *Ultrasonics Sonochemistry*, *73*, 105542.

Zheng, X. W., & Han, B. Z. (2016). Baijiu (白酒), Chinese liquor: History, classification and manufacture. *Journal of Ethnic Foods*, *3*(1), 19–25.

2

Traditional vs. Modern Usage of Nutri-Cereals

P. S. Gaikwad, Nalla Bhanu Prakash Reddy, Sandhya K., and B. K. Yadav

Contents

DOI: 10.1201/9781003251279-2

2.1 Introduction

Millets are a group of small-seeded cereal grains that are a rich source of nutrition and widely grown around the world for consumption by humans and animals. Millets can grow in dry places as they require less water, fertilizer and pesticides for proper development (Birania et al., 2020). Most of the millet grains are a native crop of India and classified based on size, such as major millets and minor millets. The major millets are sorghum and pearl millet, and minor millets are barnyard millet, brown top millet, finger millet, foxtail millet, kodo millet, little millet and proso millet (Birania et al., 2020; Sathish, 2018).

Millets are known to be the oldest "poor man's grain," but their importance and cultivation were ignored owing to modern food habits. In India, before the Green Revolution, millet production was around 40 percent more than the total cereals, including rice and wheat. However, after the revolution millets were replaced with rice and wheat and millet cultivation decreased to below 20 percent (Singh et al., 2020). Therefore, millets are popular in India as the lesser-known grain.

Millets are more nutritious compared to rice and wheat, and recently the world has recognized their importance to a healthy life. Millets are rich in calcium, protein, minerals, vitamins, iron and fiber. They are also a good source of phenols that act as antioxidants to treat cancer and heart diseases. They are gluten-free and low in the glycemic index (GI), which helps diabetic patients by indicating whether carbohydrate-containing foods will raise blood glucose levels rapidly, moderately, or slowly (Birania et al., 2020). Therefore, the Government of India promoted millets as nutri-cereals under the National Food Security Act, particularly for rural areas. And millets became more popular in urban areas owing to their nutritional, nutraceutical and health promising properties.

Traditionally millets are consumed as a staple food in India. Some of the typical value-added millet dishes in India are *roti, bread, biscuit, papad, pancake, vermicelli, idli, dosa, vada, uttapam,* fermented malt and beverages and so forth (Birania et al., 2020). Millets are gluten-free and do not produce good leavened bread; however, they are combined with other milled cereal flour (Sathish, 2018). Nowadays modern processing technologies are utilized to increase the functional aspect of millets. However, food industries adopted various modern processing, including extrusion, flaking, malting, parboiling and popping to produce various snacks and fast food from millets. India is producing a greater number of extrusion-based millet products such as pasta, vermicelli, pet food, instant beverage products and so forth (Amadou, 2019). This shows the diversification of food industries and shifts towards modern millet-based healthy food products. This chapter covers traditional and modern nutri-cereals based food products from breakfast to the main course. Also, this chapter covers the consumption pattern of Nutri-cereals along with their health benefits to human beings.

2.2 Traditional usage of nutri-cereals

Millets are popularly known as traditional food grains and are also termed by the Government of India "nutri-cereals" comprising amaranth, barnyard millet, black heat, finger millet, foxtail millet, kodo millet, little millet, pearl millet, proso millet and sorghum. Traditionally, before consumption of millets, they are processed with milling, parboiling, puffing, malting and fermented food products and converted into the different food products delineated in this section.

2.2.1 Milling

Traditionally, the decorticating of millets was performed using hand pounding and later milling machinery (Birania et al., 2020). Generally, milling technology is utilized to remove the outer layer of millets and later pulverize into flour. Before milling, all foreign materials are removed and later passed through abrasive milling, where bran and aleurone layers of millets are removed and hard in texture. Afterwards, millets are pulverized into flour and further utilized for the preparation of flatbreads, bakery products, extruded products, and so forth (Gull et al., 2016). The milling process also helps to remove the antinutrients present in the outer layer of millets and thereby improve metabolic activity. Tiwari et al. (2014) reported that pre-milling treatment of inferred (110° C for 60 s) on pearl millet helps to increase the shelf life of flour by reducing antinutrients and free fatty acid content.

2.2.2 Parboiling

The parboiling process increases the hardness of the endosperm and improves the decortication yield by decreasing the sliminess of millets. Generally, the parboiling process consists of soaking, steaming or boiling and drying that improve milling efficiency and quality of millets (Birania et al., 2020). The parboiling process is also responsible for modifying carbohydrates, lipids, phenolics and protein content in millets (Bora et al., 2019). Bora et al. (2019) studied parboiling (soaking and boiling) of pearl and proso millets and observed a significant increase in decorticated yield and phenolic content and a decrease in glycemic index (GI).

2.2.3 Multi-grain flour/composite flour

Millet flour is produced in the milling process and later utilized in various products including *roti, bread, biscuit, idli, dosa, vada, laddu* and *kheer*. Milling of millet grains is performed by cleaning of grains then pulverizing them into the fine flour using a hammer mill. The storage life of millet grains flour is limited to one to two weeks at room temperature owing to the presence of fat, lipase activity and tannins (Yadav et al., 2012a). The above issue can be improved by blending millet grains flour with cereals flour, and it is termed as multi-grain flour that has a storage life of up to one month in low-density polyethylene (LDPE) packaging (Balasubramanian et al., 2021). The use of multi-grain flour is not new to the world. Traditionally it is utilized for the preparation of *roti*, bakery products, extruded products, fermented products and snack products (Balasubramanian et al., 2021; Birania et al., 2020; Verma and Patel, 2013). According to Birania et al. (2020), multi-grain *roti* made from millet grains flour with a combination of pulses are rich in nutrients, including dietary fiber, minerals, proteins and vitamins. Also, roti made from sorghum flour provide a good taste with improved nutritional quality (Rao et al., 2014). Jaybhaye et al. (2014) reported that the amount of millet grain flour increased in the blends decreased the gluten content and increased the ash content. The dry heating method can increase the pasting viscosity and storage life of multi-grain flour (Sun et al., 2014; Tiwari et al, 2014). Yadav et al. (2012a; 2012b) successfully utilized microwave heating and hydrothermal treatments for the reduction of lipase activity and tannins that improved the storage life of pearl millet flour by more than a month at ambient conditions. Akinola et al. (2017) utilized different processing techniques, including blanching, debranning, fermentation and malting for the improvement of quality and shelf life of millet flour. They identified the fermentation process and improved the pasting viscosity by reducing the bulk density of pre-processed millet grain flour. Also, they noted the malting process improved solubility and protein content of millet flour.

2.2.4 Flatbreads and pancakes

Traditionally flatbreads are categorized into two groups as fermented and unfermented. Flatbreads are fermented flatbreads that are traditionally made from pearl millet and kodo millet as a substituent of rice is known *dosa* and popularly consumed in southern India as a breakfast or dinner (Taylor and Duodu, 2019). Generally, non-fermented flatbreads are made from the flour of sorghum, finger millet and pearl millet, and are popularly known as *roti* in India and consumed as a staple food (Taylor and Duodu, 2019; Taylor and Emmambux, 2008). Other non-fermented flatbreads popular in India are *chapati*, *bakri* and *rotla* (Taylor and Duodu, 2019).

2.2.4.1 Roti/chapati

Roti is traditional fermented flatbread prepared from millet grain flour, and it is also popular as *chapatti* all over India. Also, it is known as an unleavened pancake. The dough of *roti* is prepared by mixing warm water into millet grain flour and kneading by hand. Then the dough is flattened by beating with hands or rolled out to a thin sheet of 1.3–3.0 mm with a diameter of 12–25 cm. Afterwards, the flattened sheet is baked on metal pan at a higher temperature to get a smooth puffed texture. The prepared *roti* is served with *sabjee*, chutney or pickles (Taylor and Duodu, 2019; Taylor and Emmambux, 2008). Gull et al. (2014) prepared multi-grain *roti* by a combination of wheat flour and finger millet flour (in 7:3 ratio) that reduced gluten content, and prepared *roti* were rich in fiber. Also, multi-grain *roti* has a good taste, but the color of *roti* turned to slightly dark brown compared to *roti* made from cereal flour (Verma and Patel, 2013).

2.2.4.2 Thalipeeth

Thalipeeth is a non-fermented flatbread popular in India as a multi-grain savory. It is prepared by a combination of pearl millet flour and rice flour with other ingredients such as onion, garlic, green chilies, coriander, ajwain, salt and oil. Afterwards, the dough is prepared by mixing all the ingredients and turning them into small balls. The balls are flattened and shallow fried in hot oil until they turn golden brown (Rao et al., 2016; Gupta and Paul, 2012). The prepared hot *Thalipeeth* is served with chutney, pickles or sauce. According to Nambiar and Patwardhan (2015), the *Thalipeeth* prepared from the millet grains show a lower glycemic index (GI) and glycemic load (GL).

2.2.4.3 Papad

Papad is popular Indian snack food owing to its high nutritional content. It is also consumed in many parts of Asia. Traditionally it is prepared from legumes and millets or rice as a core ingredient and blended with spices, salt, water and oil (Gaikwad et al., 2021). Later, the prepared dough is rolled to a flat sheet (0.3–2.0 mm thickness) and dried to preserve. Traditionally the open sun drying method is utilized for *papads*. Afterward, dried flat sheets are deep-fried in oil or toasted on direct flame and consumed with an accompanying meal (Gaikwad et al., 2021).

Papad from millets (finger millet) is more popular owing to its high nutritional profile. *Papad* from finger millet is prepared with a combination of salt and spices and cooked in water till it forms a gelatinized dough (Birania et al., 2020; Verma and Patel, 2013). Afterwards, the dough of *papad* is rolled into a flat sheet and dried conventionally or mechanically. However, fried *papad* prepared from millets is a little darker in color owing to the presence of pericarp in starch (Gull et al., 2014).

Prabhakar et al. (2016) prepared *papad* from a combination of sorghum (60%) and finger millet (40%) flours and observed a high amount of protein (8.39–12.64%) and carbohydrates (76.77–80.21%), and a lower amount of moisture (9–9.25%) and fat (0.88–1.05%) present in *papad*. To improve the physical characteristics of millet-based *papad*, Yenagi et al. (2010)

soaked a combination of millets (finger millet, foxtail millet and little millet) for three days before grinding and then frying; these fried *papads* had enhanced aroma, taste (sour) and expansion.

2.2.4.4 Khakhra

Khakhra is a crisp disc type of *roti* and a popular recipe of Gujarati and Rajasthani peoples of India that is consumed as a roasted vegetarian snack food. The main ingredient required for the preparation of traditional *khakhra* is wheat flour, spices, water and oil (Giridhar, 2019). *Khakhra* prepared with millet provides more nutrients in terms of carbohydrates, dietary fibers, minerals and protein in comparison to traditional *khakhra*. Millet based *khakhra* prepared with sorghum or pearl millet or finger millet or foxtail millet flour blended with Bengal gram (chickpeas) or black gram flour, salt, spices and oil. Combine all ingredients with hot water and knead into a soft dough. Afterwards, the dough is rolled into a flat sheet and roasted or baked from both sides (Rao et al., 2016). Umarji and Vijayalaxmi (2020) prepared *khakhra* from kodo millet (60%) flour and observed an increase in carbohydrate (57%), dietary fiber (13%), protein (12%) and calories (295 K). The microbial load of *khakhra* prepared with kodo millet is shelf-stable for up to 60 days at ambient conditions.

2.2.5 Pancake

Pancake is a round flat bread cake prepared from cereal or millet flour and popularly consumed in India. The required ingredients for making millet pancakes are sorghum flour, milk powder (non-fat), baking powder, salt, sugar, eggs, oil and warm water. Prepare separate mixture of all dry ingredients in one bowl and all wet ingredients (eggs, oil and water) in another bowl. Afterwards, add all dry ingredients into wet ingredients and mix well. Then bake until a golden brown color appears (Rao et al., 2016).

Ren et al. (2016) prepared a different combination of pancake and studied the glycaemic response of foxtail millet. In their study they prepared two pancakes, one with 100 percent foxtail millet flour and another with a combination of foxtail millet flour (75%) and extrusion flour (25%). They observed that the prepared pancake with only foxtail millet has lower glycaemic index (GI–76.2 ± 10.7) compared to pancake prepared with combination of flour (GI–83.0 ± 9.6). Foxtail millet is a good source of nutraceuticals with low GI value that help to delay the growth of type 2 diabetes.

2.2.6 Popping or puffing

Popping or puffing is a traditional method of cooking millets for the preparation of ready-to-eat (RTE) breakfast food products with desirable texture and aroma (Birania et al., 2020). In the popping process, millets are exposed to high temperatures for a short time (HTST), which gelatinizes the starch and expands the endosperm layer by breaking out the outer layer of millets. While puffing, vapor pressure is released to control the expansion of millets (Mishra et al., 2014). The popping or puffing process is carried out with cereals and millets, including, corn, wheat, sorghum, ragi and foxtail millet.

Traditionally, the popping or puffing of millet (finger millet) grain is conditioned by the addition of 18–20 percent moisture and tempered for HTST under the shed. During the popping or puffing process, the temperature is held between 230–250° C for 4–6 h (Birania et al., 2020; Gull et al., 2014; Verma and Patel, 2013). This process increases the chances of Millard reaction owing to operating at such a higher temperature that reduces the antinutritional components (phytates and tannins) and increases the bio-ability of millet (Jaybhaye et al., 2014). During the Maillard reaction, carbohydrates present in the millets react with the amino acids owing to a higher temperature and produce a distinct aroma and texture to the final product. Afterward, the vapor pressure of millet grain increases, and moisture evaporates in the form of steam. And at the end, starch becomes gelatinized and explodes. In this process, millet grains are dried to a 3–5 percent moisture level to extend the shelf life (Gull et al., 2014; Jaybhaye et al., 2014; Mishra et al., 2014; Verma and Patel, 2013).

Nowadays various modern air popping or puffing methods are utilized, such as sand and salt treatment, dry air, hot air, microwave heating, hot oil popping and gun puffing (Mishra et al., 2014). Kumari et al. (2019) developed RTE breakfast cereal product including popped pearl millet and amaranth. They observed that the developed RTE breakfast food is rich in protein, minerals, folic acid and dietary fibers, and enhanced in sensory attributes.

2.2.7 Semolina (rava/suji)

Semolina is popular as *Rava* or *Suji* in India. Semolina is the coarse product obtained during the milling of cereals and millet grain. Traditionally, semolina is prepared from sorghum, finger millet, foxtail millet and pearl millet and utilized for the preparation of RTE breakfast foods, *upma, khichdi, idli, dosa, pasta, vermicelli* and so forth. Semolina is prepared by soaking millet grains for at least 12 hours and later drying them to a moisture content of 21. Afterward, dried millet grains are roasted and then pulverized in milling to obtain semolina and flour. Generally, in the market two types of semolina are available based on their particle size. One is coarse and another fine, 1.18 mm 0.71 mm particle size, respectively (Rao et al., 2016). The recovery of semolina during milling depends on the variety of millet and utilized milling machinery. While milling semolina, the recovery of coarse ranges from 50–85 percent, and fine, 40–75 percent and leftover for flour. Traditionally, semolina is prepared from sorghum, finger millet, foxtail millet and pearl millet and utilized for the preparation of RTE breakfast foods, upma, khichdi, idli, dosa and so forth. Rao et al. (2016) reported that semolina prepared from sorghum is rich in energy (353 Kcal), carbohydrates (77 g) and Protein (6.6 g).

Dharmaraj et al. (2016) prepared semolina from pre-treated foxtail millet. They reported pre-treatment of steaming for 15 min helps to increase the yield of coarse semolina (75%). Also, they observed polishing (3%) reduced ash, dietary fiber, fat and protein content of pre-treated foxtail millet. According to Balasubramanian et al. (2014), hydrothermal treatment (soaking and steaming) helps to reduce the anti-nutrition compounds and inactivate lipase activity in pearl millet semolina.

2.2.8 Malting–weaning food

Malting is also referred to as "sprouting" of millet grains. Traditionally, millet malt is utilized for the preparation of beverages combined either with milk from lukewarm water with sugar. Finger millet malt is commonly used for weaning foods owing to its easy digestibility for the elderly and infants (Rathore et al., 2019; Gull et al., 2014). Finger millet is a good source of amino acids, calcium and amylase activity compared to sorghum and other millets (Birania et al., 2020). According to Birania et al. (2020) malting of finger millet reduces its anti-nutritional compounds and increases bio-availability digestibility and sensory attributes.

The malting process is carried out by soaking millet grains into the water to hydrate them, followed by sprouting for 4–5 days at ambient conditions. Later, sprouts are dried (< 75° C) to

10–12 percent moisture level and the roots and shoots are separated manually by rubbing with hands. Afterward, these malted grains are roasted by toasting pan at 70–80° C, and obtained malt is pulverized to a fine flour (Taylor, 2017; Gull et al., 2016; Verma and Patel, 2013). Then malted weaning food is blended with powdered milk, sugar and flavoring compounds. It is popularly known as "Ragi malt" in Karnataka and Tamil Nadu and is consumed as an energy drink.

2.2.9 Fermented foods

Fermentation is the ancient preservation technique that provides distinct aroma and taste to a particular food product. The fermentation process helps to decrease the antinutrients and improved nutritional properties such as calcium, fiber, protein and vitamin B (Birania et al., 2020). *Idli* and *dosa* are popular fermented food products consumed as breakfast as well as evening meals in many parts of India (Rathore et al., 2019). The core ingredient for the preparation of *idli* and *dosa* is rice, which can be substituted with millet according to the consumer's choice. Finger millet is widely utilized for the preparation of these kinds of fermented food products and blended with the black gram (in 3:1 ratio) (Birania et al., 2020; Gull et al., 2014). Sprouted and malted millet grains are also utilized in fermented food products to improve the taste and nutritional properties (Gull et al., 2014). Germination and fermentation of millet grains improve the protein digestibility, lysin, sugars, dietary fiber, niacin, thiamine. Jaybhaye et al. (2014) reported that germinated pearl millet was ground and fermented and observed a significant increase in protein digestibility of more than 90 percent.

2.2.9.1 Idli

Millet *idli* and *dosa* are mainly prepared from sorghum, finger millet and proso millet and combined with black gram dal. For making *idli*, millet semolina (3 parts) and black gram dal (1 part) are separately soaked in portable water for 3–4 hours at ambient conditions. Then black gram dal is wet ground till it attains a soft batter. Drain out the water from the millet semolina and blend it with a soft batter of black gram. Later, salt is added as per the requirement (2%) and allow the mixed batter to ferment overnight at ambient conditions. Afterward, pour the batter into *idli* molds and allow them to steam for 10–15 min. Then cooked *idli* is served with chutney and sambar (Rao et al., 2016).

Nazni and Shalini (2010) prepared *idli* by replacing rice with pearl millet and combined with black gram. They observed that pearl millet *idli* improved nutritional properties such as calcium, fat, fiber, iron, protein and energy.

2.2.9.2 Dosa

Dosa is prepared by soaked millet grain (3 parts) and black gram dal (1 part) and grounded together into a fine batter. Then salt is added (2%) and allows the batter to ferment overnight. Afterward, pour the fermented batter on preheated *dosa* pan and spread it into a thin round shape. The fried crispy *dosa* is obtained and served with chutney and sambar (Rao et al., 2016).

To determine the glucose level in type 2 diabetic patients, Narayanan et al. (2016) prepared foxtail millet *dosa* and compared it with rice *dosa*. They observed that the GI of foxtail millet *dosa* was lower (59.25) compared to rice *dosa* (77.96).

2.2.9.3 Vada/wada

Vada or *Wada* is popular in southern India as a breakfast food product. Millet *wada* is prepared by parboiled millet grains (3 parts) wet ground to a coarse paste. Then the prepared paste is blended with Bengal gram flour (1 part), salt, chili and spices as per the requirement. Later, small balls are prepared from the blended mixture and turn into the *Wada* by frying it in the hot oil. Excess surface oil is removed and served with chutney and sambar (Rao et al., 2016).

2.2.9.4 Uttapam

Uttapam is also one of the popular breakfast foods of southern India. Millet *uttapam* is made from millet grain (3 parts) and black gram dhal soaked separately into the portable water for 3–4 hours. Then drain out the water and wet grind together into a fine batter. Later, add salt and ginger-green chili paste as per the requirement into the blended mixture. Preheat the pan and apply oil (1/2 tsp). Then pour the batter on the pan and sprinkle chopped onions, tomatoes and coriander. Press *uttapam* lightly from both sides till it completely cooks, serve with chutney and sambar (Rao et al., 2016).

2.2.9.5 Pesarattu

Pesarattu is a crispy thin *dosa*, popularly consumed in southern India. The core ingredients required for millet *pesarattu* is pearl millet and green gram. Soaked pearl millet (1 part) and green gram (1 part) together for 5–6 hours at ambient conditions. Then water is drained out and wet ground to a smooth batter. Add salt, green chili, ginger, chopped onions and coriander leaves as per the requirement. Pre-heat the pan to medium flame and pour batter on the hot pan. Cook the *pesarattu* from both sides and serve hot with chutney and sambar (Rao et al., 2016).

2.2.9.6 Adai

Adai is a healthy protein rich *dosa* popular in southern India. The core ingredients required for the preparation of *adai* are kodo millet (2 parts), toor dal (1 part), channa dal (1 part), moong dal (1 tsp) and urad dal (1 tsp). All the core ingredients are soaked for 4–6 hours at ambient conditions. Then water is drained out and the ingredients are ground together to obtain a coarse batter. Add salt, green chili, chopped onions and coriander leaves as per the requirement. Then prepare a round shape *adai* and cook in a pre-heated pan on medium flame till it turns golden brown and crispy. Serve hot with chutney or pickle (Rao et al., 2016).

2.2.10 Alcoholic and non-alcoholic beverages

Traditionally, millet-based alcoholic and non-alcoholic beverages are popular for human consumption (Francis et al., 2017). A natural fermentation process is utilized for millet-based fermented beverages that improve aroma and taste (Amadou, 2019). During the fermentation process, carbohydrates present in millets breakdown owing to the microorganisms or enzymes and result in primary metabolites, such as carbon dioxide, alcohol, acetic acid, or lactic acid (Limaye et al., 2014). Consumption of fermented millet-based beverages helps to maintain the immune system (Takahashi et al., 2017). A few popular Indian fermented beverages are given below:

2.2.10.1 Ambali

Ambali is a popular south Indian non-alcoholic summer drink. It is prepared by cooking sorghum flour (1/2 cup) in warm water and then rice starch soup and salt are added as per requirement. Cook the mixture for 15–20 min on medium flame and later cool down to ambient conditions. Then it can be served directly with the addition of buttermilk, spices and coriander leaves as required. Also, it can be stored in an earthen pot after cooking and then allowed to ferment overnight for consumption the next day (Rao et al., 2016). The prepared *Ambali* drink is rich in carbohydrates, calcium and fiber that provides energy and maintains the immune system.

2.2.10.2 Sarbat

Millet *sharbat* is mainly prepared from sorghum flour. Take sorghum flour and barley in a 2:1 ratio and boil for 15–20 min on a medium flame. Then add lemon, sweetener and peppercorns as per the requirement and mix all the ingredients. Allow the drink to cool down in refrigerated storage and serve as a cool drink with garnishing as per the requirement (Rao et al., 2016).

2.2.10.3 Lassi

Lassi is a sweet non-alcoholic fermented drink, popular in India. The millet-based *lassi* is mainly prepared with sorghum flour. Boil milk (100 ml) and cool it down to lukewarm. Then add germinated sorghum flour (5 g) and heat on a medium flame for 10 min. Allow it to cool down to ambient conditions and add curd culture into the mixture to become curd. Then store it in refrigerator condition and served chilled. To make jeera *lassi*, add cumin powder and salt as per the requirement and stir well. To make sweet *lassi*, add sugar and stir well (Rao et al., 2016).

2.2.10.4 Koozh

Koozh is a traditional Indian drink prepared with germinated finger millet flour. The millet flour is added to water, and the prepared slurry is allowed to ferment overnight at ambient conditions. Then next day, add 20 percent of broken rice cooked in water and mixed into the fermented millet slurry. Then the mixture is cooked on a medium flame to turn into thick porridge and ferment for 24 hours at ambient conditions. The resulting product is called *Kali*. To make semi-solid porridge, add water in a 1:6 ratio and salt as required and served *Koozh*. The *Koozh* is further mixed with yoghurt to get a thin porridge consistency. The shelf life of *Koozh* is short, about 12 hours after the preparation in ambient conditions (Amadou, 2019).

2.2.10.5 Kunu

Kunu is a fermented non-alcoholic drink prepared from sorghum, pearl millet or maize and flavored with sugar and spices (Taylor and Duodu, 2019). *Kunu* is also popular worldwide as *Kunu-Zaki*. It can be prepared by two different methods.

First method: millets are soaked in portable water with species (black paper, clove or ginger) for 24 hours at ambient conditions, then wet grounded, sieved and allowed filtrate and settle to the bottom. Later, the supernatant is removed from the filtrate. Then one portion of the sediment is cooked and mixed well with the uncooked sediment portion that acts as an inoculum.

Afterward, the mixture is allowed to ferment for 8–10 hours at ambient conditions. Later, the mixture is sieved and served with sugar as per the requirement (Amadou, 2019).

Second method: malted millet flour is mixed well with the uncooked millet portion. Then the mixture is mixed well with the cooked portion and allowed to ferment overnight at ambient conditions. The next day the fermented drink is served as per the requirement (Amadou, 2019).

2.2.10.6 Mageu

Mageu is a non-alcoholic fermented drink prepared from sorghum and millets. For *Mageu*, sorghum porridge is cooked to solid consistency of about 8–10 percent. Then porridge is cooled at ambient conditions. Later, malt of sorghum or finger millet is incorporated and stirred well. The prepared mixture is allowed to ferment for 24–48 hours in ambient conditions. The obtained final product is filtered and has a sour taste owing to the development of lactic acid (Taylor and Duodu, 2019).

2.2.10.7 Beer

Beer is popularly known as an alcoholic fermented drink. Sorghum and millet-based *beer* are more popular owing to their distinct aroma and taste. Lactic acid bacteria are mainly responsible for the acidic taste of millet *beer*. The *beer* prepared from malted sorghum or finger millet or pearl millet has an alcohol content of < 4 percent (v/v). However, there are two types of *beer*, such as opaque *beer* and cloudy *beer*. Opaque *beer* is viscous and has a thin consistency owing to the presence of starch and grain particulate matter. Cloudy *beer* is viscous-free because it is clarified to remove starch particulate matter (Taylor and Duodu, 2019).

2.2.10.8 Oti-oka

Oti-oka is a popular Indian fermented alcoholic beverage prepared from sorghum and pearl millet. It looks to be opaque in color and has a good aroma and slightly sour taste (Aka et al., 2014). *Oti-oka* is prepared from malted millet flour and mixed with water in a 1:5 ratio and allowed to ferment overnight at ambient conditions. The next day filtrate the mixture and boil for about 3 hours on medium flame. Later, cool down to ambient conditions and ferment for 72 hours. The prepared *oti-oka* has a slightly acidic taste owing to lactic acid bacteria and yeast (Amadou, 2019).

2.3 Modern usage of nutri-cereals

In the traditional method of millet processing, product preparation includes processes like dehulling or decortication, milling with a wide range of technologies, different fermentation methods and brewing techniques in the case of liquid or semisolid food products. But, in the current scenario, a major section of society is being gratified by RTE food products or products that can be cooked or made edible within minutes. These products by their ability to meet the requirements onsite and at once. Not only this being one of the reasons, but also the economical situations and malnutrition in the society in countries like India, Africa, and other developing countries or underdeveloped countries led to rethinking of the inclusion of millets in the daily intake (Seth and Rajamanickam, 2012).

Complete millet-based diet or composition of millet or millet compounds in significant amounts enable the population to reach their minimal nutrient requirements. Compounding all the above-mentioned factors, RTE food products have gained more importance. These RTE food products are prepared by multiple technologies, but this chapter focuses on extruded millet-based foods (Birania et al., 2020), baked foods (Verma and Patel, 2013), and other RTE food products.

2.3.1 Extruded millet products

Extrusion cooking has been one of the modern-day food processing technologies applied in the preparation of snack foods (Sumathi et al., 2007). Extruded products such as breakfast cereal and flakes have an influence on consumers as they are easy to prepare, nutrient rich and can be consumed on daily basis by all age groups (Narang et al., 2018). Extrusion technology is a high temperature and cost-effective method wherein raw material paste is plasticized, cooked, pressurized, and made to pass through a small opening called a 'die' to be expanded and dried (Kharat et al., 2019; Narang et al., 2018). Synergistic effects of pressure and heat are predominantly responsible for the modification in starch properties (Verma and Patel, 2013). This technology is used to produce pre-fabricated foods, RTE foods that are expected to be produced with the desired size and texture, and Ready-to-Cook (RTC) foods that can be served instantly (Dhas et al., 2021). Value added traditional extruded products have a high market value due to their nutraceutical effect on consumption (Kotagi et al., 2013). The processing of millets cause a reduction in antinutrient content, and enhance digestibility and shelf-life (Patil et al., 2015). This technology is used to make *flakes*, *Muesli*, *Murukku*, noodles-*vermicelli*, *utria-pasta*, and much more. The above mentioned are discussed individually below as follows:

2.3.1.1 Flakes

Extruded products, especially millet *flakes*, are eaten with milk; sometimes these *flakes* are RTC (Ranjitha Devi et al., 2016). Due to many factors – namely economic situations, nutritional awareness, easy to eat nature, changing food habits, and so forth – millet *flakes* have earned significant attention in society. Generally, millet *flakes* are obtained by tempering the grains under pressure. *Flakes* are produced in two different ways, by the process of extrusion and by roller flaking. The roller flaking method of production involves the tempering of grains with 5–10 percent moisture. It is then pre-cooked to 180° C at 20–24lbs./psi for 20 min in a rotary steamer for controlled gelatinization. The grains are cooled to surface dryness, after which they are flaked by passing them through heavy-duty rollers with 0.25 mm aperture. The flakes are sun-dried for 4–6 hours until the moisture content of the product reaches 10.5 percent. The extrusion method of *flake* production involves tempering grains to 19–20 percent moisture and grinding them into a thick slurry. The slurry is then passed through a twin-screw extruder at 155±5° C with barrel rotating at 200 rpm. The extrudate is dried to 5 percent moisture before storage. The above said method of *flakes* production is followed for the production of pearl millet, finger millet (Zacharia et al., 2020), little millet (Patil et al., 2015; Kotagi et al., 2013), little millet in combination with garden cress seeds (Takhellambam et al., 2016), a combination of pearl millet (Adebanjo et al., 2020) and various other minor millet varieties.

Narang et al. (2018) developed pearl millet *flakes* with the help of an extruder and analyzed their functional properties. It was found that the water absorption index, which generally is regarded as an index of gelatinization, shared an inversely proportional relation with feed moisture; the water solubility index showed an increase with temperature rise during extrusion and reduced with feed moisture enhancement and, similarly, the expansion ratio showed proportional relation with temperature and moisture.

RTC millet *flakes* could help in the utilization of millet flour as a whole or as a portion of a blend with other cereal flours like rice and wheat flours that have been in use for decades, and these would encourage both farmers and food industries to reap the benefits (Ranjitha Devi et al., 2016).

2.3.1.2 Muesli

Millet *muesli* is another important extruded millet based RTC food product that can be eaten at breakfast by soaking overnight in milk in a refrigerated environment, like a pancake, by making it as a batter with a mix of milk and other ingredients according to requirement, as a dessert, and as a mid-day snack, too.

Muesli is a nutrient-loaded snack that is prepared from rolled oats, nuts, seeds and dried fruits. In the case of millet-based muesli, rolled flaked millet can be used in the place of oats to formulate a nutrient-dense snack that can be consumed raw or with the addition of milk/yogurt. Rolled oats, wheat bran, whole rye, whole barley, sorghum flakes, quinoa flakes, millet puffs and millet flakes are transferred to a bowl. Nuts (sliced almonds, walnuts, cashew, pecans, pistachio and hazelnuts) and seeds (sunflower seeds, chia seeds, poppy seeds, sesame seeds) are added to the above mixture. The ingredients are roasted for 10–12 min until a characteristic aroma is achieved. After cooling down, dried fruits (dried apricots, cherries, figs, raisins, apple chips, and so forth) are added and combined. *Muesli* made with millets or millet mix is rich in protein and dietary fiber content, and mineral content like iron, magnesium, calcium, and so forth.

2.3.1.3 Muruku

Murukku (also called *muruku* or *janthikalu* in Andhra Pradesh) is a crunchy snack generally made from rice flour or a mix of rice flour with Bengal gram or black gram flour in proportions that vary from place to place. The primary step involves mixing and roasting of grains followed by grinding of the same into a fine powder. To the ground powder, salt, chili powder and cumin seeds are added to impart taste and flavor. It is then kneaded into a hard dough with the addition of water. *Muruku* is prepared by forming the paste into small cylindrical coil-shaped products, either by hand or mostly extruded with the help of a mold, and then fried in oil (180–202 °C) to impart crunchiness, taste, and extended shelf life due to moisture removal. Foxtail millet is the most suitable millet for *muruku* making, and the process is almost similar to that of rice flour (Chera, 2017; Rao et al., 2016). When the *muruku* raw materials are mixed with Bengal gram flour, it is called *chakli*. Foxtail millet shows better sensory characteristics and maintained a crispy and crunchy nature (Namitha et al., 2019).

2.3.1.4 Noodles – vermicelli

Noodles are regarded as a global food as they are consumed by people of all age groups. It is made from gluten-based products. With the increasing consumption of millet-based products, consumers choose millet noodles in place of gluten-based products. Noodles can be prepared from more than one type of millet, such as from barnyard millet and mixed with other additives, malted finger millet flour (Lande et al., 2017), and foxtail millet (Pandey et al., 2017). *Vermicelli* is kind of very thin *Pasta* mostly in the form of thin noodles and smaller in the length comparatively. *Vermicelli*-noodles have become very popular because they are very easy to cook and convenient to handle.

These are generally made from the mix of millet and wheat flours in 1:1 ratios (Verma and Patel, 2013). Wheat flour plays an important role in giving noodles a smooth texture and helps in imparting strength while extrusion process. Millet and wheat flour with the desired proportion is weighed and sieved to obtain uniform particle size. Hydrocolloids are added at gel consistency along with the addition of 2 percent NaCl and 1 percent NaOH. Water is added at 45° C to knead the flour followed by sheeting and conditioning for 15 min. The dough is transferred to the mixing vat of a cold extruder and extruded using a 1.2 mm diameter die. The noodles are steamed at 100–102° C for 3 min or fried after extrusion. Noodle strips are dried in a tray drier at 85° C for 90 min until the moisture content reaches below 10 percent.

Pandey et al. (2017) evaluated the cooking, physico-functional and textural properties of foxtail millet-based *vermicelli* and found that overall sensory characteristics were improved with foxtail millet flour, length of *vermicelli*, and non-starchy mouth coating were also seen to be enhanced. A significant reduction in cooking time was observed and cooking losses and similar results were also observed by Shukla and Srivastava (2014) for finger millet. Noodles formed with the blend of finger millet and wheat flour mix (30:70) showed a considerable reduction in glycemic index (GI), a higher nutritional value and greater dietary fiber content (Shukla and Srivastava, 2014), iron, and calcium content (Lande et al., 2017).

2.3.1.5 Nutri-pasta

Pasta is prepared in three major steps. The first step includes cleaning, sieving, and blending of different flours, namely, wheat flour, millet flour, and so forth. The second step includes mixing with warm water, forming into the paste and extruded with the help of an extruder or hand-operated mold. The final step includes pre-cooking of extruded *pasta* for some time in warm water or steaming followed by drying before packaging (Sarojani et al., 2021).

The process of preparation of pasta involves mixing wheat semolina and millet flour (mostly in the ratio 50:50) and sieved to obtain a particle size of 425 μm. Water is added in the required amount to form a dough. The dough is then passed through a twin-screw extruder with a concentric double cylinder die to obtain cylindrical hollow pasta (penne) with a wall thickness of 0.91 mm and length of 15 mm. The extruded pasta is tray dried at 50±2° C for 2 hours to obtain 8–9 percent final moisture content (Jalgaonkar et al., 2018).

Millets provide a different dimension to the nutritional value, which caters to the need of the present situation wherein people with diabetes are striving for a comprehensive diet that provides both native tastes, fewer carbohydrates, and greater dietary fiber (Rathore et al., 2019; Sarojani et al., 2021). RTC *pasta* was prepared with tropica flour and foxtail millet at various mixture ratios wherein foxtail millet flour was taken in 50, 55, 60, 70, and 80 percent and was assessed for cooking quality, functional and physico-chemical properties along with organoleptic evaluation. It was found that 70 percent of foxtail millet yielded better results of all the above-mentioned parameters (Dhas et al., 2021).

2.3.2 Bakery products

Most of the bakery products are made from cereals, in particular from wheat flour. But the major challenges of continuing with millet flour are (i) gluten intolerance section of population and (ii) Celiac disease is caused by some parts of wheat protein. To overcome these avoidable challenges, millet flour has been seen as a decent alternative. The bakery industry includes flour-based food products such as biscuits, bread, multi-grain flour bread, cookies, cakes, pastries, pies muffins, pizza, and so forth. (Aggarwal and Sharma, 2020).

Blending of millets and millet flours in raw materials for bakery product preparation not only enhances micronutrients – the dietary fiber, calcium, iron and zinc contents – but also creates a good scope for novel and healthy products in the near future (Verma and Patel, 2013). Baking allows the biofortification of millet-based food products with a wide range of minerals from a variety of sources (Balasubramanian et al., 2021). Some of the bakery products are discussed below.

2.3.2.1 Multi-grain millet bread

Bread has been known as the most popular food product for centuries due to its high nutritive value and unique organoleptic acceptability. Wheat flour is the major ingredient in bread making and due to the above mentioned reasons, there has been an increasing need to replace the content of wheat flour or gluten with alternative sources. A portion of healthy food is well balanced in terms of quantity and quality of ingredients from a variety of food groups. But, unfortunately, a single group of food cannot provide the wholesomeness to any food product, which led to the concept of a multigrain food system. One such alternative that ideally suits the requirement is millet. Multigrain blends with millet along with other cereal grains help us to maximize the nutritional, functional, and organoleptic perception of the products (Pawar et al., 2020). Formulation of the composite flour for bread making is always a tough task (Sharma et al., 2018). Agarwal et al. (2017) attempted to develop and assess the quality of multigrain bread made with different combinations of wheat flour: buckwheat flour: pearl millet flour, in different combinations namely, 85:10:5, 80:15:5, and 75:20:5. It was found that protein and ash content were increased with buckwheat flour and millet flour.

Sarabhai et al. (2021) prepared foxtile millet bread with enzymes to impart good physical and textural properties. Protease, glucose oxidase and xylanase are added, in which protease increased the specific volume and springiness of bread due to hydrolysis of protein during batter development. Sorghum-based bread with the inclusion of sprouted millet at 5 percent had better retention of air pockets during baking and improved crumb softness (Comettant-rabanal et al., 2021).

Gluten-free bread is prepared with a wide range of millets such as pearl millet (Nami et al., 2019; Wang et al., 2019), minor millet flour (Sharma and Gujral, 2019), proso millet (Tomić et al, 2020) and sprout millet flour (Comettant-rabanal et al., 2021) having reduced GI value.

2.3.2.2 Nutri-biscuits

Biscuits are the most convenient, cheap, RTE, and most popular in the major section of society. Currently available biscuits are mostly products of wheat or refined wheat flour. These biscuits may provide easy energy for the bigger part of the community. But, these biscuits may not be suitable for the section of the population with diabetics, obesity, cardiac issues, hyperlipidemia, cancer, and hypertension (Hymavathi et al., 2020). By utilizing the mix of finger millet, pearl millet and pearl millet flours along with fortification of other necessary ingredients can increase the carbohydrate, protein, fat, and several other nutritional aspects of food (Saini et al., 2021).

Millet-based biscuits exhibit a bland taste and require further addition of flavor or taste improving substances for sensorial acceptance by the consumer. Biscuit preparation involves the blending of powdered sugar with fat and vanilla extract. The blending process is carried out for 2–3 min until uniform texture appears with the development of air pockets and left undisturbed. In a planetary dough mixer, dry ingredients such as wheat flour and millet flour in the desired proportion, skim milk powder, glucose, NaCl, baking powder, sodium and ammonium bicarbonate are mixed thoroughly. To this, fat and sugar mix is added with water to form a crumbly dough. The dough is sheeted on a clean smooth surface using a rolling pin manually or by using a dough sheeter to form flat sheets. The sheet is wire cut into biscuit shape or shape cutters can be used to form shapes of various kinds manually. The cut pieces are transferred to a baking tray and baked at 205° C for 7–8 min. after baking, the biscuits are cooled to ambient temperature and packed (Kaur and Kumar, 2020).

Hymavathi et al. (2020) developed biscuits based on different formulations with the ingredients like wheat flour, skim milk powder, fat, peanuts, coconut powder, and sugar along with pearl millet, sorghum, and finger millet flour on a commercial scale. It was found that millet-based biscuits showed high dietary fiber content and low glycemic index (Biswas, 2015).

2.3.2.3 Nutri-cookies

Besides healthy products, consumer perception has always been on their taste bud satisfying foods. Biscuits, cakes, cookies, and pastries are some of the bakery food products that fall under the interest of this section of people. Cookies are convenient, inexpensive, and RTE foods, and are widely consumed all over the world (Kulthe et al., 2017).

Cookies prepared from millet include a combination of wheat flour and composite millet flour, the most common being little millet. The initial step in the processing of cookies includes dough preparation by mixing and combining millet flour (100%), sugar (58%), shortening such as margarine (28%), common salt (0.9%), sodium bicarbonate (1%) and dextrose 13.8 ml (prepared from 8.9 g glucose in 150 ml water) with the required amount of water. The dough is mixed thoroughly in a planetary mixer and dough is flat sheeted. Then sheeted dough is cut into the desired shape and transferred to the baking tray. Baking is done in an oven that is pre-heated to 204 °C for 10 min (Rai et al., 2014).

Wandhekar et al. (2021) developed cookies with regular wheat flour, sugar, milk powder, and fortified with finger millet (ragi) flour. The optimized cookies showed greater overall acceptability and greater shelf life. Similarly, pearl millet fortified cookies were affected for some of their physical and textural properties with an increasing component of pearl millet flour (Kulthe et al., 2017).

2.3.2.4 Savory cake

Savory cake, otherwise in simple and traditional terms called idli or rice cake (Nazni and Shalini, 2010). *Savory cakes* are made with the addition of herbs, spices and cheese. They are mostly added with fried onions to impart a peculiar taste of both spiciness and sweetness on consumption. The oven is pre-heated to 180° C. Butter and cottage cheese are combined till pale smooth cream appears. To this mixture, egg, sauteed onions, cheddar cheese and herbs are added and mixed thoroughly. Desired millet flour with a small quantity of all-purpose flour, baking powder, salt and pepper are added to the mixture and combined well, then transferred to a baking pan and topped with onion rings. Baking takes place for 45 min to 1 hour at 220° C until golden brown crust appears. The baked cake is cooled and sliced before consumption. Apart from the sauteed onions, olives are also added to enhance taste and flavor. Millet-based *savory cake* is still unexplored in the market.

2.3.2.5 Chocolate pudding

Pudding is nearly always a sweet dessert of milk or some fruit juices, variously flavored with rice flour or starch, corn starch, tropica, bread, or eggs. Chocolate *pudding*s are flavored with chocolate. Other than the traditional type of *pudding* preparation, *pudding*s are either boiled or steamed or boiled then chilled, dessert, texturally a custard set with starch, and a steamed or baked version, texturally similar to cake.

In India, kodo millet is dried and ground into flour and used to make *pudding*. The preparation of millet *pudding* includes milk, sugar, wheat flour at desired quantity of millet flour, cocoa powder and vanilla essence are mixed thoroughly until no lumps remain. The mixture is heated to 75° C until the gelatinization of starch occurs and held at the same temperature for 10 min. It is then transferred to a glass container and allowed to set completely at 4° C. The method is slightly modified from that stated in Samanian and Razavi (2017).

The chocolate made from millet flour mix and vegetable milk is free from gluten and is treated as vegan food. The millet insoluble fiber acts as prebiotic food and helps in feeding helpful bacteria in our gut. Little millet can be best used in the preparation of utria chocolate *pudding* as the color of the *pudding* remains unaffected.

2.3.2.6 Muffins

There has been a consistent rise in the popularity of bakery foods such as biscuits, bread, cakes, cookies, *muffins*, and so forth. Out of these, *muffins* are sweet, high-calorie products liked by almost all age groups by their soft texture and adorable taste. Generally, the major ingredient in *muffins* is wheat flour, just as in the case of any other bakery product. Wheat has been a point of concern for the last decade or so because wheat flour, which is rich in gluten, is not recommended for gluten intolerant people, persons with celiac disease. Millets, barnyard millet, in particular, is gluten-free, thereby it can be employed in the preparation of gluten-free products that bring out new opportunities in snack, baking or similar types of food industries and ultimately help gluten intolerant people (Goswami et al., 2015).

Muffin preparation involves sifting all-purpose flour, composite millet flour, powdered sugar, salt and baking powder. Butter and powdered sugar are mixed in a planetary mixer to obtain a uniform texture. In a separate mixer, egg and essence are whipped, which is then transferred in parts to buttercream to form a homogenous mixture. To this, the desired flour, salt and baking powder are added and stirred well. The batter is then transferred to greased muffin pans and baked at 180° C for 25 min. It is then cooled to ambient temperature and packed (Jadhav et al., 2021). Millet-based *muffins* are processed from pearl millet (Hassan et al., 2020; Mcsweeney et al., 2017), and foxtail and kodo millet (Barbhai et al., 2021; Rao et al., 2016).

Jyotsna et al. (2016) attempted to prepare gluten-free *muffins* made with a mix of finger millet (ragi) flour (FMF) and whey protein concentrate (WPC) blends in different proportions, and they studied the rheological, textural, and quality characteristics of gluten-free *muffins*. It was found that FMF: WPC of 90:10 along with distilled glycerol monostearate and hydroxy-propyl-methyl-cellulose fetched significant improvement in the overall quality of *muffins*.

2.3.2.7 Pizza base

Pizza is an oven-baked flatbread topped with cheese and sometimes tomato sauce. The basic dough formulation for *pizza* making includes flour (mostly maida or refined wheat flour), sugar, water, salt, and baker's yeast. *Pizza* is often a calorie-rich food, sometimes unhealthy to those who cannot afford too many calories. For those sections of people, millet mix *pizza* could be a suitable option as the dietary fiber-rich *pizza* does not release more calories, unlike the prior one. The *pizza* base or *pizza crust* is one of the important parts of the *pizza* upon which toppings and sauce are placed.

Millet-based pizza base is prepared by combining the required proportion of refined wheat flour and millet flour, salt, yeast and oil with the required quantity of warm water until a dough consistency is achieved. The dough is left undisturbed for 10 min. Kneading with knockback action is done to remove the gas trapped inside the dough. It is then rolled into sheets of circular shape using a rolling pin (Bhavya et al., 2020). Millet inclusion into *pizza* dough results in breakage as the base becomes fragile on baking. To overcome this, gums may be added during preparation to hold the shape of dough during processing. Cheese and veggies/chopped meat are added to the base before baking in nutrition-loaded millet-based pizza.

Bhavya et al. (2020) made a *pizza* base with major ingredients as refined wheat flour, proso millet flour, yeast, sugar, baking powder, and salt. It was found that 70:30 of refined wheat flour and proso millet flour, which was baked at 200° C, yielded better results in terms of texture, appearance, and overall acceptability. With temperature increment, the quality and overall sensory acceptability had been reduced.

2.3.3 Ready-to-eat millet products

Millet, one of the most significant drought-resistant crops, is widely grown in the semiarid tropics of Africa and Asia, where it provides a key source of carbohydrates and proteins (Ahmed et al., 2013). Furthermore, due to its importance to national food security and potential health advantages, millet grain is gaining popularity among food scientists, technologists, and nutritionists. Millets are better in proteins and minerals in comparison to cereals. RTE, quick-cooking, and instant dishes have grown highly popular, owing to today's fast-paced lifestyle and desire for ready-to-serve food (Dhumal et al., 2014). These RTE and ready-to-serve food products have been proved organoleptically acceptable and microbiologically safe for periods of three months to a year (Yadav et al., 2011). Ready-to-eat products from utria-cereals are easy to prepare and contain high nutritive value. They also provide an additional benefit to the consumer on convenient cooking and help maintain millet's value chain (Onyeoziri et al., 2021).

2.3.3.1 Porridge

Porridge is a breakfast and RTE food prepared by heating the mixture of cereal flour and water, which results in gelatinization of starch and can be often consumed warm with salt, sugar, milk, or butter. *Porridge* generally prepared from rice flour, corn flour, sorghum, or pearl millet or a mix of any of these is very popular, especially in Asian and African countries (Yadav et al., 2014). In some parts of the world like Nigeria, that is, an African country, a steamed *pudding* is prepared from groundnut and Bambara and served with Okpa and Akpukpa (liquid food) (Ocheme and Chinma, 2008).

Porridge is the easiest RTE product, with very few steps in its preparation. Millet flour (30 g), either whole or composite, is added to 100 ml of water and mixed until there is no lump condition. It is then transferred to boiling water and stirred continuously until starch gelatinization occurs. The slurry is left to steam for 10 min and cooled before consumption. To enhance taste and flavor, cashews and almonds in the form of ground powder can be added during the initial step (Kumari et al., 2020). Millet-based *porridge* is consumed largely by the natives of India, Kenya and Africa owing to its rich nutrient composition and antioxidant activity (Nefale and Mashau, 2018).

Onyango et al. (2020) evaluated the physico-chemical and sensory properties of hydrothermally treated finger millet flour and cassava mix *porridge*. It was found that the protein, lipid, fiber, and mineral content of the final product were improved; it was also proved that the energy density of *porridge* was enhanced due to the fermentation process. According to (Yadav et al., 2014) 0.841 mm grits of pearl millet along with 15 percent skim milk powder yielded optimal properties in preparation, which has acceptable maximum viscosity (499±6.6 cP) during preparation and optimum viscosity (450±11.9 cP) once cooled.

2.3.3.2 Kesari

Kesari is a very popular south Indian sweet *pudding* often made from *rava* (that is, cream of wheat or semolina), ghee (a milk product), saffron, sugar, and some kind of dry fruits. In many places, its color is kept from orange to red. *Kesari* is prepared by keeping the constraints of wheat usage in society; alternate multigrain grits can be used in the place of sooji to make the *kesari* a gluten-free and vegan dessert. Millets could be the perfect replacements or suitable options for the mix. *Kesari* can be prepared with any millets of our choice. Barnyard millet, finger millet, foxtail millet, pearl millet, and kodo millets are a wide range of suitable options.

Kesari preparation involves dry roasting of millet with a tablespoon of ghee at 85° C on a wide roasting pan until the onset of characteristic aroma and color change. After cooling, grind the millet into a coarse texture, similar to semolina. In a pan, milk is boiled to which ground millet is added and cooked for 5–7 min. It is stirred continuously to avoid lumps forming. Add sugar and mix thoroughly. After cooking the millet, *kesari* color, cardamom powder and sauteed cashews are added. It is left to cool and consumed.

There is still extensive research to be done to find out the nutritional changes and physico-chemical changes happening during the preparation of *kesari*, and their effects on customer acceptability and, more importantly, on their health.

2.3.3.3 Pongal

Pongal is a very popular south Indian food especially in the state of Tamil Nadu, made with moong dal and rice. It can be of two types, first one is *khara ongal* or *ven ongal* (*ven* means white in Tamil Nadu) which is savoury in taste. The second one is sweet *ongal*, which is sweet in taste. Normally, *ongal* is prepared by tempering rice and moong dal mixture with ghee, black pepper, ginger, hing, curcumin, and curry leaves. To overcome the nutritional disadvantages offered by regular types of *Pongal*, millets like kodo millet, little millet, and barnyard millets can be used in equal amounts of rice.

Millets and moong dal are washed and soaked in water for 2 hours after which the mixture is pressure-cooked until starch gets gelatinized. To the cooked mixture, salt, curry leaves, cumin, black pepper, ginger, asafoetida and ghee roasted cashew are added. The *Pongal* is combined thoroughly before serving. The resultant *Pongal* is rich in protein and fiber, more importantly, a portion of good food for children. Millet-based *Pongal* takes less time to cook than rice-based *Pongal*. RTC millet *Pongal* is available in Indian markets.

2.3.3.4 Burfi/peda

Burfi is one of the most liked Indian-based sweet recipes made with milk and sugar as the main ingredients. There are so many barfi varieties include *kaju barfi (kaju katli), besan Burfi (barfi), pista burfi, chocolate burfi, almond, or badam barfi*. These kinds of *burfi* are calorie-rich and normally are the ones to be avoided for diabetic patients. The best alternative to that section of people is millet *burfi* which can be made with multi millet mix along with dates, the recipe includes roasting the mix of sorghum *flakes*, almonds, cashew nuts, pistachio, poppy seeds, and grinding them separately. The next step includes mixing all the powders in required proportions and cooking in ghee, followed by making cylindrical (whichever shape one likes to make) shaped foods which look like normal *burfi* (Rao et al., 2016).

2.3.3.5 Kheer/payasam

An Indian dessert comprising of rice or *vermicelli*, boiled in coconut milk or milk, flavored with cardamom, and most often containing groundnuts. It is also called *kheer* in some places. For the benefit of adiabatic patients, *payasam* can be made with little millet, milk, and jaggery, which gives an acceptable recipe and nutrition too. Millet *payasam* can also be made with foxtail millet, barnyard millet, and kodo millet. Dry roast the millet in an open pan until color changes and the millet turns aromatic. The roasted millet is cooked with milk and water with constant stirring until a thick consistency is achieved. Sugar is added and cooked until it dissolves completely (Bhosale et al., 2021). The *kheer* is then transferred to a bowl and garnished with ghee roasted nuts before consumption. Various research were conducted on millet-based *kheer* and *payasam* (Bhosale et al., 2021; Kashyap et al., 2018; Singh, 2017).

2.3.3.6 Khichadi

It can be made with sorghum (ragi) and moong dal (green gram) with various other ingredients while cooking. Sorghum rawa and moong dal are to be soaked before proceeding to cook *khichadi*. In a flat bottom pan, mustard seeds, ginger-garlic paste, chilies, cut tomato pieces, turmeric powder, and curry leaves must be fried in edible oil for 5 minutes. Soaked mix has to be added along with a sufficient amount of water and the whole mix has to be cooked at low flame until it gets cooked properly (Rao et al., 2016). *Khichadi* can also be made with pearl millet. Negi and Srihari (2021) studied the optimization of millet *khichadi* instant mix, where it was concluded that cooking of millet-based *khichadi* takes less time than conventional ingredients.

Khichadi is prepared by sputtering cumin seeds in a pan with ghee at 150° C. Turmeric, asafoetida, pounded black pepper, curry leaves, green chili, veggies (tomato, beans, carrot, peas, onion, capsicum and potato), ginger and salt are added with the addition of millet, green gram dal (dal and millet soaked for 15 min prior to addition) and water. The heating process is continued until millets are cooked completely. Negi and Srihari (2021) studied the optimization of millet *khichadi* instant mix, where it was concluded that cooking of millet-based *khichadi* takes less time than conventional ingredients.

2.3.3.7 Halwa

It's a tasty sweet dish that originated in the northern part of India, is abundant in nutrients, particularly carbohydrates and fat. Although *halwa* can be made with flour or nuts, the most frequent type is wheat *sooji* (semolina) *halwa*, which is prepared using wheat flour or wheat grits (semolina). Instant *halwa* mix, which has a one-year shelf life, has gained popularity and greater acceptability among civilians and military personnel.

Yadav et al. (2011) prepared *halwa* from pearl millet, vanaspati, sugar, and water mix. Pearl millet grains were steamed for 20 min to minimize total phenols, tannins content, phytic acid, and lipase activity. Steamed grains were dried, milled, sieved for obtaining < 0.234 mm. Optimum conditions they obtained were: vanaspati 38.6 g per 100 g of pearl millet semolina (PMS), sugar 88.7 g per 100 g of PMS, and water for rehydration was 151 mL per 100 g dry mix. Kumar et al. (2017) studied the effects of quality changes in retort processed ready-to-eat foxtail millet *halwa*.

2.3.3.8 Upma

Upma is a popular breakfast food that originated in the southern part of India, is usually made from wheat semolina. Usually, it is prepared as fresh food every time and the process generally take 15–25 min. Convenience mixes are available now in most parts of the country and across the world that can reduce the preparation time by eliminating some steps in the process of cooking.

Upma is prepared by roasting millets in a pan at low flame for 5–7 min. Mustard, curry leaves and green chili are sputtered in hot oil followed by the addition of roasted millet and water. Diced tomato and onion are added during cooking until it turns soft and mushy. The cooking takes place until desired consistency is achieved. Kavali et al. (2019) studied the effects of quinoa-based *upma* where 60 percent inclusion resulted in sensorially acceptable product.

Balasubramanian et al. (2014) optimized pearl millet semolina (PMS) based *upma* for various ingredients and found that vanaspati 46.5 g per 100 g of pearl millet semolina (PMS), citric acid 0.17 g per 100 g of PMS, and water for rehydration was 244.6 mL per 100 g dry mix. It was also found that the *upma* mix was found stable and withstood sensory acceptability for 6 months under room temperature (20–30° C) in Polyethylene (PE) pouches of 75 microns.

2.3.3.9 Appalu

Appalu, are an authentic dish southern part of India and they are quite popular for their soft texture, color, and taste. There are many varieties of *appalu* including sweet, sour, and mixed taste. Generally, these are made by frying rice flour mix with ingredients like chili powder, turmeric, ginger paste, garlic paste, sesamum, cardamom, clove powder, and many more in a variety of shapes depending on the perception of the consumer. Sweet *appalu* is the regular form of *appalu* fried. In sweet *appalu* Ghee is added to boiling water to which semolina and coarse ground millet are added with continuous stirring to avoid lump formation. After mixing, heat is turned off and the pan is left covered with a lid for 5 min to cook the ingredients. Sugar and cardamom powder is added and stirred until the sugar gets dissolved completely on mild heat. The dough is made into flat round shaped size and deep-fried in oil or ghee until golden brown color reaches. But considering the nutrition requirements, millet *appalu* are in limelight these days, these can be made with pearl millet flour, and a mix of millet and rice flour. *Chakki* falls under one category of *appalu*.

2.3.4 Healthy and functional foods

Millet based healthy functional foods are a boon to people suffering from celiac disease. Functional foods prepared from millets has great impact on anemic women in turn balancing the external nutrient requirement. It also induces a feeling of satiety on consumption owing to its slow digestibility.

2.3.4.1 Laddu

Laddu is a popular sweet delicacy in India that is prepared during all festivals and celebrations. The millet flour is dry roasted until it turns golden brown in color. Dissolved, melted and strained jaggery water mixture is boiled till a thick consistency is achieved. The jaggery paste is added to roasted millet flour with the addition of ghee, cardamom powder and roasted cashews. The contents are mixed thoroughly till the millet cooks leaving behind the raw odor to form a dough. After cooling down, ball sized *laddu* is rolled and garnished with nuts (Rao et al., 2016). Refined sugar can also be used in the place of jaggery during the preparation. Nutrient loaded *laddu* made from amaranth seeds is a popular snack in Maharashtra.

2.3.4.2 Mudde

Foxtail millet based *mudde* is a popular dish of Karnataka. The initial step in *mudde* preparation includes addition of salt and millet flour to boiling water. The slurry is whisked evenly to avoid lump formation. The contents are heated until the starch gelatinizes and forms a gel consistency. Ghee is added to provide a glaze. *Mudde* is made into balls and served with chicken or vegetable stew to provide a characteristic taste and flavor (Rao et al., 2016).

2.3.4.3 Prebiotic foods/beverages

Arya and Kumar (2021) and Dong et al. (2021) studied the effects of *prebiotic* activity in formulated millet-based beverages. The preliminary step involves roasting grains at 150° C followed by boiling it in water at 100° C for 10 min. The boiled grains are ground to a fine paste and filtered using a double layered cloth to obtain the beverage. To this, Fructo-oligosaccharides (FOS), maltitol and galacto-oligosaccharides (GOS) are added to maintain the *prebiotic* activity during storage.

2.3.4.4 Nutri-bar

Nutri-bar also known as *chikki* is a popular protein-rich snack food in India. Millets (steamed and dried) are mixed with whey protein, roasted peanuts, roasted soy nuts, raisins, honey and powdered sugar. The contents are transferred to silicone molds and refrigerated to set for 2 hours. The bars are baked for 30 min at 180° C and cut into 50 g pieces for storage. (Dong et al, 2021) studied the effects of in-vitro digestibility of protein bars made of pearl millet.

2.3.4.5 Non-dairy milk

Non-dairy milk of plant origin is recently being opted by the consumers owing to its mineral and nutrient content. Millets are soaked overnight and ground to a fine paste using colloidal ball mill. The paste is then filtered using a double-layered muslin cloth to obtain milk of homogenous texture. It is then thermally processed 85° C for 10 min (pasteurized or Ultra-high temperature treated) to enhance the shelf life and cooled to 4° C for storage (Shunmugapriya et al., 2020). The major drawback in *non-dairy milk* is phase separation due to gravitation during storage. This can be overcome with the addition of gums and stabilizers during large scale processing.

2.3.4.6 Ice cream

Ice-cream is prepared by substituting a part of cow milk with millet extract or with the addition of millet flour in premix. Milk is heated to 50° C followed by the addition of skim milk powder, cream, stabilizers and emulsifiers (sodium alginate, glycerol monostearate). Sugar and malted millet flour are added to the premix which is followed by double stage homogenization and pasteurization at 80° C. The mix is cooled to 4 C overnight to initiate aging. Flavor addition takes place followed by low temperature (-18° C) freezing (Sharma et al., 2017).

2.3.4.7 Soup

Soup is a highly nutritive thickened stew prepared from various products of plant origin. Millets are soaked, boiled and dried to prepare fine millet flour for its incorporation into the *soup* mix. The mix is prepared with the addition of millet flour, dehydrated vegetable flour, spices and herbs that are sieved to maintain a homogenous texture. It is then added to boiling water until starch gelatinization occurs. For cooking of *soup* takes 5–10 min after which it is garnished with herbs, salt and pepper. Fresh cream is added in the end to impart rich taste and flavor (Rao et al, 2016). Studies on instant millet-based *soup* mix were carried out by Tulasi et al. (2020) and Upadhyay et al. (2017).

2.3.5 Value-added snack foods

The combined effect of different foods in the diet is very important not only for taste and enjoyment of eating but also for good nutritional quality and health advantages. To bridge the growing gap between food availability and nutritional security, there has been a constant effort for developing new food products that aspire for complementary foods (Verma and Patel, 2013). Numerous studies have prioritized the creation of value-added snack foods in terms of protecting and maintaining good health, increasing consumer perception in terms of nutrition, acceptability, and adaptability in recent years (Jozinović et al., 2021). Various snack food products can be prepared from by-products during the preparation of flakes, for example, bran powder and broken parts of flakes, during baking too. Other than these, millet snack foods can be enhanced with nutritional content by fortifying with zinc-rich flour and iron-rich flour (Rao et al., 2016). In this section, some of the traditional value-added snack food products like *samosa, pakoda, bhakarwadi, cutlet, biryani, bhel* that have grabbed greater attention among people are covered.

2.3.5.1 Samosa/patties

A *samosa* is a South Asian-based fried or baked dough or batter of spice-mixed all-purpose flour called maida flour, with a savory or sweet filling like spiced onions, peas, mashed, and masala fortified potatoes, chicken, and other meats. Depending upon the perception of the consumer and region in which it is made, *samosa* can be made in different shapes including cone, triangular, or half-moon shapes. *Samosas* are served with chutney (pea or coconut). Due to the disadvantages of maida and other regular cereal flours, millet flour can be a suitable alternative for *samosa* preparation. Generally, *samosa* can be made with sorghum and proso millet (Rao et al., 2016).

2.3.5.2 Pakoda

It is generally called "*pakodi*" in rural and urban regions of India, is preferred to be taken as an evening snack. This has got more popularity during the rainy season. It is a tasty and deep-fat fried food generally prepared by frying spiced rice flour batter, chopped onions, and other necessary ingredients in hot oil. *Pakoda* can be made with pearl millet flour by mixing it with rice flour to add some nutritional benefits to the final product. It is prepared by mixing flour mix, chopped onions, salt, chili powder, dhania (coriander) powder, and water to make into a batter form. Then by gently dropping that batter in clumps in hot oil to get them deep-fried which imparts taste, nutritional, physico-chemical changes, color, and enhances the shelf life of the product (Rao et al., 2016).

2.3.5.3 Bhakarwadi

Bhakarwadi is a traditional Maharashtrian, Gujarati, and Rajasthani (mostly western part of India) spicy snack. It is a deep-fried product of pinwheels (generally called as "swiss roll") shaped snack, made of all-purpose flour (maida), whole wheat flour, and gram flour which is filled with a spicy, tangy, and sweet filling. The filling has roasted nuts (coconut, coriander, cumin, peanut, poppy, and sesame) and seeds which bring mild flavors, and adds acceptable texture, taste, and crunchiness that can be felt in every bite. Millet flours like sorghum, pearl millet, little millet, barnyard millet can also be used to replace a portion of all-purpose flour which has ill effects on our health.

2.3.5.4 Cutlet

Cutlet is a snack dish generally a fried food which is prepared with the mix of different flours, slices or mashed meat, bread, or sometimes vegetables and bread. *Cutlet* can be prepared from foxtail millet, barnyard millet. Major ingredients for making *cutlet* are millet grains, potatoes, chopped carrots, beans, green chilies, breadcrumbs, and other necessary spice ingredients like salt, chat masala, pepper, ginger-garlic paste, and finally oil. Chopped onions, chopped carrot, chopped beans, and chopped green chilies once fried in oil along with ginger-garlic paste, salt, pepper powder, then mixed with chana dal flour. To that mix, pre-cooked or steamed barnyard millet or foxtail millet (debranned) and chopped coriander leaves are to be added. After cooking the whole mix for 5–7 minutes on low flame, the mix is divided into different shapes and parts depending upon the requirement, and then it is to be deep-fried till golden brown color. Usually, the *cutlet* is served with tomato sauce or chili sauce (Rao et al., 2016).

2.3.5.5 Biryani

Biryani is a mixed rice-based dish originating from the Indian subcontinent. It is generally made with rice, spices, and meat of wide varieties like chicken, lamb, goat, fish, prawn. In some cases, instead of meat, eggs and vegetables are also being used. *Biryani* is popular throughout the sub-continent. Due to the extensive advantages offered by the millets, *biryani* (both veg and non-veg) can be prepared from foxtail millet, little millet along mushroom (called mushroom biryani). It is prepared by starting with roasting dehulled little millet for 1–2 minutes. After frying and tempering onion, ginger-garlic paste, green and red chilies, masala, coriander paste, and salt, cleaned and chopped mushroom is to be added and allowed for cooking with the mix prepared earlier. Once the mushroom and masala mix start boiling, roasted millets along with sufficient water are to be mixed. Similarly mint, coriander, and salt are to be added to get the perfect taste and allowed to cook for 25–30 minutes till it gets done. Finally, it can be served and eaten with onion raita and gravy (Rao et al., 2016).

2.3.5.6 Bhel

It can be made from a mix of overnight soaked and washed different millets like sorghum, pearl millet, foxtail millet, finger millet in more or less equal amounts. After the mix of millets steamed and roasted in oil, fortified with chat masala, chopped onions, chopped tomatoes, chopped green chilies, lemon juice sprinkle, and salt in amounts required as per optimal taste. Normally, *bhel* is served with tomato sauce (Rao et al., 2016).

2.4 Future perspective

Millets are a staple and important crop of India. After the green revolution, the importance of millets was drastically reduced and shifted toward rice and wheat. Recently, the importance of millets was recognized by the Government of India owing to its multiple benefits and National Food Security Act promoted millets as Nutri-cereals. Despite the multiple benefits offered by millets and millet products, they are not widely popular because of several reasons, they are (i). lack of suitable equipment for harvesting and handling with minimal losses; (ii). Storage issues in the case of some millets with high lipid content which results in rancidity development during storage; (iii). Cultivation in clusters or some parts of the world only. Suitable future activities to reap the eligible benefits of millets are extensive research is needed to develop specific equipment for handling with minimal losses and which is adaptable enough to different products too. There is a need to shift from traditional processing to modern processing methods. Extensive use of millet starch just like any other starch for texture modification, binding, as structuring agent, and for viscosity regulation. Not only that, but focus is also needed on scaling up of millet resistant starch and application of resistant starch other food industries as a supplement. Attention is required to find out the behavioral changes in physico-chemical, structural, and nutritional properties of millet(s) during their processing like extrusion while making RTC and RTE food products. Also, scientific intervention is required while formulating different snack foods just to make sure the end product is safe to consume and does not yield destructive changes in human health thereafter.

2.5 Conclusion

With the increase in world's population which has already crossed 7.9 billion, demand for food which generally leads to high food prices, accessibility of everyone to food, and food security has always been a point of concern, with the heavy consumption of rice and wheat, especially in Asian and African countries, malnutrition is also one of the major factors which is indirectly related to the growth of the concerned country and thereby the world. For both problems cited above, millets could be one point solution, because millets are the sixth-highest yielding food grain in the world. They also have high nutritive value, some millets are rich in resistant starch or dietary fiber, they help in the formulation of low glycemic index (GI) products. Millets are significantly rich in minerals like calcium, iron, magnesium, manganese, and phosphorus than other cereals, they help in the preparation of high-nutritious food products like porridges, ladoos, pastas, biscuits, chapatti, breads, cookies, cakes, several other fermented foods, and probiotic drinks. A wide range of extruded cereal-millet-pulse snack foods can be produced commercially for an extensive availability. Similarly, blends of cereal- millet-pulse can be used in formulation and preparation of *murukus, papads, vermicelli, noodles, macaroni*, and so forth. Multiple millet blends with wheat can be used for the preparation of baked products like *biscuits, breads, buns, cookies, cakes, muffins, chocolate puddings*, and so forth. They also help in the preparation of weaning or infant foods. Millets can be a good weapon to fight against diabetes, hypertension, cancers and other diseases.

References

Adebanjo, L. A., Olatunde, G. O., Adegunwa, M. O., Dada, O. C., and Alamu, E. O. 2020. Extruded flakes from pearl millet (*Pennisetum glaucum*)-carrot (*Daucus carota*) blended flours-Production, nutritional and sensory attributes. *Cogent Food and Agriculture* 6(1), 1733332. https://doi.org/10.1080/23311 932.2020.1733332

Agarwal, M., Singh, S. S., and Ashotosh, K. 2017. Development and quality assessment of multigrain bread prepared by using wheat flour, buck wheat flour and pearl millet flour. *The Pharma Innovation Journal* 6(6): 40–43.

Aggarwal, M., and Sharma, D. 2020. Millet flour for baked food products. *Plant Archives* 20: 458–61.

Ahmed, S. M. S., Zhang, Q., Chen, J., and Shen, Q. 2013. Millet grains: nutritional quality, processing, and potential health benefits. *Comprehensive Reviews in Food Science and Food Safety* 12(3): 281–95. https://doi.org/10.1111/1541-4337.12012

Aka, S., Konan, G., Fokou, G., Dje, K. M., and Bonfoh, B. 2014. Review on African traditional cereal beverages. *American Journal of Research Communication* 2(5):103–153.

Akinola, S. A., Badejo, A. A., Osundahunsi, O. F., and Edema, M. O. 2017. Effect of preprocessing techniques on pearl millet flour and changes in technological properties. *International Journal of Food Science and Technology* 52(4):992–999. https://doi.org/:10.1111/ijfs.13363

Amadou, I. 2019. Millet based fermented beverages processing. In *Fermented Beverages*, 433–472. Woodhead Publishing.

Arya, S. S., and Shakya, N. K. 2021. High fiber, low glycaemic index (GI) prebiotic multigrain functional beverage from barnyard, foxtail and kodo millet. *LWT – Food Science and Technology* 135:109991. https://doi.org/10.1016/j.lwt.2020.109991

Balasubramanian, S., Chakraborty, S., Kotwaliwale, N., and Kate, A. 2021. Entrepreneurship opportunities in nutri-cereal processing sector. *Indian Food Industry Magazine* 3(1):29–38.

Balasubramanian, S., Yadav, D. N. Kaur, J. and Anand, T. 2014. Development and shelf-life evaluation of pearl millet based upma dry mix. *Journal of Food Science and Technology* 51(6):1110–1117. https://doi.org/10.1007/s13197-012-0616-0

Barbhai, M. D., Hymavathi, T. V., Kuna, A., Mulinti, S., and Voliveru, S. R. 2021. Quality assessment of nutri-cereal bran rich fraction enriched buns and muffins. *Journal of Food Science and Technology* 1–12. https://doi.org/10.1007/s13197-021-05236-9

Bhavya, P., Devi, S. S., Srinath, P. J., and Sreedevi, P. 2020. Development sensory evaluation of millet based pizza base. *International Journal of Chemical Studies* 8(4):3502–3504. https://doi.org/10.22271/chemi.2020.v8.i4ar.10193

Bhosale, S., Desale, R. J., and Mukhekar, A. 2021. Physico-chemical composition of millet based kheer blended with paneer. *The Pharma Innovation Journal* 10(4):320–324.

Birania, S., Rohilla, P., Kumar, R., and Kumar, N. 2020. Post harvest processing of millets: A review on value added products. *International Journal of Chemical Studies* 8(1):1824–1829. https://doi.org/10.22271/chemi.2020.v8.i1aa.8528

Biswas, S. S. 2015. Nutri Rich Diabetic Biscuits: evaluating launching communication activities. *BRAC University*. http://dspace.bracu.ac.bd/xmlui/handle/10361/4411 (October 12, 2021).

Bora, P., Ragaee, S., and Marcone, M. 2019. Effect of parboiling on decortication yield of millet grains and phenolic acids and in vitro digestibility of selected millet products. *Food Chemistry* 274:718–725. https://doi.org/10.1016/j.foodchem.2018.09.010

Chera, M. 2017. Transforming millets: Strategies and struggles in changing taste in Madurai. *Food, Culture and Society* 20(2):303–24. http://dx.doi.org/10.1080/15528014.2017.1305830

Comettant-Rabanal, R., Carvalho, C. W. P., Ascheri, J. L. R., Chávez, D. W. H., and Germani, R. 2021. Extruded whole grain flours and sprout millet as functional ingredients for gluten-free bread. *LWT – Food Science and Technology* 150:112042. https://doi.org/10.1016/j.lwt.2021.112042

Dharmaraj, U., Rao, B. S., Sakhare, S. D., and Inamdar, A. A. 2016. Preparation of semolina from foxtail millet (*Setaria italica*) and evaluation of its quality characteristics. *Journal of Cereal Science* 68:1–7. https://doi. org/10.1016/j.jcs.2015.11.003

Dhas, T. A., Rao, V. A., Sharon, M. E. M., Manoharan, P., and Nithyalakshmi, V. 2021. Development of fibre enriched ready-to-cook pasta with foxtail millet and tapioca flour. *The Pharma Innovation Journal* 10(5):1432–35.

Dhumal, C. V., Pardeshi, I. L., and Sutar, P. P. (2014). Development of potato and barnyard millet based ready to eat (RTE) fasting food. *Journal of Ready to Eat Foods* 1(1):11–17.

Dong, G. M., Dong, J. L., Zhu, Y. Y., Shen, R. L., and Qu, L. B. 2021. Development of weaning food with prebiotic effects based on roasted or extruded quinoa and millet flour. *Journal of Food Science* 86(3):1089–1096. https://doi.org/10.1111/1750-3841.15616

Francis, J. M., Grosskurth, H., Kapiga, S. H., Weiss, H. A., Mwashiuya, J., and Changalucha, J. 2017. Ethanol concentration of traditional alcoholic beverages in Northern Tanzania. *Journal of Studies on Alcohol and Drugs* 78(3):476–477. https://doi.org/10.15288/jsad.2017.78.476

Gaikwad, P. S., Pare, A., and Sunil, C. K. 2021. Effect of process parameters of microwave-assisted hot air drying on characteristics of fried black gram papad. *Journal of Food Science and Technology* 1–11. https://doi. org/10.1007/s13197-021-05050-3

Giridhar, P. 2019. Preparation and sensory evaluation of finger millet Khakhra. *Journal of Food Nutri Science and Nutrition Research* 2(1):61–64. https://doi.org/10.26502/jfsnr.2642-1100009

Goswami, D., Gupta, R. K., Mridula, D., Sharma, M., and Tyagi, S. K. 2015. Barnyard millet based muffins: physical, textural and sensory properties. *LWT - Food Science and Technology* 64(1):374–80. https://doi.org/ 10.1016/j.lwt.2015.05.060

Gull, A., Gulzar Ahmad, N., Prasad, K., and Kumar, P. 2016. Technological, processing and nutritional approach of finger millet (*Eleusine coracana*) - a mini review. *Journal of Food Processing and Technology* 7:593 http://dx.doi.org/10.4172/2157-7110.1000593

Gull, A., Jan, R., Nayik, G. A., Prasad, K., and Kumar, P. 2014. Significance of finger millet in nutrition, health and value added products: a review. *Magnesium (mg)* 130(32):120.

Gupta, S., and Paul, V. 2012. Utilization of coarse grains for formulation of value added snacks. *Food Science Research Journal* 3(1):64–68.

Hassan, E. M., Fahmy, H. A., Magdy, S., and Hassan, M. I. 2020. Physicochemical and sensorial characterization of gluten-free cupcakes. *Egyptian Journal of Nutrition*, 35(1):33–64.

Hymavathi, T. V., Robert, P., Thejasri, V., and Jyosthna, E. 2020. Proximate composition and glycemic index profiling of differently composed nutri-cereal biscuits. *International Research Journal of Pure and Applied Chemistry* 21(21):39–46. https://doi.org/10.9734/irjpac/2020/v21i2130288

Jadhav, S., Kavinya, V., Nirmal, R. V., Shameem, H. M., and Ramalakshmi, K. 2021. Physico-sensory and Textural Properties of Composite Millet Palm Jaggery Muffins. *Journal of Natural Remedies*, 37–43.

Jalgaonkar, K., Jha, S. K., and Mahawar, M. K. 2018. Influence of die size and drying temperature on quality of pearl millet based pasta 6(6):979–984.

Jaybhaye, R. V., Pardeshi, I. L., Vengaiah, P. C., and Srivastav, P. P. 2014. Processing and technology for millet based food products: a review. *Journal of Ready to Eat Food* 1(2):32–48.

Jozinović, A., Šubarić, D., Ačkar, Đ., Babić, J., Orkić, V., Guberac, S., and Miličević, B. 2021 Food industry by-products as raw materials in the production of value-added corn snack products. *Foods* 10(5):946. https:// doi.org/10.3390/foods10050946

Jyotsna, R., Soumya, C., Swati, S., and Prabhasankar, P. 2016. Rheology, texture, quality characteristics and immunochemical validation of millet based gluten free muffins. *Journal of Food Measurement and Characterization* 10(4):762–72. https://doi.org/10.1007/s11694-016-9361-9

Kashyap, A., Mehra, M., and Kashyap, Y. 2018. Study the physico-chemical attributes of kheer mix. *Development of Food Science and Technology* 2856–2857.

Kaur, A., Kumar, K., and Dhaliwal, H. S. 2020. Physico-chemical characterization and utilization of finger millet (*Eleusine coracana* L.) cultivars for the preparation of biscuits. *Journal of Food Processing and Preservation* 44(9):e14672. https://doi.org/10.1111/jfpp.14672

Kavali, S., Shobha, D., Naik, S. R., and Brundha, A. R. 2019. Development of value added products from quinoa using different cooking methods. *The Pharma Innovation Journal* 8(7):548–554.

Kharat, S., Medina-Meza, I. G., Kowalski, R. J., Hosamani, A., Ramachandra, C. T., Hiregoudar, S., and, Girish M, G. 2019. Extrusion processing characteristics of whole grain flours of select major millets (foxtail, finger, and pearl). *Food and Bioproducts Processing* 114:60–71. https://doi.org/10.1016/j.fbp.2018.07.002

Kotagi, K., Chimmad, B., Naik, R., and Kamatar, M. Y. 2013. Nutrient enrichment of little millet (*Panicum miliare*) flakes with garden cress seeds. *International Journal of Food and Nutritional Sciences* 2(3):36.

Kulthe, A. A., Thorat, S. S., and Lande, S. B. 2017. Evaluation of Physical and textural properties of cookies prepared from pearl millet flour. *International Journal of Current Microbiology and Applied Sciences* 6(4):692–701. https://doi.org/10.20546/ijcmas.2017.604.085

Kumar, R., Harish, S., Subramanian, V., Sunny Kumar, S., and Nadanasabapathi, S. 2017. Development and quality evaluation of retort processed RTE functional gluten free foxtail millet halwa. *Croatian Journal of Food Science and Technology* 9(2):114–121. https://doi.org/10.17508/CJFST.2017.9.2.05

Kumari, D., Chandrasekara, A., Athukorale, P., and Shahidi, F. 2020. Finger millet porridges subjected to different processing conditions showed low glycemic index and variable efficacy on plasma antioxidant capacity of healthy adults. *Food Production, Processing and Nutrition* 2(1):1–11. https://doi.org/10.1186/s43014-020-00027-9

Kumari, R., Singh, K., Singh, R., Bhatia, N., and Nain, M. S. 2019. Development of healthy ready-to-eat (RTE) breakfast cereal from popped pearl millet. *Indian Journal of Agricultural Sciences* 89(5):877–881.

Lande, S. B., Throats, S., and Kulthe, A. A. 2017. Production of nutrient rich vermicelli with malted finger millet (ragi) flour. *International Journal of Current Microbiology and Applied Sciences* 6(4):702–10. https://doi.org/10.20546/ijcmas.2017.604.086

Limaye, R. J., Rutkow, L., Rimal, R. N., and Jernigan, D. H. 2014. Informal alcohol in Malawi: stakeholder perceptions and policy recommendations. *Journal of Public Health Policy* 35(1):119–131. https://doi.org/10.1057/jphp.2013.43

McSweeney, M. B., Ferenc, A., Smolkova, K., Lazier, A., Tucker, A., Seetharaman, K., ... and Ramdath, D. D. 2017. Glycaemic response of proso millet-based (Panicum miliaceum) products. *International Journal of Food Sciences and Nutrition*, 68(7), 873–880.

Mishra, G., Joshi, D. C., and Panda, B. K. 2014. Popping and puffing of cereal grains: a review. *Journal of Grain Processing and Storage* 1(2):34–46.

Nambiar, V. S., and Patwardhan, T. 2015. Assessment of glycemic index (GI) and glycemic load (GL) for selected traditional millet-based Indian recipes. *International Journal of Diabetes in Developing Countries* 35(2):157–162. https://doi.org/10.1007/s13410-014-0275-y

Nami, Y., Gharekhani, M., Aalami, M., and Hejazi, M. A. 2019. *Lactobacillus*-fermented sourdoughs improve the quality of gluten-free bread made from pearl millet flour. *Journal of Food Science and Technology* 56(9):4057–4067. https://doi.org/10.1007/s13197-019-03874-8

Namitha, M. Y., Chavan, U. D., Kotecha, P. M., and Lande, S. B. 2019. Studies on nutritional and sensory qualities of foxtail millet chakli. *International Journal of Food Science and Nutrition* 4(5):68–73.

Narang, N., Kaur, N., Singh, P., Saran, R., Singh, B, Sharma, S. 2018. Development of extruded flakes from pearl millet. *The Pharma Innovation Journal* 7(7):502–506.

Narayanan, J., Sanjeevi, V., Rohini, U., Trueman, P., and Viswanathan, V. 2016. Postprandial glycaemic response of foxtail millet dosa in comparison to a rice dosa in patients with type 2 diabetes. *The Indian Journal of Medical Research* 144(5):712–717. doi:10.4103/ijmr.IJMR_551_15

Nazni, P, and Shalini, S. 2010. Standardization and quality evaluation of idli prepared from pearl millet (Pennisetum glaucum). *International Journal of Current Research* 5:84–87.

Nefale, F. E., and Mashau, M. E. 2018. Effect of germination period on the physicochemical, functional and sensory properties of finger millet flour and porridge. *Asian Journal of Applied Sciences* 6(5):360–367.

Negi, N., Srihari, S. P., Wadikar, D. D., Sharma, G. K., and Semwal, A. D. 2021. Optimization of instant foxtail millet based khichdi by using response surface methodology and evaluation of its shelf stability. *Journal of Food Science and Technology* 1–8. https://doi.org/10.1007/s13197-020-04929-x

Ocheme, O. B., and Chinma, C. E. 2008. Effects of Soaking and Germination on some physicochemical properties of millet flour for porridge production. *Journal of Food Technology* 6(5):185–88.

Onyango, C, Susan, K. L., Guenter, U., and Norbert, H. 2020. Nutrient composition, sensory attributes and starch digestibility of cassava porridge modified with hydrothermally-treated finger millet. *Journal of Agriculture and Food Research* 2:100021. https://doi.org/10.1016/j.jafr.2020.100021

Onyeoziri, I. O., Torres-Aguilar, P., Hamaker, B. R., Taylor, J. R., and de Kock, H. L. 2021. Descriptive sensory analysis of instant porridge from stored wholegrain and decorticated pearl millet flour cooked, stabilized and improved by using a low-cost extruder. *Journal of Food Science* 86(9):3824–3838. https://doi.org/10.1111/1750-3841.15862

Pandey, P., Usha, M., Nirmala, Y., and Pushpa, D. 2017. Evaluation of physico-functional, cooking and textural quality characteristics of foxtail millet (Setaria italica) based vermicelli. *International Journal of Current Microbiology and Applied Sciences* 6(10):1323–35. https://doi.org/10.20546/ijcmas.2017.610.156

Patil, K. B., Chimmad, B. V., and Itagi, S. 2015. Glycemic index and quality evaluation of little millet (Panicum miliare) flakes with enhanced shelf life. *Journal of Food Science and Technology* 52(9):6078–6082. https://doi.org/10.1007/s13197-014-1663-5

Pawar, P. P., Sawate, A. R., Kshirsagar, R. B., and Kulkarni, A. T. 2020. Studies on development and organoleptic evaluation of porridge mix. *The Pharma Innovation Journal* 9(8):203–6.

Prabhakar, B., More, D. R., Shivashankar, S., Mallesh, S., and Babu, G. N. 2016. Physico-chemical and sensory evaluation of sorghum-finger millet papad. *International Journal of Food and Fermentation Technology* 6(2):387–395. http://dx.doi.org/10.5958/2277-9396.2016.00064.7

Rai, S., Kaur, A., and Singh, B. 2014. Quality characteristics of gluten free cookies prepared from different flour combinations. *Journal of Food Science and Technology* 51(4):785–789. https://doi.org/10.1007/s13 197-011-0547-1

Ranjita Devi, T., Bharati, V. C., and Jones, N. P. 2016. Ready-to-cook millet flakes based on minor millets for modern consumer. *Journal of Food Science and Technology* 53(2):1312–1318. https://doi.org/10.1007/s13 197-015-2072-0

Rao, B. D., Patil, J. V., Reddy, K. N., Soni, V. K., and Srivatsava, G. 2014. Sorghum: an emerging cash crop. *Foundation Books* 1–158.

Rao, B. D., Vishala, A. D., Arlene Christina, G. D., and Tonapi, V. A. 2016. *Millet recipes, a healthy choice* ICAR– Indian Institute of Millets Research, Rajendranagar, Hyderabad.

Rathore, T., Singh, R., Kamble, D. B., Upadhyay, A., and Thangalakshmi, S. 2019. Review on finger millet: Processing and value addition. *The Pharma Innovation Journal* 8(4):283–291.

Ren, X., Chen, J., Molla, M. M., Wang, C., Diao, X., and Shen, Q. 2016. *In vitro* starch digestibility and *in vivo* glycemic response of foxtail millet and its products. *Food and Function* 7(1):372–379. https://doi.org/ 10.1039/C5FO01074H

Saini, S., Saxena, S., Samtiya, M., Puniya, M., and Dhewa, T. 2021. Potential of underutilized millets as nutri-cereal: an overview. *Journal of Food Science and Technology* 58(12):4465–77. https://doi.org/10.1007/s13 197-021-04985-x

Samanian, N., and Razavi, S. 2017. Investigating the relationship between the perceived thickness of the chocolate pudding in sensory and instrumental analysis. *Iranian Journal Food Science and Technology Research* 12(6):730–741. https://doi.org/10.22067/ifstrj.v12i6.50418

Sarabhai, S., Tamilselvan, T., and Prabhasankar, P. 2021. Role of enzymes for improvement in gluten-free foxtail millet bread: It's effect on quality, textural, rheological and pasting properties. *LWT – Food Science and Technology* 137:110365. https://doi.org/10.1016/j.lwt.2020.110365

Sarojani, J. K., Suvarna, C. H., Hilli, J. S., and Sneha, S. 2021. Development of ready to cook kodo millet pasta. *The Pharma Innovation Journal* 10(6):910–15.

Sathish, G. 2018. *The story of millets*. Karnataka State Department of Agriculture, Bengaluru, India with ICAR-Indian Institute of Millets Research, Hyderabad.

Seth, D., and Rajamanickam, G. 2012. Development of extruded snacks using soy, sorghum, millet and rice blend – a response surface methodology approach. *International Journal of Food Science and Technology* 47(7):1526–1531. https://doi.org/10.1111/j.1365-2621.2012.03001.x

Sharma, B., and Gujral, H. S. 2019. Modulation in quality attributes of dough and starch digestibility of unleavened flat bread on replacing wheat flour with different minor millet flours. *International Journal of Biological Macromolecules* 141:117-124. https://doi.org/10.1016/j.ijbiomac.2019.08.252

Sharma, M., Singh, A. K., and Yadav, D. N. 2017. Rheological properties of reduced fat ice cream mix containing octenyl succinylated pearl millet starch. *Journal of Food Science and Technology* 54(6) 1638–1645. https:// doi.org/10.1007/s13197-017-2595-7

Sharma, S., Sharma, N., Sharma, R., and Handa. S. 2018. Formulation of functional multigrain bread and evaluation of their health potential. *International Journal of Current Microbiology and Applied Sciences* 7(7):4120–26. https://doi.org/10.20546/ijcmas.2018.707.480

Shukla, K., and Srivastava, S. 2014. Evaluation of finger millet incorporated noodles for nutritive value and glycemic index. *Journal of Food Science and Technology* 51(3):527–534. https://doi.org/10.1007/s13 197-011-0530-x

Shunmugapriya, K., Kanchana, S., Maheswari, T. U., Saravanakumar, R., and Vanniarajan, C. 2020. Optimization of the process parameters for extraction of millet milk. *International Journal of Biochemistry Research and Review* 39–47. https://doi.org/10.9734/IJBCRR/2020/v29i430184

Singh, A., Kumar, M., and Shamim, M. 2020. Importance of minor millets (nutri-cereals) for nutrition purpose in present scenario. *International Journal of Chemical Studies* 8(1):3109–3113.

Singh, S., and Dubey, R. P. 2017. Preparation of value added food products with incorporation of peanut milk and kodo millet. *International Journal of Applied Home Science* 4(7):467–472.

Sumathi, A., Ushakumari, S. R., and Malleshi, N. G. 2007. Physico-chemical characteristics, nutritional quality and shelf-life of pearl millet based extrusion cooked supplementary foods. *International Journal of Food Sciences and Nutrition* 58(5):350–62. https://doi.org/10.1080/09637480701252187

Sun, Q., Gong, M., Li, Y., and Xiong, L. 2014. Effect of dry heat treatment on the physicochemical properties and structure of proso millet flour and starch. *Carbohydrate Polymers* 110:128–134. https://doi.org/10.1016/j.carbpol.2014.03.090

Takahashi, R., Wilunda, C., Magutah, K., Mwaura-Tenambergen, W., Wilunda, B., and Perngparn, U. 2017. Correlates of alcohol consumption in rural western Kenya: A cross-sectional study. *BMC Psychiatry* 17(1):1–10. https://doi.org/10.1186/s12888-017-1344-9

Takhellambam, R. D., Chimmad, B. V., and Prkasam, J. N. 2016. Ready-to-cook millet flakes based on minor millets for modern consumer. *Journal of Food Science and Technology* 53(2):1312-1318. https://doi.org/10.1007/s13197-015-2072-0

Taylor, J. R. 2017. Millets: their unique nutritional and health-promoting attributes. In *Gluten-free Ancient Grains*, 55–103. Woodhead Publishing. https://doi.org/10.1016/B978-0-08-100866-9.00004-2

Taylor, J. R., and Duodu, K. G. 2019. Traditional sorghum and millet food and beverage products and their technologies. In *Sorghum and Millets*, 259–292. AACC International Press. https://doi.org/10.1016/B978-0-12-811527-5.00009-5

Taylor, J. R., and Emmambux, M. N. 2008. Gluten-free foods and beverages from millets. In *Gluten-free Cereal Products and Beverages*, 119–148. Academic Press. https://doi.org/10.1016/B978-0-08-100866-9.00004-2

Tiwari, A., Jha, S. K., Pal, R. K., Sethi, S., and Krishan, L. 2014. Effect of pre-milling treatments on storage stability of pearl millet flour. *Journal of Food Processing and Preservation* 38(3):1215–1223. https://doi.org/10.1111/jfpp.12082

Tomić, J., Torbica, A., and Belović, M. 2020. Effect of non-gluten proteins and transglutaminase on dough rheological properties and quality of bread based on millet (*Panicum miliaceum*) flour. *LWT – Food Science and Technology* 118:108852. https://doi.org/10.1016/j.lwt.2019.108852

Tulasi, G., Deepika, U., Venkateshwarlu, P., Santhosh, V., and Srilatha, P. 2020. Development of millet based instant soup mix and Pulav mix. *International Journal of Chemical Studies* 8(5):832–835. https://doi.org/10.22271/chemi.2020.v8.i5l.10400

Umarji, V. K., and Vijayalaxmi, K. G. 2020. Organoleptic, physical, nutritional characteristics and storage stability of value added Kodo masala khakhra. *Journal of Pharmacognosy and Phytochemistry* 9(5):326–333. https://doi.org/10.22271/phyto.2020.v9.i5Sf.12841

Upadhyay, S., Tiwari, R., Kumar, S., and Kohli, D. 2017. Production and evaluation of instant herbal mix soup. *International Journal of Agricultural Science and Research* 7(3):37–42.

Verma, V, and Patel, S. 2013. Value added products from nutri-cereals: finger millet (*Eleusine coracana*). *Emirates Journal of Food and Agriculture* 25(3):169–176. https://doi.org/10.9755/ejfa.v25i3.10764

Wandhekar, S. S., Bondre, N., and Swami, A. 2021. Formulation and quality evaluation of finger millet (ragi) fortified cookies. *International Journal of Innovation, Leadership, Society and Sustainability* 1(1):1–8.

Wang, Y., Compaoré-Sérémé, D., Sawadogo-Lingani, H., Coda, R., Katina, K., and Maina, N. H. 2019. Influence of dextran synthesized in situ on the rheological, technological and nutritional properties of whole grain pearl millet bread. *Food Chemistry* 285:221–230. https://doi.org/10.1016/j.foodchem.2019.01.126

Yadav, D. N., Balasubramanian, S., Kaur, J., Anand, T., and Singh, A. K. 2011. Optimization and shelf-life evaluation of pearl millet based halwa dry mix. *Journal of Food Science and Technology* 1(6):313–22.

Yadav, D. N., Anand, T., Kaur, J., and Singh, A. K. 2012a. Improved storage stability of pearl millet flour through microwave treatment. *Agricultural Research* 1(4):399–404. https://doi.org/10.1007/S40003-012-0040-8

Yadav, D. N., Kaur, J., Anand, T., and Singh, A. K. 2012b. Storage stability and pasting properties of hydrothermally treated pearl millet flour. *International Journal of Food Science and Technology* 47(12):2532–2537. https://doi.org/10.1111/J.1365-2621.2012.03131.X

Yadav, D. N., Chhikara, N., Anand, T., Sharma, M., and Singh, A. K. 2014. Rheological quality of pearl millet porridge as affected by grits size. *Journal of Food Science and Technology* 51(9):2169–2175. https://doi.org/10.1007/s13197-013-1252-z

Yenagi, N. B., Handigol, J. A., Ravi, S. B., Mal, B., and Padulosi, S. 2010. Nutritional and technological advancements in the promotion of ethnic and novel foods using the genetic diversity of minor millets in India. *Indian Journal of Plant Genetic Resources* 23(1):82–86.

Zacharia, R. K., Aneena, E. R., Panjikkaran, S. T., Sharon, C. L., and Lakshmi, P. S. 2020. Standardisation and quality evaluation of finger millet based nutri flakes. *Journal of Applied Life Sciences International*, 36–42. https://doi.org/10.9734/jalsi/2020/v23i1030191

Sorghum (*Sorghum bicolor*)
Phytochemical Composition, Bio-Functional, and Technological Characteristics

Muzamil A. Rather, Rahul Thakur, Monjurul Hoque, Rahel Suchintita Das,
Karine Sayuri Lima Miki, Barbara Elisabeth Teixeira-Costa,
Poonam Mishra, and Arun Kumar Gupta

Contents

DOI: 10.1201/9781003251279-3

3.1 Introduction

Sorghum bicolor (L.) Moench, usually called broomcorn, common wild sorghum, chicken-corn, and so forth, is a multipurpose crop which is generally utilized as food, feed, fuel, and fiber. It is a widely cultivated crop and ranks among the five top cereal crops in the world (Ananda et al., 2020; Das et al., 2023). It is an annual C$_4$ plant that is adapted to high temperatures and dry conditions. Since its efficiency in utilizing high solar radiant energy in the tropical latitudes is relatively much better than maize and rice, therefore, it is the most important food crop for the semi-arid and arid areas of the world (Sage and Zhu, 2011). The United States is the largest producer of sorghum in the world, followed by Nigeria, India, Mexico and Argentina. It has been reported that on average the United States produces 50 megatons of sorghum annually (FAOSTAT, 2019). Due to the unavailability of authentic evidence, ambiguities exist regarding the time period and exact location for the domestication of sorghum. But most reports suggest that sorghum was undoubtedly first domesticated in Africa about 5,000 years ago. It has been reported that domestication took place in the Ethiopia-Sudan region of Northeast Africa (Doggett, 1988). Although there are several reports regarding the actual domestication sites of sorghum, the most confirmed site and time is Kassala in Northeastern Sudan in the period 3500–1500 BP (Beldados et al., 2015). On the basis of end use, four major categories of sorghum are being cultivated namely, grain sorghum, forage sorghum, biomass sorghum, and sweet sorghum. The varieties are genetically different from each other and are grown as per the cultivar requirement (Mullet et al., 2014). Additionally, bicolor, guinea, caudatum, kafir, and durra are five morphologically different races of sorghum. The morphological characterization of the five major races is presented in **Table 3.1**. As indicated by genetic studies all the races of sorghum are distinct from each other (Taylor, 2019). All the domesticated races of sorghum have arisen and diffused individually and separately. It is reported that the most primitive race of sorghum is bicolor, and all other races originated from it (Kimber, 2000). Bicolor race of sorghum diffused from the center of its origins (Africa) to Ethiopia and Lake Chad in the East and West, respectively. The bicolor race diffused to India before 3,000 BP and then to Indonesia, followed by China. Additionally, guinea sorghum is well adapted to high rainfall areas also and is considered as a West African race of sorghum. Similarly, it is reported that Southeast Africa is the origin site of sorghum's kafir race, and the traders have played an important role in the diffusion of this race from east coast African ports to South India. Northern African peoples, speaking the Chari-Nile group of languages, are associated with the caudatum sorghum race, which was domesticated later than sorghum's bicolor race and guinea race. The durra race is white-seeded sorghum and shows better adaptation to low rainfall regions. This race was brought to North Africa and may have been domesticated in India (Harlan and Stemler, 2011). The genus *Sorghum* belongs to the grass family and is an important member of the tribe, Andropogoneae. The genus is comprised of 25 species and five sub-genera. *S. bicolor* belongs to sub-genera *Eusorghum* that comprises various agronomically important species (Hariprasanna and Patil, 2015). Taxonomically and genetically, sorghum is related to maize, since both are diploid and have same chromosome number (2n=20) besides sharing the same subfamily, Andropogoneae (Obilana, 2004).

Table 3.1 Characteristic Features of Different Races of *Sorghum Bicolor* (L.)

S. No	Characteristic	Bicolor	Kafir	Caudatum	Durra	Guinea
1	**Inflorescence**	Open inflorescences	Moderately compact	Open to compact	Compact	Open
2	**Glumes**	Long and clasping	Long and tightly clasping	Short	Middle-creased and lower	Long and separated
3	**Spikelets/grains**	Elliptic	Elliptic	One side curved and other flat	Flat, ovate, and sessile	Obliquely twisted

Source: Moench (Doggett, 1988)

Taxonomic position of *Sorghum bicolor* (L.) Moench:
Kingdom: Plantae
   Subkingdom: Tracheobionta
      Superdivision: Spermatophyta
         Division: Magnoliophyta
            Class: Liliopsida
               Subclass: Commelinidae
                  Order: Cyperales
                     Family: Poaceae
                        Tribe: Andropogoneae
                           Genus: *Sorghum* Moench
                              Species: *Sorghum bicolor* (L.) Moench

S. bicolor is a drought resistant plant and the adaptive features to resist high temperatures include a waxy cover over the leaves, extensive root system, C_4 anatomy, and so forth. It grows well in different soils, including light sandy soils, heavy vertisols, and sandy loams. Self-pollination is predominantly found in *S. bicolor* but anemophily is reported in some cases as well (Ohadi et al., 2018). As discussed earlier, *S. bicolor* is cultivated for grain, fodder, brooms, sugar, biofuel, and as a cover crop. The crop is grown especially for human consumption as 50 percent of *S. bicolor* is consumed by humans globally. But it has also been reported that in the United States 90 percent of the total *S. bicolor* production is consumed as livestock feed while its other fraction is employed for ethanol production (Tubeileh et al., 2014). The feeding value of *S. bicolor* is quite high. It is documented that its feeding value is more than 95 percent to that of the yellow dent maize which could be attributed to the higher percentage of condensed tannins that reduce digestibility (Rooney et al., 2016). *S. bicolor* is a rich source of B-complex vitamins, and vitamin B_{12} and traces of fat-soluble vitamins are also available. Furthermore, different minerals are also present in *S. bicolor* including phosphorous, potassium, silicon, magnesium, sulphur, calcium, iron, zinc, and so forth. (Arendt and Zannini, 2013). Sorghum is also used in weaning foods, alcoholic and non-alcoholic beverages, and breakfast foods in Southern and West Africa (Aka et al., 2014). Folk pharmaceuticals and food additives are obtained from different parts of the sorghum plant, including leaves, stalks, root, sheath, and glumes. Red and purple pigments are derived from the sheath of the plant to color different foods, including cheese and porridges. These pigments are also used to dye clothes, wood and leather in some of the Asian and African countries (Adetuyi et al., 2007).

3.2 Nutraceutical properties

Nutraceuticals are bioactive, natural chemical substances with health-promoting, disease-preventive, or medicinal characteristics. Nutraceuticals are food products that have health and medical advantages, such as illness prevention and therapy. Consistent intake of foods high in phytochemicals and dietary fiber is linked to disease prevention. Because of the components in cereal bran, particularly dietary fiber and phytochemicals, whole grain intake has been shown to help reduce the risk of heart disease, diabetes, and other chronic illnesses (Thakur et al., 2022; Awika et al., 2005; Jacobs Jr. et al., 1999).

Sorghum has been a great staple food for China, Africa and parts of India for a long period of time. The pericarp, endosperm, and germ are three different anatomical components found in sorghum grains. Between the pericarp and the endosperm, some types include a fourth component, called the testa. Several researches on sorghum for human consumption have been done in these nations because of its great nutritional and functional potential. Non-starch polysaccharides, phenolic chemicals (3-deoxyanthocyanidins, tannins, and phenolic acids), and carotenoids making sorgum an important dietary component. In the endosperm, carbohydrates, proteins, B vitamins, and minerals are present. Lipids, fat-soluble vitamins, B-complex vitamins, and minerals make up the germ. Additionally, sorghum is a rich source of bioactive chemicals that have been shown to improve human health. Compounds extracted from sorghum, primarily phenolics, have been found to cause positive changes in parameters linked to noncommunicable illnesses such as obesity, diabetes, dyslipidemia, cardiovascular disease, cancer, and hypertension in *in vitro* and animal investigations (Kamath et al., 2007; Farrar et al., 2008; Awika et al., 2009; Kim and Park, 2012)

3.2.1 Chemical composition and nutritional value

Whole sorghum is chemically comparable to rice, maize, and wheat in terms of nutritional content. Sorghum grains have an energy value of 296.1 to 356.0 kcal/100 g. Polysaccharides (starch and non-starch) are the primary constituents of sorghum, followed by proteins and lipids (Martino et al., 2012.

3.2.1.1 Polysaccharides

The genetic features and growth circumstances impact the quantity and composition of starch, sorghum's primary carbohydrate. The starch content of certain cultivars varies between 32.1 and 72.5 g/100 g and is mostly made up of amylopectin and amylose. Because of the tight connection between starch granules, proteins, and tannins, sorghum has the lowest starch digestibility of all grains (Sang et al., 2008; Singh et al., 2010).

3.2.1.2 Proteins

Prolamins, are the proteins found in sorghum. Sorghum proteins are less digestible than grains like wheat and maize, especially after cooking. The major prolamins of sorghum are kafirins. Complexation of kafirins with this phenolic compound can lower protein digestibility by up to 50 percent in tannin-rich cultivars. Other exogenous elements that contribute to poor digestibility include starch, non-starch polysaccharides, phytic acid, and lipids (Belton et al., 2006; Martino et al., 2012; Afify et al., 2012b).

3.2.1.3 Lipid

Sorghum has a low lipid content (1.24 to 3.07 g/100 g), with unsaturated fatty acids accounting for 83–88 percent of the lipid content. Polyunsaturated fatty acids (PUFA) are more abundant than monounsaturated fatty acids in most sorghum cultivars (MUFA). Linoleic (45.6–51.1%), oleic (32.2–42.0%), palmitic (12.4–16.0%), and linolenic acids (1.4–2.8%) are the most common fatty acids found in sorghum (Mehmood et al., 2008; Hadbaoui et al., 2010; Afify et al., 2012a).

3.2.1.4 Minerals and vitamins

Sorghum is a great source of minerals like phosphorus, potassium, and zinc, the amount of which varies depending on where it is grown. The bioavailability of most sorghum minerals is yet unknown. However, it is known that zinc availability varies between 9.7 percent and 17.1 percent, while iron availability varies between 6.6 and 15.7 percent. Through bio-fortification, fortification, and genetic enhancement of sorghum, studies have been undertaken to increase the content and bioavailability of iron and zinc. There is not a lot of information on the vitamin composition of sorghum. However, it contains B-complex vitamins (thiamine, riboflavin, and pyridoxine) as well as fat-soluble vitamins (D, E) (Ochanda et al., 2010; Martino et al., 2012).

3.2.2 Potential impacts of sorghum on human health

The potential functional advantages to human health of chemicals extracted from sorghum, particularly the entire grain, are currently unclear. *In vitro* and animal research have revealed that phenolics, or fat-soluble chemicals extracted from sorghum, help to regulate the gut microbiota and parameters linked to noncommunicable illnesses, including obesity, diabetes dyslipidemia, cardiovascular disease, cancer, and hypertension.

3.2.2.1 Oxidative stress

The generation of free radicals is important in the development of noncommunicable diseases because it is continuous and excessive. *In vitro* studies have shown that components extracted from sorghum can protect against oxidative damage. The phenolic chemicals are responsible for these functional advantages, which are most noticeable when extracts from black or red sorghum are utilized (Burdette et al., 2010; Moraes et al., 2012). Phenolic chemicals are derived from sorghum influence phase II enzyme expression. By continually converting highly reactive electrophilic species (RES) into nontoxic and excretable metabolites, these enzymes regulate the defiance mechanism against oxidative stress (Montilla et al., 2012).

Sorghum increases NADH: quinone oxyreductase (NQO) activity, which is the major influence on phase II enzymes. This action is ascribed to sorghum 3-deoxyanthocyanidins, and it is dependent on their profile rather than the content. Apigeninidin and luteolinidin were shown to have no substantial NQO inducer action in recent research. Their 7-methoxylated versions, on the other hand, were potent NQO inducers (Awika et al., 2009; Yang et al., 2009).

3.2.2.2 Cancer

Carcinogens like toxics, mutagenic and carcinogenic agents, which make up reactive intermediates (Reactive oxygen species and Reactive nitrogen species), damage DNA and cause the majority of cancers. The activity of the phase I (cytochrome P-450) and II enzyme systems, which also eliminated endogenous and environmental carcinogens, has a substantial influence on the carcinogen rate in people (Shih et al., 2007; Sharma et al., 2010).

However, due to a paucity of research, it is impossible to deduce the effects of sorghum on people. Its ability to prevent cancer in humans is supported by epidemiological research. In black South Africans, the replacement of maize with sorghum as a staple meal increased the prevalence of esophageal cancer. Sorghum's ability to lower the risk of esophageal cancer in people is yet unclear (Isaacson, 2005).

The early removal of malignancies is considered an important element of anticancer benefits. The findings show that phenolic compounds from sorghum, particularly 3-deoxyanthocyanidins, operate directly against cancer cells by increasing apoptosis and inhibiting the development and spread of skin melanoma, colon, esophagus, liver, breast, and bone marrow cancer cells. In addition, Sorghum 3-deoxyanthocyanidins are cytotoxic to cancer cells and the estrogenic action of sorghum flavones caused apoptosis in colon cancer cells (Shih et al., 2007; Yang et al., 2009; Awika et al., 2009; Ko et al., 2012; Park et al., 2012; Hwang et al., 2013).

The phenolic components of sorghum can inhibit the cellular DNA synthesis and metastasis of cancer cells in certain ways. Tannins, which may have anticancer properties, is extracted from different foods and found to alter regulatory enzymes, inhibit signal transduction pathways, and induce apoptosis, and thus they have received a lot of interest as a cancer therapy (Hwang et al., 2013).

3.2.2.3 Obesity and inflammations

Obesity is a worldwide epidemic that is linked to a variety of non-communicable diseases. The findings of the experiments show that tannin-rich sorghum decreases weight growth in animals. The decreased weight growth is bad in slaughter animals, but it may help prevent obesity in people.

Polymeric tannins from sorghum can naturally alter starch by reacting significantly with amylose, resulting in resistant starch, according to recent research. Because resistant starch cannot be digested in the small intestine, it reaches the big intestine and provides dietary fiber advantages. Sorghum tannins also limit starch digestion by blocking the enzymes saccharase and amylase. Additionally, the complexation of tannins with proteins, as well as the suppression of digestive enzymes may have also contributed to the reduced weight gain. Despite animal evidence, it is unclear whether sorghum (tannin-rich or not) affects human weight (Nyamambi et al., 2000; Frazier et al., 2010; Rahman and Osman, 2011; Barros et al., 2013).

Adipocytes and obesity are well documented for their impact on inflammatory mediators that signal this process. The revelation that obesity causes an inflammatory response in metabolic tissues sparked a new field of study into obesity's inflammatory processes. This breakthrough provides for a better knowledge of the role of adipocytes in health and obesity, as well as how inflammatory mediators function as signaling molecules in this process (Gregor and Hotamisligil, 2011).

The functional advantages of whole sorghum and its fractions in humans are unknown, although they might be attributed in part to increased expression of adiponectin, which suppresses this inflammatory marker. Thus, *in vitro* and animal studies show that sorghum's anti-inflammatory effects are due to its impact on enzymes, whereas 3-deoxyanthocyanidins primarily affect cytokines.

3.2.2.4 Dyslipidemia

The lipidic and phenolic fractions of sorghum have been found to alter parameters linked to dyslipidemia and the risk of cardiovascular disease in both *in vitro* and animal investigations. Phytosterols, polycosanols, and phenolic chemicals, which may regulate cholesterol absorption, excretion, and synthesis, are responsible for these effects.

The addition of sorghum lipid to the diet can decrease the hepatic and plasma cholesterol. The phytosterols (bioactive compounds), polycosanols and sorghum phenolic compounds in the lipid fraction of sorghum can help to lower cholesterol absorption. In numerous investigations, phytosterols isolated from different foods were found to limit cholesterol absorption in humans, resulting in higher fecal excretion and lower plasma LDL-c levels. The lipid fraction, on the other hand, influences cholesterol absorption through changing the gut microbiota, as well as the production and excretion of endogenous cholesterol. Thus, sorghum lipid decreases the concentration of plasma cholesterol in normolipidemic hamsters (Carr et al., 2005; Hoi et al., 2009; Martnez et al., 2009).

3.2.2.5 Diabetes

According to recent findings, the activity of phenolic chemicals in sorghum fractions modulates glucose metabolism in animals. However, whether the components extracted from sorghum, particularly the entire grain, are helpful to humans is unknown.

The hypoglycemic impact of phenolic extracts of sorghum was similar to that of glibenclamida, an antidiabetic drug utilized in the control group. The areas under the glucose and glycaemia curves were decreased in mice after extracts were consumed (Chung et al., 2011; Kim and Park, 2012). According to research published in *The Journal of Agricultural and Food Chemistry*, Saracens extracts of sorghum can slow the pace of glucose digestion by inhibiting enzymes, which could be the first action mechanism of sorghum on the human metabolism (JAC). In diabetic mice given extracts of phenolic substances, a rise in insulin concentration was found. This rise suggests improved β cell activity, which has therapeutic implications for type 2 diabetes. Through an increase in insulin sensitivity, oral administration of sorghum phenolic extracts can prevent and function as an adjuvant factor in the treatment of diabetes (Park et al., 2012; Morey et al., 2018).

3.2.2.6 Hypertension

Sorghum has recently been linked to a reduction in blood pressure in the scientific literature. In this work, an isolate of sorghum α-kafirins reduced the activity of the angiotensin I converting enzyme in both competitive and noncompetitive ways (Kamath et al., 2007).

3.2.2.7 Gut microbiota

The human gut is home to a diverse range of bacterial species that develop essential metabolic and immunological activities, all of which have a significant impact on the host's nutritional and health. The health advantages of phenolic compounds in foods may be due to direct action of bioactive compounds (and their metabolites) or indirect effects mediated by non-absorbed chemicals that alter the microbiota environment and, as a result, human metabolism. It is critical to examine these impacts during human health treatments.

Because of their quantity and the fact that they are not absorbed in the large intestine, phenolic compounds in sorghum, such as tannins, are of particular importance. Instead, the intestinal bacteria metabolizes them, producing prebiotic-like effects. Sorghum contains resistant starch and dietary fiber, both of which have the ability to alter gut flora (Scott et al., 2008).

Numerous studies have shown that phenolic chemicals, such as tannins and anthocyanins, increase Bifidobacterium spp. and Lactobacillus in the gut microbiota. Bacteroides spp., Clostridium, Propionibacterium, Salmonella typhimurium, Streptococcus mutans, and Escherichia coli are all decreasing in number. Only one research has looked at the link between bioactive chemicals in sorghum and hamster gut microbiota (Hidalgo et al., 2012).

3.3 Characterization of phytochemical compounds responsible for bioactive properties

Sorghum is a rich source of bioactive compounds, vitamins, minerals, starch, proteins, policosanols, phytosterols, high fiber content, and so forth, therefore, enlisted among the major cereal crops worldwide. Flavonoids (luteolin, catechin gallate, epigallocatechin, and apigenin), tannins, phenolic acid (tannic, p-coumaric acids, and ferulic), and lipids are the major bioactive compounds largely available in sorghum (Ghimire et al., 2021). These bioactive compounds demonstrate significant antibacterial, anticancer, antioxidant – and so forth – activities and play an important role in preventing obesity, cardiovascular diseases, diabetes, and so forth. The abundance and composition of these bioactive compounds and their associated functions may fluctuate due to the varying abiotic factors (edaphic factors, temperature, precipitation, altitude, and so forth), agronomic traits, genetic variations, geographical locations, and so forth (Teixeira et al., 2016). The scientific reports suggest that the polyphenol ingredients available in sorghum play a significant role in disease prevention. The overproduction of reactive oxygen species (ROS) via metabolic processes is the causative agent of numerous degenerative diseases, namely, Alzheimer's disease, diabetes, aging, and heart diseases (Alfadda and Sallam, 2012). The ROS are scavenged by phenolic compounds, thereby neutralize peroxide and oxygen radicals. The scavenging of ROS reduces or inhibits the oxidation process, and thus help in preventing the degradation of nutritional value and food quality. The bioactive molecules also help plants evade the extreme environmental conditions, and thus assist the plant in adapting to different stress conditions. Various synthetic antioxidants are available and used in different health and food sectors, yet emphasis is put on naturally derived compounds due to the toxicity and other health effects of the former (Shahidi and Zhong, 2010). Different analytical approaches including mass spectrometry, infrared spectroscopy, and nuclear magnetic resonance (NMR) spectrometry have been performed to evaluate the composition and availability of bioactive compounds in sorghum. Reverse phase High Performance Liquid Chromatography (HPLC) analysis demonstrated the presence of eight major phenolic acids, namely, Gallic acid, p-Hydroxybenzoic acid, Caffeic acid, p-Coumaric acid, Protocatechuic acid, Cinnamic acid, Vanillic acid, and Ferulic acid (Hahn et al., 1983). Dykes et al. (2009) analyzed the flavonoid composition of different varieties of sorghum using HPLC-DAD and demonstrated the presence of the flavanones naringenin, eriodictyol, eriodictyol glucoside, the 3-deoxyanthocyanins luteolinidin and apigeninidin, and the flavone apigenin. Similarly, Abugri et al. (2015) reported the presence of gallic acid, apigenin, naringen, epigallocatechin, and gallic acid, naringen, epigallocatechin, respectively in sorghum red flour and pale brown leaves employing HPLC and Liquid chromatography–mass spectrometry (LCMS). Borokini, (2017) reported the presence of various flavonoids and phenolic acids in the leaf extract of sorghum through reverse phase HPLC-DAD, namely, gallic acid, caffeic acid, isoquercitrin, kaempferol, catechin, ellagic acid, quercitrin, chlorogenic acid, rutin, and quercetin. It is pertinent to mention that the overall secondary metabolite (phenolic acids, condensed tannins, and lavonoids) concentration of sorghum depends solely on the external environmental factors (both biotic as well as abiotic) and the genotype of the race (Hassan, 2020). Being the dietary staple food of more than 500 million people in about 30 countries of the semiarid tropic's determination of phytochemical composition and the effect of environmental factors of sorghum has received interest of the researchers (Kumari et al., 2016). Therefore, different processing techniques like germination, physical treatments, and so forth have been employed to enhance the phytochemical composition, nutritional profile and functional properties of sorghum. Germination changes the nutritional composition as different complex entities are changed to simpler ones, which are later on transformed to essential compounds. Several studies have been carried out that demonstrated a germination induced rise of free amino acids, carbohydrates, polyphenols, and so forth (Yang et al., 2016; Donkor et al., 2012). Ultrasound treatment of sorghum seed by Hassan et al. (2020) demonstrated a superior profile of phytoconstituents (saponins, phytates, sterols, alkaloids, and phenolic profile) and their associated functions (antioxidant activity).

The crude fat content of sorghum is 3 percent, which is found to be much higher than that of rice or wheat. The fat is predominantly found in germ and aleurone layers of the seed (Rooney et al., 1991). The nutritional composition of sorghum present in 100 g edible portion is protein (10.4 g), fat (3.1), ash (1.6 g), crude fiber (2.0 g), carbohydrate (70.7 g), energy (329 kcal), calcium (25 mg), iron (5.4 mg), thiamine (0.38 mg), riboflavin (0.15 mg), and niacin (4.3 mg). Furthermore, the percentage of cellulose (insoluble fiber component) among different varieties of sorghum ranged between 1.19-5.23 percent (Kulamarva et al., 2009). It is documented that embryo, bran layers and endosperm constitute 10, 8, and 80 percent, respectively of the grain, although it varies with environmental conditions, maturity, and genetic makeup. The composition of the kernel protein, starch, and B-complex vitamins is 80 percent, 94 percent, and 50 to 75 percent, respectively. Due to the restricted accessibility of starch inflicted by endosperm proteins in addition to the high content of amylose and amylopectin, the raw starch digestibility is considered low. The low starch digestibility is considered responsible due to high content of the dietary fiber and the protein digestibility has been reported in the range of 49.5 to 70 percent. The starch digestibility has been increased by some food processing technologies like steaming, micronization, flaking, puffing, fermentation or pressure-cooking, and by addition of reducing agents like cysteine, ascorbic acid, or sodium metabisulphite (Nawar et al., 1970). As discussed earlier sorghum is a rich source of minerals, proteins, fibers, and polyphenols; several studies have been carried out to investigate the effect of peptides on biological activities. It has been demonstrated that peptides obtained by *in vitro* hydrolysis of sorghum protein acts as antiviral and antihypertensive agents (Camargo Filho et al., 2008; Kamath et al., 2007).

3.3.1 Major bioactive compounds present in sorghum
3.3.1.1 Polyphenolic compounds

Polyphenols are one of the key functional compounds in sorghum that contribute to bioactive properties such as antioxidant, anti-inflammatory, anti-hypertensive activities and thereby prevent and reduce the risk of metabolic diseases like coronary heart diseases, cancer (breast, colon), and type 2 diabetes (Espitia-Hernández et al., 2020). The major polyphenols in sorghum include flavonoids and phenolic acid (Espitia-Hernández et al., 2020; Xiong, Zhang, Luo, Johnson, & Fang, 2019).

3.3.1.1.1 Flavonoids

Flavonoids are one of the most widely distributed phenolic compounds in plant-based foods but a relatively small amount is found in the frequently consumed cereal grains (Girard &

Awika, 2018; Zamora-Ros et al., 2016). Flavonoids in sorghum are found in their outer layer (bran) in variable amounts with different profiles between genotypes, and they are responsible for pericarp color, thickness, and the existence of pigmented testa (Espitia-Hernández et al., 2020; G. Wu et al., 2013). Flavonoids have the basic flavan skeleton and are classified depending on the presence of the $C_2 - C_3$ double bond and substituent groups on the C-ring. Among the wide class of flavonoids, flavones, flavanones, and anthocyanins are dominating in sorghum.

Flavones are yellow-colored flavonoids, usually found in fruits, vegetables, tubers, legumes, and cereals. Flavones in sorghum are found in the range of 20–390 µg/g, which play a biologically important role in the human diet (Girard & Awika, 2018; Xiong, Zhang, Warner, & Fang, 2019). Few sorghum-derived flavones naturally exist as glycosides – for example, luteolin derivatives – whereas other flavones like apigenin derivatives are found in aglycones forms (Girard & Awika, 2018); whereas other cereal-based flavones usually contain C-glycosides. The basic difference between sorghum flavones and other cereal-based flavones is that aglycones are readily hydrolyzable in the acidic pH and by lactic acid bacteria and thereby offer more bioaccessible flavones (Girard & Awika, 2018). Also, the presence of the catechol group on the B-ring of luteolin derivatives may influence their bioactivity as compared to apigenin derivatives, which lack a catechol group (Girard & Awika, 2018; Wang, Li, & Bi, 2018).

Flavanones are one of the major groups of flavonoid, with naringenin and glycosides of eriodictyol being the dominant in sorghum (Xiong, Zhang, Warner, et al., 2019; Yang, Allred, Dykes, Allred, & Awika, 2015). The content of flavanones varies in the range of 0 to 1800 µg/g, where the highest content is found in sorghum phenotype having a lemon-yellow pericarp color, and the white variety has been reported to have the lowest content (Dykes, Peterson, Rooney, & Rooney, 2011; Yang et al., 2015). The amount of flavanones in sorghum is the highest among the cereals as, in general, flavanones are rare in cereals grains. Like flavones, the flavanones primarily exist as O-glycosides that are sensitive and hydrolyzable in high acidic conditions (low pH of stomach), and this results in enhanced bioavailability (Girard & Awika, 2018; Xiong, Zhang, Warner, et al., 2019).

A rare type of flavonoid almost exclusively found in sorghum is 3-deoxyanthocyanidins. The sorghum with a red pericarp but phenotypically black contains the highest amount of 3-deoxyanthocyanidins in the range of 1,790–6,120 µg/g. In general, anthocyanins are C-3-hydroxylated but in the case of sorghum, anthocyanins are C-3-deoxylated, where C-3 position lacks OH group. Thereby, this structural variation renders 3-deoxyanthocyanidins to have exceptional chemical characteristics like high stability in elevated temperature and pH and also provide color stability. The primary 3-deoxyanthocyanidins found in sorghum include apigeninidin, luteolinidin aglycones, and their methoxylated forms being present in moderate to small quantities. Rarely found pyrano and dimeric forms show higher stability and are resistant to color degradation as compared to 3-deoxyanthocyanidins and thereby offer food application especially as a natural food colorant. An exception to anthocyanin, 3-deoxyanthocyanidins with a minimum of one O-methyl substitution show potent phase II enzyme induction activity; luteolinidin is a member of the 3-deoxyanthocyanidins isolated from Sorghum bicolor, and is also reported use in the treatment of ocular diseases (Girard & Awika, 2018; Xiong, Zhang, Warner, et al., 2019).

Procyanidins in sorghum are the simple condensed or non-hydrolyzable tannins comprised of oligomers and polymers of flavan-3-ol units (Espitia-Hernández et al., 2020; L. Wu et al., 2011). The most common units of procyanidins are catechin and epicatechin (Espitia-Hernández et al., 2020). Sorghum grains are the potential source of procyanidins, containing about 10.6 to 40.0 mg/g depending on the type of varieties, and procyanidins are found in the testa of the sorghum that is a structure between pericarp and endosperm of some variety (Oliveira et al., 2017; L. Wu, Huang, Qin, & Ren, 2013; Yu, Yan, Lu, & Liu, 2018). Procyanidins show antioxidant property, anti-obesity, antiallergy, antihypertensive, and antitumor activities and therefore can be used for human health benefits (L. Wu et al., 2011).

Many researchers have extracted procyanidins from sorghum following different techniques, like L. Wu et al. (2011) who characterized procyanidin-rich extract (PARE) from the bran of *Sorghum bicolor* (L.) and Moench who determined the composition and purity of procyanidins oligomers by using normal-phase HPLC-MS. The effect of PARE was investigated against oxidative stress in mice and lung tumor inhibition of mice having Lewis's lung cancer. The analysis showed that the total procyanidin was 54.68 percent. The administration of PARE (150 mg/kg) demonstrated antioxidant activity by reversing the oxidative stress induced by D-galactose. Administration of PARE also showed antitumor activity by inhibiting the tumor growth and metastasis formation by suppressing vascular endothelial growth factor (VEGF) production. Similarly, PARE extracted from bran of three different varieties of sorghum (Jinza No. 12, Jinza No. 15, and Jinza No. 18) and characterized using HPLC-FLD showed 52.01 percent, 54.03 percent, and 52.99 percent of proanthocyanidins respectively. All the three PARE samples demonstrated antioxidant activity against hydroxyl radicals with half-maximal inhibitory concentration (IC_{50}) values of 18.32 µg/mL: Jinza No. 12; 22.31 µg/mL: Jinza No. 15 and 27.28 µg/mL: Jinza No. 18. All three PARE samples showed anti-proliferation activity against HepG2 (human liver cancer cells) with half-maximal effective concentration (EC_{50}) of 32.23 ± 5.28 µg/mL: Jinza No. 31.80 ± 6.45 µg/mL: Jinza No. 15 and 67.53 ± 11.67 µg/mL: Jinza No. 18 (Zhu, Shi, Yao, Hao, & Ren, 2017).

3.3.1.2 Phenolic acid

Phenolic acids are found abundantly in all sorghum in the range of 445 to 2,850 µg/g (Girard & Awika, 2018). The phenolic acid in sorghum is present in the testa, pericarp, and endosperm of the grain and may exist in the bound, conjugated, and unconjugated form. The bound portion of phenolic acids (70 to 95%) are covalently bonded to polysaccharides that are present in the cell wall of grains and the remaining unbound phenolic acids exist as conjugates of monomeric carbohydrate in the form of esters and aldehyde, and chiefly glycerol and very least amount remain unconjugated as free acids. The unbound and the free phenolic acid are extractable with organic solvents (Girard & Awika, 2018; Xiong, Zhang, Warner, et al., 2019). Depending on their structure, phenolic acids are classified into two groups: benzoic acid and cinnamic acid (Xiong, Zhang, Warner, et al., 2019). The types of benzoic acid reported in sorghum are gallic acid (free: 0 µg/g, bound: 12.9–46 µg/g); protocatechuic acid (free: 6.3–13 µg/g, bound: 11.5–218.3 µg/g); vanillic acid (free: 0–126.7 µg/g, bound: 0-65.6 µg/g); p-hydroxybenzoic acid (free: 2.3–10.1 µg/g, bound: 11.4–24.2 µg/g); syringic acid (total: 15.7–7.5 µg/g). Similarly, the types of cinnamic acids are ferulic acid (free: 0.63 to 26.82 µg/g, bound: 30.72 to 139.93 µg/g); sinapic acid (Total: 50 to 140), caffeic acid (free: 4.51 to 35.57 µg/g, bound: 3.66 to 35.72 µg/g); p-coumaric acid (free: 6.4 to 109.1 µg/g, bound: 38 to 138.5 µg/g) (Xiong, Zhang, Warner, et al., 2019). Due to the presence of covalent bonds in the bound phenolic acids and their resistivity towards enzymatic digestion decreases the bioavailability. (Saura-Calixto, 2011; Xiong, Zhang, Warner, et al., 2019). Phenolic acids show anti-inflammatory activity, for example, ferulic acids and gallic acid demonstrated suppression of COX-2 enzyme, while TNF-α production was inhibited by ferulic acid (Burdette, 2007).

3.3.1.3 Dietary fiber

In 2009, CODEX Alimentarius put forward the revised definition of dietary fiber which mentions inter alia that dietary fiber represents carbohydrate polymers with 10 or more monomeric units that are unable to be hydrolyzed by the endogenous enzymes present in the small intestine of human beings and can include the following groups:

1. Edible polymers of carbohydrate inherently present in food.
2. Polymers of carbohydrates, which have formed in food by physical, chemical or enzymatic methods and are known to show physiological health benefits substantiated through accepted scientific data to expert authorities.
3. Synthetic carbohydrate polymers that are known to show physiological health benefits substantiated through accepted scientific data to expert authorities.

Dietary fiber passes undigested into the large intestine wherein it is fully or partially fermented by colonic microflora leading to the generation of short-chain fatty acids that are found to prevent colon cancer (Chen et al., 2017). Other physiological benefits include reduced intestinal transit time, improved bulk of stool (laxation); reduction of postprandial blood glucose levels; reduction/normalization of LDL cholesterol levels (Kendall, Esfahani, & Jenkins, 2010).

Based on solubility in water, dietary fiber has been conventionally classified into two broad groups: soluble dietary fiber (SDF) (including pectin, pentosan, gums, and mucilage) and insoluble dietary fiber (IDF) (including cellulose, lignin, and hemicellulose) (Bader Ul Ain et al., 2019; Fuller, Beck, Salman, & Tapsell, 2016). daily intake ratio for SDF and IDF has been recommended as 1:3 respectively (Bader Ul Ain et al., 2019).

Several research groups worldwide worked to develop a method for the accurate quantification of dietary fiber. While doing so, several terminologies related to dietary fiber and its categories, apart from IDF and SDF, have come into existence, and many have been refined in the course of time, such as the following (McCleary & Cox, 2017):

1. SDFP: Dietary fiber soluble in water but insoluble in 78 percent aqueous ethanol = High molecular weight soluble dietary fiber (HMWSDF).
2. SDFS: Dietary fiber soluble in water and also soluble in 78 percent aqueous ethanol = non-digestible oligosaccharides (NDO); includes inulin/FOS, GOS, polydextrose, resistant maltodextrin (such as Fibersol 2), and xylooligosaccharides (XOS).
3. HMWDF: Higher molecular weight dietary fiber = IDF + SDFP; includes natural fibers such as cellulose, β-glucan, pectin, arabinoxylan, galactomannan, arabinogalactan, and resistant starch.
4. Total dietary fiber = HMWDF (IDF + SDFP) + SDFS.

Generally, dietary fiber content in a sample is determined using an enzymatic-gravimetric method. The defatted food sample is incubated with enzymes such as amylase, protease, and amyloglucosidase that mimic the physiological process. Digestible complex carbohydrates are hydrolyzed into simple sugars and separated by filtration after ethanol precipitation. The remaining non-digestible precipitate contains the total dietary fiber, along with protein and other inorganic matter associated, which is separately measured and then deducted from the total weight. For IDF, after filtration of the enzyme digest, the residue is collected. The filtrate is precipitated using 95 percent ethanol to obtain the SDF as the non-digestible residue which, along with the IDF residue, is dried and corrected for ash, protein, and blank in the final calculation (McCleary & Cox, 2017).

It has been demonstrated by Awika, McDonough, and Rooney (2005), that, the dietary fiber content, measured by the Prosky method, represents approximately 38–45 percent of the sorghum brans, around 4 to 6 times of the level assessed in the sorghum grain. Rooney, Rooney, and Lupton (1992) stated that, on a dry-weight basis, sorghum bran fractions had 36 to 50 percent total dietary fiber and 35 to 48 percent insoluble dietary fiber on average.

In a review, Joye (2020) cited the work of Knudsen and Munck (1985) which showed that commonly consumed Sudanese sorghum varieties have a total dietary fiber, ranging from 7.55–12.3 g/100 g with IDF and SDF being in the range of 6.52–7.90 g/100 g and 1.05–1.23 g/100 g respectively. Knudsen and Munck (1985) reported that, Dabar a low-tannin sorghum variety of Sudan, had 7.6 percent total dietary fiber content, while, Feterita, a high-tannin Sudanese variety had 9.2 percent with insoluble DF being 6.5 percent in Dabar and 7.9 percent in Feterita. The authors also found that on cooking the sorghum in the form of a porridge, energy availability decreased through formation of enzyme-resistant starch, which led to the increase in the dietary fiber content of both Dabar and Feterita. The traditional Sudanese fermentation method at pH 3.9 reduced resistant starch formation and restricted formation of lignin on cooking and also inhibited protein binding.

A study was carried out by Jood, Khetarpaul, and Goyal (2012) on water-based slurries of food mixtures containing either raw or germinated sorghum flour with tomato pulp and whey powder, which were fermented after autoclaving using a probiotic curd inoculated with 5 percent *Lactobacillus acidophilus* culture. SDF and IDF content were measured by the enzymatic method of Furda (1981). Autoclaving alone along with fermentation resulted in a significant decrease in TDF and IDF content compared to the unprocessed mixture (control). But the fermented mixture displayed increased IDF content than the autoclaved food mixture. Autoclaving significantly increased SDF content while fermentation significantly decreased the same compared to the control. A similar trend in all types of dietary fibers was recorded in the germinated mixtures but the values of DF contents were significantly less than the non-germinated food mixture. The author attributed the reduction in TDF and IDF and increase in SDF content caused by autoclaving to the conversion of IDF to shorter length units, which could have possibly been precipitated with SDF. The decrease in all the DF types by fermentation could be due to the enhanced action cellulose, α-galactosidase, and so forth, which can hydrolyze the dietary fiber and convert IDF to SDF.

In their study, Badi, Pedersen, Monowar, and Eggum (1990) developed wheat bran brown bread substituted 10 percent and 20 percent with sorghum bran. The sorghum bran incorporated breads were found to be higher in dietary fiber content (soluble and insoluble) in comparison with the wheat bran breads measured by the method of Asp, Johansson, Hallmer, and Siljestroem (1983).

SDF fibers are believed to be functionally and physiologically of more importance compared to insoluble dietary fibers because of the former's higher functionality in being rapidly fermentable by colonic microflora to generate short-chain fatty acids unlike insoluble fractions (Bader Ul Ain et al., 2019). In the work carried out by Bader Ul Ain et al. (2019), the dietary fiber ratio in sorghum was chemically modified using alkaline, acid, and consecutive acid-alkaline treatment coupled with twin-screw extrusion. TDF of treated sorghum varied from 6.70–6.90 g/100 g. The authors stated that the modification of the ratio of SDF (1.19%) and IDF (24.25%) ratio by extrusion was non-significant while acid-alkaline treatment showed highly significant results, that is, increase of SDF by 952.38 percent and 71.17 percent decrease in IDF. Hence, the authors suggested that chemical treatments singly or when combined with extrusion can significantly increase SDF in sorghum.

Fiber has been found to help in the excretion of bile acids, fats including cholesterol, and minerals by binding to them (Cummings, 1973; Suhasini & Krishna, 1991). Cholesterol binding potential has led to the acknowledgement of the ability of dietary fiber in reducing serum lipids (Trowell, 1973). In a study carried out by Suhasini and Krishna (1991) on groups of healthy human volunteers (4 males and 4 females in the age group of 23–26 years), each subject was fed with pancakes prepared from ground unrefined sorghum (100 g) daily as supper for three weeks. The authors reported significant reduction in the mean total triglycerides and cholesterol level and a marked improvement in HDL cholesterol levels.

However, the binding of dietary fiber to minerals adversely affects their bioavailability. Also, it has been pointed out by Maclean Jr, Romaña, Placko, and Graham (1981) that sorghum is not properly digested by infants and children. Suitable pre-treatments through methods such as fermentation (Graham et al., 1986) or by decortication coupled with extrusion (MacLean Jr, López de Romaña, Gastañaduy, & Graham, 1983) and/or supplementation with moderately minor quantities of lysine-rich food such as fish, milk or legumes, can encourage the use of sorghum into a satisfactory weaning food (Graham et al., 1986).

An increased intake of dietary fiber has demonstrated improvement in the control of blood glucose in subjects with impaired tolerance (Lafrance, Rabasa-Lhoret, Poisson, Ducros, & Chiasson, 1998). Decreased glucose levels were observed in serum of animals receiving a mixture of soluble and insoluble NSP (Galibois, Desrosiers, Guévin, Lavigne, & Jacques, 1994; Morgan, Tredger, Wright, & Marks, 1990). The GI and GL of sorghum-based foods such as coarse, fine semolina (upma), pasta, and flakes (poha) were reported to be significantly less compared to their corresponding, either wheat or rice-based control foods by Prasad, Rao, Kalpana, Rao, and Patil (2015).

3.3.1.4 Resistant starch

The concept of incomplete starch digestion and its absorption has led to the recognition of nondigestible starch fractions (Cummings & Englyst, 1991; Englyst, Kingman, & Cummings, 1985) that came to be referred to as "resistant starches" (RS) (Sajilata, Singhal, & Kulkarni, 2006). Resistant starch has, hence, been regarded as "the sum of starch and products of starch degradation not absorbed in the small intestine of healthy individuals" (Muir & O'Dea, 1993). Resistant starch can be categorized into (i) physically inaccessible starch found in whole or partly milled grains and seeds (RS1); (ii) ungelatinized resistant granules having a crystallinity of type B, RS granules (RS2); (iii) retrograded starch found in food products undergoing repeated treatment with moist heat (RS3); (iv) chemically modified fragments formed through cross-linking by chemical reagents (RS4) (Sajilata et al., 2006) and newly discovered amylose-lipid V-type complex (RS5) (Cui & Oates, 1999; Hasjim et al., 2010). RS is a non-caloric ingredient. RS directly passes undigested into the colon, where natural microflora can ferment it to short-chain fatty acids such as butyric acids (Baghurst, Baghurst, & Record, 1996), which has been shown to induce growth inhibition and terminal differentiation in several human colon cancer cell lines (Augeron & Laboisse, 1984; Whitehead, Young, & Bhathal, 1986). RS also shows a variety of other physiological effects in the human body similar to that of dietary fiber, such as risk reduction of coronary heart disease and diabetes (Champ, Martin, Noah, & Gratas, 1999; Ranhotra, Gelroth, & Glaser, 1996). Additionally, RS does not absorb significant amount of water and can be selected as a fiber source for formulation in low-moisture bakery products (Jyothsna & Hymavathi, 2017).

Numerous direct and indirect techniques have been studied for RS quantification reaching the large intestine. Direct methods as highlighted by Åkerberg, Liljeberg, Granfeldt, Drews, and Björck (1998) mainly comprise the analysis of starch in ileal effluents (Englyst & Cummings, 1985; Jenkins et al., 1987; Langkilde & Andersson, 1994; Muir & O'Dea, 1993; Steinhart, Jenkins, Mitchell, Cuff, & Prokipchuk, 1992), ileal intubation experiments (Noah et al., 1996; Stephen, Haddad, & Phillips, 1983), balance tests in rats having repressed hind-gut microflora (Granfeldt, Drews, & Björck, 1993) and ileal excreta analysis in colectomized rats (Marlett & Longacre, 1996). Indirect methods, as mentioned by Åkerberg et al. (1998), comprise the analysis of fermentation activity markers such as breath H_2 blood acetate in the colon (Cummings & Englyst, 1991; Muir et al., 1995).

Forty nine sorghum genotypes were analyzed for their RS content and the impact of wet and dry heat treatment on both, grain and flour of two sorghum genotypes was investigated by de Carvalho Teixeira et al. (2016). Quantification of the RS content was carried out using the RS assay kit of Megazyme International, Ireland in accordance with the methods of AACC (2001) and AOAC (2000). The mean RS values were clustered in six groups ranging from 0.31 ± 0.33 g/100 g to 65.66 ± 5.46 g/100 g sorghum flour on dry basis. Dry heat was found to cause negligible losses in the RS content retaining up to 97.19 ± 1.92 percent, while wet heat could retain a maximum 6.98 ± 0.43 percent of the RS.

Sorghum grain was soaked for different periods at 37 °C, followed by autoclaving at 120 or 130 °C, and then freeze-dried by Niba and Hoffman (2003). RS content, which was analyzed by the method prescribed by McCleary, McNally, and Rossiter (2002), involving the use of the RS assay kit by Megazyme International, Ireland, was found to increase from 6.5 g/100 g to 10.4 g/100 g after soaking. On autoclaving at 120 °C, reduction in RS content was found to be more than at 130 °C.

Various modification methods were tried on isolated white sorghum starch to enhance RS formation such as heat-moisture treatment, annealing, pullulanase debranching enzyme hydrolysis, and dual autoclaving-cooling cycles. RS content measured by RS assay kit of Megazyme International Ireland of RS ingredients was found to be 1.3–fold more than native starch. The heat-moisture treatment and annealing possibly induced changes within the crystalline and amorphous regions of starch. Pullulanase action can form several linear short-chain molecules having low molecular weight adding to the content of retrograded starch. Autoclaving-cooling cycles lead to preliminary gelatinization of starch on heating, which leaches the amylose chains out of the granules, and while cooling the amylose molecules recrystallize forming a densely packed structure showing greater tolerance to enzyme hydrolysis (Giuberti, Marti, Gallo, Grassi, & Spigno, 2019).

Tabat sorghum cultivar was naturally fermented at 37 °C for 36 h and its impact on RS content was investigated. *In vitro* resistant starch (RS) was estimated using the method of M Champ et al. (1999). During the fermentation of tabat sorghum, 77 percent of the RS was eliminated after 28 h of fermentation. After 28 h of fermentation, the increase in RS detected could be attributed to the changes in the chemical nature of the starch by the effect of high concentration of lactic acid produced affecting the enzyme action (Elkhalifa, Schiffler, & Bernhard, 2004). In a study by Shaikh, Ali, Mustafa, and Hasnain (2019), the RS content of lactic acid and citric acid along with acid-heat moisture treated starches ranged from 77.9 to 90 percent, which was significantly greater than that of the native starches.

Pasta formulated with either red sorghum flour or white sorghum flour, each substituting durum wheat semolina at 20 percent, 30 percent, and 40 percent level was developed and matched with 100 percent DWS control pasta. The incorporation of both red and white flours enhanced the RS content as analyzed by the RS assay kit of Megazyme International, Ireland. Cooking of the pasta however did not change the resistant starch content of any of the pasta formulations (Khan, Yousif, Johnson, & Gamlath, 2013).

Bran extracts of sorghum varieties, namely, black, tannin, and black with tanninn were added to porridges made of normal corn starch, enzyme resistant high amylose corn starch and ground whole sorghum flours to explore the effect on RS content determined using the method described by Goñi et al. (1996). The addition of extracts (except white bran) was found to increase the RS value of high amylose corn starch porridges. The maximum increase in RS of porridges was recorded with extract of high tannin bran, then, extracts of black with tannin bran and black sorghums (Lemlioglu-Austin, Turner, McDonough, & Rooney, 2012).

Few studies have been carried out to investigate the physiological benefits imparted by resistant starch of sorghum. To study the effects of sorghum-resistant starch (SRS) consumption on blood lipid, body weight, and intestinal microbiome in the colon of 60 high-fat diet-induced (HFD), overweight and obese rats were selected. SRS was found to help the body in reducing the risk of obesity through mechanisms involving the leptin and adiponcetin's synthesis and secretion as well as enhancement in Bifidobacterium and Lactobacillus population and reduction in the population of Enterobacteriaceae in the intestine (Shen, Zhang, Dong, Ren, & Chen, 2015). Further studies are required to substantiate these effects on humans. Recently, a human study was carried out to investigate the impact of consuming sorghum muffins on plasma glucose and insulin levels. Results showed that RDS, SDS and RS content measured by the method of H.N. Englyst et al. (1985) were significantly higher in sorghum muffins in comparison to wheat muffins (p<0.05). Plasma glucose incremental area under the curve (iAUC) was reduced by 26 percent and glucose measures at 45 to 120-minute intervals were significantly less for the sorghum muffin. Also, plasma insulin iAUC and insulin measures significantly decreased at intervals of 15–90 minutes (Poquette, Gu, & Lee, 2014).

3.3.1.5 Bioactive peptides

Bioactive peptides are sequences between 3 to 20 amino acid residues that are inert or inactive within the polypeptide sequence of their native protein. They display their biological activities once they are released following hydrolytic processes (Garcia-Vaquero, Mora, & Hayes, 2019). Biaoactive peptide has been studied for its multiple physiological functions, including antioxidative, immunomodulatory, antibacterial, antithrombotic, anti-adipogenic, and antihypertensive activities (de Castro & Sato, 2015). Their bioactivity is mainly determined by their composition and amino acid sequence (Sarmadi & Ismail, 2010; Singh, Vij, & Hati, 2014).

Production of bioactive peptides includes extraction of protein, subsequently peptide release by hydrolysis; purification of peptides; and identification and validation of their bioactivities. Proteins can be extracted using conventional chemical methods using solvents including distilled water, alkaline, acid, salt or alcoholic solution and phase partitioning process (Branland & Bancel, 2007; Harrysson et al., 2018) followed by centrifugation and recovery using techniques such as ultrafiltration, precipitation and/or (Hayes, 2018). Mass spectrophotometry (MS) is the most frequently reported peptide identification technique used. Liquid chromatography (LC) coupled with an electrospray ionization (ESI) source is often used for peptide identification. Distinct specific systems permit the peptide sequencing using tandem MS/MS. LC–ESI–MS/MS can be coupled with different mass analyzers. The most common mass analyzers are quadrupole (Q), time of flight (TOF), ion trap, Q-TOF, TOF-TOF, and Q-ion trap (Cermeño et al., 2020).

Sorghum grain represents approximately 11 percent protein, ranges between 7 and 16 percent (Serna-Saldivar, 1995), which can be classified into albumins, globulins, glutelins, and prolamins named kafirins (KAF), accounting for 70–80 percent of the total protein (Oria, Hamaker, Axtell, & Huang, 2000). Sorghum kafirin has been classified according to structure, solubility and molecular weight α- (Mr 25,000 and 20,000), β- (Mr 20,000, 18,000, and 16,000), γ-kafirins (Mr 28,000) (Shull, Watterson, & Kirleis, 1991) and δ-kafirins (Mr 14,000)(Belton, Delgadillo, Halford, & Shewry, 2006). Kafirins (KAF) extraction has recently increased due to the gluten-free nature and noteworthy scientific studies displaying the health benefits of the bioactive peptides derived from them.

α-Kafirin protein was extracted using *tert*-butanol from defatted sorghum flour which, when hydrolyzed with chymotrypsin at 37° C, yielded a hydrolysate which was further fractionated using Sephadex G-25 column, yielding four fractions with significant *in vitro* angiotensin-converting enzyme (ACE) inhibitory activity. The IC_{50} values of these four fractions varied from 1.3 to 24.3 µg/mL, two of which were found to be competitively inhibiting the enzyme, while the other two were non-competitive inhibitors (Kamath, Niketh, Chandrashekar, & Rajini, 2007).

Agrawal, Joshi, and Gupta (2017) investigated the peptides for their free radical scavenging activity from green tender sorghum (GTS). protein was first extracted using phosphate buffer at pH 6.5 (1:100, g: mL) and was then hydrolyzed with alcalde enzyme at pH 9.5 and 50 °C and further purified with ultrafiltration, gel filtration, and reverse-phase ultra-flow liquid chromatography (RP-UFLC). The sequence of seven peptide fractions purified were identified using matrix-assisted laser desorption/ionization- Time-of-flight-mass spectrometer (MALDI-TOF-MS/MS), which demonstrated antioxidant activity determined using 1,1-diphenyl-2-picrylhydrazyl (DPPH), Fe^{2+} chelating activity, 2,2'-azinobis-(3-ethyl-benzothiazoline-6-sulphonate) (ABTS), and reducing power assay.

KAF was extracted using ethanol or *tert*-butanol and sodium metabisulfite after pretreatment with amyloglucosidase. The sorghum storage proteins (α-KAF and α-KAF precursor) once identified by nano LC-MS/MS. The fragment sequences were analyzed *in silico* using the BIOPEP-UWM database to screen potential bioactive peptides including the *in silico* hydrolysis of these sequences using thermolysin, chymotrypsin, and subtilisin enzymes. The PeptideRanker ranked the predicted sequences of peptides according to their bioactivity. Different biologically active peptide sequences were discovered in the two protein sequences, including ACE-inhibitors, dipeptidyl peptidase IV (DPP-IV), and dipeptidyl peptidase III (DPP-III) inhibitors, antioxidative, hypotensive, HMG-CoA reductase inhibitor, and activating ubiquitin-mediated proteolysis. Additionally, 31 uncharted peptide sequences that could demonstrate biological activity were also identified (Castro-Jácome, Alcántara-Quintana, & Tovar-Pérez, 2020).

Bioactive peptides from the grain of white sorghum were developed and evaluated for their antioxidant, anti-inflammatory, and anti-aging capacity for potential application in the cosmetic industry. The kafirin fractions (α, β, and γ-kafirin) were extracted with ethanol containing sodium metabisulfite from α-amylase liquefied defatted sorghum grain. The kafirin was then hydrolyzed with alcalase from *Bacillus licheniformis* at 50 °C to produce crude hydrolysates which were concentrated through ultrafiltration yielding two peptide extracts. Bioassays implemented on organotypic skin cultures (not exposed and exposed to UVB) using these two extracts (PE-3 and PE-1), showed that the treatments with both could significantly decrease UVB damage through reduction of depletion of superoxide dismutase (SOD) and glutathione peroxidase (GPx) activity, as well as by retaining or enhancing catalase (CAT) activity. These two peptide extracts (PE-3 and PE-1) from white sorghum diminished the levels of pro-inflammatory factors such as interleukin 1-β (1L-β), interferon-γ (IFN-γ), and tumor necrosis factor-α (TNF-α). Also, these peptides repressed collagenase, elastase, and tyrosinase activities (Castro-Jácome et al., 2021).

3.4 Techno-functional properties

Sorghum starch and flour have been widely used as a functional ingredient for bakery products and other processed foods due to its good processing properties (Sitanggang, Budijanto, & Marisa, 2018). Some of the technological properties of sorghum flour have been studied based on its macroscopic properties, such as water holding capacity (WHC) or water absorption capacity (WAC) and swelling power, as well as its pasting properties and rheological behavior (Marchini, Arduini, & Carini, 2021). These techniques give information regarding the interactions of the flour biopolymers with water, which will strongly influence the product final properties. Particularly, the WHC is linked to ability of foods or its components, such as starch, to retain water under certain conditions (Vásquez, Verdú, Islas, Barat, & Grau, 2016).

Singh, Sodhi, & Singh (2012) studied the structure and the functional-technological properties of acetylated sorghum (cultivar M-35) starch from Hyderabad, India. These authors found that native sorghum starch presented an amylose content of 18.7 percent, whereas the acetylated starch exhibited an increased amount of amylose around 19.5 percent. For them, this increased content may be related to an acetyl group interfering with amylose and amylopectin fractions, which could alter iodine absorption during the determination technique. Furthermore, the magnitude of water interaction with starch chains within both the amorphous and crystalline domains is given by the determination of swelling power and solubility of starches. In this context, these authors have found that acetylated sorghum starch presented higher values of 16.9 g g^{-1} of swelling power and 14.7 percent of solubility, while the native starch presented values ranging around 6.2 g g^{-1} and 11.9 percent, respectively. These changes could be related to the increase of hydrophilic substituting and acetyl groups, which were able to retain more water molecules and consequently to form hydrogen bonds in the starch granules (Singh, Sodhi, & Singh, 2012).

In another work, the physicochemical properties of two Indonesian varieties of sorghum starch, Numbu and Genjah, were studied by Sitanggang, Budijanto, & Marisa (2018). These sorghum varieties presented higher amounts of starch, ranging from around 79.5 to 85 percent, when compared to the content of 71.2 percent and 68.4 percent, determined to heterowaxy and waxy sorghum grain, respectively, cultivated by the University of Nebraska Field Laboratory, Ithaca (Sang, Bean, Seib, Pedersen, & Shi, 2008). In the work of Sitanggang, Budijanto, & Marisa (2018), the Numbu variety exhibited around ~22.5 percent of amylose and 77.5 percent amylopectin content, while the Genjah showed 18.6 percent and 81.4 percent, respectively. Those authors have inferred that the amylose content depends on different factors, such as starch sources, climate, and soil cultivation conditions. Thus, as the Numbu and Genjah varieties were cultivated on Yogyakarta which, having temperature between 22 and 28°C, could be one reason for a lower amount of amylose content compared to the sorghum varieties cultivated in Sahara, Algeria, where the annual temperature ranges from 8 to 45°C (Sitanggang, Budijanto, & Marisa, 2018). The amylose content in waxy, heterowaxy, and normal sorghum starches were in the range of 0 to 23.7 percent, as determined by Sang et al. (2008), a result similar to the literature.

Sitanggang, Budijanto, & Marisa (2018) determined values of starch granule size varying from 3.8 to ~39 μm for Numbu and Genjah sorghum varieties, while the waxy, heterowaxy, and normal sorghum starches from University of Nebraska Field Laboratory, Ithaca exhibited a granule size ranging from 5 to 25 μm (Sang et al., 2008). The granule size of native white sorghum starch ranged from 7–22 μm as reported by Ali & Hasnain (2014). These authors also observed the presence of pores on the surface of the native white sorghum starch granules ranging between 178–289 nm. When acid-thinned and oxidized, this native white sorghum starch exhibited pores with higher sizes, ranging between 200–245 nm and 225–275 nm, respectively (Ali & Hasnain, 2014). Moreover, in this study Ali & Hasnain (2014) observed that the dual modification of oxidized starch via acetylation increased the pore sizes up to 438 nm.

When studying the physical–chemical properties of different flours to partially wheat flour (WF) substitution, Vásquez et al., (2016) found that the WHC increased from 70.6 percent to 71.5 percent when the WF was substituted by sorghum flour from 2.5 to 5 percent. To these authors, this result was strongly affected by the flour refinement and its particle size. For them, a reduced particle size could substantially affect the dough's functional properties, because an increased particles contact surface occurs. In this context, they have determined that the WF substituted with 10 percent of sorghum flour displayed the smallest particle size, around 106.4 μm, while the WF substituted with 10 percent of oat flour presented a value of 114.2 μm. Acid-thinned and dual modified acetylated oxidized sorghum starches showed significantly higher values of WHC at 60° C, 1.13 and 1.32 g g^{-1}, when compared to native sorghum starch, 1.06 g g^{-1} (Ali & Hasnain, 2014). While this behavior was not persistent when the temperature was at 90° C, the native sorghum starch exhibited a WHC of 7.86 g g^{-1}, a much higher value than dual modified acetylated oxidized sorghum starches, 5.99 g g^{-1}.

Proteins and lipids are important nutrient components in flours, such as those obtained from sorghum, because they can interact with the starch affecting its rheological properties, that is, viscosity. Moreover, pasting properties are influenced by the amylose content, and swelling property, microstructure, and fragmentation of starch granules (Singh, Sodhi, & Singh, 2012). The peak viscosity of wheat flour substituted with 10 percent of sorghum flour increased from 2250 to 2300 cP, as reported by Vásquez et al. (2016). In this work, these authors observed the same behavior when considering the pasting temperature, which increased from around 70 °C to almost 90 °C with the 10 percent substitution using sorghum flour. The setback viscosity is correlated to the tendency of starch retrogradation after gelatinization and cooling at 50 °C, due to the recrystallization of amylose chains resulting on the formation of a gel structure (Balet, Guelpa, Fox, & Manley, 2019). Moreover, the setback viscosity can be determined by the differences between the final viscosity and the trough viscosity (Shafie, Cheng, Lee, & Yiu, 2016; Vásquez et al., 2016). Thus, higher values of setback viscosities are an indication of high rates of starch retrogradation (Balet, Guelpa, Fox, & Manley, 2019). The wheat flour substituted with sorghum flour, in the work of Vásquez et al. (2016), displayed an increased setback viscosity, which was higher the increase of substitution percentage.

Starch retrogradation can also be measured by determining syneresis (%) during a storage time. This is an important starch property and is related to the ordered crystalline structure of gelatinized starch, which occurs due to re-aligning and re-association of the amylose and amylopectin chains (Balet, Guelpa, Fox, & Manley, 2019). Native and acetylated sorghum starches, studied by Singh, Sodhi, & Singh (2012), exhibited a significant rise in syneresis with the increase of storage time, from 24 to 120 h. These authors also noted that acetylation considerably reduced the syneresis from 26.2 percent and 37.9 percent in native to 0.95 percent and 1.03 percent in acetylated sorghum starch (6.25%) after 24 h and 120 h, respectively. The syneresis reduction in acetylated starch can be due to the incorporation of acetyl groups, which improve the water retention capacity of starch due to the inhibition of interchain interaction (Singh, Sodhi, & Singh, 2012).

When studying the effect of acetylation degree in sorghum starch, Singh, Sodhi, & Singh (2012) observed that the pasting viscosity, the hot paste viscosity, breakdown viscosity and the cool paste viscosity significantly decreased with higher percentages of acetylation (acetic anhydride). The addition of acetyl groups from acetylation process by the starch chains caused a disruption of inter and intra molecular hydrogen bonds, leading to changes in hydrophilicity and fragility of it, consequently decreasing the pasting and the breakdown viscosities, respectively. The pasting temperature also significantly decreased with the increase of acetylation, from 77.4 °C in native to 75.9 °C in 6.25 percent acetylated sorghum starches, which is useful information related to the reduced need of energy during food processing (Singh, Sodhi, & Singh, 2012).

In another work, Ali & Hasnain (2014) studied the pasting properties of native, oxidized, acetylated and dual oxide-acetylated white sorghum starch. They observed that the pasting temperature significantly reduced in treated white sorghum starch, from 73.15 to 69.2 °C, which can be associated by the replacement of bulky functional groups within the starch chains that influences the inter-molecular binding forces in it. The peak viscosity of these sorghum starches was highly influenced by the oxide- and acetylation treatment, 245.5 BU in native to 54.5 BU in oxidized sorghum starch. This decreasing behavior was also noted in other results, such as hot paste viscosity, cold paste viscosity, breakdown viscosity and setback viscosity. These authors suggested that the oxide-acetylation treatments have potential application in sorghum starch, especially to the development of food products that require low thickening temperature.

Gelatinization is a commonly used functional property of starches that causes variations in the swelling, birefringence, and viscosity of granules. Moreover, it promotes irreversible changes in functional properties of the starch molecules because of the cessation of its intermolecular structure (Balet, Guelpa, Fox, & Manley, 2019). It is reported in the literature that the gelatinization temperature of sorghum starch ranged from around 67 to 81 °C, a much higher value than wheat, which ranged from 58 to 64°C (Kulamarva, Sosle, & Raghavan, 2009; Sitanggang, Budijanto, & Marisa, 2018). A higher gelatinization temperature will influence the energy and time needed during cooking of sorghum-based food products. The gelatinization peak temperature of Numbu and Genjah varieties of sorghum starch were around 75.9 and 77.3 °C, respectively, as reported by Sitanggang, Budijanto, & Marisa (2018). For them, this result is due to a higher proportion of amylopectin long chains in the Genjah starch, which requires a superior temperature to breakdown its granules and indicates a greater stability of crystal lattices in it.

Marchini, Arduini, & Carini (2021) studied by ¹H NMR (low-resolution proton nuclear magnetic resonance) and DMA (dynamic mechanical analysis), how sprouting and drying post-sprouting affected the technological functionalities of sorghum flour. These authors found that the sprouting affected the molecular properties of sorghum flour–water systems. The sprouting is a technique that can prolong amylases and proteases activation to amylose and proteins hydrolysis, which can increase solubility, WHC and stability of flours, such as sorghum. Moreover, the treatment of sorghum with drying after sprouting have also shown a potential technique to modulate the starch and protein functionalities of its flour (Marchini et al., 2021). In this work, Marchini et al. (2021) have showed that the sprouting technique significantly increases the amylose content of sorghum treated for 12 h at 40°C, consequently, altered the amylose/amylopectin ratio and increased the WHC of it. These changes caused by sprouting can be linked to the higher hydration properties of an elevated content of polar groups in proteins and polysaccharides. Likewise, in this work, the pasting properties of the sprouted sorghum were affected by the modification of starch granule structure. The authors reported that the sprouted sorghum treated for 12 h at 40°C have displayed a decreased peak viscosity, while the samples treated for 6 h at 50 °C shown a significantly higher value of peak viscosity. Finally, the sprouting at 40°C and 12 h have favored the enzymatic activity, which increased the exposure of starch to them, resulting a higher hydrolysis level and lowering its functionality.

3.5 Potential utilization in value added food products

Sources of macro, micronutrients and bioactive compounds, sorghum grains and sorghum flour are promising foods, helping to reduce diseases when induced in a diet. Studies show that its proteins do not cause autoimmune allergic reactions and may be recommended for people with celiac disease (Ofosu et al., 2020; Park and Ren, 2012; Rashwan, Yones, Karim, Taha, & Chen, 2021). However, there are several challenges to be faced in the consumption of sorghum grains, due to the presence of multiple antinutritional components, which have the ability to reduce absorption nutrient efficiency in animals and humans, such as tannins, phytic acid, trypsin inhibitor, among others (Afify, El-Beltagi, El-Salam, & Omran, 2012; Mohapatra, Patel, Kar, Deshpande, & Tripathi, 2019; Rashwan et al., 2021; Wu, Ashton, Simic, Fang, & Johnson, 2018). These factors can be reduced when applying technological approaches of processing, such as fermentation, cooking, soaking, steaming, among others, thus improving the quality of sorghum and producing foods with high nutritional value. Among the technological approaches, the best treatment is found in the fermentation process, followed by combinations of other treatments. Therefore, the use of sorghum for many types of food becomes viable, such as in bakery products, beverages, porridge, among others. (Jafari, Koocheki, & Milani, 2017; Mohapatra et al. 2019; Palavecino, Ribotta, León, & Bustos, 2019; Xiong, Zhang, Luo, Johnson, & Fang, 2019).

3.5.1 Sorghum flour

The application of flour in food production depends on its functional properties, and these factors must be considered, first so that wheat flour can be replaced by different percentages of other types of flour, such as sorghum (Vásquez, et al., 2016). The variation in the functional properties of flours will depend on factors such as compositional structure, molecular conformation, protein quality and interaction with other food components. Therefore, these functional properties play a crucial role in choosing the types of products that can be produced using the designated flour (Ojha et al. 2018).

Sorghum flour is a gluten-free and safe food for individuals who have celiac disease or who need high nutrition, and it can be consumed by people who cannot consume wheat or other types of grains (Rashwan et al., 2021). The use of sorghum flour in food production occurs normally in traditional food products, such as porridge and bread, in addition to being used for the production of cookies and snacks (Benhur et al., 2015; Liu et al., 2012; Palavecino et al. 2019; Sun, Han, Wang, & Xiong, 2014).

Despite the great commercial trend, due to health-related benefits, whole sorghum flour-based products still have some challenges to be faced within the industry, as short shelf-life and unpleasant sensory attributes, which can limit this application (Adebowale, Taylor, & de Kock, 2020). The properties of sorghum flour can be improved and/or modified using different technological treatments, depending on its purpose and formulation, such as thermal moisture treatment, extrusion cooking and microwave treatment (Jafari, Koocheki, & Milani, 2017; Marston, Khouryieh, & Aramouni, 2016; Sun et al. 2014).

3.5.2 Sorghum bread

Foods such as bread have consistently increasing consumption around the world, especially in developing countries, due to its nutrition properties, changes in eating habits and the population growth. There is a need other types of cereals, including sorghum, for the development of novel bread products, especially to meet the consumers' needs (Rashwan et al., 2021). There are many types of bread that can be made from sorghum, such as *kissra* bread, sourdough bread, flat bread, *khamir* bread, and fried bread (Abdualrahman et al., 2019; Marston, Khouryieh, & Aramouni, 2016; Osman, 2004; Rose, Williams, Mkandawire, Weller, & Jackson, 2014. However, due to the absence of gluten protein, the dough can be weak in viscoelastic and textural properties, resulting in a bread with inferior texture, color and other post-baking quality defects (Jafari, Koocheki, & Milani, 2017; Olojede, Sanni, & Banwo, 2020).

Due to these factors, many researchers have investigated possible techniques to improve the quality of sorghum bread. Marston, Khouryieh, & Aramouni (2016) studied the effect of heat treatment on bread quality. These authors reported that the ideal temperature for improving sorghum flour was 125 °C for 30 min, which showed that this treatment could produce breads with better volume and texture and with greater consumer acceptability.

In other studies, techniques such as using functional lactic acid bacteria and yeasts (*Pediococcus pentosaceus* SA8, *Weissella confusa* SD8, *P. pentosaceus* LD7 and *Saccharomyces cerevisiae* YC1) as starter cultures, as reported by Olojede, Sanni, & Banwo (2020). This study showed that the use of starter cultures used in dough fermentation had good acceptability and can improve the structure and nutrients of gluten-free sorghum-based bread (Olojede, Sanni, & Banwo, 2020). The quality of sorghum bread can also be improved by adding other types of nutritious ingredients, such as legumes, potatoes, cassava, and so forth, and/or stabilizers, as xanthan gum, dextran, and so forth, as cited by Jafari, Koocheki, & Milani (2017); Olojede, Sanni, & Banwo (2020). Studies and analyses are always necessary to find the best suitable process for producing sorghum bread with the highest quality. Sorghum bread is a well-accepted product for many consumers because it is a gluten-free product, which makes it a great choice factor, and has a great potential application in the development of novel food products (Rashwan et al., 2021).

3.5.3 Extruded sorghum-based products

Extruded products based on sorghum flour, such as pasta, noodles and French fries have been highlighted as promising products, due to the great need in the market for novel ingredients to develop new food products (Benhur et al., 2015; Liu et al., 2012; Palavecino et al. 2019). Among some studies, the development of Chinese egg noodles by Liu et al. (2012) can be cited. These authors have used sorghum flour to produce a noodle with good physical attributes, although they have noted that the functional properties of the used sorghum flour depend on the controlled quality of its grain.

Another work, Benhur et al. (2015), analyzed pasta produced from different proportions of sorghum and wheat semolina. To them, among the mixtures studied, the combinations of 50:50, 60:40 and 70:30 of sorghum:wheat ratios had the highest acceptability. Moreover, this study shown the viability and the improved the nutritional composition of the pasta produced based on a mixture of sorghum and wheat. Palavecino et al. (2019) carried out a study with pasta using only white and brown sorghum flour. Their results showed that this pasta had high protein ($\approx$ 170 g kg^{-1}), dietary fiber ($\approx$80 g kg^{-1}), polyphenols (2, 6 g GA kg^{-1} mass), and antioxidant activity. Furthermore, the sorghum pasta in their study proved to be slower in the *in vitro* digestion than the other gluten-free pasta, with a high level of protein hydrolysis (76%).

3.5.4 Sorghum-based fermented beverages

Widely consumed in sub-Saharan Africa and other (subtropical) regions, traditional cereal-based fermented beverages, especially derived from corn and sorghum grains, are produced through a series of processes. These fermented beverages can be classified as alcoholic beverages, which are mostly sour, contain no hops and are consumed without refinement, while non-alcoholic beverages are sweeter and lighter (Ezekiel et al., 2015; Rashwan et al., 2021). For example, alcoholic beverages produced from sorghum grains are mainly 'beers', commonly called *sorghum beer* or *opaque beers*. According to locality these popular beverages can be called *pito* in Nigeria, Ghana, and Togo, *chibuki* in Zimbabwe and *tchapalo* in Côte d'Ivoire, among others (Coulibaly et al., 2020; Djameh et al., 2015; Rashwan et al., 2021). According to Djameh et al. (2015) sorghum beer can have different colors, ranging from light yellow to dark brown, and a different flavor that can vary from slightly sweet to very sour, which may vary from the region in which it is produced. To these authors, it is also a good source of vitamin B complex, including thiamine, folic acid, nicotinic acid and riboflavin.

Unfermented beverages, such as *Kunun-zaki*, a refreshing drink consumed by both adults and children, mainly in northern Nigeria, can be produced from corn grains or millet and sorghum. Due to its cheap preparation, it is an accessible drink in local markets and stores (Ezekiel et al., 2015; Ndulaka, Obasi, & Omeire, 2014). Fermented beverages based on sorghum grains have a significant position on the beverage table when it comes to many countries, this is due to their nutritional value, having a series of active compounds such as phenolic compounds (anthocyanins, phenol, flavonoids, tannins) and dietary fiber (Coulibaly et al., 2020).

3.5.5 Sorghum bean tea

Due to its antioxidant potential, cholesterol-lowering ability, anti-inflammatory and anticancer properties, there has been a large increase in interest in using sorghum grains for functional food development (Morais Cardoso, Pinheiro, Martino, & Pinheiro-Sant'Ana, 2017; Stefoska-Needham, Beck, Johnson, & Tapsell, 2015). A new trend in the market is the application of cereal grains for the production of tea beverages due to their unique flavor and health benefits – such as brown rice tea, buckwheat, sorghum, among others (Guo et al., 2017). Sorghum grain tea is considered an important health drink and is mainly available in China, where it is produced using whole grains (Wu, Huang, Qin, & Ren, 2013; Xiong et al., 2019).

Quality indicators such as aroma and flavor can affect consumer satisfaction, where they are mainly affected by the composition of volatile substances. The requirement for grain-based food production is processing, the different methods can significantly affect both the physical characteristics and chemical compositions of grains (Sun, Wang, Zhang, Ajlouni, & Fang, 2020). According to Xiong, Zhang, Johnson, Luo, & Fang (2020), a total of 63 aromatic compounds, including alcohols, aldehydes and esters in processed sorghum tea were identified, which were directly influenced according to the processing method used. However, information is still limited, and further studies are needed for the use of sorghum tea as a functional beverage.

3.5.6 Sorghum ingredients as a food additive

Sorghum ingredients can be used as an additive to aid in the development and/or improvement of functional foods or beverages; wheat grain and its derivatives have a great functional potential (Adebowale, Taylor, & de Kock, 2020). Anunciação et al. (2017), used sorghum to produce a breakfast cereal. In their sensory analysis, the sorghum-based cereal had better sensory acceptance (70.6%) than the cereal based on wheat (41.18%). Additionally, this sorghum-based cereal presented a relevant content of 3-deoxyanthocyanidin (100% higher), total phenolic compounds (98.2% higher) and antioxidant activity (87.9% higher) than the wheat breakfast cereal, showing that the use of sorghum for the production of breakfast cereals should be encouraged to promote benefits to human health. Makawi, Mustafa, Adiamo, & Mohamed Ahmed (2019), have shown that the preparation of *Kisra* bread, a fermented sorghum bread, containing high amounts of sorghum flour, improved the protein, fiber, fat, ash and mineral content and its stability. These results show that the use of sorghum flour as an additive, together with other ingredients, can significantly improve food functionality, especially in gluten-free foods.

3.6 Conclusion

Sorghum represents the significant staple food of millions of people in in the semi-arid tropics of Asia and Africa, including millets. Sorghum grows well in harsh environmental conditions, where other crops fail to grow properly, and sorghum has sustained the lives of poor rural people: therefore, it is often regarded as the poor people's crop. Sorghum is cultivated for grain, fodder, brooms, sugar, biofuel, and is a rich source of bioactive compounds with potential antioxidant, antibacterial, anticancer, and so forth activities. The genus *Sorghum* is a rich reservoir of untapped genetic resources, which could be utilized for crop improvement. Therefore, scientific advances in crop improvement, storage, and consumption of sorghum could mitigate the food related issues of low-income countries by ensuring household food security and nutrition of the people.

References

Abdualrahman, M.A.Y., Ma, H., Yagoub, A.E.A., Zhou, C., Ali, A.O. and Yang, W. 2019. Nutritional value, protein quality and antioxidant activity of Sudanese sorghum-based kissra bread fortified with bambara groundnut (*Voandzeia subterranea*) seed flour. *Journal of the Saudi Society of Agricultural Sciences*, 18(1):32–40.

Abugri, D.A., Akudago, J.A., Pritchett, G., Russell, A.E. and McElhenney, W.H., 2015. Comparison of phytochemical compositions of *Sorghum bicolor* (L) Moench red flour and pale brown leaves. *Journal of Food Science and Nutrition*, 1(003).

Adebowale, O.J., Taylor, J.R. and de Kock, H.L. 2020. Stabilization of wholegrain sorghum flour and consequent potential improvement of food product sensory quality by microwave treatment of the kernels. *LWT–Food and Science Technology*, 132:109827.

Adetuyi, A.O., Akpambang, V.O., Oyetayo, V.O. and Adetuyi, F.O. 2007. The nutritive value and antimicrobial property of *Sorghum bicolor* L. stem (Poporo) flour used as food colour additive and its infusion drink. *American Journal of Food Technology*, 2:19–86.

Afify, A.E.M.M., El-Beltagi, H.S., Abd El-Salam, S.M. and Omran, A.A. 2012. Protein solubility, digestibility and fractionation after germination of sorghum varieties. *Plos one*, 7(2):31154.

Agrawal, H., Joshi, R. and Gupta, M. 2017. Isolation and characterisation of enzymatic hydrolysed peptides with antioxidant activities from green tender sorghum. *LWT–Food Science and Technology*, 84:608–616.

Aka, S., Konan, G., Fokou, G., Dje, K.M. and Bonfoh, B. 2014. Review on African traditional cereal beverages. *American Journal of Research Communication*, 2(5):103–153.

Åkerberg, A.K., Liljeberg, H.G., Granfeldt, Y.E., Drews, A.W. and Björck, I.M. 1998. An in vitro method, based on chewing, to predict resistant starch content in foods allows parallel determination of potentially available starch and dietary fiber. *The Journal of Nutrition*, 128(3):651–660.

Alfadda, A.A. and Sallam, R.M. 2012. Reactive oxygen species in health and disease. *Journal of Biomedicine and Biotechnology*, 2012.

Ali, T.M. and Hasnain, A. 2014. Morphological, physicochemical, and pasting properties of modified white sorghum (*Sorghum bicolor*) starch. *International Journal of Food Properties*, 17(3):523–535.

Ananda, G.K., Myrans, H., Norton, S.L., Gleadow, R., Furtado, A. and Henry, R.J. 2020. Wild sorghum as a promising resource for crop improvement. *Frontiers in Plant Science*, 11:1108.

Anunciação, P.C., de Morais Cardoso, L., Gomes, J.V.P., Della Lucia, C.M., Carvalho, C.W.P., Galdeano, M.C., Queiroz, V.A.V., Alfenas, R.D.C.G., Martino, H.S.D. and Pinheiro-Sant'Ana, H.M. 2017. Comparing sorghum and wheat whole grain breakfast cereals: Sensorial acceptance and bioactive compound content. *Food Chemistry*, 221:984–989.

Arendt, E.K. and Zannini, E. 2013. Sorghum. *Cereal Grains for the Food and Beverage Industries*, 1:283–305.

Asp, N.G., Johansson, C.G., Hallmer, H. and Siljestroem, M. 1983. Rapid enzymic assay of insoluble and soluble dietary fiber. *Journal of Agricultural and Food Chemistry*, 31(3): 476–482.

Augeron, C. and Laboisse, C.L. 1984. Emergence of permanently differentiated cell clones in a human colonic cancer cell line in culture after treatment with sodium butyrate. *Cancer Research*, 44(9):3961–3969.

Awika, J.M., McDonough, C.M. and Rooney, L.W. 2005. Decorticating sorghum to concentrate healthy phytochemicals. *Journal of Agricultural and Food Chemistry*, 53(16):6230–6234.

Awika, J.M., Yang, L., Browning, J.D. and Faraj, A. 2009. Comparative antioxidant, antiproliferative and phase II enzyme inducing potential of sorghum (*Sorghum bicolor*) varieties. *LWT–Food Science and Technology*, 42(6):1041–1046.

Bader Ul Ain, H., Saeed, F., Khan, M.A., Niaz, B., Khan, S.G., Anjum, F.M., Tufail, T. and Hussain, S. 2019. Comparative study of chemical treatments in combination with extrusion for the partial conversion of wheat and sorghum insoluble fiber into soluble. *Food Science & Nutrition*, 7(6):2059–2067.

Badi, S., Pedersen, B., Monowar, L. and Eggum, B.O. 1990. The nutritive value of new and traditional sorghum and millet foods from Sudan. *Plant Foods for Human Nutrition*, 40(1):5–19.

Baghurst, P. A., Baghurst, K. I., & Record, S. J. (1996). Dietary fibre, non-starch polysaccharides and resistant starch: a review. *Food Australia: official journal of CAFTA and AIFST*.

Balet, S., Guelpa, A., Fox, G. and Manley, M. 2019. Rapid Visco Analyser (RVA) as a tool for measuring starch-related physiochemical properties in cereals: A review. *Food Analytical Methods*, 12(10):2344–2360.

Barros, F., Dykes, L., Awika, J.M. and Rooney, L.W. 2013. Accelerated solvent extraction of phenolic compounds from sorghum brans. *Journal of Cereal Science*, 58(2):305–312.

Beldados, A., D'Andrea, A.C. and Manzo, A. 2015. Filling the gap: New archaeobotanical evidence for 3rd–1st millennium BC agricultural economy in Sudan and Ethiopia. In *8th International Workshop for African Archaeobotany*, 151–154.

Belton, P.S., Delgadillo, I., Halford, N.G. and Shewry, P.R. 2006. Kafirin structure and functionality. *Journal of Cereal Science*, 44(3):272–286.

Benhur, D.R., Bhargavi, G., Kalpana, K., Vishala, A.D., Ganapathy, K.N. and Patil, J.V. 2015. Development and standardization of sorghum pasta using extrusion technology. *Journal of Food Science and Technology*, 52(10):6828–6833.

Borokini, F.B. 2017. Identification and Quantification of Polyphenols in *Sorghum bicolor* (L) Moench leaves Extract Using Reverse-Phase HPLCDAD. In *5th International Conference and Exhibition on Pharmacology and Ethnopharmacology. Clinical and Experimental Pharmacology and Physiology*, 7:2.

Branlard, G. and Bancel, E. 2007. Protein extraction from cereal seeds. In *Plant Proteomics*, 15–25. Humana Press.

Burdette, A. L. 2007. *Nutraceutical uses of sorghum bran (Sorghum bicolor)*. UGA.

Burdette, A., Garner, P.L., Mayer, E.P., Hargrove, J.L., Hartle, D.K. and Greenspan, P. 2010. Anti-inflammatory activity of select sorghum (*Sorghum bicolor*) brans. *Journal of Medicinal Food*, 13(4): 879–887.

Camargo Filho, I., Cortez, D.A.G., Ueda-Nakamura, T., Nakamura, C.V. and Dias Filho, B.P. 2008. Antiviral activity and mode of action of a peptide isolated from *Sorghum bicolor*. *Phytomedicine*, 15(3):202–208.

Carr, T.P., Weller, C.L., Schlegel, V.L., Cuppett, S.L., Guderian Jr, D.M. and Johnson, K.R. 2005. Grain sorghum lipid extract reduces cholesterol absorption and plasma non-HDL cholesterol concentration in hamsters. *The Journal of Nutrition*, 135(9):2236–2240.

Castro-Jácome, T.P., Alcántara-Quintana, L.E. and Tovar-Pérez, E.G. 2020. Optimization of sorghum kafirin extraction conditions and identification of potential bioactive peptides. *BioResearch Open Access*, 9(1):198–208.

Castro-Jácome, T.P., Alcántara-Quintana, L.E., Montalvo-González, E., Chacón-López, A., Kalixto-Sánchez, M.A., del Pilar Rivera, M., López-García, U.M. and Tovar-Pérez, E.G. 2021. Skin-protective properties of peptide extracts produced from white sorghum grain kafirins. *Industrial Crops and Products*, 167:113551.

Cermeño, M., Kleekayai, T., Amigo-Benavent, M., Harnedy-Rothwell, P. and FitzGerald, R.J. 2020. Current knowledge on the extraction, purification, identification, and validation of bioactive peptides from seaweed. *Electrophoresis*, 41(20):1694–1717.

Champ, M., Martin, L., Noah, L. and Gratas, M. 1999. Analytical methods for resistant starch. In *Complex Carbohydrates in Foods*, 173–193. CRC Press.

Champ, M., Martin, L., Noah, L., Gratas, M., Cho, S.S., Prosky, L. and Dreher, M. 1999. "Complex carbohydrates in foods." Cho, S.S., Prosky, L. (Eds).

Chen, T., Kim, C.Y., Kaur, A., Lamothe, L., Shaikh, M., Keshavarzian, A. and Hamaker, B.R. 2017. Dietary fibre-based SCFA mixtures promote both protection and repair of intestinal epithelial barrier function in a Caco-2 cell model. *Food & Function*, 8(3):1166–1173.

Chung, I. M., Kim, E.H., Yeo, M.A., Kim, S.J., Seo, M.C., & Moon, H.I. 2011. Antidiabetic effects of three Korean sorghum phenolic extracts in normal and streptozotocin-induced diabetic rats. *Food Research International*, 44(1): 127–132.

Coulibaly, W.H., Bouatenin, K.M.J.P., Boli, Z.B.I.A., Alfred, K.K., Bi, Y.C.T., N'sa, K.M.C., Cot, M., Djameh, C. and Djè, K.M. 2020. Influence of yeasts on bioactive compounds content of traditional sorghum beer (tchapalo) produced in Côte d'Ivoire. *Current Research in Food Science*, 3:195–200.

Cui, R. and Oates, C.G. 1999. The effect of amylose-lipid complex formation on enzyme susceptibility of sago starch. *Food Chemistry*, 65(4):417–425.

Cummings, J.H. 1973. Dietary fibre. *Gut*, 14(1):69.

Cummings, J.H. and Englyst, H.N. 1991. Measurement of starch fermentation in the human large intestine. *Canadian Journal of Physiology and Pharmacology*, 69(1):121–129.

Das, T., Thakur, R., Dhua, S., Teixeira-Costa, B.E., Rodrigues, M.B., Pereira, M.M., Mishra, P. and Gupta, A.K. 2023. Processing of cereals. *Cereal Grains: Composition, Nutritional Attributes, and Potential Applications*: G.A. Nayik, T. Tufail, F.M. Anjum, M.J. Ansari, (Eds.), (1st ed., pp. 195—223) CRC Press.

de Carvalho Teixeira, N., Queiroz, V.A.V., Rocha, M.C., Amorim, A.C.P., Soares, T.O., Monteiro, M.A.M., de Menezes, C.B., Schaffert, R.E., Garcia, M.A.V.T. and Junqueira, R.G. 2016. Resistant starch content among several sorghum (*Sorghum bicolor*) genotypes and the effect of heat treatment on resistant starch retention in two genotypes. *Food Chemistry*, 197:291–296.

de Castro, R.J.S. and Sato, H.H. 2015. Biologically active peptides: Processes for their generation, purification and identification and applications as natural additives in the food and pharmaceutical industries. *Food Research International*, 74:185–198.

de Morais Cardoso, L., Pinheiro, S.S., Martino, H.S.D. and Pinheiro-Sant'Ana, H.M. 2017. Sorghum (*Sorghum bicolor* L.): Nutrients, bioactive compounds, and potential impact on human health. *Critical Reviews in Food Science and Nutrition*, 57(2):372–390.

de Oliveira, K.G., Queiroz, V.A.V., de Almeida Carlos, L., de Morais Cardoso, L., Pinheiro-Sant'Ana, H.M., Anunciação, P.C., de Menezes, C.B., da Silva, E.C. and Barros, F. 2017. Effect of the storage time and temperature on phenolic compounds of sorghum grain and flour. *Food Chemistry*, 216:390–398.

Di Bernardini, R., Harnedy, P., Bolton, D., Kerry, J., O'Neill, E., Mullen, A.M. and Hayes, M. 2011. Antioxidant and antimicrobial peptidic hydrolysates from muscle protein sources and by-products. *Food Chemistry*, 124(4):1296–1307.

Djameh, C., Saalia, F.K., Sinayobye, E., Budu, A., Essilfie, G., Mensah-Brown, H. and Sefa-Dedeh, S. 2015. Optimization of the sorghum malting process for pito production in Ghana. *Journal of the Institute of Brewing*, 121(1):106–112.

Doggett, H. 1988. *Sorghum*. Longman Scientific and Technical, Essex, UK, 512.

Dolara, P., Luceri, C., De Filippo, C., Femia, A. P., Giovannelli, L., Caderni, G., ... & Cresci, A. (2005). Red wine polyphenols influence carcinogenesis, intestinal microflora, oxidative damage and gene expression profiles of colonic mucosa in F344 rats. *Mutation Research/Fundamental and Molecular Mechanisms of Mutagenesis*, 591(1–2): 237–246.

Donkor, O.N., Stojanovska, L., Ginn, P., Ashton, J. and Vasiljevic, T. 2012. Germinated grains–Sources of bioactive compounds. *Food Chemistry*, 135(3):950–959.

Dykes, L., Peterson, G.C., Rooney, W.L. and Rooney, L.W. 2011. Flavonoid composition of lemon-yellow sorghum genotypes. *Food Chemistry*, 128(1):173–179.

Dykes, L., Seitz, L.M., Rooney, W.L. and Rooney, L.W. 2009. Flavonoid composition of red sorghum genotypes. *Food Chemistry*, 116(1):313–317.

Elkhalifa, A.E.O., Schiffler, B. and Bernhard, R. 2004. Effect of fermentation on the starch digestibility, resistant starch and some physicochemical properties of sorghum flour. *Food/Nahrung*, 48(2):91–94.

Englyst, H.N. and Cummings, J.H. 1985. Digestion of the polysaccharides of some cereal foods in the human small intestine. *The American Journal of Clinical Nutrition*, 42(5):778–787.

Espitia-Hernández, P., Chávez González, M.L., Ascacio-Valdés, J.A., Dávila-Medina, D., Flores-Naveda, A., Silva, T., Ruelas Chacón, X. and Sepúlveda, L. 2020. Sorghum (*Sorghum bicolor* L.) as a potential source of bioactive substances and their biological properties. *Critical Reviews in Food Science and Nutrition*, 1–12.

Ezekiel, C.N., Abia, W.A., Ogara, I.M., Sulyok, M., Warth, B. and Krska, R. 2015. Fate of mycotoxins in two popular traditional cereal-based beverages (kunu-zaki and pito) from rural Nigeria. *LWT-Food Science and Technology*, 60(1):137–141.

FAOSTAT. 2019. Production/Yield quantities of Sorghum in World (Rome). Available at: www.fao.org/faostat/en/#data/QC/visualize (Accessed 28 August 2021).

Farrar, J.L., Hartle, D.K., Hargrove, J.L. and Greenspan, P. 2008. A novel nutraceutical property of select sorghum (Sorghum bicolor) brans: inhibition of protein glycation. *Phytotherapy Research*, 22(8):1052–1056.

Frazier, R.A., Deaville, E.R., Green, R.J., Stringano, E., Willoughby, I., Plant, J. and Mueller-Harvey, I. 2010. Interactions of tea tannins and condensed tannins with proteins. *Journal of Pharmaceutical and Biomedical Analysis*, 51(2):490–495.

Fuller, S., Beck, E., Salman, H. and Tapsell, L. 2016. New horizons for the study of dietary fiber and health: a review. *Plant Foods for Human Nutrition*, 71(1):1–12.

Galibois, I., Desrosiers, T., Guévin, N., Lavigne, C. and Jacques, H. 1994. Effects of dietary fibre mixtures on glucose and lipid metabolism and on mineral absorption in the rat. *Annals of Nutrition and Metabolism*, 38(4):203–211.

Garcia-Vaquero, M., Mora, L. and Hayes, M. 2019. In vitro and in silico approaches to generating and identifying angiotensin-converting enzyme I inhibitory peptides from green macroalga ulva lactuca. *Marine Drugs*, 17(4):204.

Ghimire, B.K., Seo, J.W., Yu, C.Y., Kim, S.H. and Chung, I.M. 2021. Comparative study on seed characteristics, antioxidant activity, and total phenolic and flavonoid contents in accessions of *Sorghum bicolor* (L.) Moench. *Molecules*, 26(13):3964.

Girard, A.L. and Awika, J.M. 2018. Sorghum polyphenols and other bioactive components as functional and health promoting food ingredients. *Journal of Cereal Science*, 84:112–124.

Giuberti, G., Marti, A., Gallo, A., Grassi, S. and Spigno, G., 2019. Resistant starch from isolated white sorghum starch: functional and physicochemical properties and resistant starch retention after cooking. A comparative study. *Starch-Stärke*, 71(7–8):1800194.

Goñi, I., Garcia-Diz, L., Mañas, E. and Saura-Calixto, F. 1996. Analysis of resistant starch: a method for foods and food products. *Food Chemistry*, 56(4):445–449.

González-Montilla, F.M., Chávez-Santoscoy, R.A., Gutiérrez-Uribe, J.A. and Serna-Saldivar, S.O. 2012. Isolation and identification of phase II enzyme inductors obtained from black Shawaya sorghum [*Sorghum bicolor* (L.) Moench] bran. *Journal of Cereal Science*, 55(2):126–131.

Graham, G.G., MacLean Jr, W.C., Morales, E., Hamaker, B.R., Kirleis, A.W., Mertz, E.T. and Axtell, J.D. 1986. Digestibility and utilization of protein and energy from Nasha, a traditional Sudanese fermented sorghum weaning food. *The Journal of Nutrition*, 116(6):978–984.

Granfeldt, Y.E., Drews, A.W. and Björck, I.M. 1993. Starch bioavailability in arepas made from ordinary or high amylose corn: concentration and gastrointestinal fate of resistant starch in rats. *The Journal of Nutrition*, 123(10):1676–1684.

Gregor, M.F. and Hotamisligil, G.S. 2011. Inflammatory mechanisms in obesity. *Annual Review of Immunology*, 29:415–445.

Guo, H., Yang, X., Zhou, H., Luo, X., Qin, P., Li, J. and Ren, G. 2017. Comparison of nutritional composition, aroma compounds, and biological activities of two kinds of tartary buckwheat tea. *Journal of Food Science*, 82(7):1735–1741.

Hadbaoui, Z., Djeridane, A., Yousfi, M., Saidi, M. and Nadjemi, B. 2010. Fatty acid, tocopherol composition and the antioxidant activity of the lipid extract from the sorghum grains growing in Algeria. *Mediterranean Journal of Nutrition and Metabolism*, 3(3):215–220.

Hahn, D.H., Faubion, J.M. and Rooney, L.W. 1983. Sorghum phenolic acids, their high performance liquid chromatography separation and their relation to fungal resistance. *Cereal Chemistry*, 60(4):255–259.

Hariprasanna, K. and Patil, J.V. 2015. Sorghum: origin, classification, biology and improvement. In *Sorghum Molecular Breeding*, 3–20. Springer: New Delhi.

Harlan, J.R. and Stemler, A. 2011. The races of sorghum in Africa. In *Origins of African Plant Domestication*, 465–478. De Gruyter Mouton.

Harrysson, H., Hayes, M., Eimer, F., Carlsson, N.G., Toth, G.B. and Undeland, I. 2018. Production of protein extracts from Swedish red, green, and brown seaweeds, *Porphyra umbilicalis Kützing, Ulva lactuca Linnaeus*, and *Saccharina latissima* (Linnaeus) JV Lamouroux using three different methods. *Journal of Applied Phycology*, 30(6):3565–3580.

Hasjim, J., Lee, S.O., Hendrich, S., Setiawan, S., Ai, Y. and Jane, J.L. 2010. Characterization of a novel resistant-starch and its effects on postprandial plasma-glucose and insulin responses. *Cereal Chemistry*, 87(4):257–262.

Hassan, O.H.A. 2020. Phytochemical screening and antibacterial activities of *Sorghum bicolor* leaves derived from *in vitro* culture. *GSC Biological and Pharmaceutical Sciences*, 10(1):065–072.

Hassan, S., Imran, M., Ahmad, M.H., Khan, M.I., Xu, C., Khan, M.K. and Muhammad, N. 2020. Phytochemical characterization of ultrasound-processed sorghum sprouts for the use in functional foods. *International Journal of Food Properties*, 23(1):853–863.

Hayes, M. ed. 2018. *Novel Proteins for Food, Pharmaceuticals, and Agriculture: Sources, applications, and advances*. John Wiley.

Hidalgo, M., Oruna-Concha, M.J., Kolida, S., Walton, G.E., Kallithraka, S., Spencer, J.P. and de Pascual-Teresa, S. 2012. Metabolism of anthocyanins by human gut microflora and their influence on gut bacterial growth. *Journal of Agricultural and Food Chemistry*, 60(15):3882–3890.

Hoi, J.T., Weller, C.L., Schlegel, V.L., Cuppett, S.L., Lee, J.Y. and Carr, T.P. 2009. Sorghum distillers dried grain lipid extract increases cholesterol excretion and decreases plasma and liver cholesterol concentration in hamsters. *Journal of Functional Foods*, 1(4):381–386.

Hwang, K. A., Park, M. A., Kang, N. H., Yi, B. R., Hyun, S. H., Jeung, E. B., & Choi, K. C. (2013). Anticancer effect of genistein on BG-1 ovarian cancer growth induced by 17 β-estradiol or bisphenol A via the suppression of the crosstalk between estrogen receptor alpha and insulin-like growth factor-1 receptor signaling pathways. *Toxicology and applied pharmacology*, 272(3): 637–646.

Isaacson, C. 2005. The change of the staple diet of black South Africans from sorghum to maize (corn) is the cause of the epidemic of squamous carcinoma of the oesophagus. *Medical Hypotheses*, 64(3):658–660.

Jafari, M., Koocheki, A. and Milani, E. 2017. Effect of extrusion cooking of sorghum flour on rheology, morphology and heating rate of sorghum–wheat composite dough. *Journal of Cereal Science*, 77:49–57.

Jenkins, D.J., Cuff, D., Wolever, T., Knowland, D., Thompson, L., Cohen, Z. and Prokipchuk, E. 1987. Digestibility of carbohydrate foods in an ileostomate: relationship to dietary fiber, *in vitro* digestibility, and glycemic response. *American Journal of Gastroenterology*, 82(8):709–717.

Jood, S., Khetarpaul, N. and Goyal, R. 2012. Effect of germination and probiotic fermentation on pH, titratable acidity, dietary fibre, β-glucan and vitamin content of sorghum based food mixtures. *Journal of Nutrition & Food Sciences*, 2(164):1–4.

Joye, I.J. 2020. Dietary fibre from whole grains and their benefits on metabolic health. *Nutrients*, 12(10):3045.

Jyothsna, E. and Hymavathi, T.V. 2017. Resistant starch: Importance, categories, food sources and physiological effects. *Journal of Pharmacognosy and Phytochemistry*, 6(2):67–69.

Kamath, V., Niketh, S., Chandrashekar, A. and Rajini, P.S. 2007. Chymotryptic hydrolysates of α-kafirin, the storage protein of sorghum (*Sorghum bicolor*) exhibited angiotensin converting enzyme inhibitory activity. *Food Chemistry*, 100(1):306–311.

Kendall, C.W., Esfahani, A. and Jenkins, D.J. 2010. The link between dietary fibre and human health. *Food Hydrocolloids*, 24(1):42–48.

Khan, I., Yousif, A., Johnson, S.K. and Gamlath, S. 2013. Effect of sorghum flour addition on resistant starch content, phenolic profile and antioxidant capacity of durum wheat pasta. *Food Research International*, 54(1):578–586.

Kim, J. and Park, Y. 2012. Anti-diabetic effect of sorghum extract on hepatic gluconeogenesis of streptozotocin-induced diabetic rats. *Nutrition & Metabolism*, 9(1):1–7.

Kimber, C.T. 2000. Origins of domesticated sorghum and its early diffusion to China and India. In *Sorghum: Origin, History, Technology, and Production*. C.W. Smith and R.A. Frederiksen, (Eds.), 3–97.

Knudsen, K.B. and Munck, L. 1985. Dietary fibre contents and compositions of sorghum and sorghum-based foods. *Journal of Cereal Science*, 3(2):153–164.

Kulamarva, A.G., Sosle, V.R. and Raghavan, G.V. 2009. Nutritional and rheological properties of sorghum. *International Journal of Food Properties*, 12(1):55–69.

Kumari, P., Arya, S., Pahuja, S.K., Joshi, U.N. and Sharma, S.K. 2016. Evaluation of forage sorghum genotypes for chlorophyll content under salt stress. *The International Journal of Environmental Science and Technology*, 5(3):1200–1207.

Lafrance, L., Rabasa-Lhoret, R., Poisson, D., Ducros, F. and Chiasson, J.L. 1998. Effects of different glycaemic index foods and dietary fibre intake on glycaemic control in type 1 diabetic patients on intensive insulin therapy. *Diabetic Medicine*, 15(11):972–978.

Langkilde, A.M. and Andersson, H. 1994. *In vivo* quantification of resistant starch in EURESTA reference materials using the ileostomy model. In *Proceedings of the Concluding Plenary Meeting of EURESTA–Including the Final Reports of the Working Groups*: 31–32.

Lee, S.C., Prosky, L. and Vries, J.W.D. 1992. Determination of total, soluble, and insoluble dietary fiber in foods-Enzymatic-gravimetric method, MES-TRIS buffer: Collaborative study. *Journal of AOAC International*, 75(3): 395–416.

Lemlioglu-Austin, D., Turner, N. D., McDonough, C. M., & Rooney, L. W. (2012). Effects of sorghum [Sorghum bicolor (L.) Moench] crude extracts on starch digestibility, estimated glycemic index (EGI), and resistant starch (RS) contents of porridges. *Molecules*, 17(9): 11124–11138.

Lin, K., Zhang, L.W., Han, X. and Cheng, D.Y. 2017. Novel angiotensin I-converting enzyme inhibitory peptides from protease hydrolysates of Qula casein: Quantitative structure-activity relationship modeling and molecular docking study. *Journal of Functional Foods*, 32:266–277.

Liu, L., Herald, T.J., Wang, D., Wilson, J.D., Bean, S.R., & Aramouni, F.M. (2012). Characterization of sorghum grain and evaluation of sorghum flour in a Chinese egg noodle system. *Journal of Cereal Science*, 55(1): 31–36.

MacLean Jr, W.C., López de Romaña, G., Gastañaduy, A. and Graham, G.G. 1983. The effect of decortication and extrusion on the digestibility of sorghum by preschool children. *The Journal of Nutrition*, 113(10):2071–2077.

Maclean Jr, W.C., Romaña, G.L.D., Placko, R.P. and Graham, G.G. 1981. Protein quality and digestibility of sorghum in preschool children: balance studies and plasma free amino acids. *The Journal of Nutrition*, 111(11):1928–1936.

Makawi, A.B., Mustafa, A.I., Adiamo, O.Q. and Ahmed, I.A.M. 2019. Quality attributes of Kisra prepared from sorghum flour fermented with baobab fruit pulp flour as starter. *Journal of Food Science and Technology*, 56(8):3754–3763.

Marchini, M., Arduini, R. and Carini, E. 2021. Insight into molecular and rheological properties of sprouted sorghum flour. *Food Chemistry*, 356:129603.

Marchini, M., Marti, A., Folli, C., Prandi, B., Ganino, T., Conte, P., Fadda, C., Mattarozzi, M. and Carini, E. 2021. Sprouting of Sorghum (*Sorghum bicolor* [L.] Moench): Effect of Drying Treatment on Protein and Starch Features. *Foods*, 10(2):407.

Marlett, J.A. and Longacre, M.J. 1996. Comparison of in vitro and in vivo measures of resistant starch in selected grain products. *Cereal Chemistry*, 73(1):63–68.

Marston, K., Khouryieh, H. and Aramouni, F. 2016. Effect of heat treatment of sorghum flour on the functional properties of gluten-free bread and cake. *LWT–Food Science and Technology*, 65:637–644.

Martino, H.S.D., Tomaz, P.A., Moraes, É.A., ConceiçãoI, L.L.D., Oliveira, D.D.S., Queiroz, V.A.V., Rodrigues, J.A.S., Pirozi, M.R., Pinheiro-Sant'Ana, H.M. and Ribeiro, S.M.R. 2012. Chemical characterization and size distribution of sorghum genotypes for human consumption. *Revista do Instituto Adolfo Lutz (Impresso)*, 71(2):337–344.

McCleary, B.V. and Cox, J. 2017. Evolution of a definition for dietary fiber and methodology to service this definition. *Luminacoids Research*, 21(2):9–21.

McCleary, B.V. and Monaghan, D.A. 2002. Measurement of resistant starch. *Journal of AOAC International*, 85(3):665–675.

McCleary, B.V., De Vries, J.W., Rader, J.I., Cohen, G., Prosky, L., Mugford, D.C., Champ, M. and Okuma, K. 2010. Determination of total dietary fiber (CODEX definition) by enzymatic-gravimetric method and liquid chromatography: Collaborative study. *Journal of AOAC International*, 93(1): 221–233.

McCleary, B.V., DeVries, J.W., Rader, J.I., Cohen, G., Prosky, L., Mugford, D.C., Champ, M. and Okuma, K. 2012. Determination of insoluble, soluble, and total dietary fiber (CODEX definition) by enzymatic-gravimetric method and liquid chromatography: collaborative study. *Journal of AOAC International*, 95(3): 824–844.

McCleary, B.V., McNally, M. and Rossiter, P. 2002. Measurement of resistant starch by enzymatic digestion in starch and selected plant materials: collaborative study. *Journal of AOAC International*, 85(5):1103–1111.

Mehmood, S., Orhan, I., Ahsan, Z., Aslan, S. and Gulfraz, M. 2008. Fatty acid composition of seed oil of different *Sorghum bicolor* varieties. *Food Chemistry*, 109(4):855–859.

Mohapatra, D., Patel, A.S., Kar, A., Deshpande, S.S. and Tripathi, M.K. 2019. Effect of different processing conditions on proximate composition, anti-oxidants, anti-nutrients and amino acid profile of grain sorghum. *Food Chemistry*, 271, 129–135.

Moraes, É.A., Natal, D.I.G., Queiroz, V.A.V., Schaffert, R.E., Cecon, P.R., de Paula, S.O., dos Anjos Benjamim, L., Ribeiro, S.M.R. and Martino, H.S.D. 2012. Sorghum genotype may reduce low-grade inflammatory response and oxidative stress and maintains jejunum morphology of rats fed a hyperlipidic diet. *Food Research International*, 49(1):553–559.

Moraes, É.A., Queiroz, V.A.V., Shaffert, R.E., Costa, N.M.B., Nelson, J.D., Ribeiro, S.M.R. and Martino, H.S.D. 2012. *In vivo* protein quality of new sorghum genotypes for human consumption. *Food Chemistry*, 134(3):1549–1555.

Morey, S.R., Hashida, Y., Ohsugi, R., Yamagishi, J. and Aoki, N. 2018. Evaluation of performance of sorghum varieties grown in Tokyo for sugar accumulation and its correlation with vacuolar invertase genes SbInv1 and SbInv2. *Plant Production Science*, 21(4):328–338.

Morgan, L.M., Tredger, J.A., Wright, J. and Marks, V. 1990. The effect of soluble-and insoluble-fibre supplementation on post-prandial glucose tolerance, insulin and gastric inhibitory polypeptide secretion in healthy subjects. *British Journal of Nutrition*, 64(1):103–110.

Muir, J.G. and O'Dea, K. 1993. Validation of an in vitro assay for predicting the amount of starch that escapes digestion in the small intestine of humans. *The American Journal of Clinical Nutrition*, 57(4):540–546.

Muir, J.G., Lu, Z.X., Young, G.P., Cameron-Smith, D., Collier, G.R. and O'Dea, K. 1995. Resistant starch in the diet increases breath hydrogen and serum acetate in human subjects. *The American Journal of Clinical Nutrition*, 61(4):792–799.

Mullet, J., Morishige, D., McCormick, R., Truong, S., Hilley, J., McKinley, B., Anderson, R., Olson, S.N. and Rooney, W. 2014. Energy Sorghum – A genetic model for the design of C4 grass bioenergy crops. *Journal of Experimental Botany*, 65(13):3479–3489.

Nawar, I.A., Clark, H.E., Pickett, R.C. and Hegsted, D.M. 1970. Protein quality of selected lines of *Sorghum vulgare* for the growing rat. *Nutrition Reports International*, 1:75–81.

Ndulaka, J.C., Obasi, N.E. and Omeire, G.C. 2014. Production and Evaluation of Reconstitutable Kunun-Zaki. *Nigerian Food Journal*, 32(2):66–72.

Niba, L.L. and Hoffman, J. 2003. Resistant starch and β-glucan levels in grain sorghum (Sorghum bicolor M.) are influenced by soaking and autoclaving. *Food Chemistry*, 81(1):113–118.

Noah, L., Guillon, F., Bouchet, B., Buleon, A., Molis, C., Faisant, N. and Champ, M. 1996. Digestion of carbohydrate components of dry beans (*Phaseolus vulgaris* L.) in healthy humans. In *Symposium*.

Nyamambi, B., Ndlovu, L.R., Read, J.S. and Reed, J.D. 2000. The effects of sorghum proanthocyanidins on digestive enzyme activity *in vitro* and in the digestive tract of chicken. *Journal of the Science of Food and Agriculture*, 80(15):2223–2231.

Obilana, A.B. 2004. Sorghum: breeding and agronomy. In: Wrigley, C., Corke, H., Walker, C.E. (Eds.), *Encyclopedia of Grain Science*, vol. 3. Elsevier, Oxford, 108–119.

Ochanda, S.O., Akoth, O.C., Mwasaru, A.M., Kagwiria, O.J. and Mathooko, F.M. 2010. *Effects of malting and fermentation treatments on group B-vitamins of red sorghum, white sorghum and pearl millets in Kenya. Journal of Applied Bioscience,* 34: 2128–2134

Ofosu, F.K., Elahi, F., Daliri, E.B.M., Yeon, S.J., Ham, H.J., Kim, J.H., Han, S.I. and Oh, D.H. 2020. Flavonoids in decorticated sorghum grains exert antioxidant, antidiabetic and antiobesity activities. *Molecules*, 25(12):2854.

Ohadi, S., Littlejohn, M., Mesgaran, M., Rooney, W. and Bagavathiannan, M., 2018. Surveying the spatial distribution of feral sorghum (*Sorghum bicolor* L.) and its sympatry with johnsongrass (*S. halepense*) in South Texas. *PloS one*, 13(4):0195511.

Ojha, P., Adhikari, R., Karki, R., Mishra, A., Subedi, U. and Karki, T.B. 2018. Malting and fermentation effects on antinutritional components and functional characteristics of sorghum flour. *Food Science & Nutrition*, 6(1):47–53.

Olojede, A.O., Sanni, A.I. and Banwo, K. 2020. Rheological, textural and nutritional properties of gluten-free sourdough made with functionally important lactic acid bacteria and yeast from Nigerian sorghum. *LWT–Food Science and Technology*, 120:108875.

Oria, M.P., Hamaker, B.R., Axtell, J.D. and Huang, C.P. 2000. A highly digestible sorghum mutant cultivar exhibits a unique folded structure of endosperm protein bodies. *Proceedings of the National Academy of Sciences*, 97(10):5065–5070.

Osman, M.A. 2004. Changes in sorghum enzyme inhibitors, phytic acid, tannins and in vitro protein digestibility occurring during Khamir (local bread) fermentation. *Food Chemistry*, 88(1):129–134.

Palavecino, P.M., Ribotta, P.D., León, A.E. and Bustos, M.C. 2019. Gluten-free sorghum pasta: starch digestibility and antioxidant capacity compared with commercial products. *Journal of the Science of Food and Agriculture*, 99(3):1351–1357.

Park, J.D. and Ren, Z. 2012. High efficiency energy harvesting from microbial fuel cells using a synchronous boost converter. *Journal of Power Sources*, 208:322–327.

Park, J.H., Darvin, P., Lim, E.J., Joung, Y.H., Hong, D.Y., Park, E.U., Park, S.H., Choi, S.K., Moon, E.S., Cho, B.W. and Park, K.D. 2012. Hwanggeumchal sorghum induces cell cycle arrest, and suppresses tumor growth and metastasis through Jak2/STAT pathways in breast cancer xenografts. *PloS one*, 7(7):40531.

Park, J.H., Lee, S.H., Chung, I.M. and Park, Y. 2012. Sorghum extract exerts an anti-diabetic effect by improving insulin sensitivity via PPAR- in mice fed a high-fat diet. *Nutrition Research and Practice*, 6(4):322–327.

Poquette, N.M., Gu, X. and Lee, S.O. 2014. Grain sorghum muffin reduces glucose and insulin responses in men. *Food & Function*, 5(5):894–899.

Prasad, M.P.R., Rao, B.D., Kalpana, K., Rao, M.V. and Patil, J.V. 2015. Glycaemic index and glycaemic load of sorghum products. *Journal of the Science of Food and Agriculture*, 95(8):1626–1630.

Prosky, L., Asp, N.G., Furda, I., Devries, J.W., Schweizer, T.F. and Harland, B.F. 1985. Determination of total dietary fiber in foods and food products: collaborative study. *Journal of the Association of Official Analytical Chemists*, 68(4):677–679.

Rahman, I.E.A. and Osman, M.A.W. 2011. Effect of sorghum type (*Sorghum bicolor*) and traditional fermentation on tannins and phytic acid contents and trypsin inhibitor activity. *Journal of Food, Agriculture and Environment*, 9(3):163–166.

Ranhotra, G. S., Gelroth, J. A., & Glaser, B. K. (1996). Effect of resistant starch on blood and liver lipids in hamsters. *Cereal chemistry*, 73(2): 176–178.

Rashwan, A.K., Yones, H.A., Karim, N., Taha, E.M. and Chen, W. 2021. Potential processing technologies for developing sorghum-based food products: An update and comprehensive review. *Trends in Food Science & Technology*, 110:168–182.

Rooney, L.W. and Serna-Saldivar, S.O. 1991. *Sorghum*. In K.J. Lorenz and K. Kulp, (Eds.). *Hand book of Cereal Science and Technology*, 233–269. New York: Marcel Dekker.

Rooney, L.W., Rooney, W.L., & Serna Saldivar, S.O. (2016) *Sorghum*. In *Reference Module in Food Science*, Elsevier.

Rooney, T.K., Rooney, L.W. and Lupton, J.R. 1992. Physiological characteristics of sorghum and millet brans in the rat model. *Cereal Foods World*, 37(10):782–786.

Rose, D.J., Williams, E., Mkandawire, N.L., Weller, C.L. and Jackson, D.S. 2014. Use of whole grain and refined flour from tannin and non-tannin sorghum (*Sorghum bicolor* (L.) Moench) varieties in frybread. *Food Science and Technology International*, 20(5):333–339.

Sage, R.F. and Zhu, X.G. 2011. Exploiting the engine of C4 photosynthesis. *Journal of Experimental Botany*, 62(9):2989–3000.

Sajilata, M.G., Singhal, R.S. and Kulkarni, P.R. 2006. Resistant starch – A review. *Comprehensive Reviews in Food Science and Food Safety*, 5(1):1–17.

Sang, Y., Bean, S., Seib, P.A., Pedersen, J. and Shi, Y.C. 2008. Structure and functional properties of sorghum starches differing in amylose content. *Journal of Agricultural and Food Chemistry*, 56(15):6680–6685.

Sang, Y., Bean, S., Seib, P.A., Pedersen, J. and Shi, Y.C. 2008. Structure and functional properties of sorghum starches differing in amylose content. *Journal of Agricultural and Food Chemistry*, 56(15):6680–6685.

Sarmadi, B.H. and Ismail, A. 2010. Antioxidative peptides from food proteins: A review. *Peptides*, 31(10):1949–1956.

Saura-Calixto, F. 2011. Dietary fiber as a carrier of dietary antioxidants: an essential physiological function. *Journal of Agricultural and Food Chemistry*, 59(1):43–49.

Scott, K.P., Duncan, S.H. and Flint, H.J. 2008. Dietary fibre and the gut microbiota. *Nutrition Bulletin*, 33(3):201–211.

Serna-Saldivar, S. 1995. Structure and chemistry of sorghum and millets. *Sorghum and Millets: Chemistry and Technology*, 69–124.

Shafie, B., Cheng, S.C., Lee, H.H. and Yiu, P.H. 2016. Characterization and classification of whole-grain rice based on rapid visco analyzer (RVA) pasting profile. *International Food Research Journal*, 23(5):2138–2143.

Shahidi, F. and Zhong, Y. 2010. Novel antioxidants in food quality preservation and health promotion. *European Journal of Lipid Science and Technology*, 112(9):930–940.

Shaikh, F., Ali, T.M., Mustafa, G. and Hasnain, A. 2019. Comparative study on effects of citric and lactic acid treatment on morphological, functional, resistant starch fraction and glycemic index of corn and sorghum starches. *International Journal of Biological Macromolecules*, 135:314–327.

Sharma, S., Kelly, T.K. and Jones, P.A. 2010. Epigenetics in cancer. *Carcinogenesis*, 31(1):27–36.

Shen, R.L., Zhang, W.L., Dong, J.L., Ren, G.X. and Chen, M. 2015. Sorghum resistant starch reduces adiposity in high-fat diet-induced overweight and obese rats via mechanisms involving adipokines and intestinal flora. *Food and Agricultural Immunology*, 26(1):120–130.

Shih, I.M. and Wang, T.L. 2007. Notch signaling, γ-secretase inhibitors, and cancer therapy. *Cancer Research*, 67(5):1879–1882.

Shull, J.M., Watterson, J.J. and Kirleis, A.W. 1991. Proposed nomenclature for the alcohol-soluble proteins (kafirins) of Sorghum bicolor (L. Moench) based on molecular weight, solubility, and structure. *Journal of Agricultural and Food Chemistry*, 39(1):83–87.

Singh, B.P., Vij, S. and Hati, S. 2014. Functional significance of bioactive peptides derived from soybean. *Peptides*, 54:171–179.

Singh, H., Sodhi, N.S. and Singh, N. 2010. Characterisation of starches separated from sorghum cultivars grown in India. *Food Chemistry*, 119(1):95–100.

Singh, H., Sodhi, N.S., and Singh, N. 2012. Structure and functional properties of acetylated Sorghum starch. *International Journal of Food Properties*, 15(2):312–325.

Sitanggang, A.B., Budijanto, S. and Marisa, 2018. Physicochemical characteristics of starch from Indonesian Numbu and Genjah sorghum (*Sorghum bicolor* L. Moench). *Cogent Food & Agriculture*, 4(1):1429093.

Stefoska-Needham, A., Beck, E.J., Johnson, S.K., and Tapsell, L.C. 2015. Sorghum: An underutilized cereal whole grain with the potential to assist in the prevention of chronic disease. *Food Reviews International*, 31(4):401–437.

Steinhart, A.H., Jenkins, D.J., Mitchell, S., Cuff, D. and Prokipchuk, E.J. 1992. Effect of dietary fiber on total carbohydrate losses in ileostomy effluent. *American Journal of Gastroenterology*, 87(1):48–54.

Stephen, A.M., Haddad, A.C. and Phillips, S.F. 1983. Passage of carbohydrate into the colon: direct measurements in humans. *Gastroenterology*, 85(3):589–595.

Suhasini, G.E. and Krishna, D.R. 1991. Influence of unrefined sorghum or maize on serum lipids. *Ancient Science of Life*, 11(1–2):26.

Sun, H., Wang, H., Zhang, P., Ajlouni, S. and Fang, Z. 2020. Changes in phenolic content, antioxidant activity, and volatile compounds during processing of fermented sorghum grain tea. *Cereal Chemistry*, 97(3):612–625.

Sun, Q., Han, Z., Wang, L. and Xiong, L. 2014. Physicochemical differences between sorghum starch and sorghum flour modified by heat–moisture treatment. *Food Chemistry*, 145:756–764.

Taylor, J.R. 2019. Sorghum and millets: Taxonomy, history, distribution, and production. In *Sorghum and Millets*, 1–21. AACC International Press.

Taylor, J.R., Belton, P.S., Beta, T. and Duodu, K.G. 2014. Increasing the utilisation of sorghum, millets and pseudocereals: Developments in the science of their phenolic phytochemicals, biofortification and protein functionality. *Journal of Cereal Science*, 59(3):257–275.

Thakur, R., Gupta, V., Dhar, P., Deka, S. C., & Das, A. B. 2022. Ultrasound-assisted extraction of anthocyanin from black rice bran using natural deep eutectic solvents: Optimization, diffusivity, and stability. Journal of Food Processing and Preservation, 46(3):e16309.

Trowell, H. 1973. Dietary fibre, ischaemic heart disease and diabetes mellitus. *Proceedings of the Nutrition Society*, 32(3):151–157.

Tu, M., Cheng, S., Lu, W. and Du, M. 2018. Advancement and prospects of bioinformatics analysis for studying bioactive peptides from food-derived protein: Sequence, structure, and functions. *TrAC Trends in Analytical Chemistry*, 105:7–17.

Tu, M., Feng, L., Wang, Z., Qiao, M., Shahidi, F., Lu, W. and Du, M. 2017. Sequence analysis and molecular docking of antithrombotic peptides from casein hydrolysate by trypsin digestion. *Journal of Functional Foods*, 32:313–323.

Tubeileh, A., Rennie, T. and Alam-Eldein, S. 2014. Biofuel research in Canada: Some results from eastern Ontario. *International Journal of Environment and Sustainability*, 3(1):50–53.

Vásquez, F., Verdú, S., Islas, A.R., Barat, J.M. and Grau, R. 2016. Effect of low degrees of substitution in wheat flour with sorghum, oat or corn flours on physicochemical properties of composite flours. *Cogent Food & Agriculture*, 2(1):1269979.

Wang, J., Yin, T., Xiao, X., He, D., Xue, Z., Jiang, X. and Wang, Y. 2018. StraPep: A structure database of bioactive peptides. *Database*.

Wang, T.Y., Li, Q. and Bi, K.S. 2018. Bioactive flavonoids in medicinal plants: Structure, activity and biological fate. *Asian Journal of Pharmaceutical Sciences*, 13(1):12–23.

Whitehead, R.H., Young, G.P. and Bhathal, P.S. 1986. Effects of short chain fatty acids on a new human colon carcinoma cell line (LIM1215). *Gut*, 27(12):1457–1463.

Wu, G., Ashton, J., Simic, A., Fang, Z. and Johnson, S.K. 2018. Mineral availability is modified by tannin and phytate content in sorghum flaked breakfast cereals. *Food Research International*, 103:509–514.

Wu, L., Huang, Z., Qin, P. and Ren, G. 2013. Effects of processing on phytochemical profiles and biological activities for production of sorghum tea. *Food Research International*, 53(2):678–685.

Wu, L., Huang, Z., Qin, P., Yao, Y., Meng, X., Zou, J., Zhu, K. and Ren, G. 2011. Chemical characterization of a procyanidin-rich extract from sorghum bran and its effect on oxidative stress and tumor inhibition in vivo. *Journal of Agricultural and Food Chemistry*, 59(16):8609–8615.

Xiong, Y., Zhang, P., Johnson, S., Luo, J. and Fang, Z. 2020. Comparison of the phenolic contents, antioxidant activity and volatile compounds of different sorghum varieties during tea processing. *Journal of the Science of Food and Agriculture*, 100(3):978–985.

Xiong, Y., Zhang, P., Luo, J., Johnson, S. and Fang, Z. 2019. Effect of processing on the phenolic contents, antioxidant activity and volatile compounds of sorghum grain tea. *Journal of Cereal Science*, 85:6–14.

Xiong, Y., Zhang, P., Warner, R.D. and Fang, Z. 2019. Sorghum grain: From genotype, nutrition, and phenolic profile to its health benefits and food applications. *Comprehensive Reviews in Food Science and Food Safety*, 18(6):2025–2046.

Yang, L., Allred, K.F., Dykes, L., Allred, C.D. and Awika, J.M. 2015. Enhanced action of apigenin and naringenin combination on estrogen receptor activation in non-malignant colonocytes: Implications on sorghum-derived phytoestrogens. *Food & Function*, 6(3):749–755.

Yang, L., Browning, J.D. and Awika, J.M. 2009. Sorghum 3-deoxyanthocyanins possess strong phase II enzyme inducer activity and cancer cell growth inhibition properties. *Journal of Agricultural and Food Chemistry*, 57(5):1797–1804.

Yang, R., Wang, P., Elbaloula, M.F. and Gu, Z. 2016. Effect of germination on main physiology and biochemistry metabolism of sorghum seeds. *Bioscience Journal*, 32(2):378–383.

Yu, J., Yan, F., Lu, Q. and Liu, R. 2018. Interaction between sorghum procyanidin tetramers and the catalytic region of glucosyltransferases-I from Streptococcus mutans UA159. *Food Research International*, 112:152–159.

Zamora-Ros, R., Knaze, V., Rothwell, J.A., Hémon, B., Moskal, A., Overvad, K., Tjønneland, A., Kyrø, C., Fagherazzi, G., Boutron-Ruault, M.C. and Touillaud, M. 2016. Dietary polyphenol intake in Europe: the European Prospective Investigation into Cancer and Nutrition (EPIC) study. *European Journal of Nutrition*, 55(4):1359–1375.

Zhu, Y., Shi, Z., Yao, Y., Hao, Y., & Ren, G. (2017). Antioxidant and anti-cancer activities of proanthocyanidins-rich extracts from three varieties of sorghum (Sorghum bicolor) bran. *Food and Agricultural Immunology*, 28(6): 1530–1543.

Zou, T.B., He, T.P., Li, H.B., Tang, H.W. and Xia, E.Q. 2016. The structure-activity relationship of the antioxidant peptides from natural proteins. *Molecules*, 21(1):72.

4

Nutraceutical Potential and Techno-Functional Properties of Pearl Millet (*Pennisetum glaucum*)

Gunjana Deka, Jyoti Goyat, and Himjyoti Dutta

Contents

DOI: 10.1201/9781003251279-4

4.1 Introduction

Millets form a class of staple cereals other than more popular grain crops like wheat, maize and rice. Pearl millet (*Pennisetum glaucum* L.) is a popular drought-tolerant millet plant producing nutritionally important grains. It is also known by several common names as per languages native to the regions where the crop is grown.[1] It is a prime member under the category of major millets accounting for about 50 percent of the total global production and global market of millets. Originating in Africa, it was domesticated about 5,000 years ago and later migrated to India, Latin America, Brazil and more than 20 other nations practicing arid and semiarid cultivations. By far, pearl millet has been identified as the world's most resilient grain crop, requiring very low rainfall (<500 mm/year) and daytime temperatures higher than 30° C. Besides, it can grow well in nutrition-depleted soil. It has a long history of use as hay crop, especially in the harsh tropical arid regions of the world. The crop holds critical importance to food security in its growing nations. [1,2] It is the major staple crop for the people residing in warm deserts of West Africa and India. India is the largest pearl millet producing nation with its 9.3 million hectares of land annually producing 8.3 million tons of the crop. The grain is known as *Bajra* in India. The crop also bears much significance for the regions facing crisis of water scarcity, population density and rapid climate change.

Pearl millet is grown for both food and dry fodder. For human consumption, grains are traditionally processed into products like leavened and unleavened flat breads and porridges. Pearl millet grain contains high levels of metabolizable energy and protein with balanced amino acid profile. It also has rich composition of thiamin, niacin, riboflavin, iron, calcium, phosphorus and zinc.[3,4,5] Consumption of pearl millet has been linked with several health benefits, such as improving digestive metabolism, cholesterol reduction, diabetes inhibition, cardiovascular and locomotive improvements and lowering risks of cancers. It is a steady source of energy and proteins for millions in marginal agricultural zones. The favorable amino acid balance, rich composition of essential amino acids and superior protein digestibility make pearl millet a well-digested nutritious cereal. This drought resistant small seeded cereal has potential health benefits and, thus, also contributes to national food security across the globe.[2,6,7] The taxonomic classification of pearl millet is as follows.

Kingdom: Plantae
Subkingdom: Tracheobionta
Superdivision: Spermatophyta
Division: Magnoliophyta
Class: Liliopsida
Subclass: Commelinidae
Order: Cyperales
Family: Poaceae-grass family
Genus: *Pennisetum*
Species: *Pennisetum glaucum*

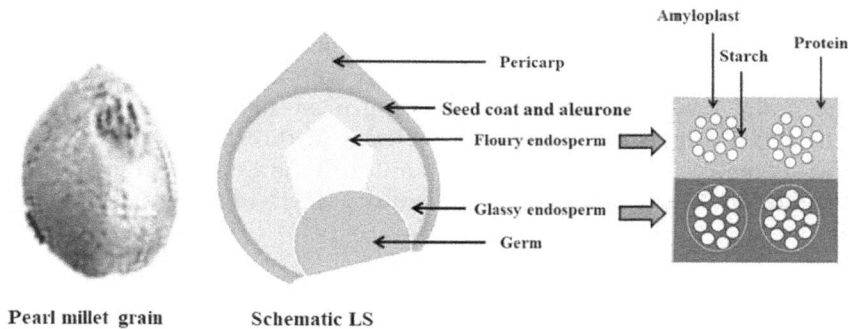

Figure 4.1 *Pearl millet grain, a schematic of longitudinal section (LS) of the grain and distribution of starch granules within proteinous matrices within "floury" and "glassy" endosperm zones of the grain.*

Source: Figure created by the authors.

The genus *Pennisetum* contains about 140 species. Four popularly cultivated forms of pearl millet are typhoides (found mainly in India and Africa), nigritarum (dominant in eastern Sahel), globosum (dominant in the western Sahel) and leonis (dominant on the West African coast).[8]

The pearl millet plant has a slender stem divided into distinct nodes. Its height can vary from 1.5–13.1 feet in height, depending on the cultivar. The annual plant is harvested after one growing season. The leaves are linear, lance-like and can grow up to more than 3 feet in length. It has a spike-like panicle inflorescence, which is further made up of many smaller spikelets. These spikelets bear the grains, when formed.[2] The grains are densely packed (~1.6 g/cm³). The grain has a "liquid drop" shape (**Figure 4.1**). It can grow up to 2 mm in length with weight varying from 3–15mg. The color of its naked kernel can range from pearly white, brown and purple. A relatively larger germ comprising about 18 percent of the matured kernel has several implications in terms of its nutrient content.[1,2,9] The intimately adhered pericarp layer (~8%, w/w) is covered on the surface by a thin waxy layer, which protects it from sudden rise or drop in humidity. The seed coat beneath the pericarp contains a single-layered aleurone. The endosperm (~75%. w/w) is mostly composed of starch with a rich proportion of protein holding the amyloplasts.[10] The outer regions of the endosperm are "glassy" with stronger protein packing. Towards the center, the endosperm is "floury" with lower protein and freer amyloplasts holding the starch granules. The ratio and distribution of the glassy and floury portions markedly affect the texture and milling characteristic of the grain. Accordingly, floury, very soft, corneous and hard endosperms of pearl millet are characterized.[5] They further relate to flour extractability and flour quality as intended for use in different processed products.

4.2 Harvesting and post-harvest processing

Pearl millet crop is harvested once the grains reach physiological maturity. Generally, it is done 40 days after flowering. Ideal moisture content for harvesting is below 20 percent to prevent integrity problems of the plant structure. At higher than 25 percent moisture, the seeds remain too soft to withstand the threshing pressure. Black spots in the hilar region of the grain can be used as indicators of attainment of maturity. Two types of methods are practiced for harvesting the crop. The first method is the one usually practiced, which is cutting the earheads followed by cutting the remaining plant. The other method is cutting the entire plant with sticks and stalking them for 5 to 7 days in sun until attainment of 14 percent moisture content in the grain. Long term storage exceeding 6 months requires further drying until achieving a 12 percent moisture content. A mechanical thresher or drawing a stone roller over the earheads or trampling with cattle separates the grain. The grains are then cleaned (winnowed) and dried. Both conventional and natural air-drying methods are used for drying the grain.[11]

Post-harvest processing of grains primarily target increasing the storage stability, usability and quality (**Figure 4.2**). Processing treatments encompass both traditional and advanced methods. Traditional methods include dehulling and decortication followed by milling to flour. Further, dry and hydrothermal treatments are used to increase the microbial stability. These treatments also help reducing the lipase activity and free fatty acid (FFA) in the grains. To obtain desired flour quality, secondary processes like germination, blanching, fermentation and so forth are also carried out. Advanced and novel thermal and non-thermal technologies have been employed to meet the gaps in nutritional and functional properties emerging from the conventional processing of pearl millet.

Traditionally, pearl millet grains are stored in mud bins, bamboo bins, straw bins, or metal bins until milled.[11] A sufficient and controlled rate of grain drying is necessary before storage to avoid mold growth. Drying is followed by bagging the grains in moisture proof bags. Bags

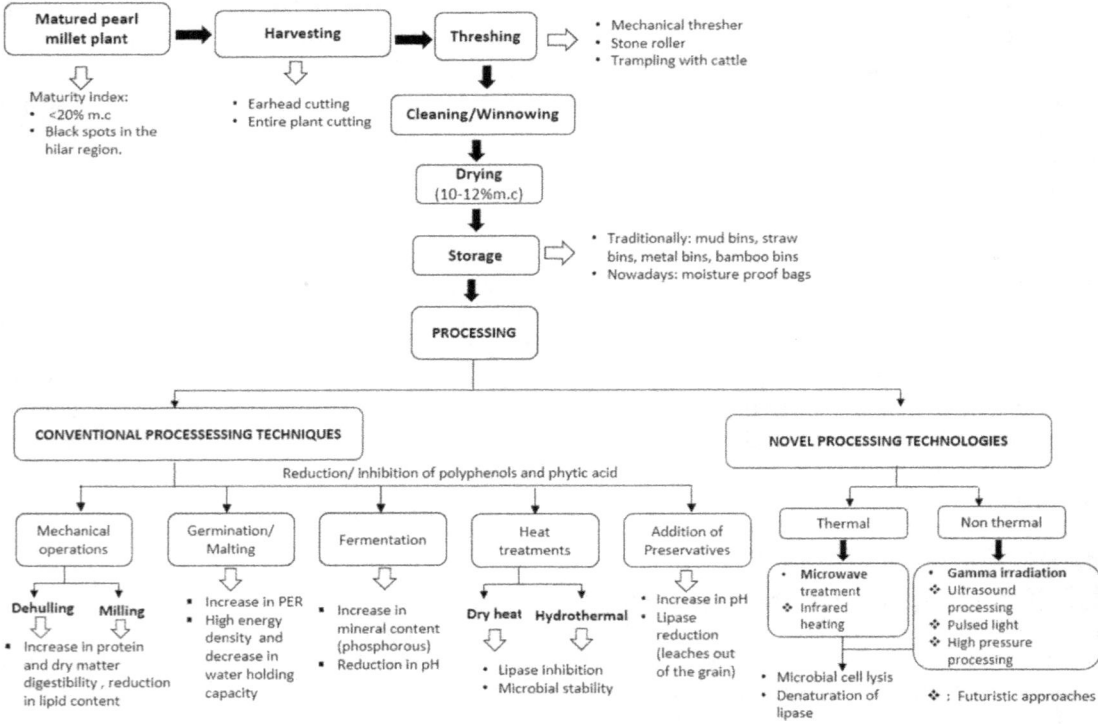

Figure 4.2 *Flowchart showing harvesting, storage and postharvest processing measures for pearl millet.*

Source: Figure created by the authors.

should be kept over wooden pallets away from walls or floor to prevent moisture absorption by the seed. Stacking more than 6 to 7 bags would exert pressure on the seeds in the basal bags and reduce their germination capacity, and should be prevented.[12]

The small kernel size with its germ firmly embedded along with the hard endosperm, make milling of pearl millet difficult. Hammer mills and roller mills are conventionally used for milling the grain to flour. Roller mills produce fine flours which can be utilized to make baked and steamed product of smooth texture. On the other hand, hammer mills produce coarser flour, with larger particle size.[4] In rural areas, hand operated non-motorized mills are used to produce domestic purpose flours. Meera et al. (2002) developed a process that was found to improve the milling characteristics of pearl millet and sorghum varieties with floury endosperm. The process is initiated by moist heating of the grains, followed by drying the grains to about 10 to 12 percent moisture and decortication. The sorghum and pearl millet flour prepared from this process had a shelf-life of about 8 to 10 months and 3 to 4 months, respectively. This process was advantageous in another context: that it also decreased the oxidative rancidity, FFA content below 10 percent, and drastically reduced the microbial load.[4,13] Owing to their high fat content and lipase activity, millets generally have a short shelf-life. Problems during storage of this grain emanates from their small size, and separating the germ from the endosperm becomes difficult. In the flour form, the lipid content in the germ of these grains gets exposed, resulting in lipolysis and oxidation of the unsaturated fatty acids. Pre-treatments and processing methods are necessary to inhibit lipase activity and improve stability of the grains. A study was carried out on the effect of decortication in pearl millet mineral and chemical composition, which revealed significant reduction in fat content due to the decortication measures taken. After 60 days of storage, a much lower value of fatty acid content was obtained.[14] However, this process considerably decreased the amount of macro and micronutrients from the grain.

To inactivate lipase and storage stability of millets, both dry heat and hydrothermal methods have been effectively applied. However, changes in certain physical and functional properties might result from this treatment that might compromise consumer acceptability. Dias-Martini et al. (2018) reported that reduction in FFA and lipase activity in dry heat-treated pearl millet. [15] A study was conducted on dehulled sorghum, foxtail millet, and pearl millet grains to study the effect of heat treatment on the total bacterial count. Heat treatment of 150–170° C for 90 seconds led to significant reduction in the total fungal count by 48.23 percent during initial treatment, and there was gradual decrease in count to 60th day of storage.[16] Hydrothermal treatment was applied by holding the grain in boiling water for 15 min followed by tray drying at 60° C for 2 h. As a result of lipase inhibition, it led to decrease in fat acidity of the flour and the shelf-life was increased up to 30 days under ambient conditions.[17]

Fermentation and germination have been found to improve the nutritional status of grains. A study was conducted on the effect of malting and fermentation on properties of pearl millet, white sorghum, and red sorghum. It was found that the total titrable acidity increased and the pH reduced in the flour sample with an increase in the fermentation time. The low pH (4.0) obtained on the second day of fermentation, made it lethal for the microorganisms, thereby increasing the storage life.[18] A decline in the fat content in fermented flour blends of pearl millet and sorghum was observed in a study. An increase in titrable acidity was observed owing to the degradation of carbohydrates by lactic acid bacteria causing acidification. A further decrease in pH was reported over prolonged process. These also led to modification of nutritional status of the flour.[19] The process of soaking prior to germination of two different varieties of pearl millets resulted in the loss of total soluble solids and fat content.[20] Results of another study revealed decrease in flour pH with consequent increase in total titrable acidity while germination was being carried out in a moistened jute sack. These variations are cumulatively brought about by changes in different complex components, namely lipids, phytin and protein getting hydrolyzed into simpler compounds at a considerable rate.[21] Pearl millet grain flour showed 73.9 percent reduction in FFA at 25–30° C after 24 h of treatment and 62.5 percent reduction after 18 h of treatment.[22]

To prolong the shelf-life of food grains, specific preservatives can be added to them. The preservatives act by inhibiting microbial contamination and decomposition. Butylated hydroxytoluene (BHT) at 0.02 percent concentration used as treatment in two cultivars of pearl millet stored for 90 days at ambient temperature showed FFA increase from 0.4 to 1.5 percent oleic acid, whereas the value in the untreated sample increased to 3.2 percent oleic acid.[23]

The disadvantages related to nutritional profile and functional changes caused by the conventional processing technologies can be mitigated by use of some advanced thermal and non-thermal treatments such as microwave processing, gamma radiation, pulsed light, ultrasonic processing and infrared heating. These treatments are mainly aimed at shelf-life stability by microbial inhibition and lipase reduction. Gamma irradiation can be used as a preserving method for food systems. A study on its applicability on fungal growth and quality characteristics on millet grains was studied that revealed a considerable reduction in FFA and fungal activity at dosage level above 0.5 kGy. The decrease in lipase content of the grain contributed to the reduction of FFA, thereby extending the shelf-life of the flour.[24] In another study, the effect of gamma radiation on different properties of pearl millet cultivars was evaluated. Results showed that intensity of color increased with the increase in radiation dosage until 4 kGy, but thereafter decreased until 8 kGy. Therefore, it was conceived as a storage-stable product and color being an index of storage.[25] In a study, microwave treatment was used to inactivate lipase enzyme in pearl millet. The treatment led to a significant decrease in the lipase activity under moisture level increment from 12 to 18 percent. Maximum reduction (92.9%) was observed at 18 percent moisture content at 100 s of treatment. Microwave exposure for 80 s duration at 18 percent moisture content was considered optimum based on lipase inactivation and pasting properties. During storage up to 30 days, significantly lowered change in FFA (20.80–22.25%, oleic acid) was observed in in flour of these grains as compared to control flour (20.11–32.43). From sensory evaluation, it was stated that the microwave-treated flour could be acceptably stored at ambient conditions for 30 days, while control flour produced unpleasant off odor and bitter taste on the 10th day of storage.[26] In another work, storage properties of millet flour were studied based on microwave treatment. The samples were treated at a constant power of 900W with different moisture contents and heating times, and it was found that the inactivation of lipase increased with increasing treatment time. During 30 days of storage, a similar decreasing trend for FFA was also recorded.[27]

4.2.1 Storehouse management

Molds, insects, rodents and other storage pests can cause spoilage in the grain and decrease the grain quality. Factors like storage temperature, moisture content of the grain, oxygen availability and presence of broken or cracked grains are responsible for safety of the stored grains. For rapid development and multiplication of the spoilage organisms, the above-mentioned factors are required. Therefore, proper management of the factors would construct unfavorable conditions for their attack. Temperatures above 42° C and below 15° C slow down reproduction and development, while above 45° C and below 10° C may kill the insects. However, heating the grains above 45° C can lose their viability. Proper drying of grains to about 10 percent moisture content can significantly prevent insect and pest attack. Rapid drying or over drying results in development of wrinkles and damage to the seeds, leading to cracks in the seed coat and destruction of vitamins and viability. Hence, cracked and broken grains are more prone to insect pest and microbial infestation. Therefore, proper screening and sorting of the grains is necessary to discard the broken grains from the lot. Use of chemicals for insect-pest management is considered the most effective one. Prophylactic and curative treatment methods are used. Since these cannot be directly sprayed on the grains, prophylactic treatment methods indirectly benefit by preventing insect infestation and cross infestation. While curative treatments use knockdown chemicals such as pyrethrum spray, lindane smoke generator or fumigant strips aimed against flying insects to immobilize or kill them. A fumigant is a chemical, which at required temperature and pressure, can exist in the gaseous state in sufficient concentrations to be lethal to a given pest organism. Infestation is a cycling problem; therefore, repeated use of chemical fumigants may cause problems which can be more severe than the insect contaminated grain itself and can cause serious residual hazards. Therefore, much care is needed in using fumigants and hazardous chemicals in storehouse facilities.[28]

4.3 Nutraceutical and therapeutic properties

As stated earlier, the occurrence of essential macronutrients and micronutrients in the pearl millet grain makes it a potential nutraceutical.[29] Pearl millet grain has been identified as a possible remedy for people with wheat allergies as it retains its alkaline properties after being cooked.[30] It is also gluten-free and ideal for use in products suitable for people suffering from celiac disease. It is high in calories and protein, which also make it favorable as a major part of staple diet, especially for pregnant women, sportspersons and infants requiring high energy diets. Its traditional usage as poultry, pig, cattle and sheep feed can be further explored and developed.[30] Whole as well as polished pearl millet grains contain high levels of a number of antioxidant compounds falling under subclasses of phenolic acids, tannins and phytates. They relate to anticancer properties and have been found to reduce risk of colon and breast cancer. Lower incidence of esophageal cancer has been reported in populations consuming sorghum and millet, compared to wheat and maize consuming communities. *In vitro* cancer initiation and progression can be prevented with the help of phenolics obtained from millets. [31,32,33] The presence of phytosterols, phytocyanins and polyphenols helps in boosting the human immune system and delaying age-related diseases. These compounds also act as detoxifying and antioxidant agents to reduce the risk of certain metabolic syndromes, including Parkinson's disease. High amounts of iron and zinc (8mg/100g and 3.1mg/100g), respectively, are present in pearl millet, which help in prevention of anemia by increasing hemoglobin.

Slow digestibility and a low glycemic index of pearl millet due to its higher dietary fibre content directs its potential consideration as an anti-diabetic crop.[34, 35] Additionally, its high amylase activity (which is about 10 times higher than that of wheat), maltose and D ribose and low fructose and glucose levels can improve the metabolic control of blood pressure and low density lipoprotein due to less pronounced insulin response.[36] These may help in managing maturity onset diabetes. It is also effective against cardiovascular diseases. Lee et al. (2010) reported that consumption of pearl millet helps in reducing plasma triglycerides in hyperlipidemic rats and hence prevented cardiovascular disease.[37] Inhibitory effects of pearl millet phenolic extracts on lipid peroxidation in *in vitro* copper-mediated low density lipoprotein (LDL) cholesterol oxidation in human and some model food systems (cooked comminuted pork and stripped corn oil) and a linoleic acid emulsion were evaluated. It was found that the extracts inhibited LDL cholesterol oxidation by 1 to 41 percent at a final concentration of 0.05mg/ml.[33] Pearl millet has better glycemic control compared to wheat and rice. Lower glycemic index (GI) is contributed by factors like the presence of slowly digestible starch and resistant starch in the genetically variable pearl millet grain. *In vivo* studies on rats demonstrated that native and treated starches from pearl millet when fed to rats showed the lowest blood glucose, serum cholesterol and triglycerides compared with rice and other minor millets. In another experiment, diabetic mice were fed diets containing added millet proteins and it was concluded that millet protein can reduce blood sugar levels as well as triglyceride levels, thereby increasing insulin sensitivity.[38] Abdelgadir et al. (2004) measured the effect of six Sudanese traditional carbohydrate rich meals on glucose and insulin response of diabetic patients and observed significant lowering of postprandial glucose and insulin responses upon consumption of a pearl millet porridge followed by wheat. However, higher postprandial glucose and insulin responses were observed upon maize porridge consumption.[39]

Omega-3 fatty acids also make pearl millet a potential source for prevention of diseases like diabetes, cardiovascular diseases, arthritis and certain types of cancer. Along with the flavonoids and phenolics, Omega-3 fatty acids help in inhibition of DNA scission, LDL cholesterol accumulation, liposome oxidation and proliferation of HT-29 adenocarcinoma cells.[40] Pearl millet contains a large amount of phosphorous, which is essential for bone growth and development and also for ATP generation in the body. The suggested therapeutic implication of pearl millet remains to be largely explored in products creating a great scope for its involvement in future nutraceutical industries.

4.4 Nutritional and phytochemical profile

4.4.1 Proximate composition

Proximate analysis of plant materials comprises determining the major classes of chemical components, which include moisture, crude protein crude fibre, ash and crude fat.[41] Pearl millet reportedly comprises dry matter (92.5%), ash (2.1%), crude fibre (2.8%), crude fat (7.8%), crude protein (13.6%), and starch (63.2%).[42] However, variations in the proximate analysis are possible due to differences in genotype, climatic conditions, soil nutrient content and type of processing. Properties of any processed or natural complex food matrix are primarily dependent on the proportion, structural state and interaction between the macro- and micro-components in it. In case of natural resources, compositions often vary with climate, soil, and growth phase among several other factors. Hence, the compositional data are seldom absolute.

Like any other cereal, the carbohydrates in pearl millet basically constitute starch (60–80%) with fractions of dietary fibre (~3%) and soluble sugars (<3%). Being the major fraction, starch mostly regulates the processed characteristics of the grain, which is thoroughly discussed later in this chapter. The apparent amylose content in starch has been reported to be within the range of 20–28 percent for different genotypes and cultivars of the grain. The starch forms a typical molecular arrangement (A-type), which is characteristic of cereals. The dietary fibre bears much importance as a chief health promoting agent for the consumer, as is also detailed later. Protein proportion ranging between 10–18 percent has been reported giving an average higher than popular grain crops such as rice, maize, sorghum and barley. Leucine, glutamic acid, proline, alanine and phenylalanine are present in the highest quantity in the pearl millet protein. As a grain protein, notably rich proportions of lysine, threonine, methionine and cysteine have been reported in it. As reported in an early study, arginine, threonine, valine, isoleucine and lysine in the protein show greater digestibility than the rest of the essential amino acids present.[43] Lipid content of 5–8 percent has been reported for pearl millet. The grain comprises higher fat content than other staple grains like wheat (<2%), rice (<1%) maize (<1%), sorghum (<2%). About 88 percent of this lipid fraction is found in the germ layer, whereas pericarp and endosperm contain almost 6 percent each. Among the whole fatty acid content, 75 percent accounts for unsaturated fat and 4 percent of the total fatty acids in its oil comprises Omega 3, linolenic acid (C18:3n-3) (LNA).[40,44,45,46] The pearl millet gives an ash content of 1.5–3.5 percent depending on the nature of soil and other stress factors. The minerals phosphorus, potassium, magnesium, calcium, sodium, iron, zinc, manganese and copper have been reported, of which the initial six are present in high proportions. The key vitamins are thiamin, niacin and riboflavin with reasonable content of tocopherol.[5,40] Pearl millet consists of 361Kcal/100g, which is higher than other commonly consumed cereals like wheat (346Kcal/100g), rice (345Kcal/100g), maize (125Kcal/100g), and sorghum (349Kcal/100g).

4.4.2 Fatty acid

This is higher than all other millets and most cereal grains. The lipids occur in free, bound, and structural forms are mostly concentrated in the germ. Of this, 85 percent are neutral (non-polar) lipids, 12 percent phospholipids and 3 percent glycolipids. Neutral lipids comprise about 85 percent triglycerides, smaller amounts of mono- and diglycerides, sterols and FFAs. Sterols comprise camesterol and stigmasterol. Lysophosphatidyl choline, phosphatidyl choline, and lysophosphatidyl ethanol amine are the major phospholipids. Sterol glycoside, sterol glucoside and mono- and digalactosyldiacyl glycerol are the major glycolipids present in this millet. Pearl millet triglycerides contain about 75 percent unsaturated fatty acids, mainly oleic (C18:1), linoleic (C18:2) and linolenic (C18:3) acids.[46,47,48] These properties of lipid in pearl millet make it almost equivalent to that of brown rice. The remaining fraction is made up of saturated fatty acid residues (palmitic (C16:0) and stearic (C18:0)). Omega-3 linolenic acid (C18:3n-3) represents 4 percent of the total fatty acids in this oil.[45,46] A higher content of n-3 fatty acids (α-linolenic acid, eicosapentaenoic acid, and docosahexaenoic acid) among cereals make pearl millet lipids physiologically advantageous. However, the presence of triglyercides imparts a negative effect to it, also affecting the shelf-life of milled four by autoxidation.[2,40]

4.4.3 Fibres

Fibres in food are important for healthy gastrointestinal activity and generation of metabolites having antioxidant and anticarcinogenic properties. Pearl millet contains 5g/100g of insoluble and 3g/100g soluble dietary fibres.[49] The process of germination increases the soluble fibre and decrease insoluble fibre content in pearl millet.[50] Insoluble (63.52g/100g) and soluble (1.63g/100g) dietary fibre are present in bran fraction as well as in endosperm fractions of pearl millet (5.47 and 1.36g/100g, respectively).[51] A study on cereals and pulses revealed that both soluble and insoluble dietary fibres affected zinc bio-accessibility in cereals, while in pulses only insoluble fibres showed the negative effect.[52]

4.4.4 Resistant starch

As cereal grains are predominantly starchy, the characteristics of starch primarily determine the processing and nutritional properties. Digestion of starch releases sugar into the blood stream. Resistant starch (RS) is the fraction of starch that cannot be hydrolyzed (digested) by the amylolytic enzymes of the human digestive system. Hence, RS does not liberate glucoses, and its presence in diet is important for designing foods for diabetic people. After surpassing the amylolytic zone, it often gets deposited in the large intestine (colon) and behaves as a dietary fibre. It is metabolized by the colon microflora. In an *in vitro* digestibility study on six Indian pearl millet varieties, RS content was found to be in the range of 6.03 to 11.06 percent.[53] In raw grains, the RS formation is generally considered to be positively correlated to the amylose content in starch. As the grain is consumed after processing, it is important to note that the RS in processed products depends on the type and severity of processing. Determination of RS in processed millet is necessary.

4.4.5 Polyphenols

Polyphenols and phytic acid content in pearl millet varies among different cultivars of the grain.[54] These compounds carry much importance for consumer nutrition as they are strong antioxidants and also act as anti-nutritional factors hindering protein digestibility and mineral absorption. Reportedly, polyphenols, tannin, phytic acid, phytates, goitrogens and oxalic acid are the antinutrients present in the millet grain. Pearl millet contains polyphenols in the range of 307–714mg/100g.[55,56] These are mainly concentrated in the bran (1.44g/100g) and also in the endosperm in lower concentrations (0.45g/100g) (Jha et al. 2015). Salar et al. (2017) studied the total phenolic content, condensed tannin content, antioxidant activity and DNA damage protection effects of 12 pearl millet cultivars from regions of North India and found that all the cultivars showed bioactive compounds with antioxidant potential when measured with different oxidant-reducing assays, namely DPPH, ABTS, FRAP, TAC and HFRSA. The highest amount of condensed tannins (138.45mg CE/100g DWB) was observed in HHB-223 cultivar, while maximum extractable total phenolic content (7.32mg GAE/g DWB) was observed in PUSA-415. The different pearl millet cultivars signified a clear correlation between total phenolic content and antioxidant activity.[57]

4.4.6 Bioactive peptides

Bioactive peptides are defined as the byproducts of biological sources that exert positive effects on the body functions and health of an individual. Fractions of albumin, globulin, cross linked prolamin, β- prolamin and glutelin are present in the protein fractions of millet grains. Additionally, the presence of high quality essential amino acids, including sulfur containing amino acids such as cysteine and methionine have been found from amino acid analysis. [58] Bioactive peptides are low molecular weight components formed by short sequence of amino acids linked together by amide or peptide bonds.[59] Bioactive millet peptides (BAMPs) exert biochemical effects, namely, antimicrobial, antioxidant, antihypertensive, ACE-inhibitory, antiproliferative/anticancer and antidiabetic effects on human health.[58] Further inhibitory effects *in vitro* have been found to be attributed to hydrolysates and bioactive peptides from millets. In a study, pepsin and pancreatic enzymes were used to hydrolyze millet protein *in vitro* and it was found that pearl millet protein hydrolysate contained higher amounts of leucine (128.04mg/g) and threonine (33.19mg/g) content.[60] In another study, trypsin enzyme was used to identify antioxidative bioactive peptide, which was SDRDLLGPNNQYLPK. These were assessed based on activity of DPPH radical, ABTS radical, hydroxyl radical, Fe^{2+} chelating ability and reducing power.[61]

4.4.7 Vitamins and minerals

Micronutrients in the form of vitamins and minerals are found in considerable amounts in pearl millet. Compared to other cereal grains, pearl millet contains comparatively higher amounts of overall mineral content (2.3mg/100g). Nutritionally important constituents like iron (0.02–0.01 g kg^{-1}), calcium (0.1–3 g kg^{-1}), phosphorus (2–3.4 g kg^{-1}) and zinc (0.3–0.6 g kg^{-1}) are present in the grain.[62] Experimental reports indicate vitamin constituents in the following range – niacin (0.9–11.1mg kg^{-1}), riboflavin (2.8–16mg kg^{-1}) and thiamine (1.5–6mg kg^{-1}).[62] Vitamin C (ascorbic acid) is absent in dried mature kernels. Since the B vitamins are concentrated in the aleurone layer and the germ, hulling by decortication reduces the contents of niacin, riboflavin, and thiamin by 50 percent. However, significant content of niacin remains after hulling, due to which consumers of millet can be observed to have not suffered from pellagra.[63] Being mainly concentrated in the germ, pericarp and aleurone layers, the mineral profile of the pearl millet gets affected during processing methods, especially milling. Whole pearl millet grains also have higher content of phytic acid due to occurrence of the intact aleurone layer.[51] Phytic acid reduces the mineral uptake in the human gut by forming complexes with the minerals and becomes insoluble at the intestine's physiological pH. Complete degradation of phytic acid resulted in a fivefold increase in mineral absorption, while a twofold increase in absorption was observed at 90 percent degradation. From their research, Hurrell et al. (2004) opined that since complete degradation of phytic acid is almost impossible, molar ratio of phytic acid to iron should be reduced from its native level to <1 or preferably <0.5 to achieve a two-fold increase in iron absorption.[64]

4.4.8 Phytochemicals

Phytochemicals present in pearl millet provide health-promoting antioxidant properties. Pearl millet contains a number of bioactive compounds like p-coumaric acid, gallic acid, ferulic acid, salicylic acid, vanillic acid, cinnamic acid, benzoic acid, catechol and so forth [62]. Present in the cell walls, p-coumaric acid is responsible for antimicrobial properties.[65] This functional compound decreases the low-density lipoproteins, free radical quenchers and antimicrobial agents.[66,67,68] Gallic acid and ferulic acid are important antioxidant constituents. Galic acid provides antiviral and antifungal properties. Ferulic acid provides anti-inflammatory properties owing to its hydrogen donating nature, making it a free radical quenching compound. Ferulic acid also helps in blood pressure management, diabetes control and protection of skin. Benzoates are popularly used as antimicrobial agents, a food preservative, flavoring agent and food additive. Ascorbic acid, the popular antioxidant is also an important cofactor for various biochemical reactions and hence provides many health benefits, including maintenance of cardiovascular and immune systems. Cinnamic acid present in bound forms in pearl millet matrix are found to be effective against diabetes and cardiovascular abnormalities. [69,70,71] Broad spectrum antibacterial properties effective against both gram-positive and gram-negative strains are provided by the extracts of syringic acids.[72] Symptoms of cellular damage and Parkinson's disease DNA damage are known to be prevented and slowed down by vanillic acid. The activity of free radicals during oxidative stress conditions can be lowered by catechols, as they possess antioxidant properties.[73,74] Therefore, pearl millet, owing to its pool of bioactive phytochemicals can be considered and distinguished as a nutraceutical grain.

4.5 Characterization of phytochemical compounds responsible for bioactive properties

The availability of macronutrients as discussed above generally makes pearl millet an nutritionally rich, commercially important grain for human consumption. Extensive research has been carried out on its nutritional attributes, and the grain has attained much global popularity in the food sector, especially in the initial decades of the 21st century. Apart from the general nutrition, the grain and its products have been found to have several nutraceutical properties owing to certain specific components present in them. However, rich but unique functional composition of millets also creates a few nutritional challenges within it and deserves obvious mention. In this section, nutraceutical components in pearl millet, their potent health promoting properties and related nutritional challenges are discussed.

4.5.1 Phenolic compounds

Phenolic compounds are plant secondary metabolites constituting phenolic acids and polyphenols. Hydroxybenzoic and hydroxycinnamic acids are the two types of phenolic acids. Flavonoids and tannins are the primary polyphenols in millets. These compounds are mostly concentrated in the pericarp, seed coat layers and the germ.[33] Diluted methanol has been most successfully used to extract phenolic compounds from millets. Hassan et al. (2020) used an advanced UPLC-MS technique to identify and quantify eight distinct phenolic compounds in methanolic extract of pearl millet flour.[75] Among these, benzoate derivatives namely protocatechuic acid (PCA) and parahydroxybenzoic acid (PHBA) and flavonoids, namely catechin, epicatechin and kaempferol glycoside were detected. All these compounds are strong antioxidants and impart related health promoting effects. Out of these, catechin was identified to be the compound with the highest quantity, followed by epicatechin. Catechin, also a major green tea and edible algae polyphenol has remarkable neuroprotective properties. It can impart pleiotropic effect at the molecular level to improve the signaling pathways. It can enter the neural cells and perform these mechanisms at the nucleic acid and cellular levels to cure neurodegenerative disorders, especially those associated with ageing.[76] Its activity as anticancer, antihypertensive, anti-inflammatory, antiproliferative, antithrombogenic, anti-hyperlipidemic and anti- several other metabolic disorders have been clinically and epidemiologically established.[77] Epicatechin is mostly known for its positive impact on endothelial cells, preventing platelet aggregation and improving vasodilation. It also maintains blood pressure and lower insulin resistance. Its strong antioxidant activity reduces inflammatory reactions in cells and inhibits abnormal growths or lysis.[78] Kaempferol modulates cellular signal transduction pathways linked to apoptosis, angiogenesis, inflammation, and metastasis. It maintains normal cell viability while inhibiting cancer cell growth.[79] PCA has been reported for its potential antibacterial, anticancer activity, antiulcer, antihyperglycemic, ageing suppressing, antifibrotic, antiviral, anti-inflammatory, anti-apoptotic, analgesic, antiatherosclerotic, cardiac and neural improvement, hepatoprotective, and nephron-protective activities.[80,81] P-hydroxybenzoic acid (PHABA) has been most commonly used as preservative with the common name "paraben" in cosmetic products like shaving cream, lubricants and toothpastes due to its antibacterial and antifungal properties.[82] Tannins and phytates, which have been known as major anti-nutritional factors present in several grains, including millets, also impart several positive effects on human health.[83,84] Apart from being carcinogen suppressants, many tannin and phyate molecules are capable of reducing mutagenic activities in the tissues, and many are antimicrobials.

Interestingly, the phenolics in pearl millet were reported to be more stable to cooking treatments than other major millets.[33] However, the overall antioxidant activities as obtained by biochemical oxidant-reducing assays were lower. Overall, the identified phenolic compounds of pearl millet may be primarily studied and targeted for their cumulative antidiabetic and anticarcinogenic properties. Enhancement of extraction efficiency by the use of modern extraction approaches like ultrasound, microwave and supercritical fluid can be used to maximize the yield. Phenolics extraction and isolation for use have been often suggested for millets. This is because their incorporation or presence in food matrices has been found to carry a number of disadvantages relating to anti-nutrition and toxicity.

Catechin and epicatechin have mild to high inhibitory effects on α-amylase hampering starch digestion.[85] Phytate can strongly chelate iron, calcium and zinc preventing them from absorption in any monogastric animal, including the human. Besides, phytate itself is not digestible by humans.[86] Tannin has a general tendency of precipitating proteins. As digestive enzymes are proteinous, tannins can highly affect the digestion of proteins, carbohydrates and lipids by inhibiting trypsin, α-amylase and lipase enzymes. PHABA has hepatotoxic effects which can be suppressed by use ginger extracts in the products containing it.[82]

4.5.2 Fatty acids

The pearl millet has a healthy fatty acid profile with about 75 percent unsaturated fatty acids in its oil extracted mainly from the germ. Of this, around 6–7 percent is omega-3 (linolenic) fatty acid which bears much significance for the nutraceutical properties of the grain. It has multifaceted biological effects towards improvement of human health. Its supplementation in oils and other products can be considered a notable approach towards value addition of processed foods across the globe. It primarily improves cardiovascular function and imparts special benefits to patients with hypertension, depression, dyslipidaemia, atherosclerosis, renal issues, schizophrenia, epilepsy, diabetes mellitus, metabolic syndrome, obesity, inflammatory diseases, neurological/ neuropsychiatric disorders and eye diseases.[87] It also lowers pregnancy risks and improves fetus development. Isolated omega-3 fatty acid has been successfully encapsulated for convenient incorporation in the food matrix.[88] Other important unsaturated fatty acids are mainly oleic (C18:1) and linoleic (C18:2) acids, which take active roles in wound healing, cancer and common inflammatory disease prevention, renal health, glycemic and obesity regulation and several other general physiological processes in the human body.[89]

4.5.3 Dietary fibre

Pearl millet contains about 5 percent insoluble and 3 percent soluble dietary fibres. They are fractions that are resistant to gastrointestinal (GI) digestive acids and enzymes. They hold the food servings longer in the upper GI tract, slow down digestive metabolism, and thereby regulate glycemic responses. In the colon, they act as prebiotic for the bacterial lining. Several resultant products of their bacterial degradation are anticarcinogens and immuno-regulatory. However, the general mechanism of their action can sometimes prove as anti-nutritional. The absorption inhibition of minerals such as iron and zinc by phytates is further enhanced by fractions of both soluble and insoluble fibres forming fibre-phytate-mineral (FPM) complex precipitates. Fibres are more concentrated in the pericarp where phytates and these two minerals are also primarily located, naturally increasing the risks of FPM complex formation. [35] Calcium is lesser affected due to its higher concentration in the endosperm region.

4.5.4 Resistant starch

Starch is the major component in pearl millet. Amylolytic digestive enzymes hydrolyze starch into glucose. Glucose is metabolized in the cells to liberate energy. Excessive glucose liberation due to surplus intake of starch, coupled with lower metabolic rate induces most diabetic and obesity conditions. Regulation of starch metabolism is therefore critical. Resistant starch (RS) are fractions within the starch matrix that are not hydrolyzed by the amylolytic enzymes due to limited accessibility into their complex molecular arrangements. The presence of RS in food thereby lowers glucose liberation and absorption by variable extents. RS has been often considered as a part of dietary fibre due to similar non-digestible and prebiotic characteristics.

Millets, including pearl millet, have been identified as a major alternative sustainable source for starch and starchy food industries.[90] RS content in raw Indian pearl millet starches have been found to range from 2–12 percent, which can further enhance, after controlled heat-moisture processing involving starch gelatinization and retrogradation.[53,91] These values are relatively higher than the common cereal starches. RS enhancement can be also achieved by selected chemical modification of starch. The higher concentration of proteins, lipids and polyphenolic compounds in pearl millet also delay starch digestion by inhibiting amylolytic enzymes.[92] These significantly relate to its lower glycemic index and usability in food products meant for diabetic patients or persons requiring delayed and limited postprandial blood glucose concentration.[93]

4.5.5 Bioactive peptides

Considering the issues related with trypsin inhibition resulting in lower pearl millet protein digestibility, extraneous extraction of bioactive peptides from its protein can be considered a promising solution. Bioactive peptides are enzymatically degraded polypeptide fractions containing 3–20 amino acid residues. They can be absorbed through the GI lining, surviving further enzymatic degradation and impart a wide range of functionalities. Some of their health enhancing functions include anti-obesity, antimicrobial, anthypertensive, antithrombotic and hypocholesterolemic and cyto- and immune-modulatory activities. The sequence of amino acids and their structure primarily determine their specific role as a functional compound.[94] High quality bioactive peptides from pearl millet protein have been extracted by pepsin and pancreatin hydrolysis. The peptides were further encapsulated in eudragit S100 microspheres for better applicability and efficacy of controlled release within the GI environment.[61]

4.5.6 Vitamins and minerals

The general need of vitamins and minerals for functioning of metabolic activities is well known. Like other cereal grains, pearl millet grain is also an important source of thiamin, niacin, and riboflavin.[9] Because of its high oil content, pearl millet is also a good source of lipid soluble vitamin E. However, riboflavin has been linked with lipid deterioration in the presence of light.[95] Thus, it may also be a potential enhancer of the deterioration of pearl millet triglycerides. Regarding mineral composition, phosphorus (~3.338mg/kg dry basis), potassium (~3.932mg/kg dry basis) and magnesium (~1.333mg/kg dry basis) are found in considerable quantities, while minerals such as calcium (~300mg/kg dry basis), iron (~18mg/kg dry basis) and zinc (~43mg/kg dry basis) are in relatively lower quantities than other common cereals.[96,97,98] Biofortification strategies have been implemented for enhancing content and bioavailability of zinc and iron in pearl millet.[99]

4.6 Techno-functional properties

Cereal grains are almost exclusively processed in kitchens and industries where thermal processing techniques are often involved. Techno-functional properties of grains are therefore of utmost importance for both processing and marketability of the grains. These properties also relate to consumer perceptions towards the raw material and its products.

4.6.1 Physical characteristics

Cultivated pearl millet grains generally have a moisture content of 10±2 percent which, how-ever, may vary with geographical and climatic conditions. Physical properties of two pearl millet varieties, namely Ex-Borno and SOSAT C88, were assessed.[100] For grains with 10 per-cent moisture, average ranges of grain length (3–4mm), width (2–3mm), thickness (1–2.5mm), sphericity (0.7), 1000 seed weight (7–10 g), bulk density (8–8.5 kg/m^3), true density (940–1030 kg/m^3), porosity (15.0–17.5 %) and angle of repose (30°) were reported. The authors also reported changes in these dimensional and other physical properties under different extents of hydration ranging from 10–20 percent for both the pearl millet varieties. Bulk density of pearl millet flour particles with size lower than 210 μ size were reported to range between 800–900 kg/m^3. The density values bear significance in packaging and transportation of grains and their flours.[101]

4.6.2 Color

A differential range of flour color profile were reported for five pearl millet varieties grown in India.[101] While the lightness (L*) values did not show much variation, the value of redness (a*) exhibited significant variation (p <0.05) among the varieties. This indicated differences in the pigment contents. Different levels of yellowness (b* = 9.9 to 13.5) were reported in the five varieties of the grain. Accordingly, the overall color difference (ΔE) values ranged between 52.5 to 75.1, suggesting appreciable color differences in the Indian pearl millet flours.

4.6.3 Water and oil absorption

The capacity of absorbing water and oil directly relates to any flour's process economy, besides various factors relating to its process efficiency. These are primarily governed by the quantita-tive and qualitative status of starch and protein in the grain and flour matrices. Water absorp-tion capacity of 153–157 percent and oil absorption capacity of 104–124 percent were reported for different pearl millet flours.[101] The latter is probably suggestive of different protein proportions and distribution in the grains. This was also supported by distinct variations on emulsifying and foaming capacities of the flours when their slurries were sufficiently agitated and intended for emulsification with addition of calculated amounts of oil.

4.6.4 Rheological

Flour and paste rheology of cereal flours directly reflect the conditions for their manual and machine handling during transportation and processing, respectively. Rheological studies measure the displacement behavior of substances on application of force. A number of instruments are used for measuring rheological characteristics of foods, including cereals. However, millets being more recently added to the list of popular cereals being scientific-ally investigated, thorough rheological investigation of the grains and flour are awaited. In one study, four different millet types including pearl millet were investigated using rotational viscometry.[102] The flour produced Non-Newtonian shear thinning type paste with higher viscosity than wheat flour taken as reference. Rudra et al. (2018) carried out on-line rheology measurement of two pearl millet flours during extrusion processing.[103] Direct indications of effect of amylose content in starch (24.42% and 15.73%) and crystallinity on shear thinning and consistency upon processing were reported. Stress-strain relationships of flour paste system also significantly rely on protein fractions among the non-starch components. This is particu-larly appropriate for millets as they have notably higher non-starch components. In a rapid viscosity analysis performed with slurry of pearl millet flour sized lower than 250 µ, pasting temperature, peak viscosity, trough viscosity, break down viscosity, final viscosity and setback viscosity values of 89.6° C, 429 cP, 390 cP, 39 cP, 983 cP and 593 cP were reported, respectively. [104] All the values were lower than that of finger millet flour taken for the same study.

4.6.5 Thermal

Thermal properties of any material that is subjected to thermal processing is of utmost importance. Studies using differential scanning calorimetry (DSC) and thermogravimetry (TGA) target identifying heat migration and molecular disassembly-induced weight loss in materials, respectively. Melting and gelatinization are the key thermal changes occurring in starchy materials and products. The gelatinization endotherm for pearl millet flour has been reported to exist between 71.35° C and 85.22° C with maximum intensity at 76.15° C.[105] A second transition peak, possibly representing melting of starch-lipid complexes was also observed. However, gelatinization transition peak positions in DSC are also dependent on the extent of moistening and other factors.[106] In TGA analysis, three distinct stages of weight loss with increasing temperature were reported.[107]

In the first stage (room temperature to 252° C) mass loss up to 8.31 percent was recorded. This is due to moisture evaporation. The second stage showed the largest mass loss of up to 52.26 percent, largely attributed to cellulose degradation and gasification of volatile components. The third stage (700–1035° C) designated lignin degradation and biochar formation to an extent of 49.26 percent with weight loss up to 4.86 percent. The degradation rates were, however, largely dependent on the specific heating rates applied by the researchers.

4.6.6 Structural

All macro-components in foods, including carbohydrates, proteins and fats, are polymeric in nature. They often exhibit a semicrystalline nature owing to variable arrangements of the polymeric chains in ordered (crystalline) and non-ordered (amorphous) fashions. Being the major component, starch plays the key role in deciding crystallinity of any cereal grain matrix and regulates several techno-functional attributes of the flour. The extent and pattern of crystalline distribution within the starch matrix is commonly measured by the X-ray diffraction (XRD) technique. The pearl millet starch exhibits an A-type diffraction pattern with major diffraction angles at and near 2θ values of 15, 17, 18 and 23, representing the typical pattern for cereal starches. However, hydrothermal processes can lead to alteration of this native crystallinity with development of V-type crystalline structures showing a diffraction peak at 2θ near 20.[108] Fourier transform infrared (FTIR) spectroscopy has been largely used to determine the types and intensity of bonds existing in food compounds. Gull et al. (2016) reported characteristic IR vibrational bands at positions of 3466.95 cm^{-1}, 2925.84 cm^{-1} and near 1661 cm^{-1} in the spectrum of pearl millet flour, which represent O-H, C-H and amine 1 group in its chemical structure.[104]

4.7 Potential utilization in value added food products

Nutrient digestibility and their bioavailability are equally important for deciding the selection of a food for inclusion in diet. As mentioned earlier, researches have sufficiently evidenced that the presence of phytates, tannins and polyphenols in pearl millet greatly lower the bioavailability of proteins and minerals and are also responsible for and inherent bitterness.

Besides, the triglycerides and fatty acids with high inherent lipase activity result in rancidity development in stored flour and products. Suitable physical, chemical and biological processes (**Table 4.1**) have been developed for suppressing the activities of anti-nutritional factors and lipase and enhance nutrient bioavailability and storability.[109]

4.7.1 Germination and malting

Germination and malting (controlled germination) result in enzymatic breakdown of starch into dextrins and simple sugars, which enhances the nutrient contents, including fibre, crude fat, vitamins B and C, minerals in millets and their bioavailability.[110,111] Malting also results in a higher protein efficiency ratio, reduction in soluble oxalates (0.502% to 0.068%) and increased soluble calcium (2.4 to 14.1mg/100g) in the grain.[111,112] Hassan et al. (2006) reported reduction in phytic acid, tannins and polyphenols after germination, soaking, debranning and dry heating of millets.[113] Reduced water holding capacity and high energy density of the flour, the factors required for the production of infant foods, weaning foods and enteral foods have been found to be present in germinated millet flours, making it a potential ingredient in such formulations. Cakes, confectionary and milk-based beverages can also be produced from malted pearl millet flours.[15]

4.7.2 Blanching

Blanching is considered as one of the basic and effective processing techniques to enhance the shelf-life of pearl millet, since it reduces the enzyme activity keeping the nutritional value intact.[109] It is done by submerging the grains in hot water for a specified short period, followed by drying.[15] In one study, blanching temperature and a time combination of 98° C and 10 seconds before milling effectively enhanced the shelf-life by 25 days by retarding fat acidity development.[113] High heat treatments result in slow lipase activity and reduction of lipids during storage. Exposure of pearl millet grains to dry heat treatments above 90° C could reduce the polyphenols and phytic acid contents.[114]

4.7.3 Fermentation

The fermentation process begins with malting followed by souring with mixed cultures of yeast and Lactobacilli.[15] The process decreases the antinutrients level and increases the nutritive value and *in vitro* protein digestibility in millets.[112] Enzymes help in degradation of the starch and soluble sugars from the grain and fermenting media in pearl millet, leading to reduction in phytic acid and increase in phosphorous content of the grain.[115] Fermentation of pearl millet flour using mixed cultures of *Saccharomyces diastaticus*, *Saccharomyces cerevisiae*, *Lactobacillus brevis* and *Lactobacillus fermentum* resulted in improved bioavailability, protein efficiency ratio, apparent protein digestibility, feed efficiency ratio, true protein digestibility, net protein utilization, net protein retention, protein retention efficiency, and utilizable protein values in the case of pure-culture-fermented pearl millet flour.[116]

4.7.4 Product profile and quality modulation in multi-grain systems

Various types of food products, traditional and non-traditional, are prepared using processed pearl millet grains and their meals (**Table 4.2**). Thick porridge, thin porridge, steam cooked products, fermented and unfermented breads, boiled rice-like products, alcoholic and non-alcoholic beverages and snacks are the nine major classes of food.[124] Ready-to-eat (RTE) products can be prepared using pearl millet grits or flour with optional coating to prepare

Table 4.1 Process Induced Inactivation of Antinutrients in Pearl Millet

Processing Type	Processing conditions	Substance inhibited/ reduced	Extent of inhibition	Reference
Dry heat treatment	150-170°C, for 90s	Total fungal count	during initial treatment: 48.23% and gradual decrease up to 60th day of storage	[117]
Hydrothermal treatment	100°C for 120min	Free fatty acid (FFA)	63.58% reduction	[22]
	Holding in boiling water for 15 min, tray drying at 60°C, 2h	Lipase activity (LA) inhibited	11.5% reduction in fat acidity	[17]
	Cooking at 90°C–95°C, 120 min	Total viable count	23.7% reduction	[118]
Germination and malting	Soaking in 0.1% formaldehyde solution, for 6h at 25–30°C	Free fatty acid (FFA)	66.25% reduction	[22]
	Steeping in 0.1 % formaldehyde water solution for 36h	Free fatty acid (FFA)	Below 1% FFA reduction up to 8th day of storage	[119]
Fermentation	Malting: steeping 24h at 25°C	Tannins, phytates	Phytate reduction: 20–21% with 2 days fermentation;	[18]
	Germination: 72h at 25°C; Natural fermentation, 25°C, 48h	pH	19–33% with 3 days malting	
	Steeping in water, 4 days fermentation Inoculum: *Lactobacillus plantarum*,	Lipid content, pH	For starter culture sample: Lipid content - 97% reduction; pH: 3.64 to 3.48; For naturally fermented sample: Lipid content- 93% reduction; pH: 4.94 to 4.53	[120]
Preservative addition	Acid treatment with 0.2N HCl, 24 h at 25–30°C	Free fatty acid (FFA)	73.97% reduction	[22]
Microwave treatment	900W, 100s	Lipase activity (LA)	92.9% reduction	[26]
Gamma- irradiation	Heat treatment combined with irradiation dose at 2.5kGy	Total fungal count	90.56% reduction	[117]

Source: Table created by authors.

Table 4.2 Reported Usage of Pearl Millet in Food Systems

Type of Food	Composition and Incorporation levels	Key changes/ findings	Reference
Ready-to-eat (RTE) breakfast cereal	Popped pearl millet (100g), popped amaranth, puffed wheat, flax seeds, sunflower seeds, raisins, honey, sugar, oil, and water	High amount of folate High amount of dietary fibre (6.13±0.14) Highly acceptable sensory attributes	[121]
Cookies	Semi refined pearl millet flour (SRPMF), sugar fat, vanilla powder. Control: Refined wheat flour	Higher protein, ash and mineral (iron, calcium and phosphorous); and better sensory attributes compared to control	[20]
Instant Kheer mix	20g pearl millet, 15 g sugar, 30g dairy whitener	Consistency: 3241.7 g.sec; Cohesiveness: 262.96g; Viscosity index: 241.07 g.sec. Overall acceptability: 7.66	[122]
Chicken nuggets	0, 10 and 20 per cent of Bajra (pearl millet) flour	Cooking yield higher: bajra flour containing sample Increased level of bajra flour: increased emulsion stability 10 %level of inclusion: optimum for preparation; better texture, juiciness and overall acceptability	[123]
Beef sausage	Roasted pearl millet flour (RoPMF) at 0%, 5%, 10% and 15% to beef meat.	Higher amounts of Fe, Mg and Ca in RoPMF sausage compared to control. Highest Zn in 5% RoPMF sausage Reduced production cost.	[124]
Oti-oka like beverage (fermented)	400g pearl millet: dried, germinated, milled, diluted with water (1:5w/v), sieved, boiled (100°C, 3h) Divided into 4 portions. 3 portions inoculated with starter culture (L. fermentum and S. cerevisiae), 1 portion allowed for natural fermentation (25 ± 2 °C for 72)	Protein content: 7.50% and calcium content 8.76mg/100ml. Phytate content reduced from 1.13±0.15mg/100ml to 0.60±0.10mg/100ml; Polyphenol content reduced from 1.33±0.64mg/100ml to 0.50±0.10mg/100ml; Tannin content reduced from 1.83±0.17mg/100ml to 1.00±0.11mg/100ml. Better taste and aroma	[125]

Source: Table created by authors.

sweet or savory and spicy snacks with crunchy texture. Blends of other protein rich ingredients along with millet flour can be prepared to produce nutritionally balanced food upon extrusion.[125] An RTE breakfast cereal was developed by mixing and baking followed by cooling of popped pearl millet, popped amaranth, puffed wheat, flax seeds, sunflower seeds, raisins, honey, sugar, oil and water.[119] This RTE product was nutritionally rich and contained total dietary fibre (6.13±0.14), calcium (62.79±0.06mg/100g), phosphorous (239.53±0.88mg/100g), iron (5.02±0.02mg/100g), folic acid (103.21±0.04mg/100g) and zinc (3.24±0.03mg/100g) in significant amounts. Brenan et al. (2012), developed an extruded breakfast cereal product using pearl millet, amaranth and buckwheat as a replacement for wheat and maize flour.[126] Significant reduction in readily digestible carbohydrate and slowly digestible carbohydrates was seen in the extruded products from pseudo cereals during *in vitro* glycemic profiling. In another study, a blend of 30 percent grain legume flour or 15 percent defatted soybean along with pearl millet was used to prepare extruded products. The products showed 14.7 percent and 16.0 percent protein, and 2.0 and 2.1 protein efficiency ratio respectively with a shelf-life of

6 months.[127] The results also indicated that protein digestibility of foods could be enhanced by extrusion cooking.[125] Development of multigrain gluten free chapati (flat bread) using pearl millet and rice along with addition of soybean (for protein content) and amaranth (for calcium and iron content) showed pearl millet as a potential grain to be incorporated in product designed for diabetic patients. This was confirmed by a considerable decrease in the blood glucose levels upon consumption of multigrain chapati.[128] Supplementation of the hydrated, dried pearl millet flour with 0.6 percent unrefined soy lecithin was reported to produce cookies with acceptable spread characteristics.[4] Blanched and malted pearl millet can be used to produce various types of biscuits and cakes. Developed products have been found to be organoleptically acceptable with storage life of at least three months.[129] Pearl millet can be used for producing noodles, macaroni and pasta-like extruded products. A process was developed to prepare noodles from sorghum and pearl millet.[130] Both sorghum and pearl millet noodles were highly acceptable in savory and sweet formulations. The solid loss of the noodles was found to be less than 6 percent, which was considered as highly acceptable. [131, 132, 133]

Formulations of pearl millet with meat have been studied. Pearl millet flour was used at 0 percent, 10 percent and 20 percent to produce chicken nuggets. pH, cooking yield, emulsion stability and sensory properties of the prepared noodles were recorded. Compared to the control, cooking yield was significantly higher at 20 percent level of incorporation. A gradual decreasing nonsignificant trend was seen for values of color and appearance and flavor at 10 percent level of incorporation, whereas texture juiciness and overall acceptability were higher. Hence, it was stated that addition of 10 percent pearl millet flour over and above the chicken meat is preferable for preparing chicken nuggets.[119] Yaro et al. (2021) observed improved mineral contents in beef sausages, when blended with roasted pearl millet flour (RoPMF).[120] Blending was carried out at 0 percent, 5 percent, 10 percent and 15 percent. RoPMF sausages showed higher amounts of iron, magnesium and calcium compared to control, while 5 percent and 15 percent showed higher potassium content. This product was considered to be high in mineral content, low in cost and sensorily acceptable.

Fermented beverages have been also prepared using pearl millet. "Oti-oka," a traditional sorghum-based beverage was prepared by substituting sorghum with pearl millet using combined starter culture of *L. fermentum* and *S. cerevisiae*. Reduction in phytate content from 1.13±0.15mg/100ml to 0.60±0.10mg/100ml, polyphenol content from 1.33±0.64mg/100ml to 0.50±0.10mg/100ml and tannin content from 1.83±0.17mg/100ml to 1.00±0.11mg/100ml were recorded. The beverage had a protein content of 7.50 percent and a calcium content of 8.76mg/100ml and was assumed to be acceptable in terms of nutritive, anti-nutritive contents and sensory properties.[121]

4.8 Conclusion

Traditionally used for feed and forage, pearl millet has gained acceptance in modern human diets as a nutrient-rich cereal. Its potential to survive high temperatures and drought makes it more useful in areas where other crops fail to persist. It is rich in several macro- and micronutrients such as carbohydrates, dietary fibres, proteins, phytochemicals, vitamins and minerals and has been regarded as a "nutri-cereal." Apart from these nutrients, the presence of a few anti-nutritional factors like phytate, tannins and polyphenols limit its utilization by reducing the availability of minerals. Additionally, the presence of the oil rich germ layer hampers the storage stability of the grain and its flour, resulting in production of off flavors under exposure to high moisture and oxygen. Therefore, specific processing treatments have been designed and applied to improve its oxidative, storage and microbial stability. Processing the grain encourages commercial utilization, thereby ensuring food security to the consumers. Pearl millet being a gluten free and nutrient rich grain provides various health benefits to consumers. Since it retains its alkaline properties after being cooked, it becomes an ideal food for people with celiac disease. The presence of antioxidant phenolic compounds in the grain provides multidimensional health benefits relating to oxidant-suppression. Due to its high iron and zinc content in the grain, it can prevent anemia and other mineral deficiency diseases. The presence of slowly digestible starch and resistant starch contribute to the lower glycemic index in the grain, which helps in lowering blood glucose, serum cholesterol and triglycerides, which are beneficial factors for diabetic patients. Inhibition of tumor development has also been found due to the presence of certain phenolic compounds like flavonoids. Additionally, various diseases like cardiovascular diseases, diabetes, arthritis and certain types of cancer can be potentially prevented due to the presence of omega-3 fatty acids in the grain. The techno-functional flour properties have been explored creating scope for future researches on its applicability in processed food systems.

References

1. Taylor, John R. N., and M. Naushad Emmambux. 2008. "Gluten-Free Foods and Beverages from Millets." In *Gluten-Free Cereal Products and Beverages*, edited by Elke K. Arendt and Fabio Dal Bello, 119–V. San Diego, CA: Elsevier. DOI: 10.1016/b978-012373739-7.50008-3
2. Punia, Sneh, Anil Kumar Siroha, Kawaljit Singh Sandhu, Suresh Kumar Gahlawat, and Maninder Kaur, eds. 2020. *Pearl Millet: Properties, Functionality and Its Applications*. CRC Press. DOI: 10.1201/9780429331732
3. Pattanashetti, Santosh K., Hari D. Upadhyaya, Sangam Lal Dwivedi, Mani Vetriventhan, and Kothapally Narsimha Reddy. 2016. "Pearl Millet." In *Genetic and Genomic Resources for Grain Cereals Improvement*, 253–89. Elsevier. DOI: 10.1016/b978-0-12-802000-5.00006-x
4. Rai, K. N., C. L. L. Gowda, B. V. S. Reddy, and S. Sehgal. 2008. "Adaptation and Potential Uses of Sorghum and Pearl Millet in Alternative and Health Foods." *Compr Rev Food Sci Food Saf* 7: 340–352. DOI: 10.1111/j.1541-4337.2008.00049.x
5. Hassan, Z. M., N. A. Sebola, and M. Mabelebele. 2021. "The Nutritional Use of Millet Grain for Food and Feed: A Review." *Agriculture and Food Security* 10 (1): 16. DOI: 10.1186/s40066-020-00282-6
6. Yadav, O. P., and K. N. Rai. 2013. "Genetic Improvement of Pearl Millet in India." *Agricultural Research* 2 (4): 275–92. DOI: 10.1007/s40003-013-0089-z
7. Saleh, Ahmed S. M., Qing Zhang, Jing Chen, and Qun Shen. 2013. "Millet Grains: Nutritional Quality, Processing, and Potential Health Benefits: Millet Grains...." *Comprehensive Reviews in Food Science and Food Safety* 12 (3): 281–95. DOI: 10.1111/1541-4337.12012
8. Claessens, Geert. 2014. "Pearl Millet." *Cgiar.Org*. September 20. https://cropgenebank.sgrp.cgiar.org/index.php/regeneration-guidelines-of-crops/pearl-millet-mainmenu-409.
9. Taylor, J. R. N. 2004. "MILLET | Pearl." In *Encyclopedia of Grain Science*, 253–61. Elsevier. DOI: 10.1016/b0-12-765490-9/00097-5
10. Kaur, Harpreet, Balmeet Singh Gill, and Brij Lal Karwasra. 2018. "In Vitro Digestibility, Pasting, and Structural Properties of Starches from Different Cereals." *International Journal of Food Properties* 21 (1): 70–85. DOI: 10.1080/10942912.2018.1439955
11. Khairwal, I. S., S. K. Yadav, K. N. Rai, H. D. Upadhyaya, D. Kachhawa, B. Nirwan, R. Bhattacharjee, B. S. Rajpurohit, C. J. Dangaria, and Srikant. 2007. "Evaluation and Identification of Promising Pearl Millet Germplasm for Grain and Fodder Traits." *Journal of Semi-Arid Tropical Agricultural Research* 5 (1): 1–6.
12. Satyagopal, K., S. N., P. Sushil, G. Jeyakumar, O. P. Shankar, S. K. Sharma, D. R. Sain, et al. 2014. "AESA Based IPM Package for Pearl Millet." Department of Agriculture and Cooperation, Ministry of Agriculture, Government of India.
13. Meera, M. S., S. Z. Ali, H. V. Narasimha, M. K. Bhashyam, A. Srinivas, and B. V. Sathyendra Rao. 2002. "A Process for the Preparation of Sorghum and Pearl Millet Grain-Based Product Having Enhanced Shelf Life." *Parent File in India and Egypt. Indian Patent* 306 (DEL/02).

14. Babiker, Elfadil, Babiker Abdelseed, Hayat Hassan, and Oladipupo Adiamo. 2018. "Effect of Decortication Methods on the Chemical Composition, Antinutrients, Ca, P and Fe Contents of Two Pearl Millet Cultivars during Storage." *World Journal of Science Technology and Sustainable Development* 15 (3): 278–86. DOI: 10.1108/wjstsd-01-2018-0005

15. Dias-Martins, Amanda M., Kênia Letícia F. Pessanha, Sidney Pacheco, José Avelino S. Rodrigues, and Carlos Wanderlei Piler Carvalho. 2018. "Potential Use of Pearl Millet (Pennisetum Glaucum (L.) R. Br.) in Brazil: Food Security, Processing, Health Benefits and Nutritional Products." *Food Research International* (Ottawa, Ont.) 109: 175–86. DOI: 10.1016/j.foodres.2018.04.023

16. Kumar, Ashwani, Vidisha Tomer, Amarjeet Kaur, Vikas Kumar, and Kritika Gupta. 2018. "Millets: A Solution to Agrarian and Nutritional Challenges." *Agriculture and Food Security* 7 (1). DOI: 10.1186/s40066-018-0183-3

17. Jalgaonkar, K., S. K. Jha, and D. K. Sharma. 2016. "Effect of Thermal Treatments on the Storage Life of Pearl Millet (Pennisetum Glaucum) Flour." *Indian Journal of Agricultural Sciences* 86 (6): 762–767.

18. Onyango, C. A., S. O. Ochanda, M. A. Mwasaru, J. K. Ochieng, F. M. Mathooko, and J. N. Kinyuru. 2013. "Effects of Malting and Fermentation on Anti-Nutrient Reduction and Protein Digestibility of Red Sorghum, White Sorghum and Pearl Millet." *Journal of Food Research* 2 (1): 41. DOI: 10.5539/jfr.v2n1p41

19. Ojokoh, A., and B. Bello. 2014. "Effect of Fermentation on Nutrient and Anti-Nutrient Composition of Millet (Pennisetum Glaucum) and Soyabean (Glycine Max) Blend Flours." *Journal of Life Sciences* 8 (27): 3566–3570. DOI: 10.17265/1934-7391/2014.08.005

20. Suma, P. Florence, and Asna Urooj. 2014. "Influence of Germination on Bioaccessible Iron and Calcium in Pearl Millet (Pennisetum Typhoideum)." *Journal of Food Science and Technology* 51 (5): 976–81. DOI: 10.1007/s13197-011-0585-8

21. Owheruo, Joseph O., Beatrice O. T. Ifesan, and Ayodele O. Kolawole. 2019. "Physicochemical Properties of Malted Finger Millet (Eleusine Coracana) and Pearl Millet (Pennisetum Glaucum)." *Food Science and Nutrition* 7 (2): 476–82. DOI: 10.1002/fsn3.816

22. Bhati, Dashrath, Vibha Bhatnagar, and Vibha Acharya. 2016. "Effect of Pre-Milling Processing Techniques on Pearl Millet Grains with Special Reference to Iron Availability." *Journal of Dairying, Foods and Home Sciences* 35 (1). DOI: 10.18805/ajdfr.v35i1.9256

23. Abdalgader, Mosab Abbas Ahmed, Syed Amir Ashraf, and Amir Mahgoub Awadelkareem. 2019. "Effect of Natural and Synthetic Antioxidant on Shelf Life of Different Sudanese Pennisetum Glaucum L. Flours." *Bioscience Biotechnology Research Communications* 12 (3): 652–57. DOI: 10.21786/bbrc/12.3/15

24. Mahmoud, N. S., S. H. Awad, R. M. A. Madani, F. A. Osman, K. Elmamoun, and A. B. Hassan. 2016. "Effect of γ Radiation Processing on Fungal Growth and Quality Characteristics of Millet Grains." *Food Sciences and Nutrition* 4 (3): 342–347. DOI: 10.1002/fsn3.295

25. Falade, K. O., and T. A. Kolawole. 2013. "Effect of γ-Irradiation on Colour, Functional and Physico-Chemical Properties of Pearl Millet [Pennisetum Glaucum (L) R." Br.] *Cultivars. Food and Bioprocess Technology* 6 (9): 2429–2438. DOI: 10.1007/s11947- 012-0981-8

26. Yadav, Deep N., Tanupriya Anand, Jaspreet Kaur, and Ashish K. Singh. 2012. "Improved Storage Stability of Pearl Millet Flour through Microwave Treatment." *Agricultural Research* 1 (4): 399–404. DOI: 10.1007/s40003-012-0040-8

27. Kashaninejad, M. 2019. "Increasing the Shelf-Life of Millet Flour by Using Heat-Moisture and Microwave Treatments." *Food Science and Technology* 16 (86): 83–93.

28. Sharma, H. C., Alur, A. S., Ravinder Reddy, C., Jayaraj, K., Varaprasad, V. J., Varaprasad Reddy, K. M., Reddy, B. V. S., and Rai, K. N. 2007. "Management of Sorghum and Pearl Millet Pests in Bulk Storage." In *Patancheru.* 502–324, Andhra Pradesh, India: 20 pp.

29. Prasad, Madhulika Esther, Ayyanadar Arunachalam, and Pankaj Gautam. 2020. "Millet: A Nutraceutical Grain That Promises Nutritional Security." *Eco. Env. & Cons.* 26 (October Suppl. Issue): S1–6.

30. Mishra et al. 2016 Mishra Shaivya, Mishra Sunita. 2016. "Nutritional analysis of value-added product by using pearl millet, quinoa and prepare ready-to-use upma mixes." *International Journal of Food Science and Nutrition*; 1(5): 41–43.

31. Graf, E., and J. W. Eaton. 1990. "Antioxidant Functions of Phytic Acid." *Free Radical Biology and Medicine* 8 (1): 61–69. DOI: 10.1016/0891-5849(90)90146-a

32. Sharma, A., and A. C. Kapoor. 1996. "Levels of Antinutritional Factors in Pearl Millet as Affected by Processing Treatments and Various Types of Fermentation." *Plant Foods for Human Nutrition* 49 (3): 241–52. DOI: 10.1007/bf01093221

33. Chandrasekara, Anoma, Marian Naczk, and Fereidoon Shahidi. 2012. "Effect of Processing on the Antioxidant Activity of Millet Grains." *Food Chemistry* 133 (1): 1–9. DOI: 10.1016/j.foodchem.2011.09.043

34. Jiri, Obert, P. L. Mafongoya, and P. Chivenge. 2017. "Climate Smart Crops for Food and Nutritional Security for Semi-Arid Zones of Zimbabwe." *African Journal of Food Agriculture Nutrition and Development* 17 (03): 12280–94. DOI: 10.18697/ajfand.79.16285

35. Krishnan, Rateesh, and M. S. Meera. 2018. "Pearl Millet Minerals: Effect of Processing on Bioaccessibility." *Journal of Food Science and Technology* 55 (9): 3362–72. DOI: 10.1007/s13197-018-3305-9

36. Oshodi, A. A., H. N. Ogungbenle, M. Oladimeji, and O. 1999. "Chemical Composition, Nutritionally Valuable Minerals and Functional Properties of Benniseed, Pearl Millet and Quinoa Flours." *Int. J. Food Sci. Nutr* 50: 325–331. DOI: 10.1080/096374899101058

37. Lee, Sun Hee, Ill-Min Chung, Youn-Soo Cha, and Yongsoon Park. 2010. "Millet Consumption Decreased Serum Concentration of Triglyceride and C-Reactive Protein but Not Oxidative Status in Hyperlipidemic Rats." *Nutrition Research* 30 (4): 290–96. DOI: 10.1016/j.nutres.2010.04.007

38. Nishizawa, Naoyuki, Tubasa Togawa, Kyung-Ok Park, Daiki Sato, Yo Miyakoshi, Kazuya Inagaki, Norimasa Ohmori, Yoshiaki Ito, and Takashi Nagasawa. 2009. "Dietary Japanese Millet Protein Ameliorates Plasma Levels of Adiponectin, Glucose, and Lipids in Type 2 Diabetic Mice." *Bioscience, Biotechnology, and Biochemistry* 73 (2): 351–60. DOI: 10.1271/bbb.80589

39. Abdelgadir, M., M. Abbas, A. Jarvi, M. Elbagir, M. Eltom, and C. Berne. 2004. "Glycaemic and Insulin Responses of Six Traditional Sudanese Carbohydrate-Rich Meals in Subjects WithType 2 Diabetes Mellitus." *Diabet Med* 22: 213–7. DOI: 10.1111/j.1464-5491.2004.01385.x

40. Nambiar, V., Dhaduk, JJ., Sareen, N., and Desai, T. (2011). "Potential Functional Implications of Pearl millet (Pennisetum glaucum) in Health and Disease." *Journal of Applied Pharmaceutical Science* 01 (10); 2011: 62–67 ISSN: 2231-3354. 1.

41. Bhattacharya, Leena, and Mukta Arora. 2005. "Nutritional Composition of Winged Bean, Psophocarpus Tetragonolobus (L) DeCandole." *Asian Journal of Dairy and Food Research* 24: 11–15.

42. Ali, Maha A. M., Abdullahi H. El Tinay, and Abdelwahab H. Abdalla. 2003. "Effect of Fermentation on the in Vitro Protein Digestibility of Pearl Millet." *Food Chemistry* 80 (1): 51–54. DOI: 10.1016/s0308-8146(02)00234-0

43. Ejeta, G., M. M. Hassen, and E. T. Mertz. 1987. "In Vitro Digestibility and Amino Acid Composition of Pearl Millet (Pennisetum Typhoides) and Other Cereals." *Proceedings of the National Academy of Sciences of the United States of America* 84 (17): 6016–19. DOI: 10.1073/pnas.84.17.6016

44. Abdelrahman, A., R. C. Hoseney, and E. Variano-Marston. 1983. "Milling Process to Produce Low-Fat Grits from Pearl Millet." *Cereal Chem* 60 (3): 189–191.

45. Burton, Glenn W., A. T. Wallace, and K. O. Rachie. 1972. "Chemical Composition and Nutritive Value of Pearl Millet (Pennisetum Typhoides (Burm.) Stapf and E. c. Hubbard) Grain 1." *Crop Science* 12 (2): 187–88. DOI: 10.2135/cropsci1972.0011183x001200020009x

46. Rooney, L. W. 1978. "Sorghum and Pearl Millet Lipids." *Cereal Chem* 55 (5): 584–590.

47. Lai, C. C., and E. Variano-Marston. 1980. "Changes in Pearl Millet during Storage. 57: 275." *Cereal Chem* 57 (4): 275–77.

48. Kapoor, R., and A. Kapoor. 1990. "Effect of Different Treatments on Keeping Quality of Pearl Millet Flour." *Food Chemistry* 35 (4): 277–86. DOI: 10.1016/0308-8146(90)90017-x

49. Nandini, Chilkunda D., and Paramahans V. Salimath. 2001. "Carbohydrate Composition of Wheat, Wheat Bran, Sorghum and Bajra with Good Chapati/Roti (Indian Flat Bread) Making Quality." *Food Chemistry* 73 (2): 197–203. DOI: 10.1016/s0308-8146(00)00278-8

50. Pushparaj, Florence Suma, and Asna Urooj. 2011. "Influence of Processing on Dietary Fiber, Tannin and in Vitro Protein Digestibility of Pearl Millet." *Food and Nutrition Sciences* 02 (08): 895–900. DOI: 10.4236/fns.2011.28122

51. Jha, Neha, Rateesh Krishnan, and M. S. Meera. 2015. "Effect of Different Soaking Conditions on Inhibitory Factors and Bioaccessibility of Iron and Zinc in Pearl Millet." *Journal of Cereal Science* 66: 46–52. DOI: 10.1016/j.jcs.2015.10.002

52. Hemalatha, Sreeramaiah, Kalpana Platel, and Krishnapura Srinivasan. 2007. "Influence of Heat Processing on the Bioaccessibility of Zinc and Iron from Cereals and Pulses Consumed in India." *Journal of Trace Elements in Medicine and Biology: Organ of the Society for Minerals and Trace Elements* (GMS) 21 (1): 1–7. DOI: 10.1016/j.jtemb.2006.10.002

53. Sandhu, Kawaljit Singh, and Anil Kumar Siroha. 2017. "Relationships between Physicochemical, Thermal, Rheological and in Vitro Digestibility Properties of Starches from Pearl Millet Cultivars." *Lebensmittel-Wissenschaft Und Technologie [Food Science and Technology]* 83: 213–24. DOI: 10.1016/j.lwt.2017.05.015

54. El Hag, Mardia E., Abdullahi H. El Tinay, and Nabila E. Yousif. 2002. "Effect of Fermentation and Dehulling on Starch, Total Polyphenols, Phytic Acid Content and in Vitro Protein Digestibility of Pearl Millet." *Food Chemistry* 77 (2): 193–96. DOI: 10.1016/s0308-8146(01)00336-3

55. Abdelrahman, S. M., H. B. Elmaki, W. H. Idris, E. E. Babiker, and Tinay AHEl. 2005. "Antinutritional Factors Content and Minerals Availability of Pearl Millet (Pennisetum Glaucum) as Influenced by Domestic Processing Methods and Cultivation." *J Food Technol* 3: 397–403.

56. Chowdhury, S., and D. Punia. 1997. "Nutrient and antinutrient composition of pearl millet grains as affected by milling and baking." *Die Nahrung* 41 (2): 105–07. DOI: 10.1002/food.19970410210

57. Salar, Raj Kumar, and Sukhvinder Singh Purewal. 2017. "Phenolic Content, Antioxidant Potential and DNA Damage Protection of Pearl Millet (Pennisetum Glaucum) Cultivars of North Indian Region." *Journal of Food Measurement and Characterization* 11 (1): 126–33. DOI: 10.1007/s11694-016-9379-z

58. Majid, Abdul, and Poornima Priyadarshini C G. 2020. "Millet Derived Bioactive Peptides: A Review on Their Functional Properties and Health Benefits." *Critical Reviews in Food Science and Nutrition* 60 (19): 3342–51. DOI: 10.1080/10408398.2019.1686342

59. Ofosu, Fred Kwame, Fazle Elahi, Eric Banan-Mwine Daliri, Akanksha Tyagi, Xiu Qin Chen, Ramachandran Chelliah, Joong-Hark Kim, Sang-Ik Han, and Deog-Hwan Oh. 2021. "UHPLC-ESI-QTOF-MS/MS Characterization, Antioxidant and Antidiabetic Properties of Sorghum Grains." *Food Chemistry* 337 (127788): 127788. DOI: 10.1016/j.foodchem.2020.127788

60. Agrawal, Himani, Robin Joshi, and Mahesh Gupta. 2016. "Isolation, Purification and Characterization of Antioxidative Peptide of Pearl Millet (Pennisetum Glaucum) Protein Hydrolysate." *Food Chemistry* 204: 365–72. DOI: 10.1016/j.foodchem.2016.02.127

61. Agrawal, Himani, Robin Joshi, and Mahesh Gupta. 2020. "Functional and Nutritional Characterization of in Vitro Enzymatic Hydrolyzed Millets Proteins." *Cereal Chemistry* 97 (6): 1313–23. DOI: 10.1002/cche.10359

62. Sandhu, Kawaljit Singh, Anil Kumar Siroha, Sneh Punia, and Manju Nehra. 2020. "Effect of Heat Moisture Treatment on Rheological and in Vitro Digestibility Properties of Pearl Millet Starches." *Carbohydrate Polymer Technologies and Applications* 1 (100002): 100002. DOI: 10.1016/j.carpta.2020.100002

63. Léder, Irén. 2004. "Sorghum and Millets." In *Cultivated Plants, Primarily as Food Sources*, edited by G. Fuleky. Encyclopedia of Life Support Systems (EOLSS).

64. Hurrell. 2004. "Phytic Acid Degradation as a Means of Improving Iron Absorption." *International Journal for Vitamin and Nutrition Research. Internationale Zeitschrift Fur Vitamin- Und Ernahrungsforschung. Journal International de Vitaminologie et de Nutrition* 74 (6): 445–52. DOI: 10.1024/0300-9831.74.6.445

65. Chandrasekara, Anoma, and Fereidoon Shahidi. 2011. "Bioactivities and Antiradical Properties of Millet Grains and Hulls." *Journal of Agricultural and Food Chemistry* 59 (17): 9563–71. DOI: 10.1021/jf201849d

66. Boz, Hüseyin. 2015. "P-Coumaric Acid in Cereals: Presence, Antioxidant and Antimicrobial Effects." *International Journal of Food Science and Technology* 50 (11): 2323–28. DOI: 10.1111/ijfs.12898

67. Gani, A., S. M. Wani, F. A. Masoodi, and G. Hameed. 2012. "Wholegrain Cereal Bioactive Compounds and Their Health Benefits: A Review." *Journal of Food Process and Technology* 3: 146. DOI: 10.4172/2157-7110.1000146

68. Roy, Abhro Jyoti, and P. Stanely Mainzen Prince. 2013. "Preventive Effects of P-Coumaric Acid on Cardiac Hypertrophy and Alterations in Electrocardiogram, Lipids, and Lipoproteins in Experimentally Induced Myocardial Infarcted Rats." *Food and Chemical Toxicology: An International Journal Published for the British Industrial Biological Research Association* 60: 348–54. DOI: 10.1016/j.fct.2013.04.052

69. Alam, Md Ashraful, Nusrat Subhan, Hemayet Hossain, Murad Hossain, Hasan Mahmud Reza, Md Mahbubur Rahman, and M. Obayed Ullah. 2016. "Hydroxycinnamic Acid Derivatives: A Potential Class of Natural Compounds for the Management of Lipid Metabolism and Obesity." *Nutrition and Metabolism* 13 (1). DOI: 10.1186/s12986-016-0080-3

70. Loader, Tara B., Carla G. Taylor, Peter Zahradka, and Peter J. H. Jones. 2017. "Chlorogenic Acid from Coffee Beans: Evaluating the Evidence for a Blood Pressure-Regulating Health Claim." *Nutrition Reviews* 75 (2): 114–33. DOI: 10.1093/nutrit/nuw057

71. Meng, Shengxi, Jianmei Cao, Qin Feng, Jinghua Peng, and Yiyang Hu. 2013. "Roles of Chlorogenic Acid on Regulating Glucose and Lipids Metabolism: A Review." *Evidence-Based Complementary and Alternative Medicine: ECAM* 2013: 801457. DOI: 10.1155/2013/801457

72. Manuja, R., S. Sachdeva, A. Jain, and J. Chaudhary. 2013. "A Comprehensive Review on Biological Activities of P-Hydroxy Benzoic Acid and Its Derivatives." *International Journal of Pharmaceutical Sciences Review and Research* 22: 109–115.

73. Li, Maoguo, Fang Ni, Yinling Wang, Shudong Xu, Dandan Zhang, Shuihong Chen, and Lun Wang. 2009. "Sensitive and Facile Determination of Catechol and Hydroquinone Simultaneously under Coexistence of Resorcinol with a Zn/Al Layered Double Hydroxide Film Modified Glassy Carbon Electrode." *Electroanalysis* 21 (13): 1521–26. DOI: 10.1002/elan.200804573

74. Salar, Raj Kumar, and Sukhvinder Singh Purewal. 2016. "Improvement of DNA Damage Protection and Antioxidant Activity of Biotransformed Pearl Millet (Pennisetum Glaucum) Cultivar PUSA-415 Using Aspergillus Oryzae MTCC 3107." *Biocatalysis and Agricultural Biotechnology* 8: 221–27. DOI: 10.1016/j.bcab.2016.10.005

75. Hassan, Zahra Mohammed, Nthabiseng Amenda Sebola, and Monnye Mabelebele. 2020. "Assessment of the Phenolic Compounds of Pearl and Finger Millets Obtained from South Africa and Zimbabwe." *Food Science and Nutrition*, no. fsn3.1778. DOI:10.1002/fsn3.1778

76. Andrade, José Paulo, and Marco Assunção. 2015. "Green Tea Effects on Age-Related Neurodegeneration." In *Diet and Nutrition in Dementia and Cognitive Decline*, 915–24. Elsevier. DOI: 10.1016/b978-0-12-407824-6.00084-7

77. Zanwar, Anand A., Sachin L. Badole, Pankaj S. Shende, Mahabaleshwar V. Hegde, and Subhash L. Bodhankar. 2014. "Antioxidant Role of Catechin in Health and Disease." In *Polyphenols in Human Health and Disease*, 267–71. Elsevier. DOI: 10.1016/b978-0-12-398456-2.00021-9

78. Abdulkhaleq, Layth Abdulmajeed, Mohammed Abdulrazzaq Assi, Mohd Hezmee Mohd Noor, Rasedee Abdullah, Mohd Zamri Saad, and Yun Hin Taufiq-Yap. 2017. "Therapeutic Uses of Epicatechin in Diabetes and Cancer." *Veterinary World* 10 (8): 869–72. DOI: 10.14202/vetworld.2017.869-872

79. Chen, Allen Y., and Yi Charlie Chen. 2013. "A Review of the Dietary Flavonoid, Kaempferol on Human Health and Cancer Chemoprevention." *Food Chemistry* 138 (4): 2099–2107. DOI: 10.1016/j.foodchem.2012.11.139

80. Kakkar, Sahil, and Souravh Bais. 2014. "A Review on Protocatechuic Acid and Its Pharmacological Potential." *ISRN Pharmacology* 2014: 952943. DOI: 10.1155/2014/952943

81. Semaming, Yoswaris, Patchareewan Pannengpetch, Siriporn C. Chattipakorn, and Nipon Chattipakorn. 2015. "Pharmacological Properties of Protocatechuic Acid and Its Potential Roles as Complementary Medicine." *Evidence-Based Complementary and Alternative Medicine: ECAM* 2015: 593902. DOI: 10.1155/2015/593902

82. Shivashankara, A. R., R. Haniadka, R. Fayad, P. L. Palatty, R. Arora, and M. S. Baliga. 2013. "Hepatoprotective Effects of Zingiber Officinale Roscoe (Ginger): A Review." In *Bioactive Food as Dietary Interventions for Liver and Gastrointestinal Disease*, 657–71.

83. Chung, K. T., T. Y. Wong, C. I. Wei, Y. W. Huang, and Y. Lin. 1998. "Tannins and Human Health: A Review." *Critical Reviews in Food Science and Nutrition* 38 (6): 421–64. DOI: 10.1080/10408699891274273

84. Markiewicz, Lidia Hanna, Anna Maria Ogrodowczyk, Wiesław Wiczkowski, and Barbara Wróblewska. 2021. "Phytate and Butyrate Differently Influence the Proliferation, Apoptosis and Survival Pathways in Human Cancer and Healthy Colonocytes." *Nutrients* 13 (6): 1887. DOI: 10.3390/nu13061887

85. Ayim, Ishmael, Haile Ma, Evans Adingba Alenyorege, and Yuqing Duan. 2019. "In Vitro Inhibitory Effect of Tea Extracts on Starch Digestibility." *Journal of Food Process Engineering* 42 (4): e13023. DOI: 10.1111/jfpe.13023

86. Sorour, M., A. Mehanni, E. Taha, and A. Rashwan. 2017. "Changes of Total Phenolics, Tannins, Phytate and Antioxidant Activity of Two Sorghum Cultivars as Affected by Processing." *Journal of Food and Dairy Sciences* 8 (7): 267–74. DOI: 10.21608/jfds.2017.38699

87. Yashodhara, B. M., S. Umakanth, J. M. Pappachan, S. K. Bhat, R. Kamath, and B. H. Choo. 2009. "Omega-3 Fatty Acids: A Comprehensive Review of Their Role in Health and Disease." *Postgraduate Medical Journal* 85 (1000): 84–90. DOI: 10.1136/pgmj.2008.073338

88. Vieira, Maiana da Costa, Karstyn Kist Bakof, Natielen Jacques Schuch, Jovito Adiel Skupien, and Carina Rodrigues Boeck. 2020. "The Benefits of Omega-3 Fatty Acid Nanocapsulation for the Enrichment of Food Products: A Review." *Revista de Nutrição* 33. DOI: 10.1590/1678-9865202033e190165

89. Sales-Campos, Helioswilton, Patrícia Reis de Souza, Bethânea Crema Peghini, João Santana da Silva, and Cristina Ribeiro Cardoso. 2013. "An Overview of the Modulatory Effects of Oleic Acid in Health and Disease." *Mini Reviews in Medicinal Chemistry* 13 (2): 201–10. DOI: 10.2174/1389557713804805193

90. Kaimal, Admajith M., Arun S. Mujumdar, and Bhaskar N. Thorat. 2021. "Resistant Starch from Millets: Recent Developments and Applications in Food Industries." *Trends in Food Science and Technology* 111: 563–80. DOI: 10.1016/j.tifs.2021.02.074

91. Sharma, Monika, Deep N. Yadav, Ashish K. Singh, and Sudhir K. Tomar. 2015. "Rheological and Functional Properties of Heat Moisture Treated Pearl Millet Starch." *Journal of Food Science and Technology* 52 (10): 6502–10. DOI: 10.1007/s13197-015-1735-1

92. Annor, George Amponsah, Massimo Marcone, Eric Bertoft, and Koushik Seetharaman. 2014. "Unit and Internal Chain Profile of Millet Amylopectin." *Cereal Chemistry* 91 (1): 29–34. DOI: 10.1094/cchem-08-13-0156-r

93. Punia, Sneh, Manoj Kumar, Anil Kumar Siroha, John F. Kennedy, Sanju Bala Dhull, and William Scott Whiteside. 2021. "Pearl Millet Grain as an Emerging Source of Starch: A Review on Its Structure, Physicochemical Properties, Functionalization, and Industrial Applications." *Carbohydrate Polymers* 260 (117776): 117776. DOI: 10.1016/j.carbpol.2021.117776

94. Varelis, Peter, Laurence Melton, and Fereidoon Shahidi. 2018. *Encyclopedia of Food Chemistry.* Netherlands: Elsevier.

95. Hamilton, R. J. 1999. "The Chemistry of Rancidity in Foods." In *Rancidity in Foods*, edited by J. C. Allen and R. J. Hamilton, 1–21. Gaithersburg, MD: Aspen Publishers.

96. Ragaee, S., E. Abdelaal, and M. Noaman. 2006. "Antioxidant Activity and Nutrient Composition of Selected Cereals for Food Use." *Food Chemistry* 98 (1): 32–38. DOI: 10.1016/j.foodchem.2005.04.039

97. Saldivar, S. O. Serna. 2016. "Cereals: Dietary Importance." In *Encyclopedia of Food and Health*, 703–11. Elsevier. DOI: 10.1016/b978-0-12-384947-2.00130-6

98. Taylor, J. R. N. 2016. *Millet Pearl: Overview, Encyclopedia of Food Grains.* 2nd Edn). 694. Oxford: Academic Press.

99. Ullah, Asmat, Ashfaq Ahmad, Tasneem Khaliq, and Javaid Akhtar. 2017. "Recognizing Production Options for Pearl Millet in Pakistan under Changing Climate Scenarios." *Journal of Integrative Agriculture* 16 (4): 762–73. DOI: 10.1016/s2095-3119(16)61450-8

100. Ojediran, J. O., M. A. Adamu, and D. L. Jim-George. 2021. "Some Physical Properties of Pearl Millet (Pennisetum Glaucum) Seeds as a Function of Moisture Content." *African Journal of General Agriculture* 6 (1): 39–46.

101. Siroha, Anil Kumar, Kawaljit Singh Sandhu, and Maninder Kaur. 2016. "Physicochemical, Functional and Antioxidant Properties of Flour from Pearl Millet Varieties Grown in India." *Journal of Food Measurement and Characterization* 10 (2): 311–18. DOI: 10.1007/s11694-016-9308-1

102. Chakraborty, Subir Kumar, Nachiket Kotwaliwale, and Surekha Ashok Navale. 2018. "Rheological Characterization of Gluten Free Millet Flour Dough." *Journal of Food Measurement and Characterization* 12 (2): 1195–1202. DOI: 10.1007/s11694-018-9733-4

103. Rudra, Shalini Gaur, Srikrishna Nishani, and B. P. Singh. 2018. "Online Rheology of Pearl Millet Flours during Extrusion: Effect of Native Amylose." *Journal of Food Process Engineering* 41 (8): e12924. DOI: 10.1111/jfpe.12924

104. Gull, Amir, Kamlesh Prasad, and Pradyuman Kumar. 2016. "Evaluation of Functional, Antinutritional, Pasting and Microstructural Properties of Millet Flours." *Journal of Food Measurement and Characterization* 10 (1): 96–102. DOI: 10.1007/s11694-015-9281-0

105. Kharat, Swapnil, Ilce G. Medina-Meza, Ryan J. Kowalski, Arunkumar Hosamani, Ramachandra, Sharanagouda Hiregoudar, and Girish M. Ganjyal. 2019. "Extrusion Processing Characteristics of Whole Grain Flours of Select Major Millets (Foxtail, Finger, and Pearl)." *Food and Bioproducts Processing* 114: 60–71. DOI: 10.1016/j.fbp.2018.07.002

106. Roy, Mrinmoy, Himjyoti Dutta, Rangarajan Jaganmohan, Monisha Choudhury, Nitin Kumar, and Avinash Kumar. 2019. "Effect of Steam Parboiling and Hot Soaking Treatments on Milling Yield, Physical, Physicochemical, Bioactive and Digestibility Properties of Buckwheat (Fagopyrum Esculentum L.)." *Journal of Food Science and Technology* 56 (7): 3524–33. DOI: 10.1007/s13197-019-03849-9

107. Boubacar Laougé, Zakari, and Hasan Merdun. 2020. "Kinetic Analysis of Pearl Millet (Penissetum Glaucum (L.) R. Br.) under Pyrolysis and Combustion to Investigate Its Bioenergy Potential." *Fuel*, London, 267 (117172): 117172. DOI: 10.1016/j.fuel.2020.117172

108. Jideani, V. A., and D. J. Scott. 2012. "Hydrothermal Characteristics of Pearl Millet (Pennisetum Glaucum) Flour during Cooking into 'Fura.'" *African Journal of Agricultural Research* 7 (18): 2751–2763. DOI: 10.5897/AJARX11.935

109. Rani, Savita, Rakhi Singh, Rachna Sehrawat, Barjinder Pal Kaur, and Ashutosh Upadhyay. 2018. "Pearl Millet Processing: A Review." *Nutrition and Food Science* 48 (1): 30–44. DOI: 10.1108/nfs-04-2017-0070

110. Suma, F. and Urooj, A. (2017). "Impact of household processing methods on the nutritional characteristics of pearl millet (*Pennisetum typhoideum*): A Review." *MOJ Food Processing and Technology.* 4: 28–32. DOI: 10.15406/mojfpt.2017.04.00082

111. Kumari, S. and Srivastava, S. (2000). "Nutritive Value of Malted Flours of Finger Millet Genotypes and Their Use in the Preparation of Burfi." *Journal of Food Science and Technology.* 37. 419–422.

112. Rao, N. B. S. 1987. "Nutritional implications of millets and pseudo millets." *Nutr News* 8(5):1–3.

113. Opoku, A. R., S. O. Ohenhen, and N. Ejiofor. 1981. "Nutrient Composition of Millet (Pennisetum Typhoides) Grains and Malt." *Journal of Agricultural and Food Chemistry* 29 (6): 1247. DOI: 10.1021/jf00108a036

114. Hassan, Amro B., Isam A. Mohamed Ahmed, Nuha M. Osman, Mohamed M. Eltayeb, Gammaa A. Osman, and Elfadil E. Babiker. 2006. "Effect of Processing Treatments Followed by Fermentation on Protein Content and Digestibility of Pearl Millet (Pennisetum Typhoideum) Cultivars." *Pakistan Journal of Nutrition:* PJN 5 (1): 86–89. DOI: 10.3923/pjn.2006.86.89

115. Kadlag, R. V., J. K. Chavan, and D. P. Kachare. 1995. "Effects of Seed Treatments and Storage on the Changes in Lipids of Pearl Millet Meal." *Plant Foods for Human Nutrition* 47 (4): 279–85. DOI: 10.1007/bf01088264

116. Ndiaye, Mame Mor Diarra, Babacar Diallo, Said Abboudi, and Dorothé Azilinon. 2018. "Theoretical and Experimental Study of a Double Air-Pass Solar Thermal Collector with an Insulating Rod of Millet." *Energy and Power Engineering* 10 (03): 106–19. DOI: 10.4236/epe.2018.103008

117. Khetarpaul, Neelam, and B. M. Chauhan. 1990. "Effect of Germination and Fermentation on in Vitro Starch and Protein Digestibility of Pearl Millet." *Journal of Food Science* 55 (3): 883–84. DOI: 10.1111/j.1365-2621.1990.tb05261.x

118. Dikkala, P. K., T. V. Hymavathi, P. Roberts, and M. Sujatha. 2018. "Effect of Heat Treatment and Gamma Irradiation on the Total Bacterial Count of Selected Millet Grains (Jowar, Bajra and Foxtail." *International Journal of Current Microbiology and Applied Sciences* 7 (2): 1293–1300. DOI: 10.20546/ijcmas.2018.702.158

119. Akinola, Stephen A., Adebanjo A. Badejo, Oluwatooyin F. Osundahunsi, and Mojisola O. Edema. 2017. "Effect of Preprocessing Techniques on Pearl Millet Flour and Changes in Technological Properties." *International Journal of Food Science and Technology* 52 (4): 992–99. DOI: 10.1111/ijfs.13363

120. Tiwari, Ajita, S. K. Jha, R. K. Pal, Shruti Sethi, and Lal Krishan. 2014. "Effect of Pre-Milling Treatments on Storage Stability of Pearl Millet Flour: Pearl Millet Flour, Storage, Phytic Acid, Polyphenol." *Journal of Food Processing and Preservation* 38 (3): 1215–23. DOI: 10.1111/jfpp.12082

121. Onuoha, Elizabeth Chinenye. 2017. "Effect of Fermentation (Natural and Starter) on the Physicochemical, Anti-Nutritional and Proximate Composition of Pearl Millet Used for Flour Production." *American Journal of Bioscience and Bioengineering* 5 (1): 12. DOI: 10.11648/j.bio.20170501.13

122. Kumari, Ritu, Karuna Singh, Rashmi Singh, Neelam Bhatia, and M. S. Nain. 2019. "Development of Healthy Ready-to-Eat (RTE) Breakfast Cereal from Popped Pearl Millet." *Indian Journal of Agricultural Sciences* 89 (5): 877–81.

123. Bunkar, D. S., Jha, A., and Mahajan, A. (2014). "Optimization of the formulation and technology of pearl millet based 'ready-to-reconstitute' kheer mix powder." *Journal of Food Science and Technology*, 51(10), 2404–2414. DOI: 10.1007/s13197-012-0800-2

124. Yaro, Joseph, Frederick Adzitey, and Nurul Huda. 2021. "Roasted Pearl Millet Flour (RoPMF) Improved the Mineral Composition of Beef Sausages." *International Journal on Advanced Science, Engineering and Information Technology* 11 (2): 769. DOI: 10.18517/ijaseit.11.2.13703

125. Temitope, Ogunbanwo Samuel, and Ogunsanya Bashirat Taiyese. 2012. "Quality Assessment of 'Oti-Oka' like Beverage Produced from Pearl Millet." *Journal of Applied Biosciences* 51: 3608–3617.

126. Murty, D. S., and K. A. Kumar. 1995. *Traditional Uses of Sorghum and Millets*. Edited by D. A. V. Dendy, Sorghum Ed., and Millets. Chemistry and Technology, American Association of Cereal Chemists, St. Paul.

127. Malleshi, N. G., N. A. Hadimani, R. Chinnaswamy, and C. F. Klopfenstein. 1996. "Physical and Nutritional Qualities of Extruded Weaning Foods Containing Sorghum, Pearl Millet, or Finger Millet Blended with Mung Beans and Non-Fat Dried Milk." *Plant Foods Human Nutr* 49: 181–89.

128. Brennan, Margaret A., Carine Menard, Gaëlle Roudaut, and Charles S. Brennan. 2012. "Amaranth, Millet and Buckwheat Flours Affect the Physical Properties of Extruded Breakfast Cereals and Modulates Their Potential Glycaemic Impact." *Die Starke* 64 (5): 392–98. DOI: 10.1002/star.201100150

129. Sumathi, A., S. R. Ushakumari, and N. G. Malleshi. 2007. "Physico-Chemical Characteristics, Nutritional Quality and Shelf-Life of Pearl Millet Based Extrusion Cooked Supplementary Foods." *International Journal of Food Sciences and Nutrition* 58 (5): 350–62. DOI: 10.1080/09637480701252187

130. Johari, Aanchal, and Asha Kawatra. 2016. "Development of Value-Added Pearl Millet and Rice-Based Gluten Free Porridge." *Journal of Food Research and Technology* 04 (02): 47–51.

131. Singh, G. 2003. "*Development and Nutritional Evaluation of Value-Added Products from Pearl Millet (Pennisetum Glaucum).*" Hisar, Haryana, India: CCS Haryana Agricultural Univ.

132. Sowbhaghya, C. M., Ali, S. Z. 2001a. "Vermicelli noodles and their quality characteristics." *J Food Sci Technol* 38:423–32.

133. Sowbhaghya, C. M., Ali, S. Z. 2001b. "A process for manufacturing high fibre low fat vermicelli noodles from millets." *Indian Patent* 1128/DEL/99.

5

Finger Millet (*Eleusine coracana*)
Phytochemical Profile, Potential Health Benefits, and Techno-Functional Properties

T. Tamilselvan, Shivani Sharma, and P. Prabhasankar

Contents

5.1 Introduction

Finger millet (*Elusine coracana* Gaertn L.), belonging to the family of genus Eleusine, is a cereal crop grown for food in Africa and Southern Asia. As per archaeological evidence, the origin and domestication of finger millet have roots about 5,000 years ago in Africa. Finger millet was brought to India from Africa around 3,000 years ago and is traditionally a staple millet popularly known as Ragi and Madua in India. Finger millet belongs to the Poaceae family under the subfamily Chloridoideae. It is an annually grown crop that matures in around 3 to 4 months and is 40 to 150 cm tall. Finger millet has spikes ranging from 3 to 20, containing approximately 70 spikelets. Its resemblance with the fingers of the hand gave this millet its common name, "Finger Millet." Each spikelet of finger millet carries around 4 to 7 seeds with varying diameters of 1 to 2 mm (Dida and Devos 2006). It is an adaptable crop grown from 500 to 2400 meters above sea level and on a wide range of tropical soils with annual rainfall as low as 500 mm, thus making it a famine crop. It is also known as a poor man's crop and can be stored for years without insect damage. Among many other finger millets, red-colored finger millets are the most widely cultivated ones throughout the world.

India, Niger and China are the world's top millet producers, contributing 55 percent of global millet production. India alone accounts for 37 percent of the total world production making it the largest producer of millet. Considering India's production area, finger millet is ranked sixth after wheat, maize, sorghum, rice, and bajra (Rathore et al. 2019). Among the various minor millets produced, finger millet (ragi) accounts for about 85 percent of production in India. The production of finger millet in India is 1.98 million tons over an area of 1.19 million hectares giving average productivity of 1661 kg per ha. In India, Ragi is extensively cultivated in Karnataka, Andhra Pradesh, Tamil Nadu, Uttarakhand, Orissa, Maharashtra, and Gujarat. The area and production of finger millet in India is led by Karnataka, which accounts for 56.21 and 59.52 percent, followed by Tamil Nadu (9.94% and 18.27%), Uttarakhand (9.40% and 7.76%), and Maharashtra (10.56% and 7.16%), respectively (Sakamma et al. 2018).

Finger millet is also considered a crop of great importance because of its high nutritional and nutraceutical significance. Finger millet has multilayer (5 layers) testa present, making it unique compared to foxtail, pearl, kodo, and proso millets, and it could also be the reason for the high dietary fiber content of the millet. The dietary fiber content of finger millet is 18g/100g. Protein content of finger millet varies considerably, depending upon its variety, and ranges from 7g/100g to 13g/100g (Chandra et al. 2016). Finger millet contains about 65–75 percent carbohydrates, 1.3–2 percent fat, and 2.5–3.5 percent minerals. In terms of minerals, protein, and carbohydrates, the nutritional value of finger millet is comparable with other staple crops such as wheat and rice. Finger millet has an excellent micronutrient profile with mineral and essential amino acid content better than major crops like wheat and rice. Finger millet has a uniquely high calcium content (344mg/100g) among other cereals. Finger millet contains phytates (0.48%), polyphenols, tannins (0.61%), and trypsin inhibitory factors, which are responsible for the nutraceutical potential of finger millet. Usually finger millet is milled with its seed coat intact, which is a good source of phytochemicals such as dietary fiber and polyphenols (0.2–3.0%). Also, seed coat contributes to the free radical scavenging activity of the millet, which is an essential factor in health, aging, and metabolic diseases (Devi et al. 2011). Finger millet has low glycemic index and therefore is considered suitable for diabetic population. Also, finger millet is gluten-free, making it an excellent wheat substitute for those who have celiac disease and individuals sensitive to gluten. The nutritional composition shows that finger millet has a great nutraceutical potential and can help combat many health issues like diabetes, hypertension, celiac disease, obesity and so forth. Despite being an economically important crop with well-documented nutraceutical potential, this crop is underutilized; still, it has a great potential to fight food insecurity and hidden hunger in the future (Gupta et al. 2017). Finger millet also has the potential to be utilized in different food formulations for value addition.

5.2 Nutraceutical properties

5.2.1 Antioxidant property

Antioxidant activity is defined as limiting or inhibiting nutrient oxidation by restraining oxidative chain reactions (Guclu, Kelebek, and Selli 2021). Oxidative stress occurs when free radicals and innate antioxidant capacity are imbalanced. Antioxidants can alleviate or neutralize free radicals, which are naturally unstable molecules produced in the body during energy production. Free radicals can quickly attack macromolecules, nucleic acids, and DNA, resulting in the primary cause of various disorders and chronic diseases. To avoid oxidative stress, increasing the dietary consumption of antioxidants is the best way to maintain balance. Different phytochemicals such as polyphenols, phenolic acids, flavonoids, tocopherols, and ascorbic acid have antioxidant activity. In recent times, consumer attention towards millets as an essential source of vital nutrients and phytochemicals helped gain commercial interests. In comparison to other major cereals, finger millet has rich phytochemical profile. Finger Millet phenolics can act as an antioxidant in both food and biological systems. It serves this purpose by various mechanisms such as scavenging the free radicals. acting as reducing agents, and as metal ion chelating agents (Chandrasekara and Shahidi 2010; Kumari, Madhujith, and Chandrasekara 2019). Due to its smaller size, it is widely consumed without dehulling. Also, the whole grain flour and dark brown hull of finger millet are rich in polyphenols. Finger millet seed coat can be used as a source of natural antioxidant as it is concentrated in polyphenol content and has significantly higher antioxidant activity than whole flour (Chethan and Malleshi 2007; Viswanath, Urooj, and Malleshi 2009). Chandrasekara and Shahidi (2012) studied the antioxidant potential of seven millet varieties, including finger millet. They observed that all millet extracts inhibit the peroxidation of LDL cholesterol from 1 to 41 percent. Subba Rao and Muralikrishna (2002) compared native and malted finger millet in terms of antioxidant activity using autoxidation of beta-carotene and linoleic acid emulsion and changes in the free and bound phenolic acid content. During the 96 h malting period, free phenolic acids showed significantly higher antioxidant activity than bound phenolic acids. Nassarawa and Sulaiman (2019) found higher reducing power in finger millet extracts and suggested regular consumption of millet varieties in the human diet. In support of the above study, Hegde, Rajasekaran, and Chandra (2005) proved that the use of whole-grain finger millet flour reduces oxidative stress by inhibiting lipid peroxidation in alloxan-induced diabetic rats. Millet consumption increases the production of antioxidant enzymes such as catalase, peroxidase, and reductase. Yadav et al. (2013) found that using bioprocessing techniques such as solid-state fermentation of finger millet grain with *Rhizopus oryzae* drastically improves the phenolic content and antioxidant property compared to enzymatic treatment. Treatment of finger millet grains with non-thermal technologies such as gamma irradiation, when used at different levels of dosage (2–15 kGy), improved the concentration of antioxidant enzymes catalase and superoxide dismutase by 72 percent and decreased the lipoxygenase activity by 35 percent (Reddy and Viswanath 2019). Finger millet can be used as a natural antioxidant in foods and as a nutraceutical and functional food ingredient to prevent oxidative stress.

5.2.2 Antidiabetic activity

The finger millet phenolics are distributed mainly in the seed coat, and the remaining are present as cell wall constituents in the endosperm. Dietary intake of polyphenols and antinutrient content decreases carbohydrate digestion and absorption resulting in lower glycemic response (Nithiyanantham et al. 2019). Hence, limiting the blood glucose level after the diet is crucial for treating diabetes and other related complications. Extensive research has been done on the inhibitory effect of polyphenols on the activity of digestive enzymes such as amylase, pepsin, glucosidase, trypsin, and lipases. Chethan, Sreerama, and Malleshi (2008) reported that the crude polyphenol extract and isolated phenolic compounds from finger millet flour extracts could decrease hydrolysis of starch and act as amylase inhibitors. They proposed that it may be used as a cost-effective nutrient to reduce diabetes mellitus complications. Kunyanga et al. (2012) suggested the combination of processing methods such as soaking and cooking of finger millet grains can be used compared to blanching and roasting to preserve the phenolic compounds present. Besides reducing oxidative stress, whole finger millet flour reduces blood glucose levels by 36 percent in alloxan-induced diabetic rats fed with a finger millet enriched diet (Hegde, Rajasekaran, and Chandra 2005). Shobana et al. (2010) proved the hypoglycemic effect of finger millet husk in the diabetic animal model. They further mentioned that dietary fiber, phytate, and polyphenols present in the seed coat matter provide health benefits. The complex phenolics mixture present in the seed coat matter acts as natural inhibitors of hydrolyzing enzymes alpha-glucosidase and pancreatic amylase, thus helps in managing the postprandial hyperglycemia (Shobana, Sreerama, and Malleshi 2009). Similarly, beta-glucan isolated from *finger millet* flour showed an antidiabetic effect by inhibiting alpha amylase (IC_{50} – 1.23 μg/ml) and alpha-glucosidase activity (IC_{50} – 1.42 μg/ml). Okoyomoh et al. (2013) found that addition of seed coat matter of finger millet in the diet of streptozotocin induced diabetic rats significantly lowered the blood glucose levels. Feeding with seed coat matter increases the regeneration of beta cells in islets of pancreas compared to control group. Lakshmi Kumari and Sumathi (2002) proved the effect of finger millet consumption on glycemic response in type 2 diabetic human subjects. Due to antinutritional content, consumption of whole finger millet flour triggers lower glycemic response and plasma glucose levels compared to germinated finger millet flour. It can be used as an essential part of the diabetic patient's diet. Kunyanga et al. (2011) found that acetonic extract of condensed tannins present in finger millet showed higher antioxidant and antidiabetic activity by reducing the rate of glucose absorption.

5.2.3 Antiobesity and hypolipidemic activity

Obesity is a chronic disorder characterized by high cholesterol, fatty acids, desensitization of insulin, excessive adipose tissue, and elevated blood pressure levels. Studies show that obesity leads to secondary complications such as retinopathy, neuropathy, chronic kidney diseases, coronary heart diseases, and insulin resistance. Alternative approaches are necessary to antiobesity medications due to their side effects. Murtaza et al. (2014) studied the effect of 10 percent finger millet bran supplementation in high fat diet fed to LACA mice for a period of 12 weeks. Finger millet bran feeding prevented the increase in serum total cholesterol, cholesterol ester, LDL/VLDL-cholesterol ratio. It also controls body weight gain, adipose tissue inflammation, and ectopic fat deposition. They concluded the study by suggesting the inclusion of finger millet bran has more beneficial effects than whole grain flour. Hegde, Rajasekaran, and Chandra (2005) fed alloxan-induced diabetic rats with 55 percent finger millet flour in their basal diet for 28 days. Reduction in total cholesterol level was observed in finger millet-fed rats (74 mg/dL) compared to alloxan induced diabetic rats (85 mg/dL). This hypolipidemic effect might be due to the synergistic effect between dietary fiber and phenolic compounds present. In the same way, Bhandari et al. (2021), who supplemented whole finger millet flour and hydroalcoholic extract (200 and 400 mg/kg) in high-fat diet-fed rats reported hypolipidemic and antiobesity effects. Aqueous finger millet seed extract showed a hypolipidemic effect by significantly reducing serum total cholesterol and LDL levels (Kareem et al. 2019). Shobana et al. (2010) observed the effects of supplementing 20 percent finger millet seed coat matter in the diet of streptozotocin-induced diabetic rats. Seed coat matter-fed rats showed a significant reduction in total cholesterol levels (43%), decrease in LDL+VLDL cholesterols (LDL+VLDL): HDL ratio and serum triacylglycerols (62%), increment in HDL cholesterol levels compared to diabetic rats. The lipid-lowering effect might be due to anthocyanins present in the seed coat.

Similarly, the obesity preventive effect of polyphenol-rich extract from finger millet and kodo millet was observed by Khare et al. (2020). Arabinoxylan extracted from finger millet supplementation (0.5 and 1 g/kg body weight) with a high-fat diet-fed to Swiss albino LACA strain mice significantly reduced the weight gain, serum triglycerides, serum total, LDL and VLDL cholesterol levels, decreased the visceral adipose tissue weight compared to a high-fat diet. But, the level of serum HDL cholesterol not increased (Sarma et al. 2018). The use of germinated finger millet as a substrate for solid-state fermentation using *Monascus purpureus* (red yeast rice) produces antihypercholesterolemic metabolites such as statin and sterol with better yield. It helps in value addition and increases the utilization of finger millet to develop functional foods (Venkateswaran and Vijayalakshmi 2010).

5.2.4 Antimicrobial activity

The secondary metabolites such as phenolics, flavonoids present in millets are potential inhibitors of micro-organisms growth. Polyphenolic compounds are mainly present in the pericarp, seed coat, and testa. Tannins present in the outer layers of finger millet grain provide resistance against fungal attack. Different parameters such as constituents and thickness of pericarp, texture, and proteins in the endosperm, polyphenols such as ferulic acid and hydroxycinnamic acid act in defense against pests and pathogens (Chandrashekar and Satyanarayana 2006). Banerjee et al. (2012) reported that acidic methanol extract of finger millet polyphenols showed inhibition against various bacteria such as *E. coli*, *B. cereus*, *S. aureus*, *L. monocytogenes*, *S. pyogenes*, *P. mirabilis*, *P. aeruginosa*, *S. marcescens*, *K. pneumonia* and *Y. enterocolitica*. Quercetin showed the highest antimicrobial activity. Similarly, Singh et al. (2015) compared the antimicrobial properties of finger and pearl millet. Finger millet showed effectiveness against *E. coli*, *S. aureus*, *P. mirabilis*, *P. aeruginosa*, *S. marcescens*, *K. pneumoniae*, *S. dysenteriae*, *Enterococcus sp.*, and *Salmonella sp.* Another study by Siwela et al. (2010) mentioned that higher phenolic content in finger millet grain showed resistance to fungal infection and was also responsible for better malt quality compared to low phenolic type malts. The mechanism involved in the antifungal activity are phenolic compounds in finger millet acts as free radicals, oxidizing the microbial membrane and cell constituents, inactivating fungal proteins, and reduced its function, forming a complex between tannins and substrates, making it inaccessible to microbes. Viswanath, Urooj, and Malleshi (2009) isolated phenolics from finger millet seed coat and whole flour using acidic methanol. Compared to whole flour, higher antimicrobial activity was observed in seed coat extract against *B. cereus and A. flavus* due to their higher polyphenol content. The lowest inhibition concentration of microbes was 30 percent of the seed coat. They concluded finger millet phenolics are responsible for antibacterial and antifungal activity. Aqueous extract of germinated finger millet seed showed antibacterial activity against *E. coli* (Vora, Rane, and Jadhav 2014). Fermentation of finger millet flour inhibits *S. typhimurium and E. coli*. The inhibitory effect might be due to metabolites produced by the fermenting microbes (Antony, George Moses, and Chandra 1998). High shelf life and storage characteristics of finger millet might be attributed to its polyphenol content. The phenolics present in finger millet can be used as a natural alternative to synthetic antimicrobials.

5.2.5 Wound healing property

Wound healing is a complicated process that involves interaction between different factors such as dermal cells, cellular matrixes, blood, cytokines, and growth factors. For wound management, various treatment options using synthetic drugs are available. Nonetheless, most of these treatments give undesirable side effects (Nagar et al. 2016). Diabetic patients and elderly people suffer from prolonged wound healing processes due to the production of higher oxidative stress (Davison et al. 2002). Antioxidants, antimicrobials, and bioactive can be included in the proper management of wounds. Rajasekaran et al. (2004) studied the wound healing activity of whole finger millet flour feeding (50%) in the diet of alloxan-induced type 1 diabetic rats. In this study, skin antioxidant status, production of nerve growth factor, and wound healing parameters were observed. They found that finger millet feeding increases the rate of wound contraction compared to the diabetic control group. Feeding the diabetic rats for four weeks with finger millet controlled the rate of glucose levels in blood plasma and increased the antioxidant status, which helped in the wound healing process. Histological and electron microscopical examinations of finger millet-fed rats showed improved epithelialization, collagen fibers, fibrillar arrangement, and mast cells in comparison to the diabetic group. Hastened wound healing process and protection of beta cells from alloxan mediated cell damage and was credited to the antioxidant property of selenium, structure of various phenolic compounds, and synergism between compounds present in finger millet. Finger millet can be included in the diet of diabetic patients to improve wound healing activity. Hegde, Chandrakasan, and Chandra (2002) prepared an aqueous paste of finger millet (300 mg) and kodo millet (300 mg), which was topically applied to the wound in Wistar rats for 16 days. Protein, DNA, and collagen content were significantly increased, and lipid peroxides decreased. The rate of wound contraction and days to complete closure of wounds were 90 percent and 13 days in finger millet, 88 percent and 14 days in Kodo millet paste applied to rats compared to 75 percent and 16 days in the control group. They reported that higher calcium content in finger millet and substantial quantity of methionine and cysteine in both millets present might be responsible for faster wound healing.

5.2.6 Anti-cataractogenic activity

Cataract causes a reduction in the transparency of the eye lens because of optical dysfunction. Chronic hyperglycemia and advanced glycation-end products (AGE) are the leading factors responsible for developing cataracts (Lim et al. 2001). Aldose reductase, a key enzyme, is the primary cause of various diabetic complications. Natural inhibitors of aldose reductase and antioxidants can be beneficial to diabetic patients without having adverse effects. Chethan, Dharmesh, and Malleshi (2008) evaluated the impact of finger millet seed coat fraction polyphenols extracted using HCl-methanol on the aldose reductase inhibitory activity. Among the finger millet phenolics, quercetin is the potent inhibitor of aldose reductase. Quercetin behaves as a non-competitive inhibitor and prevents cataractogenesis by blocking the polyol pathway. They suggested the use of finger millet seed coat polyphenols as a natural alternative to synthetic inhibitors. Shobana et al. (2010) studied the effect of finger millet seed coat matter (20%) supplementation in the diet of streptozotocin-induced diabetic rats. They concluded that the diabetic experimental group had immature cataracts with mild lenticular opacity compared to the diabetic control group. In diabetic cataracts, chronic hyperglycemia in blood resulted in increased accumulation of AGE, causing opacification. The activity of aldose reductase was decreased by 25 percent, and serum AGE level significantly reduced from 20.47 to 16.46 in the seed coat matter-fed group. Feeding of millet phenolics blocks protein glycation and delays cataractogenesis in diabetic rats. Thus, finger millet seed coat matter can be used as a functional ingredient in the development of functional foods for the diabetic population. Other proven health benefits of finger millet were given in **Table 5.1**.

Table 5.1 Health Benefits of Finger Millet

Health benefits	Extract used	Key findings	Reference
Anticancer activity	Ethanolic and methanolic extract of three-finger millet varieties (Ravi, Rawana, and Oshadha)	- Dose-dependent inhibition of Glutathione S-transferase and β-Glucuronidase enzymes indicate the anticancer potential. - Methanolic extract of Oshadha showed higher cytotoxicity against human breast cancer cells (MCF-7)	(Jayawardana et al. 2020)
Antiproliferative activity	Crude phenolic extracts from seven millets, including two-finger millet varieties, were used	- Finger millet (FM) extracts have antiproliferative activity against HT-29 human colon adenocarcinoma cells at a concentration of 0.5 mg/ml on day 4 - FM extracts prevented total DNA scission inhibition (97%) using peroxyl radicals compared to other millets - FM has a higher inhibitory effect against liposome peroxidation	(Chandrasekara and Shahidi 2011)
Prebiotic activity	Xylooligosaccharides (XOS) isolated from water-extractable xylan of finger millet seed coat	- XOS derived from seed coat showed dose-dependent antioxidant activity of 75.2 %. - L. plantarum showed higher growth by utilizing XOS as substrate, which is similar to the control dextrose. - Prebiotic activity score of seed coat XOS was similar to that of commercially available XOS.	(Palaniappan, Balasubramaniam, and Antony 2017)
Immunomodulatory activity	Arabinoxylans (AX) from finger millet bran extracted using alkali barium hydroxide, potassium hydroxide. Purified arabinoxylan was extracted using NaOH.	- Significant mitogenic activity and macrophages activation, including phagocytosis, were showed by extracted AXs. - Barium hydroxide extracted AX has higher lymphocyte proliferation due to the presence of higher ferulic acid content. - Alkali extracted AXs can be used as a potent immunomodulator than purified AXs.	(Savitha Prashanth, Shruthi, and Muralikrishna 2015)

(continued)

Table 5.1 *(Continued)*

Health benefits	Extract used	Key findings	Reference
Free-radical scavenging activity	Methanol extract of finger millet flour	- Brown finger millet has 96% higher polyphenolic content than the white variety - Fermentation and germination of finger millet decreased its free radical scavenging activity from 94% in raw to 22 and 25%. - Commonly available brown finger millet has more free radical scavenging ability.	(Sripriya et al. 1996)
Anti-inflammatory activity	Finger millet AX orally supplemented (doses 0.5 and 1.0 g/kg body wt.) to high-fat diet-fed Swiss albino mice for ten weeks	- AX supplementation downregulated the expression of the inflammatory markers compared to high-fat diet-fed mice. - AX helps in the improvement of low-grade inflammation in visceral white adipose tissue and liver.	(Sarma et al. 2018)
Collagen crosslinking and glycation	Methanolic extracts of finger and kodo millet flour	Methanolic extracts of both millets showed antiglycation activity similar to synthetic agents. Collagen treated with methanolic extracts showed lower viscosity at 35°C, similar to untreated collagen. It indicates lesser crosslinking.	(Hegde, Chandrakasan, and Chandra 2002)
Collagen glycation inhibitory activity	Finger millet and kodo millet (55% of basal diet) fed to alloxan-induced diabetic rats, respectively	Rat tail tendon collagen glycation was inhibited by 40% and 47% when diabetic rats were fed by finger and kodo millet compared to diabetic control.	(Hegde, Rajasekaran, and Chandra 2005)

5.3 Characterization of phytochemical compounds responsible for bioactive properties

5.3.1 Bioactive peptides

Bioactive or functional peptides can be described as amino acid residues with size ranges from 2 to 20 that are encrypted within the parent protein sequence and may be released during gastrointestinal digestion and food processing to achieve their specific bioactive roles (Harnedy and FitzGerald 2012). The function and efficacy of bioactive peptides depend on various factors such as their absorption, bioavailability, composition, sequence, and hydrolysis rate. Also, peptides have the advantage of higher biological activity and lower toxicity, and get metabolized easily. Millet bioactive peptides can exert various health effects such as antioxidant, antihypertensive, antimicrobial, antidiabetic, ACE-inhibitory effect, and anticancer activity (Majid and Poornima Priyadarshini 2019). Several bioactive peptides had been identified from cereal proteins; however, peptides from finger millet are yet to be identified and studied for their health benefits. Bisht, Thapliyal, and Singh (2016) isolated antimicrobial proteins/peptides from finger millet seed. It has the highest antibacterial activity against the strains *P. aeruginosa* and *S. entrica,* when compared to the barnyard and proso millet. Agrawal, Joshi, and Gupta (2019) isolated and studied the antioxidant potential of two bioactive peptides from finger millet protein hydrolysate. These peptides have higher free radical scavenging activity due to the presence of aromatic and hydrophobic amino acids in sequences. They suggested that the bioactive peptides from finger millet can be utilized as a natural antioxidant.

5.3.2 Polyphenolic compounds

In cereals, secondary metabolites such as phenolic acids and tannins were present in significant quantities, whereas flavonoids were present in minor amounts. The distribution of phenolic compounds in the finger millet grain is not uniform. About 60 percent of phenolics are mainly stored in the aleurone layer, testa, and pericarp, which are the main components of bran, and seed mass accounts for 12 percent (**Figure 5.1**). Phenolics exist in different forms, such as free, soluble, and insoluble. According to Subba Rao and Muralikrishna (2002), in finger millet flour protocatechuic acid (45 mg/100g) is the majorly present free phenolic acid. In bound phenolic acids, ferulic acid (18.60 mg/100g) is present in large quantities. Caffeic acid and coumaric acid are also present in bound form. Soluble and insoluble bound phenolics have higher antioxidant activity, reducing power, and metal chelating effect. Viswanath, Urooj, and Malleshi (2009) reported the presence of the highest concentrations of gallic acid (10.9 mg/100g) in finger millet husk and syringic acid (11.3 mg/100g) in whole flour. Other phenolic acids such as 4-Hydroxybenzoic acid, coumaric acid, ferulic acid, and vanillic acid was also identified.

Flavonoids are an important class of polyphenols with stronger antioxidant activity than vitamins C, and E. Total flavonoid content in pigmented finger millets such as open red, fist brown varieties (202.94 and 197.1 mg catechin equivalents/100g) are significantly higher than white variety (90.24 mg catechin equivalents/100g). Flavonoids such as catechin, quercetin, apigenin, and epicatechin are also present, whereas catechin and epicatechin are the majorly present flavan-3-ols (Xiang et al. 2019). Chethan, Sreerama, and Malleshi (2008) extracted polyphenols from finger millet flour using HCl-Methanol. They reported that 85 percent of total phenol compounds are benzoic acid derivatives (gallic acid, protocatechuic acid, p-hydroxybenzoic acid, syringic and vanillic acid) and remaining by cinnamic acid derivatives (trans-cinnamic acid, p-coumaric and ferulic) and flavonoid (quercetin).

Proanthocyanidins are high molecular weight polyphenols, which were reported in significant proportions only in finger millet. It is biologically active and might reduce minerals'

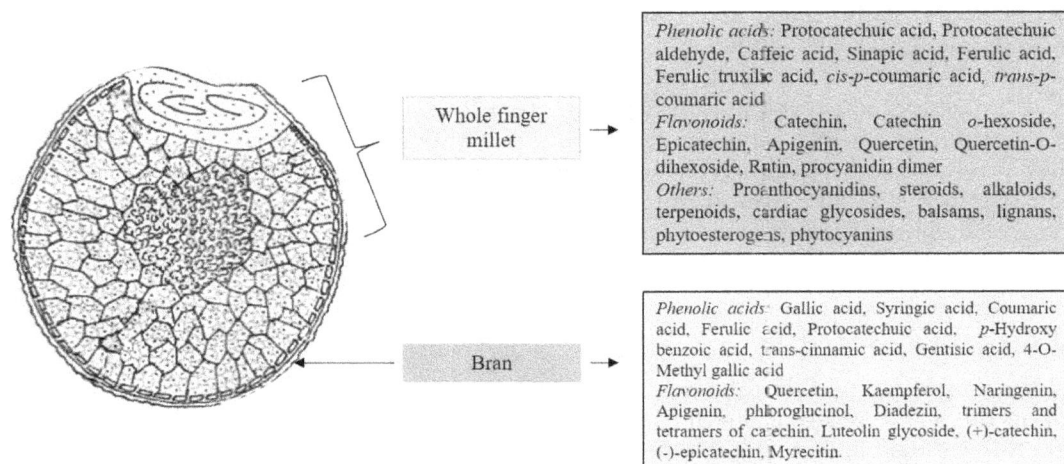

Whole finger millet →

Phenolic acids: Protocatechuic acid, Protocatechuic aldehyde, Caffeic acid, Sinapic acid, Ferulic acid, Ferulic truxilic acid, cis-p-coumaric acid, trans-p-coumaric acid
Flavonoids: Catechin, Catechin o-hexoside, Epicatechin, Apigenin, Quercetin, Quercetin-O-dihexoside, Rutin, procyanidin dimer
Others: Proanthocyanidins, steroids, alkaloids, terpenoids, cardiac glycosides, balsams, lignans, phytoesterogens, phytocyanins

Bran →

Phenolic acids: Gallic acid, Syringic acid, Coumaric acid, Ferulic acid, Protocatechuic acid, p-Hydroxy benzoic acid, trans-cinnamic acid, Gentisic acid, 4-O-Methyl gallic acid
Flavonoids: Quercetin, Kaempferol, Naringenin, Apigenin, phloroglucinol, Diadezin, trimers and tetramers of catechin, Luteolin glycoside, (+)-catechin, (-)-epicatechin, Myrecitin.

Figure 5.1 *Phytochemicals reported in finger millet.*

nutritional value and bioavailability, especially iron and proteins. Compared to white finger millet varieties, condensed tannins or proanthocyanidins is highly present in the pigmented testa of the dark brown or red color finger millet varieties (Dykes and Rooney 2006). Siwela et al. (2007) reported pigmented finger millet had condensed tannin content of 0.60–2.08 catechin equivalents/100 mg, compared to white grains 0.07 catechin equivalents/100g. Tannins are found to have numerous health benefits such as lipid peroxidation prevention, antioxidant and antimicrobial activity that are proved by various *in vivo and in vitro assays*. Studies have shown that finger millet has higher free radical scavenging activity than wheat, rice, and sorghum due to the presence of polyphenols. Some of the reported phytochemicals present in finger millet presented in **Figure 5.1**.

5.3.3 Dietary fiber

Dietary fiber (DF) intake prevents the incidence of several disease complications by influencing the gastrointestinal tract. Regular consumption of DF helps in improving the serum lipids level, blood glucose control, and immune function, helps in weight loss, reduces blood pressure, delayed absorption, and increases fecal bulk. Based on solubility, DF can be divided into two groups. Water-soluble fibers include non-starch polysaccharides such as pectins, gums, inulin, and beta-glucans. Water-insoluble fibers include lignin, hemicellulose, and resistant starch. Non-starch polysaccharides in the millets seed coat are the significant source of soluble and insoluble dietary fiber. Soluble dietary fiber can reduce cholesterol absorption and lowers the plasma cholesterol level. It is entirely fermented in the gut microbiota, produces short-chain fatty acids, mainly propionic acid. Also, the absorption of calcium and magnesium from the cecum is improved (Lopez et al. 1999).

On the other hand, insoluble fiber increases fecal bulk, fastens intestinal transit time, and binds any carcinogen present. Dharmaraj and Malleshi (2011) observed decortication alters the dietary fiber composition in terms of a significant increase in the soluble fiber content from 1.40 to 2.30 g/100g and decreased the insoluble and total dietary fiber content from 15.70 and 17.10 in native millet to 7.80 and 10.10 g/100g in decorticated finger millet, respectively. They also mentioned that finger millet decortication removes the dietary fiber and phenolics rich bran material, which in turn improves the digestibility of starch and protein and calcium bioavailability. Shukla and Srivastava (2011) reported finger millet flour has a significantly higher total, soluble and insoluble dietary fiber content than refined wheat flour. Noodles developed using 30 percent finger millet flour substitution showed a hypoglycemic effect in human subjects. Due to its higher DF content, ingestion of finger millet-based diets significantly reduces the plasma glucose levels, area under the curve, and mean rise in peak compared to other major cereals. DF delays the gastric emptying time, and forms complexes with carbohydrates available make it unabsorbable (Lakshmi Kumari and Sumathi 2002). Regular consumption of dietary fiber helps in the treatment of diabetes.

5.3.4 Resistant starch

Resistant starch (RS) refers to starch and its products that resist hydrolyzing into D-glucose units in the small intestine but are later fermented in the large intestine (Fuentes-Zaragoza et al. 2010). Consumption of RS confers various health benefits such as increased production of short-chain fatty acids, lowered plasma cholesterol level, increased fecal output, reduced postprandial glycemia, and influences immune function. With respect to RS content, finger millet and kodo millet are better than other millets. Mangala et al. (1999) found that different use of processing methods increased the resistant starch content of finger millet. Crude and partially purified resistant starch content significantly increased by up to five-fold when the number of autoclave cycles increased. This is because of increased resistant starch crystallinity by cooling in between repeated cycles. They also confirmed that RS from finger millet is an enzyme-resistant portion of dietary starch. The RS fractions isolated by either *in vitro* enzymatic method or by *in vivo* showed to have similar molecular weights. Similarly, Roopa and Premavalli (2008) reported that treatment of finger millet with pressure cooking and roasting increased the formation of resistant starch. Factors such as the type of starch present, pH, and processing methods used affect the physical changes of the starch molecule. During amylose retrogradation, linkages resistant to enzyme hydrolysis are formed. This linkage is associated with RS formation. RS content varies vastly between the millet species, with amylose content being the major affecting factor.

5.4 Techno-functional properties

5.4.1 Pasting properties

The starch pasting profile is an essential characteristic of cereals that determines how they are used in the food business. Pasting characteristics are the viscosity profile of flour (starch) as a function of temperature variation. The peak viscosity of the finger millet starch is slightly higher than the millet flour. Compared to rice starch, ragi starch has lower hot paste viscosity, owing to its amylose content. Also, qualitative, and quantitative variation in protein and lipid constituents affects its viscosity. Changes in pasting profile are also observed with the processing of finger millet, such as malting, popping, and extrusion. Popped finger millet showed lower peak and set back viscosities due to the damage of starch granules while exposed to high temperatures. Similarly, malting of ragi releases amylases, which hydrolyses the starch molecules (near the granule periphery) and thereby decreases the viscosity and consistency. Finger millet flour showed complete gelatinization at higher temperature (71.4° C) compared to rice (63.2° C), whereas, for extruded ragi, complete gelatinization takes place at low temperatures (50–55°C) (Mangala et al. 1999).

5.4.2 Color profile

Depending on different varieties, finger millet is found in yellow, white, tan, brown, violet, and red color. Finger millet grain is highly variable in color due to the presence of anthocyanin pigments and tannins present in the tesla layer of a few varieties (Naushad Emmambux and Taylor 2013). The grains and flour of finger millet have significant color variation. Native millet flour is lighter in color and has a better (L) lightness value (67.11) than grain (23.8). The flour had a lower redness than the grains but a higher yellowness (Dharmaraj et al. 2002). Among all the varieties, the red-colored ones are extensively cultivated throughout the world. Finger millet is used for fortification and value addition in many food products like biscuits, bread, noodles, pasta, and so forth. The disadvantage of finger millet substitution in food is that it negatively affects the color profile and, therefore, food appearance. The seat coat of the grain is responsible for imparting dark color to the grain and, consequently, decorticated grain is lighter in color when compared to native grain. The color of decorticated finger millet is a light cream with an opaque appearance. Shobana and Malleshi (2007) reported the whiteness percent for native and decorticated finger millet 3.2 (brown color) and 10.8 (light cream), respectively. The use of decorticated finger millet in value-added products allows for a higher percentage of incorporation without altering the color, therefore increasing the product's nutritional value without compromising its sensory acceptability.

5.4.3 Water absorption capacity (WAC) and water solubility index (WSI)

WAC shows the amount of water that starch can hold in relation to its weight. Higher values of WAC and WSI in millet flour suggests that it can be used in food products where good viscosity is important. Panda et al. (2019) observed that sprouting of finger millet significantly increased the WAC from 20.10 percent in raw to 23.34 percent and decreased the WSI from 11.47 percent in raw to 10.37 percent. Ramashia et al. (2018) evaluated WAC in three different finger millet cultivars. They found that absorption is higher in milky cream variety (1.23 mL/g) and lowest in the black variety (0.93 mL/g). Water absorption value is higher in loose starch polymers with high concentration of hydrophilic polysaccharides and lower value denotes compact structure.

5.4.4 Oil absorption capacity (OAC)

This is an important indicator of palatability, flavor and shelf stability of a food product, and also, shows the emulsifying capacity of the starch. Roopa and Premavalli (2008) compared puffed and raw finger millet flour in ten different varieties. They found that fat absorption capacity significantly increased after puffing. Gowthamraj, Jubeena, and Sangeetha (2021) treated finger millet flour using gamma radiation at varying doses (2.5 to 10 kGy). OAC increased due to starch cross linking and protein unfolding, which holds more fatty acids. Higher water and oil absorption capacity positively affects the product characteristics and prevents moisture and fat loss of food.

5.4.5 Foaming capacity and foaming stability

This parameter is important for the development of leavened products such as bread, cakes, biscuits, and muffins. For good foaming capacity, the proteins must denature and aggregate during whipping at the air–water interface. It also depends on its protein surface active properties, pH, salts, sugars, lipids, and temperature. Panda et al. (2019) reported that sprouting of finger millet significantly increases the foaming capacity (2.30 to 5.95%) and foaming stability (11.80 to 19.25%). Similarly, Kumar et al. (2021) also observed germination duration of finger millet impacts the foaming capacity and foaming stability.

5.4.6 Swelling index and solubility

Higher swelling index values denotes the higher water absorption capacity, which can be useful in low calorie foods. Swelling and solubility of starch granules depends on the interaction between amorphous and crystalline regions of the starch. Nefale and Mashau (2018) reported that germination of finger millet significantly decreased its swelling power from 4.83 g/g in control to 3.17 g/g in 72 h germinated seeds. Solubility of flour increased when the germination time increased (1.82 to 8.32%). This might be due to action of amylases and protease on the starch and proteins. Nazni and Bhuvaneswari (2021) found that finger millet (5.05% and 5.5%) has lower swelling index and solubility than little millet (6.71% and 30.8%), respectively. These lower values of finger millet showed higher starch crystallinity and poor water percolation within starch molecules.

5.4.7 Emulsion capacity (EC) and Emulsion stability (ES)

EC denotes the ability and capacity of protein present to form an emulsion. The quality and the quantity of soluble proteins influences the emulsion capacity. The mechanism involves proteins present in flour reduces the surface tension of the oil droplets, while increasing the electrostatic repulsion on the surface. Afolayan (2014) reported the emulsion activity of finger millet starch is 46 percent. Funmilayo Abioye, Oluwatoyin Ogunlakin, and Taiwo (2018) reported that germination of finger millet increased the emulsion capacity from 4.78 percent in raw flour to 8.47 percent in 96h germinated samples.

5.5 Potential utilization in value added food products

5.5.1 Malting and weaning food

Traditionally millet malt is utilized as weaning food for infants in combination with either milk or warm water. Finger millet's malting reduces the antinutrients and improves its digestibility and nutritional quality, making it reasonably popular in Tamil Nadu and Karnataka. Compared to other cereals, it is a suitable grain for malt meals because of its inherent characteristics like fungal resistance, alpha and beta amylase elaboration during germination, and desirable aroma while roasting. Malted weaning food prepared with finger millet and a blend of green and Bengal grams are nutritionally rich in protein and calcium. Ragi malt is used as a health drink, and about 5 percent of ragi malt these days is blended with energy foods to improve texture and mouthfeel (Verma and Patel 2013).

5.5.2 Fermented foods

Finger millet is widely utilized as essential ingredients in many fermented food products, which enhances the taste and enriches the content of fiber, calcium, and protein in food. Depending on taste and preference, fermented dishes can also be made using malted or sprouted finger millet grains. Fermented foods like idli and dosa are widely consumed as breakfast in many parts of India, predominantly South India. The partial replacement of black gram with finger millet in idli preparation significantly improved idli's fiber and calcium content (Ghodke and Deshpande 2020). Basappa et al. (2009), in their study, prepared a beverage (chhang) using fermented ragi, which was nutritionally more beneficial than the unfermented ragi. The beverage prepared was safe and healthy for the economically weaker section as it could provide a significant amount of B complex vitamins, calories, essential amino acids, and bioavailable minerals. Palanisamy et al. (2011) also reported that dosa (Indian breakfast food) prepared using co-fermentation of finger millet with horse gram had improved protein content with better essential amino acid profile. As a result, finger millet-based dosa might be utilized to address the problem of protein-energy malnutrition since it is an inexpensive protein-rich meal. In conclusion, various studies suggest that the addition or replacement of finger millet can be used for the value addition of fermented foods.

5.5.3 Baked foods

Bakery foods are prevalent because of their availability, ready-to-eat, and good shelf life. Bread, biscuit, cookie, muffins, cakes, and so forth are everyday bakery products consumed worldwide. Bakery products, mainly bread, are consumed daily and hence contribute a lot to human nutrition. Therefore, many studies and research are done to enhance the nutritional and functional profile of bakery products. Recently finger millet has gained lots of attention as means of value addition for development of bakery products like biscuits, bread, and muffins. Malted finger millet flour supplementation improves the nutritional quality of cakes concerning its minerals and fiber content (Verma and Patel 2013). Several studies demonstrated the use of finger millet as the main ingredient for improving the nutritional quality of bakery products such as bread, cookies, and biscuits. Finger millet substitution up to 20 percent enriches the nutritional quality of the wheat bread without affecting the physical and rheological attributes. It thus can be utilized as an ingredient for the value addition of bread (Devani et al. 2016). Finger millet seed coat is also a good source of minerals, dietary fiber, and phytochemicals and can be utilized for value addition of products. Krishnan et al. (2011) formulated finger millet seed coat (FMSC) biscuits by substituting up to 20 percent of wheat flour with FMSC. The finger millet seed coat-based biscuits have higher levels of protein, dietary fiber, and zinc.

5.5.4 Extruded Foods

The changing food habits and growing demand for Ready to Eat (RTE) and Ready to Cook (RTC) foods have increased the demand for extruded foods like noodles and pasta. Pertaining to finger millet's nutritional and nutraceutical potentials, its utilization in value addition of extruded products has gained much attention. Extruded products like pasta are usually high in starch content and low in dietary fiber and minerals. Replacement of malted finger millet flour with semolina improves the protein content of pasta and enhances the calcium content of pasta, as studied by (Tripathi et al. 2015). The malted finger millet-based pasta has higher antioxidant activity when compared with semolina pasta. Roberts Pradeepa, Thejasri, and Hymavathi (2019) developed nutrient rich noodles by blending finger millet with wheat and found that adding finger millet improved dietary fiber, calcium, and iron content of the noodles. Finger millet is also an important ingredient for the formulation of gluten-free pasta. Gluten-free foods are generally expensive, and inclusion of underutilized millets in these food products can provide an economical and affordable alternative for all sections of society. Radhika et al. (2019) developed gluten-free pasta using finger millet, pearl millet, and soybean. The formulation with 35 percent of each ingredient was most accepted and was found to have increased essential nutrients like protein, fiber, and mineral ash content. Various researches have concluded that finger millet can be used as a good source of value addition and for developing functional foods like low glycemic index and gluten-free formulations. Other value-added products developed using finger millet are elaborated in **Table 5.2**.

Table 5.2 Utilization of Finger Millet in Value Added Food Products

Product	% Incorporation	Product quality characteristics	References
Nutri cereal and milk-based beverage	(40 to 60) % malted finger millet with double toned milk	- Increased content of fiber, calcium, iron, zinc, and phosphorus with malt addition - Finger milled based drink has low fat (1.26%) and cholesterol (3.26%) levels. - Replacement of cow's milk with ragi malt drink can result in increased per capita availability, thus providing a healthier and economical option.	(Kumar et al. 2020)
Probiotic beverage	Not specified	- This probiotic beverage has good sensory acceptability up to 5 weeks in refrigerated condition. - Finger millet-based beverage has better survivability of *Lactobacillus casei 431*. There was 2.21×10^8 CFU ml^{-1} after fermentation, and the reduction rate of *L. casei431* was considerably low up to 5 weeks. - Finger millet-based probiotic beverage can be a functional beverage up to 5 weeks when stored at $5°C \pm 1$.	(Weerahewa, Fasreen, and Perera 2017)

Table 5.2 *(Continued)*

Product	% Incorporation	Product quality characteristics	References
Weaning food	Not specified	- Protein availability and starch digestibility improved by malting of finger millet - Germination of finger millet increases the protein content by 17%. - It can be helpful for optimal mental development of infants as a result of increased essential fatty acids, such as linoleic acid content of weaning food. - Malted finger millet and amaranth can be considered suitable for gluten-free weaning food. Also, malting and germination of finger millet can improve the nutritional value of infant food.	(Hejazi and Orsat 2016)
Bread	70: 30 Wheat – Finger millet	- Substitution of hydrothermally treated (HTT) finger millet in bread gives better specific volume and softer crumb comparable to 100% wheat bread. - HTT substituted finger millet bread showed higher fiber, phenolic and phytate content.	(Onyango et al. 2020)
Cookie	30% and 50%	- Finger millet can be replaced successfully up to 50% level for the development of nutritious cookies - Finger millet addition improved the fiber, calcium, iron, phosphorus, copper, and zinc content.	(Sinha and Sharma 2017)
Biscuits	0, 25, 50 and 75%	- Calcium and crude fiber significantly increased in biscuits incorporated with finger millet. - The fat content of the biscuit decreased with an increased percentage of finger millet in the biscuit. - Replacement of finger millet flour at 50% level was most suitable to prepare biscuits with improved nutritional quality and acceptable sensory acceptability.	(Kaur, Kumar, and Dhaliwal 2020)
Sponge cake **Rice noodles**	20–80 % decorticated finger millet 0, 10, 20, and 30%	- 80% replacement of refined wheat flour with decorticated finger millet flour is suitable for preparation of good sponge cake - Sponge cake prepared using decorticated finger millet flour has high dietary fiber, ash content - Calcium content increased from 18.56% to 206.48%. Also, the iron content improved from 2.13 to 11.61 %. - Rice noodles with 20% incorporation of finger millet have the best sensory acceptability - Rice noodles with 20% finger millet replacement have high protein, dietary fiber, calcium, potassium, magnesium, and phosphorus content and low cooking loss and breaking rate. - The resistant starch content of FM replaced rice noodles was 16.22-fold higher than rice noodles, thereby reducing the risk of postprandial blood glucose.	(Patel and Thorat 2019) (Chen et al. 2021)
Noodles	30 to 50 %	- 30% finger millet incorporated noodles were best in terms of sensory acceptability - Finger millet incorporated noodles have a high amount of fiber, fat, ash, calcium, and iron - Protein content reduced with an increased percent of finger millet. - Noodles with 30% incorporation of finger millet had low glycemic index (45.13), compared to rice noodles (62.59). - Finger millet incorporated noodles showed a hypoglycaemic effect.	(Shukla and Srivastava 2011)
Pasta	10 to 30 % sprouted ragi flour	- The pasta with a combination of 15% ragi flour and 15% banana flour was highly potent nutritionally and was also accepted in cooking quality and sensory. - Mineral and dietary fiber content of pasta were considerably increased with the addition of ragi and banana flour. - Pasta substituted with 20% ragi has the highest antioxidant activity	(Krishnan and Prabhasankar 2010)

5.6 Conclusion

Finger millet was considered to be a poor man's crop for a very long time. Due to its high nutritional and nutraceutical profile, finger millet is no longer seen as a coarse cereal. Instead, it is known as Nutri cereals, which can overcome the challenges of malnutrition and hidden hunger worldwide. It also exhibits excellent nutraceutical potential, which offers several health benefits such as antidiabetic, antioxidant, hypocholesterolemic, antiobesity and antimicrobial effects. Many studies suggest that regular consumption and utilization of finger millet in food can help manage chronic conditions such as diabetes, obesity, and cardiovascular diseases. This chapter also highlights the techno-functional properties of finger millet and its possible utilization as a functional ingredient for value addition in the food industry. Value addition of various foods with finger millet can help promote millet consumption and thereby nutritional intake of the population. In conclusion, it can be stated that finger millet can be used in formulation of value-added products, due to its protein profile, mineral content, and gluten-free nature. Also, the utilization of finger millet in the food industry can raise the profitability of its cultivators and help in combating the challenge of food security in the world.

References

Afolayan, Michael. 2014. "Physicochemical, Proximate and Phytochemical Evaluation of Starch from Eleusine Coracana L. (Finger Millet)." *Article in International Journal of Advanced Chemistry* 2 (4): 98–104. doi:10.14419/ijac.v2i2.3496

Agrawal, Himani, Robin Joshi, and Mahesh Gupta. 2019. "Purification, Identification and Characterization of Two Novel Antioxidant Peptides from Finger Millet (Eleusine Coracana) Protein Hydrolysate." *Food Research International* 120 (June). Elsevier: 697–707. doi:10.1016/J.Foodres.2018.11.028

Antony, U., L. George Moses, and T.S. Chandra. 1998. "Inhibition of Salmonella Typhimurium and Escherichia Coli by Fermented Flour of Finger Millet (Eleusine Coracana)." *World Journal of Microbiology and Biotechnology 1998 14:6* 14 (6). Springer: 883–86. doi:10.1023/A:1008871412183

Banerjee, S, K Sanjay, S Chethan, and N Malleshi. 2012. "Finger Millet (Eleusine Coracana) Polyphenols: Investigation of Their Antioxidant Capacity and Antimicrobial Activity." *African Journal of Food Science* 6 (13). Academic Journals. doi:10.5897/AJFS12.031

Basappa, S. C., D. Somashekar, Renu Agrawal, K. Suma, and K. Bharathi. 2009. "Nutritional Composition of Fermented Ragi (Chhang) by Phab and Defined Starter Cultures as Compared to Unfermented Ragi (Eleucine Coracana G.)." Http://Dx.Doi.Org/10.3109/09637489709028577 48 (5). Taylor & Francis: 313–19. doi:10.3109/09637489709028577

Bhandari, Upma, Lata Bisht, Sweta Joshi, Priyanka Uniyal, Veerma Ram, and Mamta Singh. 2021. "Modulatory Effect of Whole Flour and Hydroalcoholic Extract of Finger Millet (Elusine Coracana) on the Abnormalities Associated with Metabolic Syndrome in Hyperlipidemic Diabetic Rats." *University Journal of Phytochemistry and Ayurvedic Heights* 1 (1): 39–48. http://ujpah.in/wp-content/uploads/2021/07/article-doi-6.pdf

Bisht, Anjali, Manisha Thapliyal, and Ajeet Singh. 2016. "Screening and Isolation of Antibacterial Proteins/Peptides from Seeds of Millets." *International Journal of Current Pharmaceutical Research* 8 (3): 96–99. www.researchgate.net/publication/316888972

Chandra, Dinesh, Satish Chandra, Pallavi, and A. K. Sharma. 2016. "Review of Finger Millet (Eleusine Coracana (L.) Gaertn): A Power House of Health Benefiting Nutrients." *Food Science and Human Wellness* 5 (3). Elsevier: 149–55. doi:10.1016/J.FSHW.2016.05.004

Chandrasekara, Anoma, and Fereidoon Shahidi. 2010. "Content of Insoluble Bound Phenolics in Millets and Their Contribution to Antioxidant Capacity." *Journal of Agricultural and Food Chemistry* 58 (11). American Chemical Society: 6706–14. doi:10.1021/JF100868B

Chandrasekara, Anoma, and Shahidi, Fereidoon. 2011. "Antiproliferative Potential and DNA Scission Inhibitory Activity of Phenolics from Whole Millet Grains." *Journal of Functional Foods* 3 (3). Elsevier: 159–70. doi:10.1016/J.JFF.2011.03.008

Chandrasekara, Anoma, and Shahidi, Fereidoon. 2012. "Antioxidant Phenolics of Millet Control Lipid Peroxidation in Human LDL Cholesterol and Food Systems." *Journal of the American Oil Chemists' Society* 89 (2). John Wiley: 275–85. doi:10.1007/S11746-011-1918-5

Chandrashekar, A., and K. V. Satyanarayana. 2006. "Disease and Pest Resistance in Grains of Sorghum and Millets." *Journal of Cereal Science* 44 (3). Academic Press: 287–304. doi:10.1016/J.JCS.2006.08.010

Chen, Jiali, Le Wang, Panfei Xiao, Chiling Li, Hui Zhou, and Dongmin Liu. 2021. "Informative Title: Incorporation of Finger Millet Affects in Vitro Starch Digestion, Nutritional, Antioxidative and Sensory Properties of Rice Noodles." *LWT* 151 (November). Academic Press: 112145. doi:10.1016/.LWT.2021.112145

Chethan, S., and N. G. Malleshi. 2007. "Finger Millet Polyphenols: Optimization of Extraction and the Effect of PH on Their Stability." *Food Chemistry* 105 (2): 862–70. doi:10.1016/J.Foodchem.2007.02.012

Chethan, S., Shylaja M. Dharmesh, and Nagappa G. Malleshi. 2008. "Inhibition of Aldose Reductase from Cataracted Eye Lenses by Finger Millet (Eleusine Coracana) Polyphenols." *Bioorganic and Medicinal Chemistry* 16 (23). Pergamon: 10085–90. doi:10.1016/j.bmc.2008.10.003

Chethan, S., Y. N. Sreerama, and N. G. Malleshi. 2008. "Mode of Inhibition of Finger Millet Malt Amylases by the Millet Phenolics." *Food Chemistry* 111 (1). Elsevier: 187–91. doi:10.1016/ J.Foodchem.2008.03.063

Davison, Gareth W., Lindsay George, Simon K. Jackson, Ian S. Young, Bruce Davies, Damian M. Bailey, John R. Peters, and Tony Ashton. 2002. "Exercise, Free Radicals, and Lipid Peroxidation in Type 1 Diabetes Mellitus." *Free Radical Biology and Medicine* 33 (11). Pergamon: 1543–51. doi:10.1016/S0891-5849(02)01090-0

Devani, Bansee. M., Bhavesh. L. Jani, Mansukhlal. B. Kapopara, Devendra. M. Vyas, and Manda Devi Ningthoujam. 2016. "Study on Quality of White Bread Enriched with Finger Millet Flour." *International Journal of Agriculture, Environment and Biotechnology* 9 (5). New Delhi Publishers: 903. doi:10.5958/2230-732X.2016.00116.9

Devi, Palanisamy Bruntha, Rajendran Vijayabharathi, Sathyaseelan Sathyabama, Nagappa Gurusiddappa Malleshi, and Venkatesan Brindha Priyadarisini. 2011. "Health Benefits of Finger Millet (Eleusine Coracana L.) Polyphenols and Dietary Fiber: A Review." *Journal of Food Science and Technology 2011 51:6* 51 (6). Springer: 1021–40. doi:10.1007/S13197-011-0584-9

Dharmaraj, Usha, and N. G. Malleshi. 2011. "Changes in Carbohydrates, Proteins and Lipids of Finger Millet after Hydrothermal Processing." *LWT - Food Science and Technology* 44 (7). Academic Press: 1636–42. doi:10.1016/J.LWT.2010.08.014

Dharmaraj, Usha, M. S. Meera, S. Yella Reddy, and Malleshi Nagappa Gurusiddappa. 2002. "Decorticated Finger Millet (Eleusine Coracana) and a Process for Preparing the Decorticated Finger Millet." *Journal of Food Science and Technology 2011* 52 (3): 1361–71.

Dida, Mathews M., and Katrien M. Devos. 2006. "Finger Millet." *Finger Millet: A Valued Cereal*, March. Springer, Berlin, Heidelberg, 333–43. doi:10.1007/978-3-540-34389-9_10

Dykes, Linda, and Lloyd W. Rooney. 2006. "Sorghum and Millet Phenols and Antioxidants." *Journal of Cereal Science* 44 (3). Academic Press: 236–51. doi:10.1016/J.JCS.2006.06.007

Fuentes-Zaragoza, E., M. J. Riquelme-Navarrete, E. Sánchez-Zapata, and J. A. Pérez-Álvarez. 2010. "Resistant Starch as Functional Ingredient: A Review." *Food Research International* 43 (4). Elsevier: 931–42. doi:10.1016/J.Foodres.2010.02.004

Funmilayo Abioye, Victoria, Grace Oluwatoyin Ogunlakin, and Ganiyat Taiwo. 2018. "Effect of Germination on Anti-Oxidant Activity, Total Phenols, Flavonoids and Anti-Nutritional Content of Finger Millet Flour." *Journal of Food Processing and Technology* 9 (2): 1–4. doi:10.4172/2157-7110.1000719

Ghodke, S. V., and H. S Deshpande. 2020. "Studies on Standardization and Development of Innovative Finger Millet Fortified Idli" 7. JETIR. www.jetir.org

Gowthamraj, G., C. Jubeena, and Narayanasamy Sangeetha. 2021. "The Effect of γ-Irradiation on the Physicochemical, Functional, Proximate, and Anti-Nutrient Characteristics of Finger Millet (CO14 & CO15) Flours." *Radiation Physics and Chemistry* 183 (June). Pergamon: 109403. doi:10.1016/J.RADPHYSCHEM.2021.109403

Guclu, Gamze, Hasim Kelebek, and Serkan Selli. 2021. "Antioxidant Activity in Olive Oils." In *Olives and Olive Oil in Health and Disease Prevention*, 313–25. Academic Press. doi:10.1016/B978-0-12-819528-4.00031-6

Gupta, Sanjay Mohan, Sandeep Arora, Neelofar Mirza, Anjali Pande, Charu Lata, Swati Puranik, J. Kumar, and Anil Kumar. 2017. "Finger Millet: A 'Certain' Crop for an 'Uncertain' Future and a Solution to Food Insecurity and Hidden Hunger under Stressful Environments." *Frontiers in Plant Science* (April). Frontiers: 643. doi:10.3389/FPLS.2017.00643

Harnedy, Pádraigín A., and Richard J. FitzGerald. 2012. "Bioactive Peptides from Marine Processing Waste and Shellfish: A Review." *Journal of Functional Foods* 4 (1). Elsevier: 6–24. doi:10.1016/J.JFF.2011.09.001

Hegde, Prashant S., Gowri Chandrakasan, and T. S. Chandra. 2002. "Inhibition of Collagen Glycation and Crosslinking in Vitro by Methanolic Extracts of Finger Millet (Eleusine Coracana) and Kodo Millet (Paspalum Scrobiculatum)." *Journal of Nutritional Biochemistry* 13 (9): 517–21. doi:10.1016/S0955-2863(02)00171-7

Hegde, Prashant S., Namakkal S. Rajasekaran, and T. S. Chandra. 2005. "Effects of the Antioxidant Properties of Millet Species on Oxidative Stress and Glycemic Status in Alloxan-Induced Rats." *Nutrition Research* 25 (12). Elsevier: 1109–20. doi:10.1016/J.NUTRES.2005.09.020

Hejazi, Sara Najdi, and Valérie Orsat. 2016. "Optimization of the Malting Process for Nutritional Improvement of Finger Millet and Amaranth Flours in the Infant Weaning Food Industry." Http://Dx.Doi.Org/10.1080/09637486.2016.1261085 68 (4). Taylor & Francis: 429–41. doi:10.1080/09637486.2016.1261085

Jayawardana, S A S, J K R R Samarasekera, G H C M Hettiarachchi, M J Gooneratne, M I Choudhary, R Imad, and A Naz. 2020. "Glutathione S-Transferase and β-Glucuronidase Enzymes Inhibitory and Cytotoxic Activities of Ethanolic and Methanolic Extracts of Sri Lankan Finger Millet (Eleusine Coracana (L.) Gaertn.) Varieties | South Asian Research Journal of Natural Products." *South Asian Journal of Natural Products* 3 (2): 1–9. https://journalsarjnp.com/index.php/SARJNP/article/view/30100

Kareem, Adepoju M., Shuaibu O. Bello, Emmanuel Etuk, Sarafadeen A. Arisegi, and Muhammad T. Umar. 2019. "Evaluation of the Hypoglycemic and Hypolipidemic Effects of Aqueous Eleusine Coracana Seed Extract." *International Archives of Medicine and Medical Sciences* 1 (3): 35–41. doi:10.33515/iamms/2019.032/32

Kaur, Arshdeep, Krishan Kumar, and Harcharan Singh Dhaliwal. 2020. "Physico-Chemical Characterization and Utilization of Finger Millet (Eleusine Coracana L.) Cultivars for the Preparation of Biscuits." *Journal of Food Processing and Preservation* 44 (9). John Wiley: e14672. doi:10.1111/JFPP.14672

Khare, P., R. Maurya, R. Bhatia, P. Mangal, J. Singh, K. Podili, M. Bishnoi, and K. K. Kondepudi. 2020. "Polyphenol Rich Extracts of Finger Millet and Kodo Millet Ameliorate High Fat Diet-Induced Metabolic Alterations." *Food & Function* 11 (11). Food Funct: 9833–47. doi:10.1039/D0FO01643H

Krishnan, Murali, and Prabhasankar, Pichan. 2010. "Studies on pasting, microstructure, sensory, and nutritional profile of pasta influenced by sprouted finger millet (eleucina coracana) and green banana (musa paradisiaca) flours." *Journal of Texture Studies* 41 (6). John Wiley: 825–41. doi:10.1111/J.1745-4603.2010.00257.X

Krishnan, Rateesh, Usha Dharmaraj, R. Sai Manohar, and N. G. Malleshi. 2011. "Quality Characteristics of Biscuits Prepared from Finger Millet Seed Coat Based Composite Flour." *Food Chemistry* 129 (2). Elsevier: 499–506. doi:10.1016/J.FOODCHEM.2011.04.107

Kumar, Ashwani, Amarjeet Kaur, Kritika Gupta, Yogesh Gat, and Vikas Kumar. 2021. "Assessment of Germination Time of Finger Millet for Value Addition in Functional Foods." *Current Science* 120 (2): 406–13.

Kumar, Ashwani, Amarjeet Kaur, Vidisha Tomer, Prasad Rasane, and Kritika Gupta. 2020. "Development of Nutricereals and Milk-Based Beverage: Process Optimization and Validation of Improved Nutritional Properties." *Journal of Food Process Engineering* 43 (1). John Wiley: e13025. doi:10.1111/JFPE.13025

Kumari, K. D. D., W. M. T. Madhujith, and G. A. P. Chandrasekara. 2019. "Millet Phenolics as Natural Antioxidants in Food Model Systems and Human LDL/VLDL Cholesterol in vitro." *Tropical Agricultural Research* 30 (3). Postgraduate Institute of Agriculture (PGIA): 13. doi:10.4038/TAR.V30I3.8316

Kunyanga, C. N., J. K. Imungi, M. Okoth, C. Momanyi, H. K. Biesalski, and V. Vadivel. 2011. "Antioxidant and Antidiabetic Properties of Condensed Tannins in Acetonic Extract of Selected Raw and Processed Indigenous Food Ingredients from Kenya." *Journal of Food Science* 76 (4). J Food Sci. doi:10.1111/J.1750-3841.2011.02116.X

Kunyanga, Catherine N., Jasper K. Imungi, Michael W. Okoth, Hans K. Biesalski, and Vellingiri Vadivel. 2012. "Total Phenolic Content, Antioxidant and Antidiabetic Properties of Methanolic Extract of Raw and Traditionally Processed Kenyan Indigenous Food Ingredients." *LWT – Food Science and Technology* 45 (2). Academic Press: 269–76. doi:10.1016/J.LWT.2011.08.006

Lakshmi Kumari, P., and S. Sumathi. 2002. "Effect of Consumption of Finger Millet on Hyperglycemia in Non-Insulin Dependent Diabetes Mellitus (NIDDM) Subjects." *Plant Foods for Human Nutrition 2002 57:3* 57 (3). Springer: 205–13. doi:10.1023/A:1021805028738

Lim, S. S., S. H. Jung, J. Ji, K. H. Shin, and S. R. Keum. 2001. "Synthesis of Flavonoids and Their Effects on Aldose Reductase and Sorbitol Accumulation in Streptozotocin-Induced Diabetic Rat Tissues." *The Journal of Pharmacy and Pharmacology* 53 (5): 653–68. doi:10.1211/0022357011775983

Lopez, Hubert W., Marie Anne Levrat, Christine Guy, Arnaud Messager, Christian Demigné, and Christian Rémésy. 1999. "Effects of Soluble Corn Bran Arabinoxylans on Cecal Digestion, Lipid Metabolism, and Mineral Balance (Ca, Mg) in Rats." *The Journal of Nutritional Biochemistry* 10 (9). Elsevier: 500–509. doi:10.1016/S0955-2863(99)00036-4

Majid, Abdul, and C G Poornima Priyadarshini. 2019. "Millet Derived Bioactive Peptides: A Review on Their Functional Properties and Health Benefits." *Critical Reviews in Food Science and Nutrition* 60 (19). Taylor & Francis: 3342–51. doi:10.1080/10408398.2019.1686342

Mangala, Silanere L., Nagappa G. Malleshi, Rudrapatnam N. Tharanathan, and Mahadevamma. 1999. "Resistant Starch from Differently Processed Rice and Ragi (Finger Millet)." *European Food Research and Technology 1999 209:1* 209 (1). Springer: 32–37. doi:10.1007/S002170050452

Murtaza, Nida, Ritesh K. Baboota, Sneha Jagtap, Dhirendra P. Singh, Pragyanshu Khare, Siddhartha M. Sarma, Koteswaraiah Podili, et al. 2014. "Finger Millet Bran Supplementation Alleviates Obesity-Induced Oxidative

Stress, Inflammation and Gut Microbial Derangements in High-Fat Diet-Fed Mice." *British Journal of Nutrition* 112 (9). Cambridge University Press: 1447–58. doi:10.1017/S0007114514002396

Nagar, Hemant Kumar, Amit Kumar Srivastava, Rajnish Srivastava, Madan Lal Kurmi, Harinarayan Singh Chandel, and Mahendra Singh Ranawat. 2016. "Pharmacological Investigation of the Wound Healing Activity of Cestrum Nocturnum (L.) Ointment in Wistar Albino Rats." *Journal of Pharmaceutics* 2016 (February). Hindawi: 1–8. doi:10.1155/2016/9249040

Nassarawa, Sanusi Shamsudeen, and Salamatu Ahmad Sulaiman. 2019. "Comparative Analyses on the Chemical Compostion, Phytochemcimal and Antioxidant Properties of Selected Milled Varieties (Finger and Pearl Millet)." *International Journal of Food Sciences* 2 (1): 25–42. doi:10.47604/IJF.997

Naushad Emmambux, M., and John R.N. Taylor. 2013. "Morphology, Physical, Chemical, and Functional Properties of Starches from Cereals, Legumes, and Tubers Cultivated in Africa: A Review." *Starch/Staerke* 65 (9–10): 715–29. doi:10.1002/STAR.201200263

Nazni, P., and J. Bhuvaneswari. 2021. "Analysis of Physico Chemical and Functional Characteristics of Finger Millet (Eleusine Coracana L) and Little Millet (P. Sumantranse)." *International Journal of Food and Nutritional Sciences* 4 (3). Medknow Publications: 109. www.ijfans.org/article.asp?issn=2319-1775;year= 2015;volume=4;issue=3;spage=109;epage=114;aulast=Nazni;type=0

Nefale, Fulufhelo E, and Mpho E Mashau. 2018. "Effect of Germination Period on the Physicochemical, Functional and Sensory Properties of Finger Millet Flour and Porridge." *Asian Journal of Applied Sciences*, 2321–0893. www.ajouronline.com

Nithiyanantham, Srinivasan, Palanisamy Kalaiselvi, Mohamad Fawzi Mahomoodally, Gokhan Zengin, Arumugam Abirami, and Gopalakrishnan Srinivasan. 2019. "Nutritional and Functional Roles of Millets – A Review." *Journal of Food Biochemistry* 43 (7). Blackwell Publishing. doi:10.1111/JFBC.12859

Okoyomoh, K., Okere, O. S., Olowoniyi, O. D., and Adejo, G. O. 2013. "Antioxidant and antidiabetic properties of eleucine coracana (l.) geartn. (finger millet)seed coat matter in streptozotocin induced diabetic rats." *ASJ International Journal of Advances in Herbal and Alternative Medicine (IJAHAM)* 1 (1): 1–09. www.academiascholarlyjournal.org/ijaham/index_ijaham.htm

Onyango, Calvin, Susan Karenya Luvitaa, Guenter Unbehend, and Norbert Haase. 2020. "Physico-Chemical Properties of Flour, Dough and Bread from Wheat and Hydrothermally-Treated Finger Millet." *Journal of Cereal Science* 93 (May). Academic Press: 102954. doi:10.1016/J.JCS.2020.102954

Palaniappan, Ayyappan, V. Geetha Balasubramaniam, and Usha Antony. 2017. "Prebiotic Potential of Xylooligosaccharides Derived from Finger Millet Seed Coat." *Food Biotechnology* 31 (4). Taylor & Francis: 264–80. doi:10.1080/08905436.2017.1369433

Palanisamy, Bruntha Devi, Vijayabharathi Rajendran, Sathyabama Sathyaseelan, Rajeev Bhat, and Brindha Priyadarisini Venkatesan. 2011. "Enhancement of Nutritional Value of Finger Millet-Based Food (Indian Dosa) by Co-Fermentation with Horse Gram Flour." Http://Dx.Doi.Org/10.3109/09637486.2011.591367 63 (1). Taylor & Francis: 5–15. doi:10.3109/09637486.2011.591367

Panda, Debabrata, N. Hema Sailaja, Bandana Padhan, and Kartik Lenka. 2019. "Sprouting-Associated Changes in Nutritional and Physico-Functional Properties of Indigenous Millets from Koraput, India." *Proceedings of the National Academy of Sciences, India Section B: Biological Sciences 2019 90:1* 90 (1). Springer: 79–86. doi:10.1007/S40011-019-01085-X

Patel, Priya, and S. S. Thorat. 2019. "Studies on Chemical, Textural and Color Characteristics of Decorticated Finger Millet (Eleusine Coracana) Fortified Sponge Cake." ~ 64 ~ *The Pharma Innovation Journal* 8 (3): 64–67. www.thepharmajournal.com

Radhika, Amreen Virk, Manpreet Kaur, Priyanka Thakur, Divya Chauhan, Qurat Ul Eain Hyder Rizvi, Sumaira Jan, and Krishan Kumar. 2019. "Development and Nutritional Evaluation of Multigrain Gluten Free Cookies and Pasta Products." *Current Research in Nutrition and Food Science* 7 (3). Enviro Research Publishers: 842–53. doi:10.12944/CRNFSJ.7.3.23

Rajasekaran, N. S., M. Nithya, C. Rose, and T. S. Chandra. 2004. "The Effect of Finger Millet Feeding on the Early Responses during the Process of Wound Healing in Diabetic Rats." *Biochimica et Biophysica Acta* 1689 (3). Biochim Biophys Acta: 190–201. doi:10.1016/J.BBADIS.2004.03.004

Ramashia, S. E., E. T. Gwata, S. Meddows-Taylor, T. A. Anyasi, and A. I.O. Jideani. 2018. "Some Physical and Functional Properties of Finger Millet (Eleusine Coracana) Obtained in Sub-Saharan Africa." *Food Research International* 104 (February). Elsevier: 110–18. doi:10.1016/J.FOODRES.2017.09.065

Rathore, Teena, Rakhi Singh, B Dinkar Kamble, Ashutosh Upadhyay, S. Thangalakshmi.. 2019. "Review on Finger Millet: Processing and Value Addition." ~ 283 ~ *The Pharma Innovation Journal* 8 (4): 283–91. www.thepharmajournal.com

Reddy, Chagam Koteswara, and Kotapati Kasi Viswanath. 2019. "Impact of γ-Irradiation on Physicochemical Characteristics, Lipoxygenase Activity and Antioxidant Properties of Finger Millet." *Journal of Food Science and Technology 2019 56:5* 56 (5). Springer: 2651–59. doi:10.1007/S13197-019-03753-2

Roberts Pradeepa, Tp, V. Thejasri, and T. V. Hymavathi. 2019. "Enhancing Cooking, Sensory and Nutritional Quality of Finger Millet Noodles through Incorporation of Hydrocolloids." www.researchgate.net/publication/332142609

Roopa, S., and K. S. Premavalli. 2008. "Effect of Processing on Starch Fractions in Different Varieties of Finger Millet." *Food Chemistry* 106 (3). Elsevier: 875–82. doi:10.1016/J.Foodchem.2006.08.035

Sakamma, S., K. B. Umesh, M. R. Girish, S. C. Ravi, M. Satishkumar, and Veerabhadrappa Bellundagi. 2018. "Finger Millet (Eleusine Coracana L. Gaertn.) Production System: Status, Potential, Constraints and Implications for Improving Small Farmer's Welfare." *Journal of Agricultural Science* 10 (1). doi:10.5539/jas.v10n1p162

Sarma, Siddhartha Mahadeva, Dhirendra Pratap Singh, Paramdeep Singh, Pragyanshu Khare, Priyanka Mangal, Shashank Singh, Vandana Bijalwan, et al. 2018. "Finger Millet Arabinoxylan Protects Mice from High-Fat Diet Induced Lipid Derangements, Inflammation, Endotoxemia and Gut Bacterial Dysbiosis." *International Journal of Biological Macromolecules* 106 (January). Elsevier: 994–1003. doi:10.1016/J.IJBIOMAC.2017.08.100

Savitha Prashanth, M. R., R. R. Shruthi, and G. Muralikrishna. 2015. "Immunomodulatory Activity of Purified Arabinoxylans from Finger Millet (Eleusine Coracana, v. Indaf 15) Bran." *Journal of Food Science and Technology* 52 (9). J Food Sci Technol: 6049–54. doi:10.1007/S13197-014-1664-4

Shobana, S., and N. G. Malleshi. 2007. "Preparation and Functional Properties of Decorticated Finger Millet (Eleusine Coracana)." *Journal of Food Engineering* 79 (2). Elsevier: 529–38. doi:10.1016/J.JFOODENG.2006.01.076

Shobana, S., M. R. Harsha, K. Platel, K. Srinivasan, and N. G. Malleshi. 2010. "Amelioration of Hyperglycaemia and Its Associated Complications by Finger Millet (Eleusine Coracana L.) Seed Coat Matter in Streptozotocin-Induced Diabetic Rats." *The British Journal of Nutrition* 104 (12). Br J Nutr: 1787–95. doi:10.1017/S0007114510002977

Shobana, S., Y. N. Sreerama, and N. G. Malleshi. 2009. "Composition and Enzyme Inhibitory Properties of Finger Millet (Eleusine Coracana L.) Seed Coat Phenolics: Mode of Inhibition of α-Glucosidase and Pancreatic Amylase." *Food Chemistry* 115 (4). Elsevier: 1268–73. doi:10.1016/J.FOODCHEM.2009.01.042

Shukla, Kamini, and Sarita Srivastava. 2011. "Evaluation of Finger Millet Incorporated Noodles for Nutritive Value and Glycemic Index." *Journal of Food Science and Technology 2011 51:3* 51 (3). Springer: 527–34. doi:10.1007/S13197-011-0530-X

Singh, Nidhi, G. Meenu, A. Sekhar, and J. Abraham. 2015. "Evaluation of Antimicrobial and Anticancer Properties of Finger Millet (Eleusine Coracana) and Pearl Millet (Pennisetum Glaucum) Extracts." *The Pharma Innovation Journal* 3 (11): 82–86.

Sinha, Rekha, and Bindu Sharma. 2017. "Use of Finger Millet in Cookies and Their Sensory and Nutritional Evaluation." doi:10.18805/ajdfr.v36i03.8962

Siwela, Muthulisi, John R. N. Taylor, Walter A. J. de Milliano, and Kwaku G. Duodu. 2010. "Influence of Phenolics in Finger Millet on Grain and Malt Fungal Load, and Malt Quality." *Food Chemistry* 121 (2). Elsevier: 443–49. doi:10.1016/J.FOODCHEM.2009.12.062

Siwela, Muthulisi, John R. N. Taylor, Walter A. J. De Milliano, and Kwaku G. Duodu. 2007. "Occurrence and Location of Tannins in Finger Millet Grain and Antioxidant Activity of Different Grain Types." *Cereal Chemistry* 84 (2): 169–74. doi:10.1094/CCHEM-84-2-0169

Sripriya, G., K. Chandrasekharan, V. S. Murty, and T. S. Chandra. 1996. "ESR Spectroscopic Studies on Free Radical Quenching Action of Finger Millet (Eleusine Coracana)." *Food Chemistry* 57 (4). Elsevier: 537–40. doi:10.1016/S0308-8146(96)00187-2

Subba Rao, M. V., and G. Muralikrishna. 2002. "Evaluation of the Antioxidant Properties of Free and Bound Phenolic Acids from Native and Malted Finger Millet (Ragi, Eleusine Coracana Indaf-15)." *Journal of Agricultural and Food Chemistry* 50 (4). J Agric Food Chem: 889–92. doi:10.1021/JF011210D

Tripathi, Jaya, Ranu Prasad, Alka Gupta, and Vinita Puranik. 2015. "Development of Value Added Pasta with Incorporation of Malted Finger Millet Flour." *Journal of Applied and Natural Science* 7 (2). ANSF Publications: 598–601. doi:10.31018/JANS.V7I2.651

Venkateswaran, V., and G. Vijayalakshmi. 2010. "Finger Millet (Eleusine Coracana) – an Economically Viable Source for Antihypercholesterolemic Metabolites Production by Monascus Purpureus." *Journal of Food Science and Technology 2010 47:4* 47 (4). Springer: 426–31. doi:10.1007/S13197-010-0070-9

Verma, Veenu, and S. Patel. 2013. "Value Added Products From Nutri-Cereals: Finger Millet (Eleusine Coracana)." *Emirates Journal of Food and Agriculture* 25 (3): 169/176-169/176. doi:10.9755/EJFA.V25I3.10764

Viswanath, Varsha, Asna Urooj, and N. G. Malleshi. 2009. "Evaluation of Antioxidant and Antimicrobial Properties of Finger Millet Polyphenols (Eleusine Coracana)." *Food Chemistry* 114 (1). Elsevier: 340–46. doi:10.1016/J.FOODCHEM.2008.09.053

Vora, Jyoti, Ashwini Ganesh Rane, and Priyanka Jadhav. 2014. "Biochemical, Antimicrobial and Organoleptic Studies on the Germination Profile of Finger Millet (Eleusine Coracana) - Volume 3, No.4, October, 2014 - IJLBPR." *International Journal of Life Science Biotechnology and Pharma Research* 3 (4): 123–28.

Weerahewa, H. L. D., M. M. F. Fasreen, and Odan Perera. 2017. "Development of Finger Millet Based Probiotic Beverage Using Lactobacillus Casei431." *Journal* 12 (1): 128–38. doi:10.4038/ouslj.v12i1.7384

Xiang, Jinle, Franklin B. Apea-Bah, Victoria U. Ndolo, Mangani C. Katundu, and Trust Beta. 2019. "Profile of Phenolic Compounds and Antioxidant Activity of Finger Millet Varieties." *Food Chemistry* 275 (March). Elsevier: 361–68. doi:10.1016/J.FOODCHEM.2018.09.120

Yadav, Geetanjali, Anshu Singh, Patrali Bhattacharya, Jude Yuvraj, and Rintu Banerjee. 2013. "Comparative Analysis of Solid-State Bioprocessing and Enzymatic Treatment of Finger Millet for Mobilization of Bound Phenolics." *Bioprocess and Biosystems Engineering 2013 36:11* 36 (11). Springer: 1563–69. doi:10.1007/S00449-013-0924-4

Nutraceutical and Functional Attributes of Foxtail Millet (*Setaria italica*)

L. Karpagapandi, A. Kalaiselvan, and V. Baskaran

Contents

DOI: 10.1201/9781003251279-6

6.1 Introduction

Foxtail millet (*Setaria italica* L.) is an essential ancient staple food of dry land agriculture, able to withstand climate change and provide food and nutrition security. It is an annual grass from the *Poaceae* family and has a "self pollinating" C4 crop, adapted to hot and dry environments and good as food for human consumption and feed for cattle, poultry and caged birds. Due to its poor crossover fertility, distinctive floral shape and antithesis behavior, it also possesses fodder species. Due to this, foxtail millet is one of the species that is hardest to cross-pollinate (Murugan and Nirmalakumari, 2006). The prominent attributes of foxtail millet – including a tiny genome and a brief life cycle, abiotic stress tolerance and an in-breeding nature – have emphasized foxtail millet as a novel model food grain for researching C4 photosynthesis, stress biology and biofuel characteristics (Yang et al., 2020b).

Foxtail millet, also known as Italian millet (or German or Chinese or Hungarian millet) is widely considered to be the first whole grain produced by humans. It is a native crop of China. It is currently the sixth highest grain yielding crop in the world and is grown in 26 Asian, African, and American subcontinent countries (Dayakar Rao et al., 2017). Among the millets, foxtail millet is the second most cultivated millet after pearl millet and continues to have an important place in the world agriculture, providing approximately six million tons of food to millions of people, mainly to the poor. Additionally, it is grown in marginal soils in Southern Europe as well as in temperate, subtropical, and tropical Asia (Dayakar Rao et al., 2017).

6.1.1 Origin and distribution

Foxtail millet is thought to have originated in China around 2700 BC (8,000 years ago) and was domesticated in Central Asia's highlands. During China's Neolithic era, it was the most important plant food. Foxtail millet has been planted in Russia since ancient times, with evidence dating back as far as 1,500 years. From China or Central Asia, it spread to India and European countries. However, a multiple domestication hypothesis has been widely accepted because of its cultivation about 4,000 years ago at Europe (Hariprasanna, 2016). Among the famous Tamil literature, "Thirukural," which was written 2,500 years ago by the poet Thiruvalluvar, used the name *thinai* meaning foxtail millet in five of his *kural* (Thirukural

No. 104, 144, 433, 931 and 1282). This shows that foxtail millet has been cultivated in India for the past 2,500 years.

Foxtail millet can grow in altitudes of 2,000 m above sea level. It is a drought tolerant crop, but it cannot tolerate water logging conditions due to its shallow root system. It is, nevertheless, adapted to grow in a wide range of altitudes, soils, and temperatures. Due to the planting of several cash crops, like sunflower and soybean, on black soils throughout the 1990s, the area of foxtail millet cultivation in India has been reduced by more than half. The well-known Old World species of foxtail millet among human food is *Setaria italica*. On the other hand, foxtail species have been farmed and utilized in Europe, Africa, and Asia. Until the early twentieth century, *Setaria italica* was a popular cuisine in southern Europe. Its importance has waned since then, although it is still a necessary diet in portions of India, China, Korea, and Japan.

In India, foxtail millet cultivation predates that of China, but archaeological sites in the states of Tamilnadu, Karnataka, and Andhra Pradesh include seeds collected from wild plants that date to about 2800 cal BC. They date to 2400 BC in the Harappan region of northern India and Pakistan, making them almost as old. The "Navadhanyam" ("Nine Grains") of India, which include foxtail millet, are even less well-known in the West than the "Five Grains of China." In both India and the United States, this grain blend is used in regular diets and offered as *pujas* (offerings for worship) at shrines and temples.

In Europe and the Middle East, foxtail millet is a somewhat more recent crop. Grains have been found in Greece, Italy, Switzerland, Turkey, and Yugoslavia, among other places, that belong to the Bronze Age (2000 BC). In contemporary hearth sweepings and domestic detritus, Miller and Smart (1984) discovered the foxtail millet species *Setaria italica* in the villages of Milyan, Iran. Additionally, they found *Setaria's* remains in burned dung in the "prehistoric" settlement of Milyan. Foxtail millet was a popular cuisine during the Roman era and it was especially popular in Pontus, Thrace, the Po Valley in northern Italy, and Elymian Sicily.

6.1.2 Botanical information

Foxtail millet belongs to the *Setaria* genus, has over 125 species, including *Setaria italica*, and is extensively dispersed throughout warm and temperate regions of the world. The green foxtail (*Setaria viridis* (L.) P. Beauv) is a wild ancestor of foxtail millet *S. italica*. Biologically these two species are same and *S. viridis* is probably an ancestor of domesticated foxtail millet. This genus is a member of the tribe *Paniceae* and subfamily *Panicoideae*. They contain weed, grain and wild species with various life cycles, ploidy levels and breeding strategies. It is an erect annual grass with slim, vertical, leafy stems that can reach a height of 90–220 cm, fast-growing and tufted. The seedhead is a dense, hairy panicle 5–30 cm. It has a tiny seed with a diameter of about 2 millimeters that is covered in a papery, thin shell that is simple to remove during threshing. The colour of the seed varies greatly and mainly depends on varieties. Its root system is thick, with thin adventitious roots. The stems are straight and slender, with a tillering base. The blades are lanceolate and serrated, the leaves are 15–50 cm long and 0.5–4 cm wide. An erect or pendulous spike-like bristly panicle, 5–30 cm long x 1–5 cm wide, with between 6 and 12 spikelets, forms the inflorescence. It is fairly tolerant to drought and hence safe from droughts because of early maturity. Due to its fast growth can be grown as a short-term catch crop. It is adapted to a wide range of elevations, soils and temperatures and harvested within 90 days after sown. Moreover, it is a spring season crop.

6.1.3 Production

As mentioned in earlier section, for the purpose of grains or fodder, foxtail millet is widely grown in Asia, Europe, North America, Australia, and North Africa. Foxtail millet is one of the staple grains consumed in China, India, Korea and Japan. With an annual cultivating area of over two million hectare and an annual total grain production of roughly six metric tones, China ranks first in the world for foxtail millet production (80%). With 10 percent of the global production, India is the second-largest producer of foxtail millet. One of the principal food crop in Northern China is foxtail millet, which is mostly grown in arid and semi-arid regions in the Northern section of India (Zhang et al., 2017). It is a significant crop that is farmed for silage and hay in South and North America and is used as a staple food in many countries throughout the world, including China, India, and Japan.

Since it was first cultivated in India in antiquity, foxtail millet has had the longest history of all millet varieties. In India it is mainly grown in the states of Andhra Pradesh, Telangana, Karnataka, Tamilnadu, Maharashtra, Rajasthan, Madhya Pradesh and Uttar Pradesh and to a small extent in the northeastern states. In southern states like Tamilnadu the flour of foxtail millet mixed with honey (*Thaen Thinai Laddu*) and given as *Prasadham* (holly food) to the devotees in temples of Lord Murugan which shows other than food purpose it has spiritual value too. Like other millets, foxtail millet can be cultivated in harsh climates like high temperatures with minimum rainfall. In various regions of India, foxtail millet is referred to as several names, including *Kangni, Kakum* (Hindi), *Kanguni* (Sanskrit), *Kang* (Gujrati), *Kaon dana* (Bengali), *Kanghu, Kangam, Kora* (Oriya), *Kangni* (Punjabi), *Kang, Rala* (Marathi), *Navane* (Kannada), *Korra, Korralu* (Telugu), *Kavalai, Tenai* (Tamil) and *Thina* (Malayalam).

Foxtail millet was a staple food grain of rural people in the central and southern states of India till the introduction of subsidized rice schemes in the early 1990s like public distribution systems. When people switched over to rice the demand of foxtail millet started declining. But, currently, due to increase in awareness and knowledge about its nutritional and nueutraceutical values the demand for this millet is increasing and owing to it is slightly expensive than rice. In India, state agricultural universities have introduced different varieties of foxtail millet with improved yield potential and nutritional values. The following is a list of the most recent and popular foxtail millet cultivars that are suggested for various states (**Table 6.1**).

Table 6.1 State-wise Popular Foxtail Millet Cultivars

State	Popular varieties in the states of India
Uttarakhand	PS 4, PRK 1, Sreelaxmi, SiA 326, SiA 3156, SiA 3085
Uttar Pradesh	PRK 1, PS 4, SiA 3085, SiA 3156, Sreelaxmi, Narasimharaya, S-114,
Bihar	RAU-1, SiA3085, SiA 3088, SiA 3156, PS 4
Rajasthan	Prathap Kangani 1 (SR 51), SR 1, SR 11, SR 16, SiA 3085, SiA 3156, PS 4
Andhra Pradesh	SiA 326, SiA 3085, SiA 3088, SiA 3156, Lepakshi, Narasimharaya, Krishnadevaraya, PS 4
Telangana	SiA 326, SiA 3085, SiA 3088, SiA 3156, Lepakshi
Tamil Nadu	TNAU 43, TNAU-186, Co 1, Co 2, Co 4, Co 5, Co (Te) 7, K2, K3, SiA 3085, SiA 3088, SiA 3156, PS 4
Karnataka	DHFt-109-3, HMT 100-1, SiA 326, SiA 3085, SiA3088, SiA 3156, PS 4, Narasimharaya

Source: www.millets.res.in/technologies/2-Recommended_Package_of_Practices-Foxtail_Millet.pdf and Hariprasanna et al., ICAR-IIMR, Extension Folder 4, 2018.

6.1.4 Physical, nutritional and therapeutic properties

Foxtail millet grains have a length of around two millimetres and the glumes might be red, white, yellow, brown, or black. Tannins were discovered in both white and red variants. The weight of 1,000 grains after husk and bran removal is around 2.6 g, and the milling yield is approximately 77 percent. Although polygonal forms have also been discovered, the starch granules in the floury endosperm of foxtail millet are typically spherical (Taylor and Emmambux, 2008). The pericarp layer, aleurone layer, and germ of foxtail millet make up around 8–10 percent of millet quality and are the principal by-products of foxtail millet processing. Previously, foxtail millet bran was widely used as animal feed in China, but it has recently gained popularity due to the discovery that millet bran oil is high in polyunsaturated fatty acids, particularly linoleic acid and tocopherol, and phytosterol composition of millet bran oil is also available (Slama et al., 2020).

Foxtail millets are nutritious, gluten-free and non-acid forming foods. It is easy to digest and also a good source of energy, fat, protein, fatty acids, vitamins, minerals and dietary fibre (See Table 6.2). Foxtail millets contain moderate protein (11–12%) and crude fibre (above 8%). The starch content is about 50–55 percent, which is quite low for a cereal, though some varieties with about 70 per cent starch content. Muthamilarasan et al. (2016) reported that foxtail millet seeds comprise 12.5 percent protein, 8 percent crude fibre, 4.3 percent fat, 3.3 ash and 61 percent carbohydrates. The digestible portion of foxtail millet is approximately 80 percent, with the remaining 20 percent consisting of dietary fibres, bioactive elements, anti-nutritional factors, and other indigestible oligo-polysaccharides that aid in the digestive process by causing laxative effect and reducing transit time (Sharma and Niranjan, 2018).

The amount of amylose and amylopectin in the starch varies depending on the variety of foxtail millet used. There are three types of foxtail millet: waxy (high amylopectin), regular (low amylose), and non-waxy (high amylose). Generally, the amylose content of foxtail millet is 33 percent. Except for lysine and threonine, foxtail millet proteins are rich providers of essential amino acids, however they have a lot of sulphur-containing amino acids (methionine and cysteine) (Huchchannanavar et al. 2019). Prolamin accounts for roughly 40 percent of total extractable nitrogen, while 20 percent is extracted using a reducing agent, showing that foxtail millet, like most other cereals, is high in prolamin protein with a large proportion of disulfide linkages.

Foxtail millet is also rich in minerals, like calcium, iron, zinc, manganese, potassium, copper, and phosphorus (Kola et al., 2020). Millets have a mineral composition comparable to sorghum, but calcium and manganese levels are extremely high. Common, little, foxtail, kodo, and barnyard millets had higher total mineral matter or ash content than most regularly consumed cereal grains, including sorghum. Compared with rice, foxtail millet has two-fold higher proteins content, four-fold minerals and fat and three-fold content of calcium. Ample evidence shows increasing consumption of foxtail millet, and its bioactivy is associated with a lower risk of diabetes and obesity (Anitha et al., 2021). As a result, foxtail millet is gaining popularity among customers. In general, the yellow colour of the de-husked grain is the most visible indicator of millet quality for consumers.

6.1.5 Utilization method

Traditionally, the method of utilization of cereal in general or millet grains are for foods and another for beverages. Further, it is categorized into three groups (i) grain; (ii) meal and (iii) flour, and the latter non-alcoholic and alcoholic beverages. For preparation of food from

Table 6.2 Nutrient Composition of Foxtail millet (*Setaria italica*) Grain (per 100 g)

Component	Quantity
Proximate composition	
Carbohydrate (g)	60.9
Protein (g)	12.3
Fat (g)	4.3
Energy (KCal)	331
Crude fibre (g)	8.0
Mineral matter (g)	3.3
Starch (g)*	57.9
Amylose (%)	17.5
Amylopectin (%)	82.5
Total Sugars (g)*	1.5
Essential amino acid (per g of N)	
Arginine (mg)	220
Histidine (mg)	130
Lysine (mg)	140
Tryptophan (mg)	60
Phenylalanine (mg)	420
Methionine (mg)	180
Cystine (mg)	100
Threonine (mg)	190
Leucine (mg)	1040
Isoleucine (mg)	480
Valine (mg)	430
Fatty acids	
Palmitic acid (mg)	6.4
Stearic acid (mg)	6.3
Oleic acid (mg)	13.0
Linoleic acid (mg)	66.5
Vitamins	
Thiamine, B_1 (mg)	0.59
Riboflavin, B_2 (mg)	0.11
Niacin, B_3 (mg)	3.2
Pantothenic acid, B_5 (mg)	0.82
Vitamin A (Carotene)	32
Vitamin E	31
Folic acid	5
Minerals	
Calcium (mg)	31
Phosphorus (mg)	290
Iron (mg)	2.8
Zinc (mg)	2.4
Magnesium (mg)	81
Sodium (mg)	4.6
Potassium (mg)	250
Copper (mg)	1.4
Manganese (mg)	0.6
Chromium (mg)	0.03
Chlorine (mg)	37

Source: Gopalan et al., 2015 Nutritive Value of Indian Foods, NIN; *Heuzé et al. (2015).

the grains of millets especially with foxtail millet, seven subcategories have been recorded, namely boiled grain, gruel, mochi (cake made from waxy starch), porridge and dumplings, flour porridge and bread.

Foxtail millet is a staple meal in northern China, where it is commonly used as a nourishing gruel or soup for expectant and nursing mothers as well as for food therapy (Jali et al., 2012). It is frequently combined with pulses and boiled, or the flour is used with other grains to make bread or noodle dough. Sprouted foxtail millet also supplemented the Chinese diet. In India, the husked grain is cooked as rice; the flour is made into porridge or mash and baked into unleavened bread.

Foxtail cultivars with yellow seeds are used as astringents, digestives, emollients and stomachics in medicine. Additionally, dyspepsia, poor digestion, and stagnant food in the abdomen are all treated with it. White seeds are refrigerant and used to treat cholera and fever, whereas green seeds are diuretic and boost virility (Dias-Martins et al., 2018). In Asia, Southeastern Europe, and Africa, the husked grains of foxtail millet are used as food. The flour is used for making cakes, porridges and puddings. It is also used to make beer and alcohol in Russia and Myanmar, as well as vinegar and wine in China. It is predominantly grown as bird feed, hay and silage in Europe and the United States. In China, the straw is an important fodder. Due to its diuretic, appetite stimulant, emollient, and digestive characteristics, foxtail millet is widely utilized as a geriatric food in Korea, China, India, and Japan (Sireesha et al., 2011; Vetriventhan et al., 2016; Bembem et al., 2020).

Foxtail millet is considered to be an easily digestible food grain, being gluten-free like buckwheat. It also contains more proteins, fibre, and minerals, and because of its high lysine level, it can be used as a functional food ingredient as a supplementary protein source for most cereals. Nowadays because of these qualities, medical practitioners now recommend it in Ayurvedic and Unani products (Indian traditional medicine). Due to its short growing season, foxtail millet is well adapted to agricultural area with low rainfall. Even though the yield of the grain is very low it is widely cultivated in India due to its medicinal qualities. Though foxtail millet has lost its importance as a food crop due to competition from rice, wheat, maize, and sorghum, it still has its own place in world agriculture and market value in terms of output and productivity.

6.2 Nutraceutical properties of foxtail millet

The "Concepts of Food" are shifting from a previous emphasis on survival, hunger satisfaction, absence of adverse effects on health and health maintenance to a current emphasis on the use of nutraceuticals in foods, which promise to promote better health and well-being, thus helping to reduce the risk of chronic life-style illnesses such as cardiovascular diseases, some cancers, diabetes, obesity, and so on. Because they are high in health-promoting elements and phytochemicals, epidemiological studies indicate that diets rich in plant foods, especially whole grains, protect men from noncommunicable diseases. Millets are considered as functional foods/food grains since they are rich in highly effective health-promoting phytochemicals (Himanshu et al., 2018). The health benefits of millets are dependent upon their metabolic profiles, including the types and amounts of natural phenolic compounds present (Pradeep and Sreerama, 2018).

Millets being a rich source of micronutrients and bioactive phytochemicals reportedly exhibit antioxidant, anti-cancerous, anti-diabetic, anti-aging, anti-hypertensive, cardio protective, and many other health attributes (Pradeep and Sreerama, 2018). These compounds include phenolic (phenolic acids, flavonoids, and tannins), carotenoid and tocopherols, phytosterols, policosanols and bioactive peptides. Studies revealed that these bioactive compounds have the potential to combat diet-related non-communicable diseases such as cardiovascular disease, diabetes, cancer, and so forth (Duodu and Awika, 2019; Shah et al. 2021).

Foxtail millet contains fatty acids and vitamins such as oleic acid, linoleic acid, tocopherol, and phytosterol, as well as tannins, phytate, alkaloids, polyphenols, flavonoids, and triterpenoids, all of which have anti-oxidative, anti-inflammatory, anti-carcinogenic, antimicrobial, hepatoprotective, hypocholesterolemic/hypolipidemic and anti-hyperglycemic activities, thereby reducing the risk of several degenerative chronic complications. Free radical scavengers, reducing agents, radical quenchers, and metal chelating agents have been reported to act as phytochemical substances in foxtail millet, preventing the generation of reactive oxygen species (ROS), hydroxyperoxides, and singlet oxygen molecules (Kumari et al., 2016). The consumption of foxtail millet is thought to be anti-cholera and anti-fever. It is commonly used as an astringent and stomachic, as well as for boosting virility. This seed extract is thought to provide a gluten-free treatment for celiac disease (Amadou et al., 2013b) and reduce type II diabetes (Zhang and Liu, 2015). The hypoglycemic effect of foxtail millet was first documented more than 400 years ago in traditional Chinese medicine literature. Studies have also suggested that foxtail millet protein, peptide fractions, and phenols have a hypolipidemic impact. In this chapter, the various health benefits based on the nutraceutical properties of phytochemical components of foxtail millets are detailed.

6.2.1 Antioxidant properties

Antioxidants, which are needed to minimize the risk of chronic illness and to keep foods from spoiling, have received a lot of attention in recent years because of their capacity to neutralize free radicals' effects. Free radicals are potentially hazardous chemicals produced by a variety of natural processes in the body and the food system that are linked to cell and tissue ageing and food rotting. The accumulation of free radicals reported to aggravate cardiovascular disease, cancer, diabetes, arthritis and various neurodegenerative disorders. Interestingly, antioxidants (phenolics and carotenoids) found in the foxtail millet grains have potential to modulate the free radicals inside the body (Zhang and Liu, 2015) and foods (Suma and Urooj, 2011). According to Suma and Urooj (2011), bran-rich foxtail millet fraction and whole flour had stronger radical scavenging activity (44.62 percent and 51.80 percent, respectively) and reduction power (0.381 and 0.455, respectively) at the 2 mg level. The free radical quenching potential of foxtail millet crude extract and fractionated protein hydrolysate were revealed to have significant antioxidant activity by scavenging free radicals.

Studies revealed relatively high antioxidant activity of organic solvents and aqueous extracts of foxtail millet against stable DPPH radical cations. Bangoura et al. (2013a) reported the inhibition activity of solvent extracts of two species of foxtail millet's fibres against ABTS$^+$ radical cations. Sharma, S. et al. (2018) reported that the total antioxidant activity and reducing power of germinated foxtail millet found to be higher due to a quantitative increase in polyphenol content of foxtail millet. They also described that *in-vitro* antioxidant capacities of foxtail millet were improved by the germination process as measured by metal chelating activity (34.92 to 57.38 µM of ferrous equivalent Fe (II) per gram of phenolic extract), reducing power assays (0.76 (µg/ml)), DPPH free radical scavenging activity and hydrogen peroxide scavenging activities (35.44 to 63.07 mM trolox/g) and total antioxidant capacity (29.0–45.23 mgAAE/g) by phosphomolybdinum reduction assay.

6.2.2 Anti-obesity effects

Obesity leads to improper lipid and glucose metabolism, which increases the risk of cardiovascular disease and type II diabetes. Obesity-related type II diabetes is characterized by insulin action and secretion abnormalities, resulting in changes in lipid and glucose metabolism. Insulin resistance, hyperinsulinemia, hypertension, and dyslipidemia are the most common causes of metabolic syndromes. It is widely understood that an increase in abdominal visceral adipose tissue leads to a decrease in physical activity and the development of obesity. Adipose tissue has also just been discovered to express genes for a number of adipocytokines and release anti-atherogenic and anti-insulin resistance proteins like tumor necrosis factor, interleukin-6, resistin, leptin, and adiponectin in addition to serving as a tissue for energy storage (Sharma and Baskaran, 2021). Adiponectin levels in the blood decrease with the advent of obesity and type II diabetes.

Proteins purified from foxtail millet are reported to serve as a food supplement for the obesity-related diseases such as type II diabetes mellitus and cardiovascular diseases. Studies have found that dietary protein from foxtail millet, compared with casein-based diet, increased the plasma concentrations of High-Density Lipoprotein (HDL) cholesterol and adiponectin and greatly decreased the concentration of insulin protein. In young healthy men, serum adiponectin concentration was found to be positively connected to HDL-cholesterol and apolipoprotein-A or Low-Density Lipoprotein (LDL) levels and negatively related to triglyceride and Apolipoprotein B levels. Increases in insulin sensitivity were more closely linked to plasma adiponectin concentration than changes in body weight or plasma glucose concentration. Hypoadiponectinemia in obesity and type II diabetes is thought to be influenced by hyperinsulinemia and insulin resistance (Kazumi et al., 2004). The presence of high dietary fibre and tryptophan in foxtail millet is found to help control the body weight and slow down the digestion process and appetite that prevents intake of extra calories. The fibre-rich foxtail millet-based diet gives a feeling of fullness, which helps to control excessive food consumption. Besides, changing eating habits, exciting physical activity with enough sleep may also facilitate overcoming diabetes and obesity of all age groups.

6.2.3 Anti-inflammatory properties

Free radicals are normal metabolic products in the body. However, over-production of reactive oxygen species (ROS) is far beyond the body's scavenging capacity of antioxidant defense mechanisms that lead to oxidative stress. Many degenerative disorders, including cancer, stroke, myocardial infarction and inflammation, have been linked to the formation of ROS. When compared to other natural substances, plant-derived polyphenolic compounds are the most potent anti-oxidant and anti-inflammatory agents. According to numerous studies (Naveen et al., 2021). In LPS-induced HT-29 cells and nude mice, the bound polyphenols of the inner shell of foxtail millet bran were found to have anti-inflammatory properties (Shi et al., 2017). Bioactive peptides such as Pro-Phe-Leu-Phe (PFLF) and Ile-Ala-Leu-Leu-Ile-Pro-Phe (IALLIPF) separated from foxtail millet prolamin using enzymatic hydrolysis, exhibited the capacity to scavenge free radicals (Ji et al., 2019). Peptides derived from foxtail millet prolamins substantially reduced ROS production, inhibited malondialdehyde synthesis and increased glutathione (GSH) levels in H_2O_2-induced HaCaT cells, indicating promising anti-oxidant bioactivity (Ji et al., 2020).

6.2.4 Antihypertensive properties

Hypertension is one of the potent risk factors for cardiovascular disease (CVD), which includes coronary heart disease, heart failure, myocardial infarction, stroke, end-stage renal disease and erectile dysfunction (Kjeldsen, 2018). Cardiovascular illnesses are the major cause of death worldwide and while their causes are complex, eating habits have been linked to their progression. The role of nutraceuticals and their supplementations are found to help in treatment of CVD (Sosnowska et al., 2017).

The foxtail millet protein hydrolysates may ameliorate hypertension and alleviate related cardiovascular diseases. Chen et al. (2017) reported that the ingestion of 200 mg peptides per kg of body weight of foxtail millet protein hydrolysates (raw and extruded hydrolysates) for 4 weeks, blood pressure was lowered significantly compared to fermented samples; the extruded and uncooked samples performed better. The treatment groups' blood levels of ACE activity and angiotensin II were considerably lower than those of the control group. The high amount of α-glucans in foxtail millet increases sugar and cholesterol metabolism, resulting in the effects of hypoglycemia and hypocholesterolemia that are beneficial for diabetes and cardiovascular disease prevention. Because of these benefits, foxtail millet is used in the formulation of low GI diets for the treatment of diabetes, especially type II diabetes, and also cardiovascular diseases (Hariprasanna, 2016). The possible anti-hypertensive effect of foxtail millet was studied through dietary intervention. The daily consumption of whole foxtail millet (50 g) in the form of bread and pancake significantly reduced blood pressure, body mass index, body fat percentage and fat mass in untreated mildly hypertensive subjects without affecting their dietary habits. This positive effect of millet is postulated to the combined properties of dietary fiber, protein, minerals and micronutrient present in the foxtail millet, as well as the decreased intake of fat and cholesterol. Based on these findings, it is currently advised that daily consumption of whole foxtail millet lower blood pressure in people with mild hypertension (Hou et al., 2018).

6.2.5 Anticancer properties

Dietary habits have repeatedly been identified as one of the most important factors of chronic diseases such as cancer in epidemiological research. Antioxidant consumption should be raised in the diet to minimize the possibility of cancer. Functional foods, nutraceuticals and additional micronutrients may inhibit the development of cancer cell, proliferation and induce apoptosis in cancer cells (Choudhari et al., 2020). A novel 35 kDa protein, named fibroin-modulator-binding protein (FMBP) extracted and purified from foxtail millet bran reported to exhibit anti-tumor effect implicating that FMBP had the ability to suppress xeno grafted tumor growth in mice (Shan et al., 2014). Through the onset of G1 phase arrest and loss of mitochondrial transmembrane potential, FMBP can inhibit the proliferation of colon cancer cells in a way that triggers caspase-dependent death. Normal colon epithelial cells are substantially less harmful to this new protein. As a result, FMBP from foxtail millet bran has specific anti-colon cancer actions and may treat the disease (Shan et al., 2014). Anti-tumor potency of bound polyphenol found in the inner shell of foxtail millet bran (BPIS) was reported in female BALB/c mice (Zhang et al., 2021a). Studies have found that BPIS is composed of 12 phenolic compounds that suppressed breast cancer cell proliferation specifically. The results suggested that vitexin and syringic acid were the main active components in BPIS for inhibiting the growth of breast cancer cells.

6.2.6 Antidiabetic properties

As mentioned elsewhere in this chapter, diabetes mellitus is a metabolic disease caused by insulin resistance or decreased insulin production. Hyperglycemia is another medical term used to name the condition of high blood glucose. It is a prevalent disease as a result of an unhealthy diet and lack of physical activity. Cardiovascular diseases, tumor growth, retinopathy, nephropathy, neuropathy, polydipsia, polyphagia, polyuria, and weight loss are all complications of diabetes, as are gene-environment interactions and downstream genetic and epigenetic repercussions (Deshpande et al., 2008). The understanding of molecular mechanisms for glycemic control aids in the introduction and implementation of effective bioactive components for diabetes treatment. Despite the availability of various pharmaceutical medications on the market, the scientific development of foods with therapeutic benefits is a promising alternative for diabetes therapy (Said et al., 2008; Das et al., 2022).

Foxtail millet grains have a low glycemic index (50.8) and a higher proportion of non-starchy polysaccharides and dietary fibre content. It is used as diabetics' food due to their complex carbohydrate that causes slow release of glucose into the blood streams and thus prevent an abrupt rise in glycemic value (Ren et al., 2018). The insoluble fibre and phenolics of foxtail millet has more of a positive effect on glucose diffusion than that of commercial soy insoluble fibres (Geetha, et al., 2020). The protein concentrates of Korean foxtail millet significantly elevate the plasma adiponectin and HDL cholesterol levels and causes decreases in insulin levels relative to a casein diet in type II diabetic mice. Through a rise in adiponectin levels, foxtail millet protein may enhance insulin sensitivity and lipid metabolism. As a result, the protein found in foxtail millet would be an advantageous food component in the management of diabetes type II, obesity, and cardiovascular illnesses.

Several epidemiological studies have demonstrated that foxtail millet consumption lowers the risk of lifestyle problems. Foxtail millet reduced fasting blood glucose in diabetic rats with a daily dosage manner (300 mg per kg body weight) (Sireesha et al., 2011). Foxtail millet is a promising functional food because of the antioxidants, vitamins, minerals, phytochemicals, and other bioactive compounds it contains. Laxation, lowered blood cholesterol, and reduced blood sugar are just a few of the advantageous physiological benefits of dietary fibres. Bangoura et al. (2013b) also found that insoluble dietary fibres generated from foxtail millet could slow glucose diffusion and improve absorption in the gastrointestinal tract. It also inhibits the activity of α-amylase, delaying carbohydrate digestion and slowing glucose release. The quantity and kind of fatty acids in foxtail millet may further contribute to the starches' limited digestion.

Foxtail millet is an alternative diet for diabetics as low GI diet to reduce blood glucose levels, cholesterol and glycosylated hemoglobin. It also has β-glucans (water soluble gums) that are helpful in improving glucose metabolism. Glycemic index of foxtail millet dosa is lower (59.5) than that of rice (77.96). Therefore, foxtail millet can be substituted with rice in food processing for managing diabetes. Ren et al. (2016) reported the influential effect of different processing methods on starch digestibility and glycemic responses of foxtail millet. According to their theories, heating and deproteinization processes increase the synthesis of resistant and slowly digestible starch, whereas extrusion, steaming, and boiling cooking techniques favour the formation of digestible starch. These findings show that foxtail millet can be used to generate a variety of low-glycemic index dishes when cooked properly, resulting in a significant improvement in diabetes and cholesterol control over time.

The interactions between endogenous proteins and lipids with starch play a significant role in the hypoglycemic properties of foxtail millet. In foxtail millet flour, the protein matrix not only acts as a physical barrier between starch and α-amylase, but it also partially sequesters α-amylase, slowing starch breakdowns. Additionally, foxtail millet starches can form amylose–lipid complexes with free fatty acids, which can reduce the rate of enzymatic-hydrolysis of starch. Therefore, consumption of whole foxtail millet grain that has a complex matrix may help to achieve nutritional and health benefits due to the unique digestive properties of these foods (Jin et al., 2019). Many studies found that consuming a foxtail millet-based diet resulted in significantly lower blood glucose levels, as well as that refined wheat flour-based food products that were swapped with foxtail millet flour resulted in a considerable reduction in the glycemic index

It is important to debate the role of gut microbiota on digestion and fermentation of millet dietary fibre and diabetes (Yang et al. 2021). The gut microbiota plays an important role in the onset and progression of type II diabetes (Gurung, et al., 2020). The gut microbiome reacts quickly to dietary changes. Type II diabetes is thought to be a result of gut microbial dysbiosis brought on by an imbalanced diet (Zhang et al., 2019). Studies demonstrated that the whole grain of foxtail millet has a prebiotic impact on the human gut bacteria. The gut microbiota disturbance caused by colitis-associated carcinogenesis has been reported to be restored by bound polyphenol from foxtail millet bran (Yang et al. 2020a). Foxtail millet supplementation may improve blood glucose metabolism by (i) modulating the structure of the gut microbiota, especially by increasing the relative abundance of Lactobacillus; (ii) inhibiting gluconeogenesis, stimulating glycolysis, and repairing fatty acid synthesis via the insulin-mediated PI3K/AKT signaling a pathway; and (iii) reducing inflammation via the NF-B signaling pathway (Ren et al., 2021). The link between whole-grain foxtail millet and gut bacteria in controlling glucose metabolism was also described (Ren, et al., 2022). Foxtail millet supplementation, namely 30 and 48 percent, considerably decreases the concentrations of fasting blood glucose, 60 min blood glucose and blood triglycerides. The findings revealed that consuming whole grains of foxtail millet might be beneficial to individuals suffering from type II diabetes.

6.2.7 Hypolipidemic effect

Hyperlipidemia denotes the raised serum levels of total cholesterol (TC), low-density lipoprotein cholesterol (LDL-C) and triglycerides (TG). Similar to hyperglycemia, the condition of hyperlipidemia contributes to the onset of chronic diabetes. Other components of foxtail millet, in addition to dietary fibres, have a significant impact on plasma levels of lipid, glucose, insulin, and adiponectin, all of which are linked to risk factors such as dyslipidemia, raised LDL cholesterol, decreased HDL cholesterol and hyperglycemia. The protein content of foxtail millet has been linked to a reduction in the risk of cardiovascular disease (Chen et al., 2017). Foxtail millet contains antioxidants that have been demonstrated to lower triglycerides, C-reactive protein and LDL peroxidation levels, all of which are linked to inflammation responses, implying that these grains could be used to treat cardiovascular disease. The consumption of quinoa and foxtail millets was found to lower TC, TG, LDL-C, VLDL-C and HDL-C. It is suggested that consumption of foxtail millets contributes to the risk reduction of diabetics and lipids in human subjects (Anusha et al., 2018).

6.2.8 Gastro-protective effect

One of the most common digestive illnesses in the twenty-first century is the peptic ulcer. This is also known as gastric ulcers or gastric mucosal injuries due to the imbalance in the mucous membrane's defensive and aggressive components. Peptic ulcers are long-term solitary lesions that can appear anywhere in the digestive tract (Karampour et al., 2019). Although the mortality of stomach ulcers is decreased by utilizing antibiotics, proton pump inhibitors, prostaglandin analogues and H2 receptor blockers, efforts must be made to find new medications with lower costs and fewer side effects. Natural products with lower side effects, ease of accessibility and affordability are a valuable source for stomach ulcer prevention and therapy. In experimental stress-induced gastric ulcer models, foxtail millet has been shown to have gastroprotective properties. Lin et al. (2020) suggested that millet and adlay diets endorse ulcer protection by the decrease in the ulcer index, and TBARS values, and increase in NPSH concentrations. Further, millet and adlay diets hold the benefit of being a natural product that may guard the gastric mucosa against ulceration.

The gastro-protective effect was observed in stress-induced gastric ulcer rats' administrated diet with foxtail millet exhibited major reduction in the severity and quantity of stomach lesions when compared to the control. Foxtail millet's anti-ulcer response and comprehensive antioxidant defense system are important in preventing the occurrence of severe gastric epithelium injury (Lin et al., 2020). Mice fed on diets with foxtail millet (fermented) had the lowest plasma IL-6 levels and claudin2 expression in the colon, representing reduced systemic inflammation and enhanced gut barrier function (Zhang et al., 2021d). In traditional Chinese medicine, foxtail millet has long been used to treat spleen and stomach vacuity heat, stomach reflux vomiting, reduced food intake with abdominal distention, and diabetes mellitus.

6.2.9 Antimicrobial properties

Antimicrobial and anti-oxidative peptide production has been discovered as a widespread technique utilized by plants, animals and microorganisms to battle pathogenic germs and stress. It is widely assumed that a protein that is resistant to proteolytic digestion in the digestive tract preserves enough structural integrity to boost the likelihood of the organism's natural defense against invading pathogens (Amadou et al. 2013a). But the problem is with the allergenic effect of peptides when ingested. Studies have described that the fermented foxtail millet meal extracts by *L. paracasei* Fn032 are shown to inhibit *Escherichia coli* growth. Natural antioxidants and antibacterial activity resistant to enzymes from fermented foxtail millet meal were both shown by Amadou et al. (2013b). The purified peptide fractions (FFMp) of Tyrosine/Leucine-rich (FFMp4=756.84, FFMp6=678.74 and FFMp10=678.87Da) from foxtail millet meal fermented by *Lactobacillus paracasei* Fn032, showed significant scavenging activities for superoxide anion radicals and DPPH. The Escherichia coli ATCC 8099 growth has been moderately inhibited by the produced FFMp peptides. The findings showed that fermented foxtail millet meal can be used to produce natural antioxidants, that possess antibacterial activity and enzyme resistance.

In foxtail millet seeds, an antifungal peptide with a molecular mass of 26.9 kDa has been found (Bisht et al., 2016). These millets types are high in polyphenolic content and are regarded as the most important phytochemicals in millets, with antioxidant activity, anti-inflammatory, anticarcinogenic, antibacterial, anti-diarrheal, anti-ulcer, and anti-cardiovascular capabilities. Ghimire et al. (2019) described that the antimicrobial activity of 15 accessions of foxtail millet against five different pathogenic bacteria using the minimum inhibitory concentration (MIC). The extracts of all accessions of foxtail millet are more sensitive towards *Escherichia coli* and *Staphylococcus aureus*. The presence of phenolic chemicals in foxtail millets, such as naringenin, chlorogenic acid, and myricetin, may help foxtail millet accessions have better antibacterial action. Banerjee et al. (2012) demonstrated proliferation inhibitory activities of millet polyphenols on *Escherichia coli, Bacillus cereus, Staphylococcus aureus, Streptococcus pyogenes, Listeria monocytogenes, Pseudomonas aeruginosa, Proteus mirabilis, Serratiamarcescens, Klebsiella pneumonia* and *Yersinia enterocolitica*. Among polyphenols, quercetin displayed higher inhibition of the growth of these pathogenic bacteria suggesting the potential nutraceutical functionality of the millet polyphenols. Millet polyphenols could be employed as a natural source of antioxidants, especially for reducing the risk of diseases caused by oxidative degradation, based on the findings.

6.2.10 General health benefits

- Foxtail millet is helpful to maintain good nervous health. Currently neural ailments are a serious concern for many people including adults. The daily consumption of foxtail millet-based diet can manage the condition of the common nervous system ailments namely, nervous weakness, diabetic neuropathy which involves tingling sensations all over the body, burning sensations in the feet during night times and so forth.
- Consumption of foxtail millets slows down Alzheimer's disease in humans. The presence of Vitamin B1 in foxtail millet can slow the progression of Alzheimer's disease. It also improves the concentration plus memory power of individuals.
- Foxtail millet grains are an excellent source of iron and calcium, which play a crucial part in maintaining the health of bones and muscles. A regular diet with foxtail millet can meet the calcium and phosphorous needs of the body for combating brittle bones, inflammation and other bone related chronic conditions like osteoporosis, arthritis, spondylitis and so forth. Sprouted foxtail millet helps to fight diseases like osteoporosis and also reduces the risk of fracture.
- Foxtail millet contains high dietary fibre that prevents constipation and aids in digestion. Thus, foxtail millet, particularly cooked millet, has the potential to be a source of nutraceuticals and functional foods that can help to prevent the onset of chronic diseases (Ren et al., 2022).

6.3 Characterization of phytochemical compounds responsible for bioactive properties

Plant primary and secondary phytochemicals are important for plant growth and development, as well as for colouring in flowers and fruits; and protecting plants from microbes and animals, thereby defenses against abiotic stresses and so forth. Flavonoids are the well-known component in secondary metabolites of plants, which provide a variety of biological tasks, including UV protection, phytopathogens as well as male fertility (Li et al., 2018). The long-term dietary administration of flavonols provides cardio-protection and anthocyanin gives protection against cancer; they also inhibit the tumor development during the initiation and progression stages. The carbohydrates, amino acids, fatty acids, nucleotides, and organic acids are the primary metabolites, whereas phenolic acids, flavonoids, phytosterols, alkaloids, polyamines, lignans, phytohormones, phyto-oestrogens, phytocyanins, and vitamins are secondary metabolites.

Foxtail millet is high in a variety of nutrients that are good for one's health – namely, starch, resistant starch, oligosaccharides, lipids, protein, dietary fibres, vitamins, minerals and antioxidants, including flavonoids, lignans, avenanthramides, and phytosterols – and thought to be responsible for a number of health benefits. These compounds' major roles as antioxidants, immunological modulators, detoxifying agents and protect against age-related degenerative diseases such as CVD, diabetes and cancer, among others. In addition to their established roles of avoiding nutritional deficiency disorders, some well-known nutrients, such as vitamins, minerals and essential fatty acids, have been shown to have therapeutic benefits in the prevention of degenerative diseases. Foxtail millets are suitable for celiac disease sufferers because they are non-glutinous. Because of this rich nutritional composition and the presence of functional components in foxtail millet, it is considered as a neutraceutical grain (Huchchannanavar et al. 2019). Suma and Urooj (2011) reported that alkaloids, phenolics, reducing sugars, and flavonoids were found in methanolic and aqueous extracts; tannins and terpenoids were found in all solvent extracts of foxtail millet whole flour and bran rich fraction. The characters of phytochemical compounds in foxtail millet, which are responsible for bioactive properties, are explained here.

6.3.1 Starch and resistant starch

The chemical composition, shape, architectures and physicochemical properties of starch in different varieties of millet vary naturally. In terms of morphology, starch is a significant component of all kinds of cereals and millets, such as foxtail millet, and it is crucial for post-processing quality as well as other functional factors. A millet-based product's starch quality is determined by its gelatinization parameters, molecular make-up, amylose and amylopectin content, gel-forming properties, and digestion and absorption capacities. Starch is one of the main carbohydrate components in foxtail millet, accounting for around 61 percent of the dry content. It is made up of two primary components, amylose and amylopectin, which are connected to the creation and products quality of foxtail millet in the weight ratio of 25:75. These amylose and amylopectin control the biosynthesis, physical granular structure, functionality, and potential applications of the starch. The starch granules of foxtail millet are typically tiny or large polygonal (particularly pentagonal) and rarely elliptical in shape, with round edges and some surface indentations induced by the protein bodies. In comparison with the starch granules of other millets, foxtail millet starch granules are the largest and the structure greatly resembles that of maize.

Any functional component's digestion and subsequent absorption by the human body plays a vital role in the component's health promoting effects. Heat processing makes starch more digestible, although some of the starch is still indigestible by mammalian digestive enzymes. This portion of starch, referred to as "resistant starch," has a major positive effect on health, especially in the management of chronic diet-related illnesses, glucose metabolism, mental performance and weight control. Foxtail millet starch has a lower digestibility than wheat flour (Ren et al., 2016). Human pancreatic amylase resists the digestion of resistant starch in the small intestine and it is fermented by the gut microbiota in the large intestine, resulting in a gradual and persistent release of glucose. Resistant starch's mode of digestion not only promotes gut health, obesity and insulin resistance, but it also reduces the risk of cardiovascular disease and colon cancer.

Kumari and Thayumanavan (1998) investigated the functional properties of foxtail millet resistant starch as a faecal bulking, hypoglycemic, and hypolipidemic drug in rats. The synthesis of resistant starch in native and processed foxtail millet was favorably linked with its absolute amylose concentration, according to the findings. Native resistant starch from foxtail millet dramatically reduced blood sugar, cholesterol and triglyceride levels in rats while increasing body weight and digestibility. According to Bangoura et al. (2012), heating foxtail millet reduced the resistant starch content while increasing the amylose concentration, giving the starch a significant level of syneresis and gel consistency. Additionally, the swelling capacity and solubility of foxtail millet starch enhanced with rising temperature while cooking due to the retrogradation of the amylose and amylopectin chains.

6.3.2 Dietary fibre

Dietary fibre, also known as roughage, is the portion of a food generated from plants that cannot be fully broken down by digestive enzymes in humans and it found to be fermented totally or partially in the large intestine. It is a carbohydrate-based polymer with beneficial physiological properties and it includes polysaccharides, oligosaccharides, lignin and associated plant substances. Dietary fibre comes from a variety of cereals, millet, legumes, vegetables and fruits. It is ascribed to its properties of shortening chyme transit time in the small intestine and increasing stool volume, as well as lowering blood total cholesterol, LDL cholesterol, blood sugar and insulin levels. Foxtail millet bran is a good source of dietary fiber. The purity of foxtail millet bran dietary fibre is about 79.22 per cent and it mainly consists of a high amount of insoluble dietary fibre (77.57%) and a lesser amount of soluble dietary fibre (1.75%). The cellulose and hemicellulose accounted for a large part of the foxtail millet bran dietary fibre (Zhu et al., 2018). Wankhede et al. (1979) reported that the pentosans content in foxtail millet is about 5.5 per cent, which bind water and contribute to the formation of viscous dough in the bakery industry.

The water holding capacity and swelling power of foxtail millet bran dietary fibre are 3.24 g/g and 2.06 ml/g respectively. This fibre has good adsorption capacities to lipophilic substances such as lard (3.34 g/g), peanut oil (2.32 g/g) and cholesterol (5.19 mg/g). The bile salts adsorption capacity of foxtail millet bran fibre indirectly reflected its cholesterol lowering capacity due to rich hemicelluloses content. The chemical composition and microstructure studies of foxtail millet bran dietary fibre revealed that this dietary fibre has a great potential to be used as functional ingredient in food products (Zhu, et al., 2018). The insoluble fibres in foxtail millet also exhibited antioxidant activity. The yellow variety has higher amounts of antioxidants when compared to the white variety of foxtail millet. The antioxidant potential of insoluble fibres may be linked to their phytochemical composition. In a basic pH solution, polysaccharides from foxtail millet showed a better ability for scavenging DPPH and hydroxyl radicals.

6.3.3 Protein

The majority of the dry mass of millet seed is made up of protein, with the biochemical component varying depending on the kind of millet. Foxtail millet protein is easily digested and has a high net utilization rate in the body. It is a good and reasonably inexpensive protein source for adding value to functional foods. In foxtail millet, alcohol soluble prolamines are the most common protein storage type. The characterization of protein in this millet explains about the protein concentrate, which is rich in lysine, is a potent bioactive food ingredient that is used as an additional source of protein in most cereals. The majority of proteins in foxtail millet, which has an average total protein concentration of 11.54 g/100 g, are albumin, gliadin, globulin, glutein, and other proteins (Jing-ke et al., 2014b). The size and chemical makeup of amino acids present in foxtail millet play a crucial role in influencing the functioning of proteins. Foxtail millet protein contains large amounts of glutamic acid, leucine, alanine, aspartic acid, and lysine. The essential amino acid content of foxtail millet was found to be in compliance with WHO and FAO criteria, as well as comparable to soy protein concentrate.

The protein concentrate from foxtail millet is superior in quality than that of protein concentrate from other millets due to its higher amino acid content. The functional properties of foxtail millet protein concentrate obtained by protease enzyme shows that the antioxidant properties of the protein hydrolyzates are similar to α-tocopherol and have the activity of metal chelating. Peptides could be added to functional foods and nutraceuticals to minimize oxidative stress-related damage in a variety of human illnesses. In radical scavenging, amino acids including histidine, leucine, glycine, and proline are important. The PR-HPLC purified peptide fractions of fermented Tyrosine/Leucine rich foxtail millet were found to have considerable radical scavenging activity (Amadou et al., 2016).

6.3.4 Phytochemicals

Phytochemicals are non-nutritive plant bioactive constituents that are biologically active organic substances of plant origin having disease preventing and health promoting properties. These substances are isolated as nutrients, dietary supplements and specific diets to genetically modified foods, herbal products and beverages. Polyphenols, flavonoids, isoflavonoids, anthocyanidins, phytoestrogens, terpenoids, carotenoids, limonoids, phytosterols, glucosinolates and fibre are all phytochemicals. Phytochemical screening in foxtail millet revealed the presence of polyphenols, flavonoids, phenolics, and tannins along with other components that are potent antioxidants.

6.3.4.1 Phenolic compounds

The qualities of antioxidants – including phenolics (phenolic acids and flavonoids), carotenoids, α-tocopherol, and certain polysaccharides – are widely acknowledged as the primary cause of foxtail millet's health advantages. According to Pradeep and Sreerama (2018), the main flavonoids in foxtail millet's free fractions are luteolin and kaempferol, whereas the main phenolic acids are ferulic, caffeic and sinapic acids. A variety of phenolic compounds found in dehulled foxtail millet could be exploited to produce functional meals with unique antioxidant properties. According to Xiang *et al.* (2019), four ferulic acid dimers (DFAs) and twenty-three phenolic acid derivatives were found in the bound fraction with ferulic acid as well as twenty-one phenolics, including a series of nine hydroxycinnamic acid spermidines, in the free fraction of foxtail millet. Foxtail millet was found to have nine hydroxycinnamic acid spermidines for the first time. In addition to being shown to considerably block HIV-1 protease, which is necessary for the HIV life cycle, hydroxycinnamic acid amides are a group of important plant secondary metabolites that are often concentrated in the floral sections of the plants.

The methanolic and aqueous extracts of foxtail millet whole flour and bran rich fraction contain phytochemicals like alkaloids, phenolics, reducing sugars and flavonoids, while terpenoids and tannins are found in all solvent extracts of whole flour and bran rich fractions (Suma and Urooj, 2011). The recovery of phytochemicals using solvents extraction is more from the coloured foxtail millet (yellow species) than that of the white species (Bangoura et al., 2013a). The higher concentration of phytochemicals such as total phenol and total flavonoid content is positively associated with the colour of the grain. Fourteen various types of phenolic compounds, including ρ-hydroxybenzoic acid, chlorogenic acid, o-coumaric acid, ferulic acid, naringin, hesperetin, myricetin, catechin, caffeic acid, syringic acid, salicylic acid, t-cinnamic acid, quercetin and naringenin were identified and quantified in 15 foxtail millet accessions. Myricetin was the most common flavonoid among the phenolic chemicals found in the accessions. Higher antioxidant properties are presumably due to the synergy of phenolic chemicals found in plant extracts (Ghimire et al., 2019).

Flavonoids found in foxtail millets, such as flavone, flavanone, flavonols, isoflavone and anthocyanins, may offer health benefits such as antioxidation, anti-inflammatory, anticholinesterase, anticancer and heart disease prevention, and antibiotic resistance prevention. Zhang et al. (2021c) described that flavonoid metabolite in the examined foxtail millet varieties varied substantially, and each one accumulated its own set of metabolites, suggesting that these variants may have medical applications. They discovered 116 flavonoid compounds in foxtail millet and 33 flavonoid metabolites that differed between good and low eating quality foxtail millet cultivars. The total polyphenolic content of foxtail millet was dramatically impacted during the germination process due to the release and production of polyphenolic compounds by endogenous enzymes that were activated during germination and increased the total polyphenolic compounds. As a result, the foxtail millet's functional qualities may be affected (Sharma, S. et al., 2018).

6.3.4.2 Tocopherol

Tocopherol (also known as Vitamin E or alpha-tocopherol) is a nutrient that is important for many body processes. It acts as an antioxidant, anti-inflammatory, anti-atherosclerotic compound and decreases superoxide production in mitochondria and helps the nerves and muscles work well, prevents blood clots and boosts the immune system. Millet seeds have a very low α-tocopherol concentration, and tocopherols are predominantly found as the γ-isomer in millets. When compared to other tocopherols, α-tocopherol has an extremely high vitamin activity. Liang et al. (2010) reported the tocopherol content of foxtail millet bran oil to be about 64.83 ± 0.83 mg/100 g which consisted mainly of γ-tocopherol (48.79 mg/100 g) and α-tocopherol (15.53 mg/100 g).

6.3.4.3 Phytosterols

Phytosterol is a significant component of vegetable oils that has strong nutritional and biological importance. Phytosterols are bioactive molecules which are essential to reduce blood cholesterol levels and cardiovascular disease incidence. The most prevalent phytosterols are sitosterol, stigmasterol, and campesterol. Furthermore, cereal crops constitute the primary source of phytosterol in the human diet, outnumbering oilseeds in daily consumption. Because of the rising interest in phytosterols, the creation of functional foods with phytosterols has developed. Foxtail millet is an excellent source of beneficial natural fatty acids and phytosterols including stigmasterol, campesterol and sitosterol which help to reduce the risk of diseases and promote a healthy population (Sharma and Niranjan, 2018). According to Tsuzuki et al. (2018), the steryl ferulates content of whole foxtail millet was 28.9 mg-oryzanol equivalent/100 g, indicating that the steryl ferulates content in whole foxtail millet is almost 80 percent of oryzanol in brown rice. Steryl ferulates are one of the important phytochemicals which exhibit antioxidant activities through scavenging of free radicals.

6.3.4.4 Fatty acid profile

Foxtail millet is an excellent source of fatty acids and the main components of foxtail millet are palmitic acid, stearic acid, oleic acid, linoleic acid, and linolenic acid (Li et al., 2021a). However, the alteration in quality parameters can take place during cooking due to the oxidation of fatty acids. The change in aldehyde content is considerable to the quality of the porridge because aldehydes contribute to the aroma of foxtail millet porridge. The volatiles produced during porridge cooking are mainly due to the oxidation of stearic acid. During cooking of foxtail millet flour into porridge, the free fatty acid content increased 0.8–1.2 fold, whereas the total fatty acid content decreased 0.1–0.4 fold.

Millets have a fat content ranging from 1 to 5 percent, which is dispersed in the bran as well as the endosperm. The high fat content is present in pearl, foxtail and proso millets (5 percent) where as low in finger and kodo millet (1 percent). Millets have a high percentage (60 percent) of unsaturated fatty acids, such as linolenic acid, in their fat composition. The fat in millets generally consists of more than 60 percent unsaturated fatty acids including linolenic acid. Kim et al. (2010) reported that the presence of essential fatty acids, namely, linoleic acid and linolenic acid in foxtail millet along with palmitic acid, stearic acid and oleic acid are very important to the human immune system to regulate blood pressure. The foxtail millet bran oil is rich in linoleic acid (66.5–71.2 percent), which has a major contribution in fatty acid profile of foxtail millet than oleic acid (13.0–16.11 percent) (Liang et al. 2010). Thus, the foxtail millet could be an alternative source of edible oil due to presence of all saturated and unsaturated fatty acids that are required for human health.

Another study revealed that the presence of essential fatty acids in 35 foxtail millets including seven varieties planted in five different regions of China (Zhang et al. 2015). The fat content of foxtail millets ranged from 3.38 to 6.49 percent. The major fatty acids in foxtail millets are linoleic acid (66.68%), oleic acid (16.11%), palmitic acid (7.42%), stearic acid (6.84%) and linolenic acid (2.48%). Pang et al. (2014) identified eight major components of fatty acids: that includes four saturated fatty acids (SAFA), two monounsaturated fatty acids (MUFA) and two polyunsaturated fatty acids (PUFA) in foxtail millet bran oil. The most common SAFAs among them were palmitic acid C16:0 (10.23%) and stearic acid C18:0 (3.73%), the most common MUFA was oleic acid C18:1 (19.65%), and the most common PUFAs were linoleic acid C18:2 (58.85%) and linolenic acid C18:3 (3.96%). This suggests that foxtail millet oil, which is high in linoleic acid, linolenic acid, and tocopherols, could be a suitable source of natural oil (Liang, et al., 2010).

6.3.4.5 Bioactive peptides

Bioactive peptides are protein fragments that have a beneficial effect on physiological functions or situations that may affect health. These are organic substances formed by amino acids, whereas proteins are polypeptides with a greater molecular weight. Bioactive peptides and proteins play important roles in the metabolic functions by affecting the digestive, endocrine, cardiovascular, immune and nervous systems. The growing interest in bioactive peptides has incentivized the scientific community and the food industry to explore the development of new food additives and functional foods based on these peptides.

The trypsin resistance of peptides produced from fermented foxtail millet meal by *L. paracasei* Fn032 and their potential to prevent the development of E. coli ATCC 8099 are discussed in (Amadou, et al., 2013b). The results demonstrate that short peptides from foxtail millet meal with 2–10 amino acids have better antioxidant activity and other functional properties than their parent native proteins or large polypeptides. Fermented foxtail millet peptides have significant antioxidant and antibacterial properties. The inclusion of numerous functional components with bioactive qualities distinguishes foxtail millet from other cereals. Numerous studies have been conducted throughout the years to recognize, quantify, examine and evaluate the key elements of foxtail millet. The existence of these numerous nutritional and health-related functional components demonstrates how versatile foxtail millet is as an ingredient.

6.4 Techno-functional characteristics of foxtail millet

The techno-functional qualities of cereal and millet grains are significant in the formulation of processed and convenient food products as well as the creation of machinery for processing and storing crops before and after harvest. Under these following topics, the techno-functional properties of grains and flours of foxtail millet that are suitable for processing are discussed, that are (i) physical; (ii) thermal; (iii) gelatinization; (iv) rheological; (v) hydration behavior; (vi) colour and flavour profile and (vii) other functional characteristics.

6.4.1 Physical characteristics

Techno-functional benefits of millets having noteworthy role in processing of various pharmacological and functional products to enhance their use is growing currently. Functional and organoleptic characteristics of foods are generally influenced by physical properties of millets. Physical characteristics of food grains generally influence the functional and organoleptic characteristics. Physical properties of the millets are evaluated in terms of length, breadth, thickness, breadth ratio, size, colour, geometric mean diameter, arithmetic mean diameter, thousand grain volume and weight, true density, bulk density, grain hardness and so forth. The physical and functional properties of millets are also essential factors for the design of equipment used for harvesting, processing, handling and storage. Bulk density is an important parameter that helps to determine the packaging requirement of a product. As a functional attribute, the mass of a thousand grains can be used to estimate weight loss during storage. Grain hardness aids in grain classification and has an impact on milling and end product quality.

Many studies have undertaken to determine the physical and functional characteristics of raw and cooked forms of foxtail millet grains and flours. For foxtail and other minor millets, the thousand grain weight varies linearly with moisture content, whereas bulk density, true density, porosity and grain hardens inversely with moisture content. Foxtail millet grain is spheroid in shape with a size and roundness of 2.78 mm and 0.23 cm. Roundness value is indicating the maturity stage of foxtail millet grain during harvesting. Sunil et al. (2016) reported that physical properties of foxtail millet variety (HMT 1001). Individual foxtail millet grain length, breadth, and thickness are 2.169, 1.590, and 1.454 mm, respectively, according to the data. The average values for geometric mean diameter, weight of hundred grains, bulk density, actual density, and porosity of foxtail millet grain are 1.703 mm, 0.245 g, 737.127 kg.m-3, 1260.132 kg.m-3, and 41.471 percent respectively. The sphericity is determined by measuring the projected area of the grain particle. The mean sphericity value is 0.785, with sphericity values ranging from 0.659 to 0.881. The mean value of 1000 grain weight of foxtail millet is 2.36 g (Gouda et al., 2019). The thousand grain volume of foxtail millet is 2.96 ml. Depending on the variations of moisture content, the grain volumes can change significantly and most often regularly. The knowledge of physical properties of foxtail millet grain and flour is necessary while formulating products and to design machinery for harvesting, processing and storage.

Various cooking methods such as boiling, steaming, drying, frying, microwave cooking, and so forth, can alter the physical and functional properties of food grains. The effect of microwave pre-treatment prior to milling of foxtail millet on physical and functional characteristics of flours, as well as their quality, was studied by Rao et al. (2021). When compared to the untreated sample, the colour values of microwave treated foxtail millet flour are minimally altered within an acceptable range. The bulk density of the boiling, pressure cooked and roasted samples are significantly higher than that of raw foxtail millet flours, however the bulk density of the germinated foxtail millet flour is lower. It signifies that the flour has a low porosity or air spacing, which means that autooxidation, is unlikely.

6.4.2 Thermal characteristics

Thermal characteristics are critical engineering and functional considerations in millet processing equipment design. For the grains and flours of foxtail millet in the moisture content range of 10–30 percent, several thermal parameters such as specific heat, thermal conductivity, and thermal diffusivity were investigated (Subramanian and Viswanathan, 2003). They have also found that specific heat increases (1.33–2.40 kJ kg^{-1} K^{-1}) for both grains and flours with an increase in moisture content. Thus, the specific heat of grains and flours of foxtail millet has a positive association with moisture content. The thermal conductivity of foxtail millet grains and flours can also increase (0.119–0.223 W m^{-1} K^{-1} and 0.026–0.128 W m^{-1} K^{-1} respectively) by the 60 percent with moisture content in the same range (10–30%). It indicates that flour's heat conductivity was significantly lower than that of grains. Among the millets, foxtail flour has higher value of thermal conductivity than the other millets' flours followed by proso millet, kodo millet, finger millet, little millet and barnyard millet. With moisture content in the same range, the thermal diffusivity of grains (0.731×10$^{(-3)}$ m2 h^{-1}) and flours (0.820×10$^{(-3)}$ m2 h^{-1}) decreases slightly. This tendency might be a result of the components and ratios in foxtail millet grains and flours changing as the moisture content rises. Even though, the thermal processing can affect not only the content of bioactive compounds like phenolics in millets but also affect their beneficial biological properties.

Thermal characteristics such as onset temperature, peak temperature and enthalpy of gelatinization of native flours, starches and gel flours of foxtail millet have been estimated and compared with green gram native flours, starches and gel flours by Nagaprabha et al. (2018). The onset temperatures of the foxtail millet native flour, starch and gel flour are 70.7, 70.4 and 55.4°C respectively. It was slightly higher than the green gram samples. The peak temperature of the native flour of foxtail millet (75.7°C) is significantly greater than their corresponding starch and gel flour samples (75.0 and 58.4°C respectively). The conclusion temperature of the foxtail millet native flour (81.2°C) and starch (79.4°C) samples are higher than that of native green gram samples. The gelatinization temperature range of the native flours of the foxtail millet 10.6°C is significantly higher than the starch sample. Due to complete gelatinization of the gel flour, no gelatinization peaks were found; instead, retrogradation peaks were observed. The presence of other components such as protein and fat may be the influential factors to affect the phenomenon of gelatinization of the crystalline region of the flour, and the gelatinization temperature can be decreased by the removal of the other components. High gelatinization temperatures indicate that more energy is needed to start the starch gelatinization process, and these thermal features can also affect the textural aspects and other physical parameters of the gels generated.

6.4.3 Gelatinization characteristics

Starch gelatinization is a complex process that involves the slow expansion of intact starch granules in the presence of water and heat, allowing the hydrogen bonding sites to detect more water and rapidly increasing the apparent viscosity of the paste. Then, the starch granules are broken down, leading to an obvious drop in viscosity. During processing, the millet flours that are suitable for industrial scale utilization in the manufacture of food items like baby foods, snack foods and dietary food, undergo the process of gelatinization. Additionally, procedures like baking bread and cakes, extruding cereal-based products, thickening and gelling sauces, and making pie fillings call for the proper starch gelatinization method. Cooked suspensions of cereal flours are frequently used in the food business for specialized meals such as sauce, soup, heat-set gel, porridge, instant powders, modified flour, and starches. Foxtail millet flour is structurally more stable and resistant to gelatinization when compared to the other millet flours like little, barnyard, finger, kodo and proso millet. Sharma and Niranjan (2018) also reported the higher gelatinization range in foxtail millet (28.4°C) followed by finger millet (17°C) and little millet (12°C) as lower range.

Shinoj et al. (2006) and Annor et al. (2014) reported the values of gelatinization properties such as onset, peak/maximum, conclusion temperatures and enthalpy of gelatinization (ΔH) of foxtail millet types. It is revealed that there are slight differences in the gelatinization temperatures and enthalpies in different varieties of foxtail millet. The variation in the values of gelatinization temperature and other related properties among the varieties and other grains may be attributed to the grain's size, shape, type, concentration and distribution of starch granules, amylose-amylopectin ratio and nutritional composition. The melting temperature of foxtail millet is narrower than that of other millets, indicating that the amylopectin crystals in the starch granules are homogeneous and uniform in quality. Furthermore, there are numerous other factors that influence the degree of starch gelatinization, such as sample characteristics, water absorption capacity, and the presence of other components such as protein and fat, all of which reduce the percentage of starch gelatinization even at high pressures.

The wet cooking methods like germination or germination with soaking have positive correlations with the improved gelatinization properties of foxtail millet for commercial application. Similarly, when germinated and non-germinated foxtail millet grains were soaked at high pressures and temperatures, different degrees of starch gelatinization were found. The enthalpy of gelatinization revealed the loss of molecular order of starch granules due to hydrogen bond breaking, whereas the gelatinization temperature defines the structural stability of starch molecules. Non-germinated samples had higher gelatinization temperatures and enthalpies than germinated samples, showing that germination reduces resistance to high pressure induced gelatinization of foxtail millet grains due to deformation of starch granule structure during high pressure treatment (Liu et al., 2009). The findings of gelatinization qualities of control and high pressure-soaked foxtail millet samples were published by Sharma, N. et al. (2018). In comparison to all the high pressure treated samples, the natural starch of foxtail millet had greater gelatinization temperatures and enthalpies. It shows that germinated foxtail millet grains have more starch gelatinization than non-germinated foxtail millet grains. This could be related to increased grain permeability following high-pressure soaking and germination. When compared to non-germinated foxtail millet grains, high pressure soaking of germinated foxtail millet grains can provide better partially cooked grain or flour for formulation of functional goods.

6.4.4 Rheological characteristics

Rheological or viscoelastic parameters are the most practical indicators of food product quality and texture. Rheological characteristics such as elasticity, viscosity, extensibility, swelling capacity, water absorption, flow property, mixing property, rolling quality and so forth are significant in the milling and baking industries for predicting dough processing parameters and end product quality. The rheological behavior of doughs show how they deform, flow, or rupture when subjected to stress and can be used to help choose and specify the right basic materials. However, there have been few investigations on the qualities of millet flours and doughs. The understanding of fundamental rheological properties of millet flours and doughs can help to predict how the dough will respond under different processing circumstances. They are important in product formulation and optimization, raw material quality control, dough machining qualities, process scale-up, automation and consumer-determined quality evaluation by comparing rheological measurements with sensory tests.

Dynamic rheometry is a method for figuring out the elastic and flow characteristics of substances, and it has been extensively used to explain the rheological characteristics of foods, including doughs. The energy retained during deformation is represented by the storage modulus, which is proportional to the elastic energy of the sample, whereas the energy lost during deformation is represented by the loss modulus, which is proportional to the viscose energy of the sample. The storage and loss moduli (G' and G"), along with the phase angle (δ), may give a clear indicator of the extensibility and stiffness of the dough. A stiff dough has a high G' value and a low G", whereas a softer and more extensible dough has a lower G'. Shinoj et al. (2006) reported that increased viscosity of all millet flours during heating started decreasing just before holding and increased again during the cooling cycle. The viscosity values for the foxtail millet flour ranged from 0.055 to 0.220 Pa. On cooling, the viscosity of starch rises due to the high retrogradation tendency of the amylose fraction. The storage modulus and the loss modulus ranged from 2.55 to 20.8 Pa and 0.06 to 2.9 Pa for the flours of foxtail millets, respectively. It indicates the stiffness of the foxtail millet dough. For foxtail millets flour, the phase angle values, which indicate the phase shift throughout the heating/cooling cycle process, exhibited a wide range of values ranging from 1.1 to 12.8 degrees.

Based on the consumers' interest on millet-based healthy foods, many food industries and food enterprises are providing the millet-based products in the form of ready-to-eat convenience food with improved acceptability. Of late, millets are incorporated in various bakery products. Many researches have suggested that millet flour can be combined with wheat flour in various quantities to make bread, biscuits and other snacks. Despite the fact that foxtail millet is a gluten-free grain with the potential to be employed in gluten-free product development, it behaves differently from wheat and has poor rheological qualities in terms of pliability, extensibility and rollability. The viscosity of the flour/dough formulations increases with wheat flour and lowers with millet flour, according to the basic rheological measures (higher than 30 percent incorporation). Several studies approved the substitution level of millet flour up to 30 per cent in composite flour for making bakery products. High levels of millet flour that is, 40 and 50 per cent incorporation negatively influence the quality of the product. When compared to durum wheat bakery products, gluten free bakery products have a larger cooking loss and a completely different rheological behavior. To improve the quality of gluten-free products, various approaches must be used, such as (i) the addition of specific food additives and ingredients (e.g., hydrocolloids, emulsifiers, stabilizers, protein isolates, customized starches, etc.) and (ii) the use of various production processes (sour dough technology, enzymatic or heat treatments, etc.) or their combination, to improve the properties of gluten-free bakery products.

As foxtail millet flour is gluten-free, making bakery products like bread is challenging, however adding enzymes improved the rheological quality of the flour. Sarabhai et al. (2021) have revealed the potential use of enzymes for improvement in texture and rheological characters of gluten-free foxtail bread. Foxtail millet breads treated with enzymes had significantly higher specific volume and crumb springiness. When comparing protease-treated bread to glucose oxidase and xylanase-treated bread, crumb hardness, chewiness and cohesiveness all decreased dramatically. The expansion of the batter was aided by the decreased paste consistency and viscosity during proofing and baking. The addition of enzymes such as protease to foxtail millet resulted in breads with improved physical, textural and sensory quality. Adding enzymes such as protease to foxtail millet is a promising approach that helps to prepare high quality gluten free foxtail millet bread that meets consumer expectations from an industrial standpoint. The absence of water holding capacity was demonstrated by the considerable increase in baking loss of breads containing glucose oxidase 0.1 g/100 g and xylanase 0.05 g/100 g. Although the viscous modulus (G") of the foxtail millet dough batters was lower than the elastic modulus (G'), the batters displayed a solid, elastic-like characteristic.

Nazni and Devi (2016) have reported the effect of different cooking methods on rheological qualities of foxtail millet flours. The roasted foxtail millet flour has a high peak viscosity of

1143cP, indicating that it has a strong water binding capacity, causing the starch granules to swell more. This is frequently linked to the end product's quality. The starch granules' great swelling capacity is shown by their high peak viscosity. The foxtail millet raw and processed flour through viscosity ranges from 29 to 420cP, respectively. Foxtail millet flour that has been pressure-cooked and boiled (2cP) has a low breakdown viscosity, indicating that the starches are stable under high temperatures. Final viscosity is a metric for how well a starch can create a viscous paste. The raw foxtail millet flour's high final viscosity (1398cP) reveals its high shear resistance. The setback viscosity, a retrogradation index, is frequently used to assess the starch's gelling capacity or retrogradation tendency. It is the increase in viscosity brought on by the rearrangement of amylose molecules that have leached out of inflated starch granules during cooling. The lower setback value (-107cP) of roasted foxtail millet flour indicated the lowest rate of retrogradation, implying that products created with low setback viscosity flour will have a longer shelf life. These rheological metrics indicate that germinated and roasted foxtail millet flours have excellent functional and pasting capabilities, making them good basis ingredients in infant food formulation.

6.4.5 Hydration behavior

Pasting behaviors of millet flour is one of the important measuring parameters to illustrate the quality of millet starch granules. Their hydration properties before and after processing determines the application values of flours and starches. The hydration behavior is another important property of millet grains. According to Vasudeva et al. (2010), foxtail millet hydrates very slowly and takes longer (24 hours) to reach equilibrium than other millet. In comparison to other cereals and millets, foxtail millet has the lowest equilibrium moisture content, at 28.3 percent on a wet basis. The pasting qualities and hydration capacity of both the raw and processed foxtail millet flour were reported by Nazni and Devi in 2016. Temperatures of foxtail millet flour paste, both raw and processed, ranged from 48.5°C to 89.6°C, with fresh foxtail millet flour having the highest temperature (89.6°C) and boiled and pressure-cooked foxtail millet flour having the lowest temperature (48.5°C). The lowest temperature required for cooking the samples is indicated by the pasting temperature. The unprocessed foxtail millet has a hydration capacity of 2.01 g/1000 seeds and a hydration index of 76.9 percent. The swelling capacity of foxtail millet grain is about 0.21 ml / 1000 seeds with an index of 6.72 percent. The presence or inclusion of high protein, lipid, fibre, and a higher proportion of amylose-lipid complex in flour may inhibit starch granules from expanding.

The various processing methods like soaking, germination and microwave cooking of millet grain play a major role in improving the qualities of hydration behavior in terms of water absorption capacity. In both germinated and non-germinated samples, foxtail millet grains increased their water intake as soaking pressures and temperatures increased. Germination raises the grain's water intake by 3.13 percent due to the rupture of the grain's outer walls, which enhances the grain's water absorption capacity. The water absorption capacity of micro-wave treated foxtail millet is increased while oil absorption capacity is decreased. The highest water absorption capacity is observed at 840 W and 6 min treatment level. The swelling power of microwave-treated foxtail millet flour decreases as the power level and time rise, with the exception of 840 W, 2 min. Microwave treatment, on the other hand, enhanced the solubility of flour. At 840 W and 2 minutes of treatment, the maximum solubility is observed. These findings suggest that using microwave heating as a pre-treatment can successfully improve the functional qualities of foxtail millet flours in terms of hydration behavior without changing physical parameters, which could assist increasing the value of underutilized foxtail millet (Rao et al., 2021).

Nazni and Devi (2016) reported that the reduced swelling power of processed foxtail millet flours was consistent at 90°C when compared with raw samples. This flour samples' limited swelling behavior shows that they are resistant to shearing action when heated. The maximum solubility behavior was observed in the germinated foxtail millet flour. The amount of amylose leaching out of starch granules when swelling could be indicated by solubility. When boiling, pressure cooked, or roasted samples were compared to raw samples, the water absorption capacity increased dramatically; however germinated foxtail millet flours had the lowest water absorption capacity. Several factors influence the water absorption capacity of processed millets, including the number of hydration locations, the physical environment, pH, solvent and the presence of lipids and carbohydrates. The endogenous proteins in foxtail millet has an influential effect on the functional properties of foxtail millet during the hydrothermal process, especially water absorption and swelling of starch granules. A study confirms that the water absorption and swelling power of foxtail millet flour is increased, followed by lower water mobility and greater hydration of foxtail millet flour pastes only after removing endogenous proteins (Zhang and Shen, 2021).

Rehydration properties are another important behavior to test the quality of dehydrated convenient millet-based products. The rehydration capabilities and mouthfeel of dried gelatinized millet products differed significantly depending on the drying processes used. The ability of a dried food to rehydrate is widely cited as an indicator of its ready-to-eat potential. The crystallization of gelatinized millet starch could be another reason for the deteriorated hydration and swelling ability. Retrogradation can occur when gelatinized millet is exposed to drying circumstances caused by low temperature or insufficient energy input. The production of greater V-type structure, that is, stable crystalline amylose-lipid complexes, may be induced by severe hydrothermal techniques such as too high temperature or microwave power, or quick water removal. These complexes have a significant impact on the goods' rehydration qualities and mouthfeel. Wang et al. (2017) reported the drying techniques for formulating gelatinized foxtail millet (α-millet) products. Microwave and hot air drying both resulted in significant loss in eating quality; however, freeze drying procedures can maintain the desired rehydration properties and texture.

6.4.6 Colour and flavour profile

Colour plays an important role in consumer acceptability and quality assessment of any kind of food product. The most essential criterion for evaluating the foxtail millet grain quality is the golden colour of the de-husked grain. In dehulled foxtail millet, the predominant yellow pigments are carotenoids (lutein and zeaxanthin) and flavonoids. The concentration of carotenoids and flavonoids has been shown in studies to contribute to the seeds' palatability and nutritional value. A study described that the colour values of four different foxtail millet samples mainly based on the colour of the grain, that is, yellow, white and black millets. Because of the highest b* value (40.65 ± 0.29), the yellow millet looked to be the most yellowish. Yellow millet and white millet had considerably different L* values than the other two samples, showing that yellow millet and white millet had brighter colours. Among the four millet varieties, the total colour change was also significantly varied (Li et al., 2021b). Colour variation in foxtail millet grains is a significant attribute to consider when determining market value and nutritional quality.

However, the colour quality of the foxtail millet grain may be altered during thermal processing. As like other carotenoid rich foods, the colour deterioration – that is, turning into a pale colour – is a main problem with foxtail millet grains during cooking. Carotenoid degradation during cooking is responsible for the discoloration of foxtail millet grains. A study found that when compared to micro-pressure and high-pressure cooking, atmospheric pressure cooking retained more yellow colour.

Aroma is a key indicator of food quality and odour activity, which measures how much aroma molecules contribute to food flavour. Around 52 to 62 chemical components have been identified among the commercially available foxtail millet grains. The predominant volatile molecules among the components are aldehydes (Li et al., 2021a). The major flavour components of foxtail millet are composed of 18 aldehydes, 10 hydrocarbons, 10 benzene derivatives, 9 ketones, 6 alcohols, 5 acids and 4 other components. Among 30 odour characteristics, the odour activity value of 18 compounds in foxtail millet is greater than 1, including nonanal, (Z)-2-nonenal, (E,E)-2,4-nonadienal, naphthalene dibutyl phthalate, and others, which have the largest impact on millet flavour (Bi et al., 2020). Millet is irradiated by visible light during transit, storage, and marketing, resulting in a faster rate of lipid oxidation. Aldehydes, ketones, esters, alcohols, and hydrocarbons are among the volatile components found in millet that deteriorate after storage under various light conditions. To prevent the negative impacts of these volatile components during long-term storage, it is suggested that blue light be used as a clean and low-cost technique to increase the shelf life of millet while maintaining grain quality and delaying oxidative rancidity (Liang et al., 2021).

6.4.7 Other functional properties

Oil absorption capacity is another important functional characteristic of foxtail millet that is used to determine the quality of fried products. Because of their excellent oil absorption capability, the flours are ideal for enhancing flavour and mouth feel when employed in food preparations. In comparison to raw samples, boiling, germinated, pressure cooked, and roasted foxtail millet flours have a higher oil absorption capacity, whereas the decreased oil absorption capacity is observed in microwave treated foxtail millet (Rao et al., 2021). Protein concentration, degree of interaction with water and oil, and different cooking methods may all have a role in fat absorption.

6.5 Potential utilization of foxtail millet in value-added food products

Millet is currently attracting more interest from food scientists, technologists, and nutritionists around the world due to its significant contribution to global food security and potential health benefits. The fundamental reason for this is that millet grains, along with other cereals, are the world's most important source of food and play an essential role in human diets all over the world. The advancement and shift towards potential use of millets is due to climate changes, water shortage, growing populations, escalating costs of food, and various other socioeconomic impacts (Saleh et al., 2013). The utilization of millets, such as foxtail millet, as food items has the potential to alleviate not only malnutrition and nourish the general population, but also to aid in the prevention and treatment of diseases such as obesity, diabetes, and CVD. Millet grains need to be processed before consumption and for preparation of products either by traditional or industrial methods of processing techniques to improve their palatability, nutritional, sensory and functional properties. Foxtail millet contains some kind of phytochemicals with considerable antinutrient behavior that can potentially decrease the quality and bioavailability of nutrients. Decortications, soaking, milling, heating, extrusion, germination, fermentation, roasting and other millet processing techniques can be used to improve palatability, reduce anti-nutritional effects, and increase micronutrient bioavailability in foxtail millet (Saleh et al., 2013; Nazni and Devi, 2016). Even simple measures like dehulling, soaking, and cooking reduces anti-nutrients by around 50 percent, improved in-vitro protein digestibility (32.71%), and enhanced mineral bioavailability, like iron and zinc. The post harvest losses of millets could be minimized by adopting the various post harvesting systems, include drying, threshing, cleaning, dehulling, milling, grading, packaging, transportating and storaging that ensures good quality products (Beta and Ndolo, 2019).

6.5.1 Food processing techniques

Millets have three main components: pericarp, a starch-containing endosperm and a germ section that is partially segregated during processing. The basic processing procedures are used to separate the undesirable components (offal) of the grains, such as the pericarp and germ (in some cases), which are not fit for human consumption. The procedure of eliminating offal is known as dehulling or decortication. Millets' tough outer coat, corresponding unique flavour, and lack of processed products in ready-to-use form of millets similar to rice and wheat are the major reasons for their low appeal among rice and wheat eaters. To make foxtail millet into convenient form, there are many processing techniques available that is, primary and secondary processing. Primary millet processing techniques are cleaning, decortication, grading, polishing/refining, sorting, size reduction/milling, drying and storage. This phase of the process can be done at the farm or at the producer level and it enhances grain quality and turns it into a more usable form. The secondary post harvest operations such as roasting, puffing, popping, baking, milling, flaking, extrusion, germination, malting fermentation and so forth, largely involve the transformation of processed grain into value-added products intended for direct consumption. These operations are usually carried out off-farm, in either unorganized or organized sectors. During basic processing, foxtail millets are dehulled and mechanically processed to increase palatability before cooking and eating. The husk and bran of foxtail millet could be used as possible sources of phenolic antioxidants and functional components for the treatment of postprandial hyperglycemia (Zhang et al. 2021b). The various secondary processing methods such as soaking, milling, high-temperature processing, fermentation, germination and value-addition techniques that are adopted for processing of foxtail millet grains have been discussed here.

6.5.1.1 Process of soaking

Soaking grains is a common household food processing procedure for lowering antinutritional components such as phytic acid and phytase activity, improving mineral absorption, and making cooking go faster. Combining several processing methods such as dehulling, soaking, and heating reduces the amount of antinutrients like as polyphenols and phytate, increases protein digestibility in vitro and improves the bioavailability of micronutrients like iron and zinc. Sharma, N. et al. (2018) found that high pressure soaking reduced levels of anti-nutrients like phytic acid and tannin, proving that the quality of foxtail millet grains and flour can be improved by soaking at high pressures due to leaching loss in steep water.

6.5.1.2 Milling process

The dehusking and polishing of millet and cereal grains is part of the milling process. There is a significant amount of husk and bran in the foxtail millet grain which can be removed, either traditionally or by using a sheller and polisher. Due to the significant amount of phenolics, millet is employed as a potential source of antioxidants. Different milled fractions of foxtail millet, like whole, brown, polished millet and flours, have different nutritional, anti-nutritional, and functional properties. Minerals and insoluble dietary fibre, such as lignin, cellulose and hemicelluloses, are abundant in whole millet flour, whereas brown grain flour fraction contains more amounts of protein and soluble dietary fiber. Total phenols, total flavonoids, and proanthocyanidin content are higher in the whole grain fraction of foxtail millet, but phytic acid concentration is higher in the brown grain fraction. The foxtail millet is ground into flour, which contains special properties that can be used to create novel meals (Devisetti, et al., 2014). The milled fractions of foxtail millet, such as brown millet, milled millet and millet bran, contain sulfur-containing compounds, aldehydes, ketones, hydrocarbons, benzene derivatives, alcohols, acids and esters. Millet bran has a stronger aroma due to the presence of more heterocycles. Flaking, extrusion, and roller drying are utilized to further process the decorticated foxtail millet grains for the roasting and popping processes, in which native grains are employed. Popped grains have the finest nutritional profile, since the bran layer remains intact after roasting.

The processing techniques have an impact on the nutraceutical characters of foxtail millet, such as phenolics, tannins and antioxidant activity. This indicates that the suitable simple processing techniques may be adopted for improvement of the availability of nutrients from foxtail millet. Kumar et al. (2016) investigated the possibility of fractionating foxtail millet, using a roller mill without dehusking. The advantages of the roller mill without dehusking are preparation of flour from millet by bypassing dehusking and retaining specific nutrients in the prepared different fractions, which might be used in a variety of designer cuisines. As a result, foxtail millet can be easily transformed into flour without the need for dehusking and separation stages that are normally necessary during milling. This also suggests that foxtail millet, milled under ideal conditions, can have better quality attributes.

6.5.1.3 High-temperature processing methods

The grain characteristics of foxtail millet are significantly impacted by high-temperature processing techniques like roasting, popping, frying and sophisticated techniques like extrusion. *Popping* is a processing method that involves using sand as a heat transfer medium for a short period of time with high temperature, resulting in starch gelatinization and endosperm bursting open (expanded), imparting a very desirable flavour and aroma. It is sold as a ready-to-eat snack, promoting the consumption of millet grains. Since the starch granules are separated from the protein matrix during popping, the popped grains offer better nutritional qualities, including a higher protein, lower fat and increased starch and better digestibility of protein. A strong correlation was found between foxtail millet's roasting and popping characteristics and its weight to volume ratio. As starch granules undergo the same structural changes during popping and flaking, they have similar functional qualities of oil and water absorption, solubility and swelling power. *Roasting* is a high-temperature and quick-processing procedure that releases volatile aromatic molecules and odour. When roasted, foxtail millet was compared to raw foxtail millet and over 44 different aromatic components were discovered. When foxtail millet is roasted, it produces more aldehydes and heterocyclics while producing less hydrocarbons and benzene derivatives (Jing-ke et al., 2014a).

Cooking: All kinds of millets are frequently cooked to create a consumable kind of food product. It is another method that utilizes high temperatures. As with rice and other millets, foxtail millet can be cooked in water. Cooking time will vary depending on amylose and amylopectin concentration, type, gel viscosity and gelatinization temperature. Cooking quality of foxtail millet grain can be determined by evaluating factors such as cooking time, grain extension during heating, gelatinization, water uptake ratio, solid loss and organoleptic properties, all of which affects the development of food products and their use in both domestic and industrial settings. Medium to low amylose content, 90–130 g kg–1; medium to hard gel consistency, 60–75 mm; and medium to low gelatinization temperature, 58°C–63°C, all are features of a good grade of foxtail millet grain. While soaking and heating dehulled foxtail millet grains, there is a reduction in the total phenolic content, water-soluble nutrients, and antioxidant properties. An alkaline cooking technique was developed by Rajeswari et al. (2015) that involves hulling foxtail millet grains after soaking them in lime water at a boiling temperature to reduce nutrient losses. This alkali-treated foxtail millet flour has higher protein, ash, and carbohydrate contents when compared to untreated foxtail millet flour, and it has lower fat and anti-nutritional components.

Despite the fact that these ancient processing methods are popular in foxtail millet and other grains, foxtail millet is underutilized in the development of processed foods employing advanced techniques. ***Extrusion*** is one of the more sophisticated processing techniques and is frequently employed to create functional foods. When compared to other heat processing techniques, extrusion typically causes less nutritional loss. Extrusion cooking is a versatile, low-cost, and highly efficient food processing technology that involves continuous cooking, mixing, and shaping. Foxtail millet flour has good extrusion properties, resulting in a smooth surface with a crisp feel (Kadam, et al., 2018). Extrusion can effectively prevent the production of undesirable aromas in foxtail millet-based products while also extending their shelf life. Formulated foxtail millet flour was extruded in a twin-screw pilot-scale extruder in the temperature range of 150–160°C. The barrel temperature, 118.23°C; moisture level, 15.88 percent; and to a lesser extent the screw speed, 400 rpm during extrusion; affect the physical and nutritional characteristics of foxtail millet extrudates, which have greater expansion ratios, water absorption indices, water solubility indices, and overall acceptability with lower bulk density and hardness values (Kharat et al., 2015). All of these findings reveal that foxtail millet food products with good sensory and functional qualities may be made using both traditional and innovative procedures using high-temperature short-time processing. All types of cereal processing technologies might be successfully applied to foxtail millet to produce convenient foods, thereby boosting its usage.

6.5.1.4 Fermentation process

The fermentation process is widely employed in food preservation, and it produces a wide range of food products with varying aromas and textures, as well as greatly improving the nutritional values of raw food. Fermentation reduces antinutrient levels while also improving protein availability, in-vitro digestibility and a noticeable change in the chemical makeup of food. Fermentation is a quick, simple and cost-effective way to boost the nutritional value of foxtail millet while also improving its sensory and functional qualities. *Lactobacillus paracasei* Fn032; *Lactobacillus rhamnosus, Lactobacillus acidophilus, Bifidobacterium longum, Bifidobacterium bifidus* and *Saccharomyces boulardii* have all been used to ferment foxtail millet flour. According to Amadou et al. (2013b), solid state fermentation with *Lactobacillus paracasei* Fn032 and protease enhanced the physicochemical and nutritional qualities of a foxtail millet meal while also producing antioxidant and antibacterial activity as a byproduct of the synthesis of bioactive peptides.

The nutritional content of foxtail millet flour was further increased through microbial processing utilizing probiotic yeast and lactic acid bacteria. When foxtail millet was infused with a mixture of *S. boulardii* and *L. acidophilus*, the amount of crude protein increased, while the amount of fat, fibre, ash and total carbohydrates decreased. A single strain of *Lactobacillus acidophilus* boosted mineral values (iron, calcium, phosphorus, magnesium and zinc) while lowering anti-nutritional factors like phytic acid. Amadou et al. (2014) found that fermentation and high-moisture treatment had a substantial impact on the physicochemical composition, thermal characteristics, starch digestibility and pasting properties of foxtail millet flour. The tryptophan and riboflavin spectra, as well as the spectra of the rapidly digesting starch, the slowly digesting starch and the resistant starch, were all affected by the high-moisture treatment, according to fluorescence spectroscopy analysis.

According to Sharma and Sharma (2022), bioprocessed flours may be employed as potential ingredients in cereal products that have been valorized and improved in terms of techno-logical and biological functionality. The biological processing of foxtail millet resulted in less anti-nutritional compounds, a greater bioactive profile and antioxidant potential, increased techno-functional qualities and improved in-vitro protein and starch digestibility. This implies that functional foods may contain bioprocessed foxtail millet flour as a value-added ingredient. Due to its compositional, technological and bio-functional characteristics, it can be used to create baked goods and extruded foods that are nutritionally enhanced without sacrificing their technological functionality.

6.5.1.5 Germination process

A biochemical process called germination improves the functional properties of millets by converting existing chemicals and creating new ones. When grains germinate, enzymes are activated, which speeds up the decomposition of starch, protein and cell wall components, boosting nutritional digestibility and releasing bound polyphenols. The vegetative components of the grain will emerge during germination and the husk will invariably separate from the grain, allowing for better water absorption. Germination also aids in the decrease of anti-nutrients such as phytic acid, tannins and other compounds that obstruct the absorption of nutrients. The levels of thiamine, niacin, total lysine, protein fractions, carbohydrates and soluble dietary fibre improved significantly after germination and probiotic fermentation. This technique of processing produces natural, minimally processed and high-nutrient products that can assist in reducing the risk of hyperglycemia and hyperlipidemia. One of the most prevalent outcomes of millet germination is an increase in the production of water-soluble protein that inhibits hydroxyl radicals (Sharma, N. et al., 2018).

The functional components of foxtail millet seeds – such as total phenolics, antioxidants, total flavonoids, dietary fibre, proteins and minerals like phosphorus, calcium, zinc and copper – are significantly enhanced and modified during germination. Foxtail millet is one of the few grains that produce Gamma-aminobutyric Acid (GABA) during germination and can thus be used to make a variety of functional products. Considerable research has improved foxtail millet germination conditions using the physicochemical and functional properties of the product. According to Laxmi et al. (2015), the nutritional characteristics of foxtail millet were greatly increased by steeping it for 12 hours, followed by 48 hours of germination. When glutamic acid (Glu), pyridoxal-5-phosphate (PLP), and CaCl were added to the germination medium, the glutamate decarboxylase (GAD) activity was impacted and the yield of γ-aminobutyric acid (GABA) was increased. According to Pawar and Pawar (1997), the maximum malt output was obtained when grains were soaked in water for 16 hours and allowed to germinate for 48 hours. Yellow and purple cultivars of foxtail millet had their germination processes improved by Choudhury et al. in 2011. After soaking for 20 hours and germination for 72 hours at temperatures of 25 to 30°C, the foxtail millet was processed with negligibly diminished pro-tein content, noticeably lower fat and carbohydrate content, and improved starch and protein digestibility. Because of the activity of lipolytic enzymes, fat in grains is utilized to provide energy for seed expansion during germination. Consuming fat enhances the flavour of grains even when lipase is not present. Similarly to how amylase breaks down starch granules, it transforms complicated sugars into simpler, easier-to-digest forms.

The functional properties of foxtail millet are influenced by its soaking time, germination period and temperature; a favourable combination of all three can result in significant phytochemical health advantages. Because of the rupture of the outer walls of the grains during germination, the water intake of the grains increased by 3.13 percent. It is generally known that when hydrolytic enzymes are triggered during germination, cell walls are broken down, improving water absorption capacity. The ability of foxtail millet β-glucan extract to froth, bind water, swell, and maintain stability all significantly improve after germination. Enhancements to the functional and rheological characteristics of β-glucan in foxtail millet could be used in the food and pharmaceutical industries. Based on these findings, the germination process may be helpful in the creation of functional and value-added products made from foxtail millet, such as bakery goods, convenience foods, homemade goods and composite flours.

6.5.1.6 Value-addition in foxtail millet

Advanced processing techniques play a major role in diversification of foxtail millet products. This attracts great attention in all over the world where products made from foxtail millet can be found in a variety of dishes, including porridge, breakfast items, ready-to-eat snacks and beverages. The majority of processed foxtail millets are sold as rice, flour, or semolina; however, there is a greater potential for its use in the development of baked goods such as biscuits, cookies, cakes, bread, muffins, pies, pancakes, snacks, extruded foodstuffs and breakfast cereals like idli, dosa, papad and chakli, which are highly depend on its flour functionality. Milling is one of the processing procedures that results in the concentration of functional constituents that can be employed in the large-scale production of products like infant foods, snack foods, nutritious foods and traditional sweets. Due to its therapeutic potential, gluten-free foxtail millet flour can be used to partially or completely replace wheat flour in culinary products. Verma et al. (2015) reported on the nutritional value and sensory quality of food items made from foxtail and rice, such as laddus, halwas and biryani. Among these products there is no significant difference in sensory qualities. According to them foxtail millet biryani was most acceptable compared to barnyard millet and control biryani. The formulated foxtail products had higher protein, fat and fibre content than the rice products. The findings showed that foxtail millet has a higher nutritional content than rice and has the potential to be used in both traditional and functional food formulations.

The use of whole grain, under-utilized millets and pseudo cereals, as well as herbals, in composite flour blending has been raised in order to assess the viability of using locally available alternative grains and improve the qualities of the end products. Divakar and Prakash (2021) demonstrated that the composite flour made from foxtail millet along with brown rice, soybean, grain amaranth, flaxseed and black cumin was a good source of nutrients like protein, fat and dietary fibre, and bioactive components such as phenols, flavonoids and antioxidant activity. They also reported that it has a high protein digestibility along with high-quality essential amino acids and mono and polyunsaturated fatty acids. For this reason, the foxtail millet based composite flour can be recommended as a rich source of nutrients and bioactive components, which has poor digestibility of starch and can be used in development of nutritious products for all age groups.

Similarly, the current trend in the baking industry is to formulate fortified composite flour bakery products, such as biscuits and cookies. Many research studies report the incorporation of gluten free millet flour with a certain portion of refined wheat flour for formulation of bakery products. However, Marak et al. (2019) created cookies using foxtail millet (30%) and ginger powder (10%) in addition to wheat flour and they found that the protein concentration from foxtail millet is a potential functional food ingredient that can be used as an additional source of protein due to the composition of proteins and can add essential amino acid pattern (high in lysine). Furthermore, foxtail millet's high fibre and phenolic content make it a suitable ingredient for creating baked items with improved functional and nutritional qualities. These findings also showed that adding foxtail millet along with ginger powder to a product increased its nutritional and phenolic associated antioxidant potential, potentially extending its shelf life and improving health benefits through its nutraceutical and functional qualities.

Cookies made with germinated foxtail millet had the highest acceptability compared to cookies made with raw foxtail millet in a 70:20:10 ratio. This could be due to the fact that malting in germinated millet cookies reduces bitterness and improves flavour quality when compared to raw millet flour cookies. The sensory evaluation found that cookies made with germinated foxtail millet are superior to cookies made with raw ingredients (Sharma et al., 2016). Nowadays development of food additives from plant extracts is in trend. Food proteins play an important role in the encapsulation of active substances because they are good delivery mechanisms. Proline-rich prolamins (60%) with a high amount of hydrophobic amino acids and proline, which comprise fundamental delivery system properties, are the most important components of foxtail millet proteins (Chen, et al., 2022). As a result, foxtail millet prolamin, a novel encapsulant, has the potential to function as excellent carriers for lipophilic components such as curcumin encapsulation.

6.6 Conclusion

Foxtail millet is a small pointed food grain which has high economical value currently. In many parts of the world, foxtail millet is used as a staple food. It has a long history since 2700 BC. Like other millets, foxtail millet is also a non-glutinous grain and so considered as an easily digestible and non-allergenic food. Foxtail millet contains a high amount of protein, minerals, fibre and bioactive nutraceutical components, namely, phytosterol, tocopherol, phenolics, flavonoids and so forth, which attribute the wide range of biological properties, including hypoglycemic, hypolipidemic, anticancer, antioxidant and hepatoprotective activities. Due to the high lysine content, foxtail millet can be a safe and potential functional food ingredient in supplements. Conventionally, foxtail millet flour is used to prepare various foods like pan cakes, porridge and puddings, among other foods. However, in the majority of situations foxtail millet is rarely used as a food source. The functional qualities of foxtail millet play a significant role in the development of convenient and functional food products, as well as increasing their utility. Foxtail millet, like other millets, should be pre-processed before ingestion so as to improve its palatability, nutritional, sensory and functional qualities. Aside from various conventional methods, appropriate processing procedures like germination, extrusion, fermentation and diversified value-added products made from foxtail millet have had a favourable impact on nutraceutical qualities such phenolics, tannins and antioxidant activity. Considering all these nutritional and health benefits foxtail millet can be included in diet so as to improve nutritional and food security.

References

Amadou, I., Gounga, M.E., Shi, Y.H., et al. 2014. Fermentation and heat-moisture treatment induced changes on the physicochemical properties of foxtail millet (*Setaria italica*) flour. *Food and Bioproducts Processing*, 92: 38–45.

Amadou, I., Le, G.W. and Shi, Y.H. 2013a. Evaluation of Antimicrobial, Antioxidant Activities, and Nutritional Values of Fermented Foxtail Millet Extracts by Lactobacillusparacasei Fn032, *International Journal of Food Properties*, 16(6): 1179–90.

Amadou, I., Le, G.W., Amza, T., et al. 2013b. Purification and characterization of foxtail millet-derived peptides with antioxidant and antimicrobial activities. *Food Research International*, 51: 422–28.

Amadou, I., Sun, G.W., Gbadamosi, O.S., et al. 2016. Antimicrobial and Cell Surface Hydrophobicity Effects of Chemically Synthesized Fermented Foxtail Millet Meal Fraction Peptide (ffmp10) Mutants on *Escherichia Coli* ATCC 8099 Strain. *International Food Research Journal*, 23(2): 708–714.

Anitha, S., Botha, R., Kane-Potaka, J., et al. 2021. Can Millet Consumption Help Manage Hyperlipidemia and Obesity?: A Systematic Review and Meta-Analysis. *Frontiers in Nutrition*, 17(8):700778.

Annor, G.A., Marcone, M., Bertoft, E., et al. 2014. Physical and Molecular Characterization of Millet Starches. *Cereal Chemistry*, 91: 286–92.

Anusha B., Hymavathi T.V., Vijayalakshmi, V. et al. 2018. Lipid-lowering Effects of Foxtail Millet (*Setaria italica*) and Quinoa (*Chenopodium quinioa* wild) in Pre-diabetics. *Journal of Pharmaceutical Research International*. 24(5): 1–7.

Banerjee, S., Sanjay, K.R., Chethan, S., et al. 2012. Finger millet (Eleusine coracana) polyphenols: Investigation of their antioxidant capacity and antimicrobial activity. *African Journal of Food Science*, 6(13): 362–74.

Bangoura, M.L., Nsor-Atindana, J., Ming, Z.H., et al. 2012. Starch functional properties and resistant starch from foxtail millet [*Setaria italica* (L.) P. Beauv] species. *Pakistan Journal of Nutrition*, 11: 821.

Bangoura, M.L., Nsor-Atindana, J. and Z.H. Ming. 2013a. Solvent optimization extraction of antioxidants from foxtail millet species' insoluble fibers and their free radical scavenging properties. *Food Chemistry*, 141: 735–44.

Bangoura, M.L., Nsor-Atindana, J., Zhu, K., et al. 2013b. Potential hypoglycaemic effects of insoluble fibres isolated from foxtail millets (*Setaria italica* (L.) P. Beauvois). *International Journal of Food Science & Technology*, 48: 496–502.

Bembem, Khwairakpam and Dipika Agrahar-Murugkar. 2020. Development of millet based ready-to-drink beverage for geriatric population. *Journal of Food Science and Technology*, 57(9): 3278–3283.

Beta, T. and V.U. Ndolo. 2019. Postharvest Technologies. In: *Sorghum and Millets – Chemistry, Technology, and Nutritional Attributes* (2nd edn.), Ed. Taylor, J.R.N. and Duodu, K.G, 225–258. Woodhead Publishing, Elsevier, UK.

Bi, S., Wang, A., Lao, F., et al. 2020. Effects of frying, roasting and boiling on aroma profiles of adzuki beans (*Vigna angularis*) and potential of adzuki bean and millet flours to improve flavor and sensory characteristics of biscuits. *Food Chemistry*, 339: 127878.

Bisht, A., Thapliyal, M. and A. Singh. 2016. Screening and isolation of antibacterial proteins/peptides from seeds of millets. *International Journal of Current Pharmaceutical Research*, 8(3): 96–99.

Chen, J., Duan, W., Ren, X. et al. 2017. Effect of foxtail millet protein hydrolysates on lowering blood pressure in spontaneously hypertensive rats. *European Journal of Nutrition*, 56: 2129–38.

Chen, X., Zhang, T.Y., Wu, Y.C., et al. 2022. Foxtail millet prolamin as an effective encapsulant deliver curcumin by fabricating caseinate stabilized composite nanoparticles. *Food Chemistry*, 367: 130764.

Choudhari, A.S., Mandave, P.C., Deshpande, M., et al. 2020. Corrigendum: Phytochemicals in Cancer Treatment: From Preclinical Studies to Clinical Practice. *Frontiers in Pharmacology*, 11:175.

Choudhury, M., Das, P. and Baroova, B. 2011. Nutritional evaluation of popped and malted indigenous millet of Assam. *Journal of Food Science and Technology*, 48: 706–11.

Das, A., Panneerselvam, A., Yunnam, S.K., et al. 2022. Shelf-life, nutritional and sensory quality of cereal and herb based low glycaemic Index foods for managing diabetes. *Journal of Food Processing and Preservation*, 46: e16162.

Dayakar, Rao B., Bhaskarachary, K., Arlene, C.G.D., et al. 2017. *Nutritional and Health Benefits of Millets*, 1–112. ICAR Indian Institute of Millets Research (IIMR) Rajendranagar, Hyderabad.

Deshpande, A.D., Harris-Hayes, M., and Schootman, M. 2008. Epidemiology of diabetes and diabetes-related complications. *Physical Therapy*, 88(11): 1254–64.

Devisetti, R., Yadahally, S.N., Bhattacharya, S. 2014. Nutrients and antinutrients in foxtail and proso millet milled fractions: Evaluation of their flour functionality. *Food Science and Technology*, 59: 889–95.

Dias-Martins, A.M., Pessanha, K.L.F., Pacheco, S., et al. 2018. Potential use of pearl millet (*Pennisetumglaucum* (L.) R. Br.) in Brazil: Food security, processing, health benefits and nutritional products. *Food Research International*, 109:175–86.

Divakar, S.A. and Prakash J. 2021. Nutritional and Bioactive Properties of Foxtail Millet Based Composite Flour. *Indian Journal of Nutrition*, 8(1): 223.

Duodu, K.G. and Awika, J.M. 2019. Phytochemical-Related Health-Promoting Attributes of Sorghum and Millets. In: *Sorghum and Millets – Chemistry, Technology, and Nutritional Attributes* (2nd edn), Ed. Taylor, J.R.N., and Duodu, K.G, 225–258. WOODHEAD Publishing, Elsevier, UK.

Geetha, K., Yankanchi, G.M., Hulamani, S., et al. 2020. Glycemic index of millet based food mix and its effect on pre diabetic subjects. *Journal of Food Science and Technology*, 57(7): 2732–38.

Ghimire, B.K., Yu, C.Y., Kim, S.H., et al. 2019. Assessment of Diversity in the Accessions of *Setaria italica* L. Based on Phytochemical and Morphological Traits and ISSR Markers. *Molecules*, 24: 1486.

Gopalan, C., Rama Sastri, B.V. and S.C. Balasubramanian. 2015. *Nutritive Value of Indian Foods*, Printed by National Institute of Nutrition, 1–164. Indian Council of Medical Research, Hyderabad.

Gouda, G.P., Sharanagouda, H., Nidoni, U., et al. 2019. Studies on engineering properties of foxtail millet [Setaria italica (L.) Beauv.], *J. Farm Sci.*, 32(3): 340–45.

Gurung, M., Li., Z., You, H., et al. 2020. Role of gut microbiota in type II diabetes pathophysiology, *Ebiomedicinne*. 51.

Hariprasanna, K. 2016. Foxtail Millet – Nutritional importance and cultivation aspects. *Indian Farming*. 65 (12): 25–29.

Hariprasanna, K., Sangappa, Kumar, S., et al. 2018. Extension Folder – 4: Foxtail Millet. ICAR- IndianInstitute of Millets Research, Rajendranagar, Hyderabad, Telangana.

Heuzé V., Tran, G., Sauvant, D., et al. 2015. *Foxtail millet (Setaria italica),* grain. Feedipedia, a programme by INRAE, CIRAD, AFZ and FAO.

Himanshu, Chauhan, M., Sonawane, S.K., et al. 2018. Nutritional and Nutraceutical Properties of Millets: A Review. *Clinical Journal of Nutrition & Dietetics*. 1(1): 1–10.

Hou D., Chen J., Ren X., et al. 2018. A whole foxtail millet diet reduces blood pressure in subjects with mild hypertension. *Journal of Cereal Science* 84: 13–19. www.millets.res.in/technologies/2-Recommended_Package_of_Practices-Foxtail_Millet.pdf.

Huchchannanavar, S., Yogesh, L.N. and S.M. Prashant. 2019. Nutritional and Physicochemical Characteristics of Foxtail Millet Genotypes. *International Journal of Current Microbiology and Applied Science*s, 8(01): 1773–78.

Jali, M.V., Kamatar, M.Y., Jali, S.M., et al. 2012. Efficacy of value added foxtail millet therapeutic food in the management of diabetes and dyslipidamea in type 2 diabetic patients. *Recent Research in Science and Technology*, 4(7): 03–04.

Ji Z., Feng R. and Mao J. 2019. Separation and identification of antioxidant peptides from foxtail millet (*Setaria italica*) prolamins enzymatic hydrolysate. *Cereal Chemistry*. 96(6): 981–93.

Ji Z., Mao J., Chen S. and Mao J. 2020. Antioxidant and anti-inflammatory activity of peptides from foxtail millet (*Setaria italica*) prolamins in HaCaT cells and RAW264.7 murine macrophages. *Food Bioscience*. 36: 100636.

Jin, Z., Bai, F., Chen, Y., et al. 2019. Interactions between protein, lipid and starch in foxtail millet flour affect the in-vitro digestion of starch, *CyTA-Journal of Food*, 17(1): 640–47.

Jing-ke, L., Wei, Z., Shao-hui, L., et al. 2014a. Effect of roast on the volatile compounds in foxtail millet. *Food Science and Technology*, 39: 181–85.

Jing-ke, L., Yu-zong, Z., Ying-ying, L., et al. 2014b. Analysis of protein components in foxtail millet. *Food & Machinery*, 6: 39–42.

Kadam, S., Yadav, K.C. and R. Saini. 2018. Development and quality evaluation of foxtail millet (Setaria italica) based extruded product using twin screw extruder. *Journal of Pharmacognosy and Phytochemistry*, 7(4): 658–63.

Karampour N.S., Arzi A, Rezaie A, et al. 2019. Gastroprotective Effect of Zingerone on Ethanol-Induced Gastric Ulcers in Rats. *Medicina (Kaunas)*. 55(3):64.

Kasaoka, S., Oh-hashi, A., Morita, T., et al. 1999. Nutritional characterization of millet protein concentrates produced by a heat-stable alpha-amylase digestion. *Nutrition Research*, 19: 899–910.

Kazumi, T., Kawaguchi, A., Hirano, T., et al. 2004. Serum adiponectin is associated with high-density lipoprotein cholesterol, triglyceride, and low-density lipoprotein particle size in young healthy men. *Metabolism*, 53: 589–593.

Kharat, S., Hiregoudar, S., Beladhadi, R.V. 2015. Optimization of extrusion process parameters for the development of foxtail millet based extruded snacks. *Karnataka Journal of Agricultural Sciences*, 28: 301– 03.

Kim, S.N., Zhang, Q.Y., Yu, X.Z., et al. 2010. Fatty Acids Composition of Foxtail millet (*Setaria italica* BEAUVOIS) Seeds Collected in South Korea. *Korean Society of Medicinal Crop Science*, 18(6): 405–08.

Kjeldsen, S.E. 2018. Hypertension and cardiovascular risk: General aspects. *Pharmacological Research*, 129: 95–99.

Kola, G., Reddy, P.C.O., Shaik, S., et al. 2020. Variability in seed mineral composition of foxtail millet (*Setaria italica* L.) landraces and released cultivars. *Current Trends in Biotechnology & Pharmacy*, 14(3): 239–55.

Kumar, K.V.P., Dharmaraj, U., Sakhare, D.S. et al. 2016. Flour functionality and nutritional characteristics of different roller milled streams of foxtail millet (*Setaria italica*). *LWT – Food Science and Technology*, 73: 274–79.

Kumari, D., Madhujith, T., and Chandrasekara, A. 2016. Comparison of phenolic content and antioxidant activities of millet varieties grown in different locations in Sri Lanka. *Food Science and Nutrition*, 5(3): 474–85.

Kumari, K.S. and Thayumanavan, B. 1998. Characterization of starches of proso, foxtail, barnyard, kodo, and little millets. *Plant Foods for Human Nutrition*, 53:47–56.

Laxmi, G., Chaturvedi, N., Richa, S. 2015. The Impact of malting on nutritional composition of foxtail millet, wheat and chickpea. *Food Science & Nutrition*, 5: 2.

Li, S., Dong, X., Fan, G., et al. 2018. Comprehensive Profiling and Inheritance Patterns of Metabolites in Foxtail Millet. *Frontiers in Plant Science*, 9:1716.

Li, P., Zhao, W., Liu, Y., et al. 2021a. Precursors of volatile organics in foxtail millet (Setaria italica) porridge: The relationship between volatile compounds and five fatty acids upon cooking. *Journal of Cereal Science,* 100: 103253.

Li, S., Zhao, W., Liu, S. et al. 2021b. Characterization of nutritional properties and aroma compounds in different colored kernel varieties of foxtail millet (*Setaria italica*). *Journal of Cereal Science,* 100: 103248.

Liang, K., Liu, Y. and Liang, S. 2021. Analysis of the characteristics of foxtail millet during storage under different light environments. *Journal of Cereal Science,* 101:103302.

Liang, S., Yang, G. and Ma, Y. 2010. Chemical Characteristics and Fatty Acid Profile of Foxtail Millet Bran Oil. *Journal of the American Oil Chemists' Society*, 87(1): 63–67.

Lin, H.C., Sheu, S.Y., Sheen, L.Y. et al. 2020. The gastroprotective effect of the foxtail millet and adlay processing product against stress-induced gastric mucosal lesions in rats. *Journal of Traditional and Complementary Medicine.* 10: 336–44.

Liu, H.S., Yu, L., Dean, K., et al. 2009. Starch gelatinization under pressure studied by high pressure DSC. *Carbohydrate Polymers*, 75: 395–400.

Marak, N.R., Malemnganbi, C.C., Marak, C.R. et al. 2019. Functional and antioxidant properties of cookies incorporated with foxtail millet and ginger powder. *Journal of Food Science and Technology*, 56(11):5087–96.

Miller, N.F. and Smart, T.L. 1984. Intentional burning of dung as fuel: A mechanism for the incorporation of charred seeds into the archaeological record. *Journal of Ethnobiology* 4(1):15–28.

Murugan, R. and Nirmalakumari, A. 2006. Genetic divergence in foxtail millet *Setaria italica* (L.) Beauv. *Indian Journal of Genetics and Plant Breeding*, 66: 339–40.

Muthamilarasan, M., Dhaka, A., Yadav, R., et al. 2016. Exploration of millet models for developing nutrient rich graminaceous crops. *Plant Science*, 242: 89–97.

Nagaprabha, P., Devisetti, R. and Bhattacharya, S. 2018. Physicochemical and microstructural characterisation of green gram and foxtail millet starch gels. *Journal of Food Science and Technology*, 55(2): 782–91.

Naveen, J., Revathy, B., and Baskaran, V. 2021. Profiling of bioactives and in-vitro evaluation of antioxidant and antidiabetic property of polyphenols of marine algae Padinatetrastromatica, *Algal Research*, 55: 102250.

Nazni, P. and Devi, S.R. 2016. Effect of Processing on the Characteristics Changes in Barnyard and Foxtail Millet. *Journal of Food Processing & Technology*, 7(3): 566.

Pang, M, He, S, Wang, L, et al. 2014. Physicochemical properties, antioxidant activities and protective effect against acute ethanol-induced hepatic injury in mice of foxtail millet (*Setaria italica*) bran oil. *Food & Function*, 5(8):1763–70.

Pawar, V.S. and Pawar, V.D. 1997. Malting characteristics and biochemical changes of foxtail millet. *Journal of Food Science and Technology,* 34: 416–18.

Pradeep, P.M., and Sreerama, Y.N. 2018. Phenolic antioxidants of foxtail and little millet cultivars and their inhibitory effects on alpha-amylase and alpha-glucosidase activities. *Food Chemistry,* 247: 46–55.

Rajeswari, J.R., Guha, M., Jayadeep, A. et al. 2015. Effect of Alkaline Cooking on Proximate, Phenolics and Antioxidant activity of foxtail millet (*Setaria italica*). *World Applied Sciences Journal*, 33: 146–52.

Rao, M.V., Akhil, K.G, Sunil, C.K, et al. 2021. Effect of microwave treatment on physical and functional properties of foxtail millet flour. *International Journal of Chemical Studies*, 9(1): 2762–67.

Ren, X., Chen, J., Molla, M.M., et al. 2016. *In-vitro* starch digestibility and in vivo glycemic response of foxtail millet and its products. *Food & Function*, 7: 372–79.

Ren, X., Wang, L, Chen, Z., et al. 2021. Foxtail Millet Improves Blood Glucose Metabolism in Diabetic Rats through PI3K/AKT and NF-$_k$B Signaling Pathways Mediated by Gut Microbiota. *Nutrients*, 13: 1837.

Ren, X., Wang, L., Chen, Z., et al. 2022. Foxtail millet supplementation improves glucose metabolism and gut microbiota in rats with high-fat diet/streptozotocin-induced diabetes. *Food Science & Human Wellness.* 11: 119–128.

Ren, X., Yin, R., Hou, D., et al. 2018. The Glucose-Lowering Effect of Foxtail Millet in Subjects with Impaired Glucose Tolerance: A Self-Controlled Clinical Trial. *Nutrients*, 10(10): 1509.

Said, O., Fulder, S., Khalil, K. et al. 2008. Maintaining a physiological blood glucose level with glucolevel, a combination of four anti-diabetes plants used in the traditional Arab herbal medicine. *Evidence-Based Complementary and Alternative Medicine*, 5(4): 421–28.

Saleh, A.S.M., Zhang, Q., Chen, J., et al. 2013. Millet grains: Nutritional quality, processing, and potential health benefits. *Comprehensive Reviews in Food Science and Food Safety*, 12: 281–95.

Sarabhai, S., Tamilselvan, T. and Prabhasankar, P. 2021. Role of enzymes for improvement in gluten-free foxtail millet bread: It's effect on quality, textural, rheological and pasting properties. *LWT – Food Science and Technology*, 137: 110365

Shah, P., Kumar, A., Kumar, V., et al. 2021. Millets, Phytochemicals, and Their Health Attributes. In: *Millets and Millet Technology*, ed. Kumar, A., Tripathi, M.K., Joshi, D., Kumar, V, 191–218. Springer, Singapore.

Shan, S., Li, Z., Newton, I. P., et al. 2014. A novel protein extracted from foxtail millet bran displays anti-carcinogenic effects in human colon cancer cells. *Toxicology Letters*, 227(2): 129–38.

Sharma, N. and K. Niranjan. 2018. Foxtail millet: Properties, processing, health benefits, and uses. *Food Reviews International*, 34(4): 329–63.

Sharma, N., Goyal, S.K., Alam, T., et al. 2018. Effect of high pressure soaking on water absorption, gelatinization, and biochemical properties of germinated and non-germinated foxtail millet grains. *Journal of Cereal Science*, 83: 162–70.

Sharma, P.P. and Baskaran, V. 2021. Polysaccharide (laminaran and fucoidan), fucoxanthin and lipids as functional components from brown algae (Padinatetrastromatica) modulates adipogenesis and thermogenesis in diet-induced obesity in C57BL6 mice, *Algal Research*, 54:102187.

Sharma, R. and Sharma, S. 2022. Anti-nutrient & bioactive profile, in vitro nutrient digestibility, techno-functionality, molecular and structural interactions of foxtail millet (*Setaria italica* L.) as influenced by biological processing techniques. *Food Chemistry* 368: 130815.

Sharma, S., Saxena, D.C., and Riar, C.S. 2016. Nutritional, sensory and in-vitro antioxidant characteristics of gluten free cookies prepared from flour blends of minor millets. *Journal of Cereal Science*, 72: 153–61.

Sharma, S., Saxena, D.C. and Riar, C.S. 2018. Changes in the GABA and polyphenols contents of foxtail millet on germination and their relationship with in vitro antioxidant activity. *Food Chemistry*, 245: 863–70.

Shi, J., Shan, S., Li, H. et al. 2017. Anti-inflammatory effects of millet bran derived-bound polyphenols in LPS-induced HT-29 cell via ROS/miR-149/Akt/ NF-kB signaling pathway. *Oncotarget*, 8: 74582–94.

Shinoj, S., Viswanathan, R., Sajeev, M.S., et al. 2006. Gelatinisation and Rheological Characteristics of Minor Millet Flours. *Biosystems Engineering*. 95 (1): 51–59.

Sireesha, Y., Kasetti, R.B., Nabi, S.A., et al. 2011. Antihyperglycemic and hypolipidemic activities of Setaria italica seeds in STZ diabetic rats. *Pathophysiology*. 18: 159–64.

Slama, A., Cherif, A., Sakouhi, F. et al. 2020. Fatty acids, phytochemical composition and antioxidant potential of pearl millet oil. *Journal of Consumer Protection and Food Safety*, 15, 145–151.

Sosnowska, B., Penson, P., and Banach, M. 2017. The role of nutraceuticals in the prevention of cardiovascular disease. *Cardiovascular Diagnosis and Therapy, 7(Suppl 1)*, S21–S31. https://doi.org/10.21037/cdt.2017.03.20.

Subramanian, S. and Viswanathan, R. 2003. Thermal properties of minor millet grains and flours. *Biosystems Engineering*, 84: 289–296.

Suma, P.F. and Urooj, A. 2011. Antioxidant activity of extracts from foxtail millet (*Setaria italica*). *Journal of Food Science and Technology*, 49: 500–04.

Sunil, C.K., Venkatachalapathy, N., Shanmugasundaram, S., et al. 2016. Engineering Properties of foxtail millet (*Setaria italic* L): Variety – HMT 1001. *International Journal of Science, Environment and Technology*, 5(2): 632 – 37.

Taylor, J.R.N. and Emmambux, M.N.. 2008. Gluten-free foods and beverages from millets. In: *Gluten-Free Cereal Products and Beverages*, E.K. Arendt and F. D. Bello, 119–148, IV–V. A volume in *Food Science and Technology*. Elsevier Academic Press, Burlington, MA.

Tsuzuki, W., Komba, S., Kotake-Nara, E. et al. 2018. The unique compositions of steryl ferulates in foxtail millet, barnyard millet and naked barley. *Journal of Cereal Science*, 81: 153–60.

Vasudeva, S., Vishwanathan, K.H., Aswathanarayana, K.N., et al. 2010. Hydration behavior of food grains and modelling their moisture pick up as per Peleg's equation: Part I. Cereals. *Journal of Food Science and Technology*, 47: 34–41.

Verma, S., Srivastava, S. and Tiwari, N. 2015. Comparative study on nutritional and sensory quality of barnyard and foxtail millet food products with traditional rice products. *Journal of Food Science and Technology*, 52(8):5147–55.

Vetriventhan, M., Upadhyaya, H. D., Dwivedi, S. L., et al. 2016. Finger and foxtail millets. In Genetic and Genomic Resources for Grain Cereals Improvement, ed. M. Singh and H. D. Upadhyaya, 291–319. Elsevier Academic Press, USA.

Wang, R., Chen, C. and Guo, S. 2017. Effects of drying methods on starch crystallinity of gelatinized foxtail millet (α-millet) and its eating quality. *Journal of Food Engineering*, 207: 81–89.

Wankhede, D.B., Shehnaj, A. and M.R. Rao. 1979. Carbohydrate composition of finger millet (*Eleusine coracana*) and foxtail millet (*Seturia italica*). *Plant Foods for Human Nutrition*, 28:293–303.

Xiang, J.L., Zhang, M., Apea-Bah, F.B., et al. 2019. Hydroxycinnamic acid amide (HCAA) derivatives, flavonoid C-glycosides, phenolic acids and antioxidant properties of foxtail millet. *Food Chemistry* 295: 214–23.

Yang, G., Wei, J., Liu, P., et al. 2021. Role of the gut microbiota in type 2 diabetes and related diseases. *Metabolism: Clinical and Experimental*, 117, 154712.

Yang, R., Shan, S., Zhang, C. et al. 2020a. Inhibitory effects of bound polyphenol from foxtail millet bran on colitis-associated carcinogenesis by the restoration of gut microbiota in a mice model, *Journal of Agricultural and Food Chemistry*, 68(11): 3506–17.

Yang, Z., Zhang, H., Li, X. et al., 2020b. A mini foxtail millet with an Arabidopsis-like life cycle as a C4 model system. *Nature Plants* 6, 1167–1178.

Zhang, A.X., Liu, X.D., Wang, G.R., et al. 2015. Crude Fat Content and Fatty Acid Profile and Their Correlations in Foxtail Millet. *Cereal Chemistry*, 92(5): 455–59.

Zhang, C.G., Gong, W.J., Li, Z.H. et al. 2019. Research progress of gut flora in improving human wellness, *Food Science and Human Wellness*, 8(2): 102–05.

Zhang, F. and Shen, Q. 2021. The impact of endogenous proteins on hydration, pasting, thermal and rheology attributes of foxtail millet. *Journal of Cereal Science*, 100: 103255.

Zhang, L., Li, J., Han, F. et al. 2017. Effects of Different Processing Methods on the Antioxidant Activity of 6 Cultivars of Foxtail Millet. *Journal of Food Quality*. (2):1–9.

Zhang, L.Z. and Liu, R.H. 2015. Phenolic and carotenoid profiles and antiproliferative activity of foxtail millet. *Food Chemistry*, 174: 495–501.

Zhang, L., Xiaoqin L., Tian, J. et al. 2021a. The phytochemical vitexin and syringic acid derived from foxtail fillet bran inhibit breast cancer cells proliferation via GRP78/SREBP-1/SCD1 signaling axis. *Journal of Functional Foods,* 85: 104620.

Zhang, M., Xu, Y., Xiang, J., et al. 2021b. Comparative evaluation on phenolic profiles, antioxidant properties and α-glucosidase inhibitory effects of different milling fractions of foxtail millet. *Journal of Cereal Science*, 99: 103217.

Zhang, Y., Gao, J., Qie, Q., et al. 2021c. Comparative Analysis of Flavonoid Metabolites in Foxtail Millet (Setaria italica) with Different Eating Quality. *Life*. 11: 578.

Zhang, Y., Liu, W., Zhang, D., et al. 2021d. Fermented and Germinated Processing Improved the Protective Effects of Foxtail Millet Whole Grain Against Dextran Sulfate Sodium-Induced Acute Ulcerative Colitis and Gut Microbiota Dysbiosis in C57BL/6 Mice. *Frontiers in Nutrition*, 29(8):694936.

Zhu, Y., Chu, J., Lu, Z. et al. 2018. Physicochemical and functional properties of dietary fiber from foxtail millet (*Setaria italic*) bran. *Journal of Cereal Science,* 79: 456–61.

7

Proso Millet (*Panicum miliaceum*)
Bioactive Composition, Pharmacological Impact and Techno-Functional Attributes

Shubhendra Singh and Anil Kumar Chauhan

Contents

7.1 Introduction

Millets are a unique component of biodiversity in agricultural and food security systems. Millets are frequently processed into flour, rolled into large balls, parboiled, and served with milk as porridge; millets are also prepared as beverages in some cases. Farmers in Gujarat, India, have relied on roti prepared from pearl millet as their primary source of nourishment (FAO 2012). There is an increasing need for the globe to feed its expanding population. In low-income households of nations like India and the Sahel zone, plants like millets are grown and consumed widely, thus increasing the need for research on these plants (Obilana 2002). Cereals, particularly millet-based dishes and beverages, are well-known around the world and remain an important part of most African diets (Amadou et al. 2011). India has chosen 2018 as the "Year of Millets," while the United Nations Food and Agriculture Organization (FAO) has named 2023 the "International Year of Millets." The most widely grown species of millet in the

DOI: 10.1201/9781003251279-7

world is a kind known as pearl millet (*Pennisetum typhoides*), sometimes known as bulrush millet. Other major varieties of millets are foxtail millet (*Setaria italica*), proso millet (*Panicum miliaceum*), also known as common millet, Japanese barnyard millet (*Echinochloa crusgalli var. Frumentacea* or *E. colona* (*Paspalum scorbiculatum*)) (FAO 2017).

Proso millet is classified as a C_4 cereal because it absorbs more CO_2 from the atmosphere and converts it to oxygen. It is better for the environment since it uses less water, has fewer byproducts, requires less fertilizers and pesticides, and can withstand cold weather. Farmers in many regions of the country cultivate millets, although the crop's performance has fluctuated over the past decade for reasons that may be related to both external and internal agro-climatic factors (Habiyaremye et al. 2017).

Having the shortest growth season of any farmed cereal crop, at about 10–11 weeks, proso millet is well-known for its ability to thrive in arid climates. It goes by different names in various parts of the world, including Russian millet, broomcorn millet, hog millet and common millet. One of the world's oldest crops is proso millet (*Panicum miliaceum L.*), sometimes known as yellow rice. It was domesticated for the first time in China some 10,000 years ago and is now a major food crop in northwest China (Yang et al. 2012). Because of its high tolerance to stress, such as poor soil and drought, it is still commonly cultivated. Proso millet grains include a variety of nutrients, including starch, protein, dietary fiber, vitamins, and minerals, with starch accounting for the majority of the carbohydrate content (58.5–73.5 percent). People favor proso millet as diet food, functional food, and so on because of its nutritious content. The physico-chemical features of proso millet have a significant role in determining its eating quality.

Although it provides one-third of protein and energy in impoverished nations, proso millet is an underutilized crop. Birdseed industries comprise the majority of the proso millet market share (Saxena et al. 2018). Proso millet is a grain that has yet to gain widespread acceptance in the human food market despite its many potential health benefits, including being gluten-free, having a low glycaemic index, and being high in protein and fiber (**Table 7.1**). The rising popularity of gluten-free diets and general concern for one's health has renewed an interest in these long-forgotten grains. Proso millet is now widely used in a variety of breakfast cereals, breads, and fermented and alcoholic beverages. Promoting the commercialization of proso millet, conducting research to generate new kinds, and raising public awareness are all vital steps towards reviving the once-popular crop.

7.2 Nutraceutical properties

Like wheat, rice, and maize, millets are an excellent source of protein and energy. As a result of their very high levels of minerals including potassium, magnesium, calcium, iron, dietary

Table 7.1 Nutritional Composition of Small Millets (per 100g)

Crop	Dietary fiber (g)	Carbohydrate (g)	Fat (g)	Protein (g)	Mineral matter (g)	Calcium (mg)	Phosphorus (mg)	Fe (mg)
Finger millet	18.8	72.1	1.32	7.3	2.8	345	284	4.0
Kodo millet	15.0	65.2	1.43	8.3	2.7	28	189	12.3
Proso millet	14.2	70.3	3.14	12.5	2.0	15	205	10.2
Foxtail millet	14.0	61.0	4.31	12.3	3.2	32	292	5.4
Little millet	12.2	67.2	4.74	7.7	1.6	18	222	6.2
Barnyard millet	13.7	65.7	2.23	6.2	4.5	10	281	16

Source: Adapted from Saha et al. 2016.

fiber, polyphenols, phosphorus, zinc, and protein, millets are in a class of their own among grains (Baltensperger 2002). Since millet is gluten-free, it is a great option for those with celiac disease or wheat allergies, but millet flour cannot be used to produce leavened bread. Millets are easy to digest due to their low glycemic index. They are high in lecithins, which help the body repair and maintain neurons, regenerate myelin sheaths, and provide energy for brain cells. Millets also contain high amounts of beneficial micronutrients like niacin, B vitamins, and folic acid. Particularly sulfur-rich millets are praised for their abundance of all nine necessary amino acids (methionine and cysteine). For all amino acids except lysine and threonine, millets are a good source of methionine, as shown by Saleh et al. (2013). The fat content of millets is higher than that of corn, rice, and sorghum. Researchers Ravindran (1991) examined the chemical composition of a wide variety of millets, including four distinct foxtail millet varieties. Common millet, finger millet, and foxtail millet each contained 14.4 percent protein and 3.2 to 4.7 percent crude fiber. The mineral content was often high, especially in comparison to other conventional cereal grains. There was large variability in nutritional content both between and among different types of millet.

The results as a whole demonstrate the nutritional potential of millets that has yet to be fully developed. Around 11 percent (on a dry weight basis) of proso millet is protein (Kalinova and Moudry 2006). There are a greater number of essential amino acids in proso millet protein compared to wheat protein (leucine, isoleucine, and methionine). Therefore, proso had a higher quality protein (Essential Amino Acid Index, which was 51%) than wheat. Nutritional value, processing characteristics, and proximate composition of proso-millet flour were the subject of a comprehensive investigation by Lorenz and Dilsaver (1980). Obilana and Manyasa (2002) find in their studies that, when compared to wheat flours, millet flours have higher protein and less ash. Nutritionally and medicinally, millet can be used in many ways. Because of their abundance of beneficial phytochemicals, millets have gained popularity as a functional food (Pathak 2013). Proso millet, like other millets, is rich in magnesium and other minerals linked to a lower risk of developing Type 2 diabetes. Glucose and insulin synthesis are both regulated by a number of biochemical activities that require magnesium as a cofactor (Shobana and Malleshi 2007). In addition to alleviating the symptoms of migraines and heart attacks, magnesium has been shown to help patients with atherosclerosis and diabetic heart disease (Gélinas et al. 2008). The maintenance of human body's cell structure depends on phosphorus, which is abundant in millet (Laxmi Kumari and Sumathi 2002). Liang et al. (2010) and Devi et al. (2014) found that millet's phosphorus content is important since it is needed to make bone mineral matrix and is also a key component of ATP, the body's energy currency. One cup of millet, once cooked, provides roughly 24 percent of the phosphorus the body needs for a day. Furthermore, phosphorus is required for the synthesis of nucleic acids, the structural components of DNA (Laxmi Kumari and Sumathi 2002).

Possibly preventing cancer and cardiovascular disease, millets are a whole grain that is rich in fiber, antioxidants, and complex carbs. Millet is a major crop with several health benefits, according to Coulibaly et al. (2011), who point to the presence of phytochemicals in millet grains as evidence. Three varieties of proso millet were analyzed by Zhang et al. (2014) for their phytochemical profiles, antioxidant activity, and anti-proliferative effects. MDA-MB (Human Mammary Carcinoma) cells, human breast cancer cells, and HepG2 human liver cancer cells were all tested for antiproliferative properties in vitro. The results indicated that proso millet possesses a novel and, potentially, selective anti-proliferative effect. Recent studies have discovered that eating millets can lower levels of the cancer and heart disease-promoting compounds phytate and cholesterol. Millet's lignans are an important phytonutrient because they protect against breast cancer and other hormone-dependent malignancies and lower the risk of cardiovascular diseases (CVD) (Coulibaly et al. 2011). Millets are a great source of fiber and protein, and frequent consumption lowers cardiovascular disease risk factors like high blood pressure and cholesterol, especially in postmenopausal women (Shahidi and Chandrasekara 2013).

Millets may help slow down the aging process in humans and guard against age-related degenerative diseases as well (Pathak 2013; Shahidi and Chandrasekara 2013). Millet's cholesterol-lowering protein is a win-win (Nishizawa and Fudamo 1995). The effect of dietary proso-millet protein on plasma levels of HDL cholesterol in rats was studied by Nishizawa et al. (2002). They confirmed the previous findings regarding proso- millets, that increasing one's intake of millet protein led to an increase in plasma HDL-cholesterol. Millet protein has an anti-atherogenic activity, which suggests it may be useful as a novel dietary component that modulates cholesterol metabolism. It is believed that this protein can also protect the liver from harm (Nishizawa et al. 2002).

Several methods exist for preparing millets. The grains can be prepared in the same ways as rice: by boiling, steaming, or any other conventional method. Millet flour, together with a few additional ingredients, can be used to make porridge. Proso millet's subtle flavor, pale color, gluten-free quality, and promising health advantages have aroused the interest of food manufacturers in Europe and North America (Santra 2012. As a subset of the bread and grain industry, gluten-free products are experiencing rapid growth in the United States. Expectations for this market's size by 2017 are $6.6 billion, up from $4.8 billion in 2012 compound annual growth rate (CAGR). Additionally, proso millet is utilized in the production of fermented beverages and as a possible substrate in distilled spirits and beers across Africa and Asia. Therefore, proso millet may experience growth in the alcoholic beverage industry, notably in the gluten-free sector, in the United States and Europe (Santra et al. 2015).

Few people are familiar with millet, much less its culinary or commercial use (Saleh et al. 2013). As per the findings of Common Fund for Commodities (CFC) and the International Crops Research Institute for the Semi-Arid Tropics (ICRISAT), industrial millets are increasingly threatened by competition from other industrially produced grain crops in poor nations (CFC and ICRISAT, 2004). There are a few studies out there on proso millet's nutritional value, health benefits, and uses, but there is a big hole where information on the nutritional profile of different types and how they are used in processed foods is concerned. It is required to research the processing methods for producing a wide range of food products from proso millet. For proso millet to be used as a food additive, it is crucial that the results of these investigations be made available to relevant parties in the food processing sector. To develop millet cultivars that are suitable for various agroecosystems and suitable production methods, more study is required. Proso millet's nutritional characteristics and end-uses must be linked to its agronomic characteristics. Increased demand in the market may result from informing consumers about the nutritional and health advantages of millets (Amadou et al. 2013; Saleh et al. 2013). Research into the viability of proso millet as a nutritious whole grain alternative is required if it is to acquire traction in industrialized nations (Amadou et al. 2013).

The pinnacles must be subdued to gain the health advantages of little millets. Nutritional deficiencies, including both over- and under-nutrition, are now widespread due to changes in lifestyle and diet, placing an additional load on our nation. The prevalence of micronutrient deficiencies is a reason for worry even among the wealthiest parts of the population. Consumer demand for little millets in the food basket has declined over time, and it is past time to resuscitate efforts to avoid the slow elimination of indigenous crops from our diets (Mathanghi and Sudha 2012). As one of the world's leading producers and consumers of millets, we recognize the importance of creating and standardizing millet-based pre-validated functional meals to fulfil global demand. Millet's phytochemicals, polyphenols, and high concentrations of micro and macro minerals have been demonstrated to have positive effects on health. More epidemiological and experimental research is necessary, however, for more confidence (Chauhan et al. 2018).

The high concentrations of various minerals, vitamins, and trace elements, along with the abundance of numerous phytochemicals such as polyphenols, tocopherols, phytosterols, and dietary fiber is responsible for the majority of the health advantages associated with millets. Millets have been linked to a variety of health benefits, some of which are backed up by scientific evidence. Animal research provides the most compelling evidence for millets' health benefits, while human studies provide additional support (epidemiology and experimental). Millet's biodiversity and nutraceutical quality have been thoroughly established in several researches and to classify them, we can look at how well they address issues of poverty and income generation, health care, food safety, and environmental sustainability (Gupta et al. 2012; Gupta et al. 2014). Since millet is gluten-free, it is a suitable substitute for celiac disease patients who are frequently irritated by the gluten component in wheat and other commonly consumed cereal grains. People with heart conditions like atherosclerosis or diabetes benefit from it as well (Gélinas et al. 2008).

Type 2 diabetic mice had their insulin sensitivity and lipid profiles vastly improved when they were fed a diet high in protein concentrates from Korean foxtail millet and proso millet, compared to a casein diet (Choi et al. 2005). In addition, plasma levels and glycemic responses were both improved by proso millet (Park et al. 2008), see **Figure 7.1**. Moreover, the protein concentrate from proso millet prevents D-galactosamine from damaging the livers of rats (Ito et al. 2008).

Figure 7.1 *Health benefits of Proso millet.*

Source: Figure created by authors.

7.3 Characterization of phytochemical compounds responsible for bioactive properties

7.3.1 Polyphenolic compounds

According to Zhang et al. (2014) broomcorn millet possesses the phenolic acids that are commonly present in the combined form. Among the most prominent phenolic chemicals found in millet are syringic acid, gallic acid, vanillic acid, 4-hydroxyl benzoic acid, and hydroxycinnamic acid derivatives including caffeic acid, ferulic acid, and sinapic acid (Chandrasekara et al. 2010). The most abundant phenolic acids in broomcorn millet, finger millet, and foxtail millet are hydroxycinnamic acids (Hithamani et al. 2014). The broomcorn millet contains the hydroxycinnamic acids ferulic acid, chlorogenic acid, caffeic acid, and coumaric acid. Ferulic acid is bound in broomcorn millet and finger millet, whereas it is free in pearl millet (Zhang et al. 2014).

Ferulic acid, a major phenolic component, contributes greatly to the outstanding antioxidant activity of millets, as reported by Chadrasekara and Shahidi (2010). Gallic acid, p-hydroxybenzoic acid, vanillic acid, caffeic acid, sinapic acid, and syringic acid are only few of the many phenolic acids found in millets. In humans, ferulic acid has been demonstrated to reduce the risk of cancer (Rice-Evans et al. 1996), while caffeic acid preferentially reduces leukotriene synthesis and syringic acid protects the liver (Koshihara et al. 1984).

7.3.2 Factors affecting resistant starch in foods

7.3.2.1 Effect of protein

Starch digestibility may be affected by interactions with specific digestive proteins. Scientists think that enzymes are unable to access starch granules because denatured protein binds to them. The gelatinization process could be influenced by globule encapsulation in the gel formed by the hydrophilic interaction of protein and starch (Guerrieri et al. 1997). This theory was corroborated by studies on the effect of protein enrichment in proso-millet IVSD (Zheng et al. 2020). Protein supplements of zein, soy isolate, and whey isolate were added to proso millet flour to determine their effects on the flour's physicochemical properties and starch digestibility. With the addition of protein, starch digestibility was greatly enhanced. It was discovered that whey protein isolate had the greatest effect on the protein content of RS-millet flour, increasing it from 4.49 to 11.73 percent (w/w on a dry weight basis).

7.3.2.2 Effect of lipid

Studies conducted have shown Lipid-amylose form to be a complex form (Annor et al. 2014; Meenu and Xu 2019; Rattanamechaiskul et al. 2014). The complex is formed when lipids, particularly long-chain fatty acids, are trapped within the hydrophobic helical chambers of amylose. The polarity of the trapped fat molecule affects how many glucose units are present per turn in the complex. To maintain the lipid-inclusion complex, nonpolar, bulky lipids require additional glucose residues at each step. When these compounds cool, they rearrange to create a crystalline structure, as confirmed by XRD. The resistance to enzymatic hydrolysis is brought on by the combination of the hydrophobicity and crystallinity of the lipid-starch inclusion. It has also been shown that this complex can become unstable with a higher lipid concentration, which will cause amylose to leak out (Sajilata et al. 2006). These amylo-lipid complexes are also known as RS5, according to certain studies1 (Fuentes-Zaragoza et al. 2011; Srichuwong et al. 2017). An intriguing investigation on the effect of extra fatty acid on millet

IVSD revealed that fatty acids had a detrimental effect on starch digestibility. They investigated the IVSD of proso, foxtail, finger, and pearl millet flours with different amounts of various fatty acids (0.01–0.09 mmol/g starch) (palmitic acid, oleic acid, linolenic acid, eladic acid). All millet flour showed a comparable increase in RS content, with proso millet exhibiting the largest increase in RS from 13.54 percent to 27.35 percent (w/w on a dry weight basis). They also concluded that unsaturated fats generate more RS than saturated fats (Annor et al. 2015).

7.3.2.3 Effect of dietary fiber

According to a considerable number of studies, dietary fiber (DF) does not influence RS production (Bae et al. 2014; Bae et al. 2016). In research to test the physicochemical and nutritional digestibility of DF (millet bran) enriched bread, a similar observation was observed (Li et al. 2020). The study found that the concentration of RS was not noticeably altered by the addition of up to 5 percent millet flour or DF. While there was a moderate increase in RS quantity due to an increase in DF. The increased water-binding capacity of DF may be to blame, since it may prevent starch from becoming fully hydrated, leading to inefficient gelatinization and more pronounced starch crystallinity (Agama-Acevedo et al. 2012). Furthermore, higher dietary fiber concentrations may inhibit hydrolysis by retaining starch inside the pores (Oh et al. 2014). Nonetheless, augmenting necessary DF to achieve a substantial RS rise may endanger food sensory appeal to a greater extent.

7.3.3 Antioxidant properties

Both the presence of primary and secondary antioxidants as well as oxidants have an impact on gene regulation and cell signaling. Check the proso millet's lignin, phytic acid, and condensed tannin levels to see if they provide any antioxidant protection. Bound phenolics are thought to play an important part in the production of SCFs and B-complex vitamins in the colon. Prebiotics such as oligosaccharides promotes skeletal health. In terms of nutraceutical properties, melatonin, phytosterols, and para-aminobenzoic acid need to be studied further. The health benefits of proso millet, both as a cereal grain and in the form of its numerous dietary derivatives, are well established. According to Asharani et al. (2010), the total antioxidant activity of proso millet edible flours ranges between 0.5 and 5.7 mM tocopherol equivalent/g. In addition, Chandrasekara and Shahidi (2010) investigated the bioactive capabilities of proso millet in vitro chemical testing using the $DPPH_2$ radical scavenging capability and reactive oxygen species (ROS). Zhang et al. (2014) examined the phenolic acid composition and observed that the bound fraction accounted for 65 percent of total phenols. It also contains antioxidant and anti-proliferative compounds such as ferulic acid, chlorogenic acid, syringic acid, caffeic acid, and p-coumaric acid.

7.4 Techno-functional properties

7.4.1 Pasting properties

Starch granules in proso millet are typically either small spherical granules or large polygonal granules, with diameters ranging from 1.3 to 8.0 mm (Kumari et al. 1998). Proso millet had the lowest pasting temperature (PT) and setback (SB) but the highest peak viscosity (PV) and breakdown of the four grains tested (foxtail, barnyard, Kodo, and millet). Yao et al.(2009) reported that the amylose concentration in proso millet ranges from 2.85 to 19.3 percent, and that the starch granule size ranges from 1.9 to 10.0 mm. As a result of its wide range of gelatinization properties, proso millet starch is an excellent substitute for potato and maize starch when it comes to gel formation and heat and cold paste stability, respectively. Starch granule diameters in proso millet were found to range from 1.3 to 8.0 mm, with small spherical granules and large polygonal granules having bimodal distributions (Kumari et al. 1998). Pasting temperature (PT), setback (SB), the peak viscosity (PV), and breakdown were all lower for proso millet than for foxtail, barnyard, Kodo, and millet (BD). According to Yao et al. (2009) the starch granule diameters of proso millet range from 1.9 to 10.0 mm, while the amylose concentration ranges from 2.85 to 19.33 percent. The gel-forming properties of potato starch are lower than those of maize and potato starch, but it has greater thermal and cold paste stability than proso millet starch. Previous research has shown that waxy proso millet starch granules are polyhedral with smooth edges and range in size from 5.76 to 8.64 mm, making them larger than waxy rice starch granules but on a par with non-waxy millet starch granules (Wang et al. 2012; Yang et al. 2019).

7.4.2 Thermal properties

Starch undergoes gelatinization, or an irreversible phase change, when boiled in an excessive amount of water. Food manufacturers utilize gelatinization parameters to select specific starch grades for usage in cuisines preparation. Many instruments, including DSC, NMR spectroscopy, and FTIR, are utilized to evaluate gelatinization characteristics; nevertheless, DSC is the most commonly used by scientists to examine starch thermal properties (Bangar et al. 2021). Proso millet (68.4° C) exhibited the highest onset temperature of any of the millet starches studied by Annor et al. (2014), followed by foxtail millet (66.7° C), finger millet (63.9° C), and pearl millet (62.8° C). Also, the enthalpy (H) of PMS starch (13.1 J/g) was higher than that of pearl millet (12.3 J/g) and fox millet (11.8 J/g), but lower than that of finger millet (13.2 J/g). In the same way that commercial starches (such as wheat and maize) can be utilized in thermally processed foods, millets can too.

Although Annor et al. (2014) found that fox millet had the highest onset temperature (69.7° C), Wu et al. (2014) found that pearl millet, barnyard millet, and finger millet all had higher H values, and that these values varied by millet type. Higher PMS values indicate that more heat is needed to breakdown the starch intermolecular bonds. Using proso-millet starch in foods including soups, sauces, breads, and even dairy products was suggested due to the starch's improved heat stability.

7.4.3 Textural properties

The textural properties of starch are essential to the research, development, and end result of any new product. The consistency of the starch gel changes depending on a variety of circumstances. Physicochemical and textural requirements for various starch-based products led to the evaluation of 95 proso millet accessions for potential use in these industries. The cohesiveness of cooked proso millet varied from 0.122 to 0.856 depending on whether the PMS starch was waxy or non-waxy. Furthermore, it was discovered that the amylose content was highly related to the hardness of the PMS gel. Non-waxy PMS has higher hardness than waxy PMS, which supports this conclusion (Li et al. 2020). Another study discovered that uncooked waxy proso millet had higher chewiness and hardness values than cooked waxy proso millet, which had higher cohesiveness and adhesiveness (Yang et al. 2018). However, there is a dearth of writing on the topic of PMS's textural components; in the future, the scientific community can organize research initiatives centered on this component in order to create novel food products with the appropriate textural features.

7.5 Potential utilization in value-added food products

Because of its improved functioning and higher health benefits, processed food is currently gaining a lot of attention. Processing procedures are expected to have an impact on the availability of nutrients and phenolic compounds. Pre-treatment/mechanical treatment has been observed to liberate phenolic compounds from their bound state. As a result, numerous processing procedures have been employed to enhance the release and accessibility of bound phenolic chemicals in wheat grains (Wang et al. 2014).

Many studies have found that pre-treatments like dehulling or milling have a considerable favorable influence on the nutritious content of the millet flour (Taylor et al. 2015; Zhang et al. 2017). To enhance the profiles of secondary bioactive compounds, a variety of food processing methods are being used, including as thermal processing, sprouting, and fermentation (Hotz et al. 2007). Steaming and heating methods increase the phenolic content and antioxidant profile of foxtail and proso millet, according to Geetha et al. (2014) and Fei et al. (2018). Roasting is used in the food sector to increase secondary chemical bioavailability. Roasting alters the physical, chemical, and nutritional aspects of plant foods, causing both positive and unfavorable alterations (Zhang et al. 2017). Food items are subjected to roasting to remove antinutritive elements and give finished products a distinctive flavor and brown color (Han et al. 2018). Extrusion may aid in the breakdown of secondary compound's high molecular weight ingredients into smaller molecular weight constituents with greater bioavailability (Deprez et al. 2001).

Noodles are a worldwide favorite among quick-and-easy meals. Millet mixes composite flour produced noodles own a low glycemic index but take longer to cook than branded noodles (Vijayakumar et al. 2010). More recent research with similar aims and enhanced consideration of sensory, dietary, and nutritional aspects have also been reported (Collar 2017). A tiny concentration of proso millet's bran has been observed to have a greater phenolic content (117%) and dietary fiber (76%), suggesting that gluten-free bread containing millets bran could be a significant component. Its nutritional and sensory qualities, however, have not been investigated (Mustac et al. 2020).

Grain and legume flours constitute the foundation of pasta and papad, respectively; these RTC staples are then supplemented with spices and other dry ingredients. Noodles are dry, hard, and brittle items made utilizing cold extrusion technology, and are sometimes referred to as convenience foods. These noodles can be ready in just a few minutes and need no effort. Finger millet noodles, finger millet and wheat noodles at a 1:1 ratio, and finger millet noodles made with a 5:4:1 ratio of wheat and soy flour are just a few of the noodle preparations that can be made. Finger millet, refined wheat, and soy flour/whey protein concentrate composite flours created at 50 percent, 40 percent, and 10 percent (Devaraju et al. 2006) can all be used to make pasta, as can a proso millet and wheat flour blend that keeps for a long time (Devi et al. 2013).

Complaints regarding the texture and taste of biscuits and extruded snacks manufactured from a refined corn/refined proso millet mix increased as the millet content increased (from 0 to 100) (McSweeney et al. 2016). Female and male panelists did not agree completely on how to rank the products, but they did agree that bitterness was a major issue. Greasy, rancid, gritty, and crumbly were all associated with biscuits with 75 and 100 percent millet inclusion. Snacks that were extruded with a higher percentage of millet were deemed crunchier and thicker, but had a more unpleasant aftertaste. As has been suggested for other bitter grains like quinoa (Suárez-Estrella et al. 2018), pre-treatments (such as fermentation or sprouting) may be employed to improve the sensory profiles of millet products.

7.6 Conclusion

Proso millet starch has been the focus of numerous scientific investigations on its nutritional value, digestibility, temperature stability, and structural diversity. Cooked proso millets have been shown to have low protein digestibility, spurring studies into the chemical mechanisms at play and potential techniques for improving digestibility. Less research has gone into other phytochemicals, especially their fate following processing, as phenolic acids. Nutrient fiber content and lipid stability information is likewise scarce. The relationships between, say, variety, amylopectin structure, and starch functioning, or variety, prolamin composition/structure, and protein functionality, can be better understood with additional research on a larger range of variations. Collaborations between plant breeders and food scientists would help maximize the project's use of proso millet's agronomic, nutritional, and technological features.

References

Agama-Acevedo, Edith, José J. Islas-Hernández, Glenda Pacheco-Vargas, Perla Osorio-Díaz, and Luis Arturo Bello-Pérez. "Starch digestibility and glycemic index of cookies partially substituted with unripe banana flour." *LWT-Food Science and Technology* 46, no. 1 (2012): 177–182.

Amadou, I., O. S. Gbadamosi, and Guo-Wei Le. "Millet-based traditional processed foods and beverages – A review." *Cereal Foods World* 56, no. 3 (2011): 115.

Amadou, Issoufou, Mahamadou E. Gounga, and Guo-Wei Le. "Millets: Nutritional composition, some health benefits and processing – A review." *Emirates Journal of Food and Agriculture* (2013): 501–508.

Annor, George Amponsah, Massimo Marcone, Eric Bertoft, and Koushik Seetharaman. "Physical and molecular characterization of millet starches." *Cereal Chemistry* 91, no. 3 (2014): 286–292.

Annor, George Amponsah, Massimo Marcone, Milena Corredig, Eric Bertoft, and Koushik Seetharaman. "Effects of the amount and type of fatty acids present in millets on their in vitro starch digestibility and expected glycemic index (eGI)." *Journal of Cereal Science* 64 (2015): 76–81.

Asharani, V. T., A. Jayadeep, and N. G. Malleshi. "Natural antioxidants in edible flours of selected small millets." *International Journal of Food Properties* 13, no. 1 (2010): 41–50.

Bae, In Young, and Hyeon Gyu Lee. "In vitro starch digestion and cake quality: Impact of the ratio of soluble and insoluble dietary fiber." *International Journal of Biological Macromolecules* 63 (2014): 98–103.

Bae, In Young, Yujin Jun, Suyong Lee, and Hyeon Gyu Lee. "Characterization of apple dietary fibers influencing the in vitro starch digestibility of wheat flour gel." *LWT-Food Science and Technology* 65 (2016): 158–163.

Baltensperger, David D. "Progress with proso, pearl and other millets." *Trends in New Crops and New Uses* (2002): 100–103.

Bangar, Sneh Punia, Adeleke Omodunbi Ashogbon, Sanju Bala Dhull, Rohit Thirumdas, Manoj Kumar, Muzaffar Hasan, Vandana Chaudhary, and Srilatha Pathem. "Proso-millet starch: Properties, functionality, and applications." *International Journal of Biological Macromolecules* 190 (2021): 160–168.

Chandrasekara, Anoma, and Fereidoon Shahidi. "Content of insoluble bound phenolics in millets and their contribution to antioxidant capacity." *Journal of Agricultural and Food Chemistry* 58, no. 11 (2010): 6706–6714.

Chauhan, Manish, Sachin K. Sonawane, and S. S. Arya. "Nutritional and nutraceutical properties of millets: a review." *Clinical Journal of Nutrition and Dietetics* 1, no. 1 (2018): 1–10.

Choi, You-Young, Kyoichi Osada, Yoshiaki Ito, Takashi Nagasawa, Myeong-Rak Choi, and Naoyuki Nishizawa. "Effects of dietary protein of Korean foxtail millet on plasma adiponectin, HDL-cholesterol, and insulin levels in genetically type 2 diabetic mice." *Bioscience, biotechnology, and biochemistry* 69, no. 1 (2005): 31–37.

Collar, C. "Significance of heat-moisture treatment conditions on the pasting and gelling behaviour of various starch-rich cereal and pseudocereal flours." *Food Science and Technology International* 23, no. 7 (2017): 623–636.

Coulibaly, A., Brou Kouakou, and Jie Chen. "Phytic acid in cereal grains: structure, healthy or harmful ways to reduce phytic acid in cereal grains and their effects on nutritional quality." *American Journal of Plant Nutrition and fertilization technology* 1, no. 1 (2011): 1–22.

Deprez, Stephanie, Isabelle Mila, Jean-François Huneau, Daniel Tome, and Augustin Scalbert. "Transport of proanthocyanidin dimer, trimer, and polymer across monolayers of human intestinal epithelial Caco-2 cells." *Antioxidants and Redox Signaling* 3, no. 6 (2001): 957–967.

Devaraju, B., Mushtari J. Begum, Shemshed Begum, and K. Vidhya. "Effect of temperature on physical properties of pasta from finger millet composite flour." *Journal of Food Science and Technology-Mysore* 43, no. 4 (2006): 341–343.

Devi, G. Sudha, V. Palanimuthu, H. S. Arunkumar, and P. Arunkumar. "Processing, packaging and storage of pasta from proso millet." *International Journal of Agricultural Engineering* 6, no. 1 (2013): 151–156.

Devi, Palanisamy Bruntha, Rajendran Vijayabharathi, Sathyaseelan Sathyabama, Nagappa Gurusiddappa Malleshi, and Venkatesan Brindha Priyadarisini. "Health benefits of finger millet (Eleusine coracana L.) polyphenols and dietary fiber: a review." *Journal of Food Science and Technology* 51, no. 6 (2014): 1021–1040.

Fei, H., Z. Lu, D. Wenlong, and L. Aike. "Effect of roasting on phenolics content and antioxidant activity of proso millet." *International Journal of Food Engineering* 4 (2018): 110–116.

Food, F. A. O. "Agriculture Organization of the United Nations (2012) Economic and Social Department, The Statistical Division." (2012).

Fuentes-Zaragoza, Evangélica, Elena Sánchez-Zapata, Esther Sendra, Estrella Sayas, Casilda Navarro, Juana Fernández-López, and José A. Pérez-Alvarez. "Resistant starch as prebiotic: A review." *Starch-Stärke* 63, no. 7 (2011): 406–415.

Geetha, R., H. N. Mishra, and P. P. Srivastav. "Twin screw extrusion of kodo millet-chickpea blend: process parameter optimization, physico-chemical and functional properties." *Journal of Food Science and Technology* 51, no. 11 (2014): 3144–3153.

Gélinas, Pierre, Carole M. McKinnon, Mari Carmen Mena, and Enrique Méndez. "Gluten contamination of cereal foods in Canada." *International Jjournal of Ffood Sscience & and Ttechnology* 43, no. 7 (2008): 1245–1252.

Guerrieri, Nicoletta, Lucia Eynard, Vera Lavelli, and Paolo Cerletti. "Interactions of protein and starch studied through amyloglucosidase action." *Cereal Chemistry* 74, no. 6 (1997): 846–850.

Gupta, Nidhi, A. K. Srivastava, and V. N. Pandey. "Biodiversity and nutraceutical quality of some indian millets." *Proceedings of the National Academy of Sciences, India Section B: Biological Sciences* 82, no. 2 (2012): 265–273.).

Gupta, Sangeeta, S. K. Shrivastava, and Manjul Shrivastava. "Proximate composition of seeds of hybrid varieties of minor millets." *Int. J. Res. Eng. Technol* 3 (2014): 687–693.

Habiyaremye, Cedric, Janet B. Matanguihan, Jade D'Alpoim Guedes, Girish M. Ganjyal, Michael R. Whiteman, Kimberlee K. Kidwell, and Kevin M. Murphy. "Proso millet (Panicum miliaceum L.) and its potential for cultivation in the Pacific Northwest, US: A review." *Frontiers in Plant Science* 7 (2017): 1961.

Han, F., Z. Lu, W. Di, and A. Li. "Effect of roasting on phenolics content and antioxidant activity of proso millet." *Int. J. Food Eng* 4 (2018): 5006–5014.

Hithamani, Gavirangappa, and Krishnapura Srinivasan. "Effect of domestic processing on the polyphenol content and bioaccessibility in finger millet (Eleusine coracana) and pearl millet (Pennisetum glaucum)." *Food Chemistry* 164 (2014): 55–62.

Hotz, Christine, and Rosalind S. Gibson. "Traditional food-processing and preparation practices to enhance the bioavailability of micronutrients in plant-based diets." *The Journal of Nutrition* 137, no. 4 (2007): 1097–1100.

Kalinova, Jana, and Jan Moudry. "Content and quality of protein in proso millet (Panicum miliaceum L.) varieties." *Plant Foods for Human Nutrition* 61, no. 1 (2006): 43–47.

Koshihara, Neichi Tomohiro, Murota Sei-Itsu, Lao Ai-Na, Fujimoto Yasuo, and Tatsuno Takashi. "Caffeic acid is a selective inhibitor for leukotriene biosynthesis." *Biochimica et Biophysica Acta (BBA)-Lipids and Lipid Metabolism* 792, no. 1 (1984): 92–97.

Kumari, S. Krishna, and B. Thayumanavan. "Characterization of starches of proso, foxtail, barnyard, kodo, and little millets." *Plant Foods for Human Nutrition* 53, no. 1 (1998): 47–56.

Lakshmi Kumari, P., and S. Sumathi. "Effect of consumption of finger millet on hyperglycemia in non-insulin dependent diabetes mellitus (NIDDM) subjects." *Plant Foods for Human Nutrition* 57 (2002): 205–213.

Li, Kehu, Tongze Zhang, Shwetha Narayanamoorthy, Can Jin, Zhongquan Sui, Zijun Li, Shunguo Li, Kao Wu, Guoqing Liu, and Harold Corke. "Diversity analysis of starch physicochemical properties in 95 proso millet (Panicum miliaceum L.) accessions." *Food Chemistry* 324 (2020): 126863.

Li, Yunlong, Jing Lv, Lei Wang, Yingying Zhu, and Ruiling Shen. "Effects of Millet Bran Dietary Fiber and Millet Flour on Dough Development, Steamed Bread Quality, and Digestion In Vitro." *Applied Sciences* 10, no. 3 (2020): 912.

Liang, Shaohua, Guolong Yang, and Yuxiang Ma. "Chemical characteristics and fatty acid profile of foxtail millet bran oil." *Journal of the American Oil Chemists' Society* 87, no. 1 (2010): 63–67.

Lorenz, K., W. Dilsaver, and L. Bates. "Proso millets. Milling characteristics, proximate compositions, nutritive value of flours." *Cereal Chem* 57, no. 1 (1980): 16–20.

Mathanghi, SK and K. Sudha. "Functional and phytochemical properties of finger millet (Eleusine coracana L.) for health." *International Journal of Pharmaceutical, Chemical and Biology Sciences* 2, no. 4 (2012): 431–438.

McSweeney, Matthew B., Lisa M. Duizer, Koushik Seetharaman, and D. Dan Ramdath. "Assessment of important sensory attributes of millet based snacks and biscuits." *Journal of Food Science* 81, no. 5 (2016): S1203–S1209.

Meenu, Maninder, and Baojun Xu. "A critical review on anti-diabetic and anti-obesity effects of dietary resistant starch." *Critical Reviews in Food Science and Nutrition* 59, no. 18 (2019): 3019–3031.

Mustač, Nikolina Čukelj, Dubravka Novotni, Matea Habuš, Saša Drakula, Ljiljana Nanjara, Bojana Voučko, Maja Benković, and Duška Ćurić. "Storage stability, micronisation, and application of nutrient-dense fraction of proso millet bran in gluten-free bread." *Journal of Cereal Science* 91 (2020): 102864.

Nishizawa, Naoyuki, and Yoshiharu Fudamoto. "The elevation of plasma concentration of high-density lipoprotein cholesterol in mice fed with protein from proso millet." *Bioscience, Biotechnology, and Biochemistry* 59, no. 2 (1995): 333–335.

Nishizawa, Naoyuki, Daiki Sato, Yoshiaki Ito, Takashi Nagasawa, Yasuko Hatakeyama, Myeong-Rak Choi, You-Young Choi, and Yi Min Wei. "Effects of dietary protein of proso millet on liver injury induced by D-galactosamine in rats." *Bioscience, biotechnology, and biochemistry*, 66, no. 1 (2002): 92–96.

Obilana, A. B., and E. Manyasa. "Millets. In 'Pseudocereals and less common cereals: grain properties and utilization potential.'" (P. S. Belton and J. R. N. Taylor eds). (2002): 177–217.

Oh, Hannah, Hanseul Kim, Dong Hoon Lee, Ariel Lee, Edward L. Giovannucci, Seok-Seong Kang, and NaNa Keum. "Different dietary fibre sources and risks of colorectal cancer and adenoma: A dose–response meta-analysis of prospective studies." *British Journal of Nutrition* 122, no. 6 (2019): 605–615.

Park, Kyung-Ok, Yoshiaki Ito, Takashi Nagasawa, Myeong-Rak Choi, and Naoyuki Nishizawa. "Effects of dietary Korean proso-millet protein on plasma adiponectin, HDL cholesterol, insulin levels, and gene expression in obese type 2 diabetic mice." *Bioscience, Biotechnology, and biochemistry* 72, no. 11 (2008): 2918–2925.

Pathak, H. C. "Role of millets in nutritional security of India." *New Delhi: National Academy of Agricultural Sciences* (2013): 1–16.

Rattanamechaiskul, Chaiwat, Somchart Soponronnarit, and Somkiat Prachayawarakorn. "Glycemic response to brown rice treated by different drying media." *Journal of Food Engineering* 122 (2014): 48–55.

Ravindran, G. J. F. C. "Studies on millets: Proximate composition, mineral composition, and phytate and oxalate contents." *Food Chemistry* 39, no. 1 (1991): 99–107.

Rice-Evans, Catherine A., Nicholas J. Miller, and George Paganga. "Structure-antioxidant activity relationships of flavonoids and phenolic acids." *Free Radical Biology and Medicine* 20, no. 7 (1996): 933–956.

Saha, Dipnarayan, MV Channabyre Gowda, Lalit Arya, Manjusha Verma, and Kailash C. Bansal. "Genetic and genomic resources of small millets." *Critical Reviews in Plant Sciences* 35, no. 1 (2016): 56–79.

Sajilata, M. G., Rekha S. Singhal, and Pushpa R. Kulkarni. "Resistant starch–a review." *Comprehensive Reviews in Food Science and Food Safety* 5, no. 1 (2006): 1–17.

Saleh, Ahmed S.M., Qing Zhang, Jing Chen, and Qun Shen. "Millet grains: nutritional quality, processing, and potential health benefits." *Comprehensive Reviews in Food Science and Food Safety* 12, no. 3 (2013): 281–295.

Santra, D. K., R. F. Heyduck, D. D. Baltensperger, R. A. Graybosch, L. A. Nelson, G. Frickel, and E. Nielsen. "Registration of 'Plateau' Waxy (Amylose-Free) Proso Millet." *Journal of Plant Registrations* 9, no. 1 (2015): 41–43.

Santra, Dipak K. "Proso millet varieties for western Nebraska." *Horizon* 2603 (2012): 2257.

Saxena, Rachit, Sai Kranthi Vanga, Jin Wang, Valérie Orsat, and Vijaya Raghavan. "Millets for food security in the context of climate change: A review." *Sustainability* 10, no. 7 (2018): 2228.

Shahidi, Fereidoon, and Anoma Chandrasekara. "Millet grain phenolics and their role in disease risk reduction and health promotion: A review." *Journal of Functional Foods* 5, no. 2 (2013): 570–581.

Shobana, S., and N. G. Malleshi. "Preparation and functional properties of decorticated finger millet (Eleusine coracana)." *Journal of Food Engineering* 79, no. 2 (2007): 529–538.

Srichuwong, Sathaporn, Delphine Curti, Sean Austin, Roberto King, Lisa Lamothe, and Hugo Gloria-Hernandez. "Physicochemical properties and starch digestibility of whole grain sorghums, millet, quinoa and amaranth flours, as affected by starch and non-starch constituents." *Food Chemistry* 233 (2017): 1–10.

Suárez-Estrella, Diego, Luisa Torri, Maria Ambrogina Pagani, and Alessandra Marti. "Quinoa bitterness: Causes and solutions for improving product acceptability." *Journal of the Science of Food and Agriculture* 98, no. 11 (2018): 4033–4041.

Taylor, John R. N., and Kwaku G. Duodu. "Effects of processing sorghum and millets on their phenolic phytochemicals and the implications of this to the health-enhancing properties of sorghum and millet food and beverage products." *Journal of the Science of Food and Agriculture* 95, no. 2 (2015): 225–237.

Vijayakumar, T. Poongodi, Jemima Beryl Mohankumar, and T. Srinivasan. "Quality evaluation of noodles from millet flour blend incorporated composite flour." *Journal of Scientific and Industrial Research* 69 (2010): 84–89.

Wang, Tao, Fuli He, and Guibing Chen. "Improving bioaccessibility and bioavailability of phenolic compounds in cereal grains through processing technologies: A concise review." *Journal of Functional Foods* 7 (2014): 101–111.

Wang, Ying, QiuGe Yang, GuiMei Chao, PengKe Wang, JinFeng Gao, XiaoLi Gao, and BaiLi Feng. "Comparison of physicochemical properties of starches from proso millet and glutinous rice." *Journal of Northwest A & F University-Natural Science Edition* 40, no. 12 (2012): 157–163.

Wu, Yue, Qinlu Lin, Ting Cui, and Huaxi Xiao. "Structural and physical properties of starches isolated from six varieties of millet grown in China." *International Journal of Food Properties* 17, no. 10 (2014): 2344–2360.

Yang, Qinghua, Panpan Zhang, Yang Qu, Xiaoli Gao, Jibao Liang, Pu Yang, and Baili Feng. "Comparison of physicochemical properties and cooking edibility of waxy and non-waxy proso millet (Panicum miliaceum L.)." *Food Chemistry* 257 (2018): 271–278.

Yang, Qinghua, Weili Zhang, Jing Li, Xiangwei Gong, and Baili Feng. "Physicochemical properties of starches in proso (Non-waxy and waxy) and foxtail millets (Non-waxy and waxy)." *Molecules* 24, no. 9 (2019): 1743.

Yang, Xiaoyan, Zhiwei Wan, Linda Perry, Houyuan Lu, Qiang Wang, Chaohong Zhao, Jun Li et al. "Early millet use in northern China." *Proceedings of the National Academy of Sciences* 109, no. 10 (2012): 3726–3730.

Yao, Y. P., Tian, C. R., Zhang, G. Q., Analysis on physicochemical properties of proso millet starch. *Journal of the Chinese Cereals and Oils Association* 24 (2009): 45–52.

Zhang Lizhen, Ruihai Liu, and Wei Niu. "Phytochemical and antiproliferative activity of proso millet." *PloS one* 9, no. 8 (2014): e104058.

Zhang, PanPan, BaiLi Feng, PengKe Wang, Hui Song, XiaoLi Gao, JinFeng Gao, Jia Chen, and Yan Chai. "Leaf senescence and activities of antioxidant enzymes in different broomcorn millet (Panicum miliaceum L.) cultivars under simulated drought condition." *Journal of Food, Agriculture & Environment* 10, no. 2 Part 1 (2017): 438–444.

Zhang, Tongze, Kehu Li, Xinghua Ding, Zhongquan Sui, Qiong-Qiong Yang, Nagendra P. Shah, Guoqing Liu, and Harold Corke. "Starch properties of high and low amylose proso millet (Panicum miliaceum L.) genotypes are differentially affected by varying salt and pH." *Food Chemistry* 337 (2021): 127784.

Zheng, Mingzhu, Yu Xiao, Shuang Yang, Meihong Liu, Ling Feng, Yuhang Ren, Xinbiao Yang, Nan Lin, and Jingsheng Liu. "Effect of adding zein, soy protein isolate and whey protein isolate on the physicochemical and in vitro digestion of proso millet starch." *International Journal of Food Science & Technology* 55, no. 2 (2020): 776–784.

8

Kodo Millet (*Paspalum scrobiculatum*)
Bioactive Profile, Health Benefits and Techno-Functionality

Rajeev Ranjan, Shubhendra Singh, Subhamoy Dhua, Poonam Mishra,
Anil Kumar Chauhan, and Arun Kumar Gupta

Contents

8.1 Introduction

Since the dawn of history, cereal grains have provided nourishment for humanity and played a critical role in the daily diets of billions of people in the world. Still, many important cereal crops are in short supply because of climate changes (such as climate shifts, weather conditions that cause flooding and drought, declining productivity due to nutrient depletion of the soil, and harmful chemical residues spoiling soil health). Agriculture and the food industry are under immense pressure. Furthermore, there are other societal and economic impacts that have threatened global agriculture and food security. These include the fact that a growing world population is raising food prices and commodities and threatening to bring food security and global agriculture to its knees. The citizens living in arid and sub-arid climates, who are already limited in resources, will be hit the hardest (Saleh et al. 2013). The challenges of getting enough food for everyone that scientists in the food production, processing, storage, and nutrition fields face are vast. People know millets as "ancient grains" since they have been cultivated for thousands of years, emerging around the beginning of human civilization. Some archaeological finds seem to indicate millet cultivation dates back 4,000 years (Shahidi and Chandrasekara 2013). Even though millet grains have not been utilized much in the past, they have recently become more popular due to the benefits they provide. This means that millet production, which had previously been in decline, is now rising due to the growth in consumer interest. Millets, one of the most drought-resistant crops grown in Asian countries, are an important source of food for people living in rural areas (Gupta et al. 2020). These millets are widely used in nutraceuticals due to their high antioxidant content, which far exceeds that of most other cereal crops. In addition to helping reduce non-communicable disease, it is said that they are beneficial in preventing and treating cardiovascular issues. Millets contain lots of fiber, which fights against gallstone formation. One study found that every tenth of a serving of whole grains per day could help prevent type 2 diabetes, heart disease, stroke, and death. Another study found that even more frequent intake – two servings of whole grains per day – led to reduced incidences of breast cancer, childhood asthma, and premature death (Balasubramanium 2013). Millets have the added advantage of having numerous uses, including in functional foods and as a nutraceutical. Nutri-cereals, therefore, describe grains that are a significant source of nutrients for infants. In addition, millet's contribution to national food security and possible benefits of countering diseases (Shahidi and Chandrasekara 2013) has caused food processors, technologists, and nutritionists to investigate its increasing use in food processing and nutrition. Until recently, it was only available to poor people, but now it is found on rich people's plates, as they try to be healthy. An ancient millet grain called kodo millet, native to Africa, and domesticated in India about 3,000 years ago, grows in the Deccan Plateau of India (De Wet et al.1983). The kodo millet (also known as "cow grass," "rice grass," "ditch millet," "Native Paspalum," or "Indian Crown Grass") is a small-grained Asian grass. It is popular in the rural areas of Deccan Plateau (including areas of Gujarat, Karnataka, and Tamil Nadu), and many of the eastern states, the Himalayas, and the northern plains, which see it as a valuable and healthy food (Hegde and Chandra 2005). The different dialects in India refer to kodo differently: *kodo* and *kodra* in Punjabi. *Kodon* and *harka* are other common names, and *koduain* is used in Odia. Kannada uses *varagu*, Tamil *arikelu*, and Telugu *arika*. Estimates state that about 907,800 hectares of land are used to grow kodo millet, with an annual yield of 310,710 tonnes (Yadav et al. 2013). Kodo millet is heavily supported by Madhya Pradesh and Tamil Nadu, which contribute the most to its production and promotion. This crop is also in the government's plans for rural and agricultural development, where they are promoting cultivation and the marketing of it. A genus of grass called Paspalum, with more than 400 species and many cultivars, is often used as an annual crop, although many cultivars take root at the nodes and produce new culms after their flower completes its life cycle. The species has perennial characteristics (De Wet et al. 1983). A drought-resistant crop is grown in semi-arid climates, and it is not frequently inter-culturally managed. Kodo is a monocot, and the seeds are roughly 1.5 mm wide and 2 mm long, appearing light brown to dark grey. Intercropping might work well for millet with shallow roots, like kodo millet. The grain is contained within tough, hardened husks that do not erode easily (FAO 1995).

8.2 Nutritional aspects of kodo millet

Kodo millet is an excellent grain to consume, with a nourishing nutritional profile, making it a decent replacement for rice or wheat. It contains far more protein, fiber, and minerals than rice, a major grain crop. The kodo millet grain is quite high in protein, with an 8 percent composition. Kodo millet has a glutelin protein (Sudharshana et al. 1988). Compared to wheat (0.2% fiber) and rice (0.2% fiber), millet, or kodo millet, is a top-quality fiber provider (9% fiber) (1.2%). Millet in general is loaded with carbs and calories, which is why kodo millet offers 63.6 g of carbohydrates and 353 calories per 100 g of grain. It has a fat content of 1.4 percent and a mineral content of 2.6 percent. Iron is found in amounts ranging from 25.86 ppm to 39.60 ppm in kodo millet. The phosphorus concentration in it is the lowest among all the millets. Kodo millet requires a smaller amount of DPPH quenching compared to vitamin C and vitamin E, according to Hegde and Chandra (2005). The gelatinization temperature range of kodo millet flour is 13°C (76.6–90°C), which means it can be included in things like bread, cakes, and cereal-based products. Kodo millet flour can also be used in things like gravy, soup, and heat set gel. Additionally, kodo millet flour is a great choice for recipes such as porridge, instant powders, and modified flour and starches for specialty foods. A protein boost like the one used with other foods can help make the protein in kodo millet more nutritious.

8.3 Phytochemical composition of kodo millet

Millets are well-known for their high content of bioactive substances as well as their possible functional characteristics (Okwudili et al. 2017). Millets contain a lot of polyphenols (Chethan and Malleshi 2007). Gallic acid, tannins, gentisic acid, protocatechuic acid, caffeic acid, vanillic acid, syringic acid, ferulic acid, para coumaric acid, transcinnamic acid, and 5-n-alkyl-resorcinols are the most important phenolic chemicals found in millet grains (Bellato et al. 2013; Pradeep and Sreerama 2015; Nithiyanantham et al. 2019). Many phenolic compounds have the ability to scavenge free radicals and reduce ROs, making them useful in the prevention and treatment of diseases caused by free radicals (Chandrasekara and Shahid 2011b; Nithiyanantham et al. 2019). Kodo millet is a good source of phenolic compounds (Chandrasekara and Shahidi 2011b). Millets have a large amount of flavonoids, which are a type of polyphenol. Catechin, orientin, iso-orientin, quercetin, luteolin, apigenin, vitexin, myricetin, isovitexin, daidzein, sponarin, violanthin, lucenin-1, and tricin are some of the significant polyphenols found in millets (Chandrasekara and Shahidi 2011b; Pradeep and Sreerama 2015). Millet flavonoids have been found to have a wide range of therapeutic qualities and have a broad effect on disorders such as inflammation, cancer, hypertension, diuresis, and pain (Sarita and Singh 2016; Nithiyanantham et al. 2019). Because its seed coat includes pectin, cellulose, and hemicelluloses, which are difficult for digestive enzymes to break down, millet is a rich source of dietary fiber (Chethan and Malleshi 2007). Dietary fiber in millet grain seed coatings is beneficial to human health because it affects a number of metabolic and digestive processes, such as glucose absorption and cholesterol levels (Johnson, et al. 2019). It is rich in polyunsaturated fatty acids (PUFA), despite its modest fat content. Kodo millet and tiny millet are also said to contain the greatest levels of nutritional fiber among grains, at 37 to 38 percent (Malleshi and Hadimani 1993). Kodo millet has the greatest amount of DPPH free radical scavenging activity, followed by sorghum and finger millet (Deshpande et al. 2015). Kodo millet exhibited the highest DPPH free radical scavenging activity, followed by great millet (sorghum) and finger millet, according to Hegde and Chandra (2005). Because of its strong antioxidant content, it protects against oxidative stress and keeps glucose levels stable in type 2 diabetes. Dietary supplementation with phenolic acid-rich foods has been shown to have antimutagenic, antihyperglycemic, and antioxidant effects, which can be used to develop health products (Shobana et al. 2007). Refined millets are gluten-free, easy to digest, high in antioxidants, and may have anti-carcinogenic properties (Dykes and Rooney 2006).

8.4 Characterization of phytochemical compounds responsible for bioactive properties

8.4.1 Dietary fiber

Millets contain good amounts of nutrition, like 60–70 percent carbohydrate, proteins (7–11%), crude fiber (2–7%), and fat (1.5–5%). Millets are gluten-free and high in nutrients like vitamin B, calcium, and zinc. Millets are becoming more popular because they contain essential nutrients as well as bioactive components that have health-promoting properties. Millets could be used in the treatment of chronic diseases such as obesity, heart disease, and cancer as nutraceuticals (Guo et al. 2018).

When compared to other cereals, millets' digestibility rates for both dietary starch and protein have been found to be lower. Various millet properties, including polyphenols, fibers, starch-protein-lipid interactions and intrinsic structural features of starch contained in millets, are responsible for this hypoglycaemic trait (Annor et al. 2017). Consuming dietary fibers have several beneficial physiological effects, such as laxation, lowering blood cholesterol levels, and lowering blood glucose levels. Foxtail millet has been found to have properties comparable to soy insoluble fibers, such as aiding digestion and lowering the postprandial increase in serum glucose levels while having a little effect on serum cholesterol levels (Bangoura et al. 2013; Anderson et al. 2009). Taylor and Duodu (2015) discovered that the digestibility of starch and protein was altered after processing millets. This was significant since millet-based meals may be beneficial to one's health (Singh 2010). The biggest issue was dehusking kodo millet, which is a small grain. (Devi et al. 2013). Not only does the process of dehusking minor millet take a long time, but it also consumes a lot of energy. Dry milling or dehulling a fractionated seed caryopsis separates it into three fundamental components: germ, pericarp, and endosperm. Anti-nutrients, fiber, fat, mineral content, and phenolic acids were all reduced by decortication (Lestienne et al. 2007). The elimination of fat, protein, or both enhanced the in vitro digestibility of kodo millet starch, according to Annor et al. (2013). Millet germ and pericarp contain most lipids and proteins (**Table 8.1**); thus, it was hypothesized that removing the outer layers would boost starch digestibility by increasing the amount of lipids and proteins available to the body. Following decortication, the GI of kodo millet increased by 42 percent, whereas the

Table 8.1 Millets and Other Grains Nutrition Chart (per 100 g at 12% Moisture Content)

Millets	Protein (g)	Crude fat (g)	Dietary fiber (g)	Carbohydrates (g)	Minerals (mg)
Barnyard millet	6.20	4.40	13.70	65.50	4.40
Finger millet	7.20	1.90	11.20	66.80	2.70
Kodo millet	8.90	2.60	6.40	66.20	2.60
Foxtail millet	12.30	4.30	14.00	60.90	3.30
Little millet	10.40	3.90	7.70	65.60	1.50
Pearl millet	11.00	5.40	11.50	61.80	2.30
Proso millet	12.50	1.10	8.50	70.40	1.90

Source: Dayakar et al. 2017; Saleh et al. 2013.

GI of other millet types increased by less than 6 percent. Because insoluble dietary fiber, polyphenolic chemicals, and lipid content are reduced during decortication of millets, this may result in an increase in GI (Bora 2014).

8.4.2 Polyphenolic compounds

The most common phenolic acids found in millets are ferulic and *p*-coumaric acids (Kulp 2000). There is more information available about finger millet varieties than other millet grains, but it still requires an extensive research to strongly support their pharmacological properties. These chemicals are abundant in the testa of finger millets and vary according to millet variety (Subba Rao and Muralikrishna 2002). Flavones are the only flavonoids found in millets so far. Finger millets are believed to contain proanthocyanidins, also known as condensed tannins (Dykes and Rooney 2006). Hegde et al. (2005) found that whole grain meals of kodo and finger millets protect Wistar rats against hyperglycemic and alloxan-induced oxidative damage. In many developed countries, millets are under-utilized because they have a short shelf life. In poor countries, millet grains have enormous potential for processing into value-added meals and drinks. Millets are also healthy for coeliacs because they do not contain gluten. The phenolic content of the bound phenolic extract of kodo millet was 2.5 times that of the soluble equivalent. According to published studies, bound phenolic extracts of corn, wheat, oats, rice, and barley exhibited more phenolic content than their soluble counterparts (Liyana-Pathirana and Shahidi 2006; Madhujith and Shahidi 2009). The antioxidative activity of bound extracts is in the following order: kodo> foxtail> little> finger> proso> pearl> little finger > proso> pearl> little finger (local). However, with the exception of little millet, all of the other kinds studied showed significant differences (p< 0.05) in the bleaching efficiency of carotene between the two types of extracted phenols (Chandrasekara and Shahidi 2010). In addition to the phenolic content, millet phenolic extracts appear to be efficient antioxidants for a variety of reasons. Finger millet variants (finger and pearl millet) had higher antioxidant efficacy when compared to their unbound and bound counterparts while kodo millet, had a lower antioxidant efficacy (Chandrasekara and Shahidi 2011b). Millets have phenolic content that ranges from 146 to 1157 μmol of FAE (ferulic acid equivalents) per g of extract, with the greatest concentration found in kodo millet (Chandrasekara and Shahidi 2011a). Flavonoids are polyphenolic chemicals that give foods their blue, crimson, and purple hues. They are structurally defined by aromatic rings joined by a heterocyclic ring of three carbon atoms (Duodu and Awika 2019). Anthocyanins, flavones, flavanols, and proanthocyanidins are only a few of the flavonoids found in millet grains. With a total flavonoid concentration ranging from 25.15 to 120.3 μmol catechin equivalent per g of crude extract, kodo millet had the greatest flavonoid content, followed by finger millet and proso millet (Chandrasekara and Shahidi 2011a). Finger and kodo millet diets have been shown to offer antioxidant effects, which may be due to the presence of polyphenols. Millet polyphenol extracts, which Hegde et al. (2005) identified as a preventive agent against oxidative stress, have the ability to inhibit ROS generation.

Antinutritional components in cereals and legumes inhibit nutrient absorption and decrease nutrient bioavailability in the body. Phytates, tannins, alkaloids, lectins, and protease inhibitors are among the antinutritional compounds found in cereal grains (Ram et al. 2020). Millet's main antinutritive agents are tannins and phytates. Whole grains had a higher phytic acid content than polished grains (Devisetti et al. 2014). Except for proso and barnyard millet, all millets had higher levels of inositol hexaphosphates, while kodo had the highest level of condensed tannins, followed by finger and barnyard millet (Sharma and Gujral 2019). The levels of phytic acid and tannin in kodo millet differed significantly from those in other millet varieties. Phytic acid is known to impair mineral bioavailability and enzyme function due to their ability to bind and precipitate proteins and minerals.

The interactions of lipids and proteins influence the digestibility of kodo millet starch. In vitro, protein and fat removal improved kodo millet starch digestibility and predicted glycemic index. Because amylose is present in fatty acids, they can form complexes that are resistant to enzymatic degradation. These complexes hydrolyze at a slower rate than raw amylose (Sharma et al. 2021).

8.4.3 Resistant starch

Millets are also used to make a variety of other foods and beverages (Jha et al. 2021; Gupta et al. 2020). For example, flatbreads, beer, porridge, and non-alcoholic drinks are all made from millets (Amadou et al. 2011). The content, structure, characteristics, and interactions of starch, the principal component of these products, have a significant impact on their production and quality. Rapidly digestible starch (RDS), slow digestible starch (SDS), and resistant starch (RS) are all terms used to classify starch based on their hydrolysis by α-amylase (Englyst et al. 2018). Numerous health benefits can be derived from RS. In addition to creating a sense of satisfaction, RS escapes digestion and enters the colon, where it ferments and works as a prebiotic for the digestive bacteria (Fuentes-Zaragoza et al. 2011). Millets are highly preferred since they contain a high level of resistant starch (RS) (Annor et al. 2017). Kodo millet starch, on the other hand, has both high RS and low SDS (Annor et al. 2013). Millets are highly sought after because of their high level of RS. While kodo millet starch has a lower SDS than rice or wheat, it has an increased RS. According to their amylose content and their starch composition millet species differ in terms of starch. (Li et al. 2008). Amylose-rich starches are suitable for RS production (Gao et al. 2011). The lowest RS concentration was found in pearl millet (1.4–2.2%), while the highest was found in kodo millet (37.52%). Repeated autoclaving, roasting, and pressure cooking all increased the RS content of millet flour/starch (Mangala et al. 1999; Roopa and Premavalli 2008). Five millets (foxtail, proso, barnyard, kodo, and small millets) and rice were autoclaved and cooled to determine RS content. The concentration of RS in millet flour/starch increased with repeated autoclaving, baking, or pressure cooking (Kumari and Thayumanavan 1998). The native millet starches had the following RS contents (%): barnyard (0.50) > kodo (0.47), little (0.45), foxtail (0.42), and proso (0.39), while the treated millet starches had the following RS contents (%): barnyard (9.91), kodo (9.71), little (9.29), foxtail (8.89), and proso (8.45) (Mahajan et al. 2021). Enzymatic starch hydrolysis is reduced when lipids are present because they form complexes, such as amylose lipid complexes (ALC), with free fatty acids and digestion (Annor et al. 2015). Kodo millet's GI (49.4) was less than that of wheat (58.3) and rice (75.0). In the first step of amylose and lipid degradation, the amorphous region is quickly hydrolyzed and then the inclusion complex is slowly degraded. As much as

85 percent of millet's entire composition is constituted by three primary fatty acids: palmitic, oleic and linoleic. Even after boiling, millet flour retains an RS content higher than other types of flour and is much more RS than wheat and rice flour (Annor 2013).

8.5 Kodo millet nutritional and functional properties alterations during processing

Multiple important micronutrients, such as polyphenols, phosphorous, antioxidants, and phytic acids, can be found in abundance in kodo millets. Complexes formed by these nutrients with other micronutrients, such as calcium and zinc, can reduce both their bioavailability and their soluble solubility (Bunkar et al. 2021; Sharma et al. 2017). Soaking, cooking, and fermenting millet-based foods can significantly reduce tannin and phytate levels, increasing the bioavailability of these micronutrients (Bunkar et al. 2021). Essentially, the full nutritional potential of kodo millets can be unlocked by using the proper preparation techniques. Nutritional bioavailability, organoleptic qualities, and antinutritional factors are all improved through processing (Bunkar et al. 2021). In most cases, a single processing method will not suffice to accomplish this. Thus, a wide variety of processing methods are employed, including but not limited to decortication, milling, soaking, cooking, roasting, germinating, and fermenting. Antioxidants like polyphenols, flavonoids, and carotenoids can be found in kodo millets, as they can in many other whole grains (Liang et al. 2019). Research shows that these antioxidants have many positive effects on health, including lowering the risk of cancer, cardiovascular disease, and diabetes (Saleh at al. 2013). However, the millets' antioxidant activity can be diminished during the cooking and dehulling processes (Chandrasekara et al. 2012). When grains are dehulled, the protective outer layer and many of the antioxidants it contains are lost. Antioxidants are susceptible to destruction by both heat and water, which can occur during cooking (Chandrasekara et al. 2012). Antioxidant activity in processed kodo millets has been the subject of several studies. One study found that when compared to the whole grain, the antioxidant activity of kodo millets drops significantly after being dehulled and cooked. Soaking and cooking the millets reduced their antioxidant activity (Sharma et al. 2017). According to Hegde and Chandra (2005), when compared to other millets – like finger millet, small millet, foxtail millet, barnyard millet, and large millet – kodo millet has a greater ability to scavenge free radicals. Decreased activity was also observed after kodo millet husk and endosperm were separated. According to Chandrasekara et al. (2012) the antioxidant activity of dehulled kodo millet is only 6.06 percent as high as that of whole kodo, while it drops from 32.4 percent to 6.86 percent in boiled kodo. Whole kodo millet starch has a lower expected glycemic index (AGI) than rice starch (Annor et al. 2017). The starch and AGI (amyloglucosidase) digestibility of kodo grain is lower than that of decorated grains (Yadav et al.2013). Starch digestibility refers to the body's capacity to metabolise and absorb starch from food. Amyloglucosidase (AGI) is an enzyme that facilitates starch digestion. Lower AGI digestibility indicates that the starch in kodo grain is less easily digested by the body, which can affect nutrient absorption and energy levels. Notable influences on the digestibility of starch and AGI in grains include cooking methods, processing, and other components present in the grain (Annor et al. 2017). From this, it can be inferred that different processing methods have different effects on the nutritional and functional properties of kodo millets.

8.5.1 Decortication

The pericarp and outer covering of the grain, known as the hull and shell, are removed in this process. Utricles have been discovered in a wide range of grains and millet species, implying that the "hull" must be removed before the grain can be further processed (Sharma and Niranjan 2018). As a result, only 12 to 30 percent of the kernel extracted during dehulling may be used in the recipe. When grains are dehulled beyond this point, dietary fiber, some beneficial chemicals, and around 80 percent of phenolic content are lost (Akanbi et al. 2019). Dehulling of millet grains has no effect on moisture content. The current decorticated green cultivar, on the other hand, had lower levels of antinutrients, phosphorus, and iron (Babiker et al. 2018). After decortication, there was a significant decrease in protein, lipid content, insoluble dietary fiber, ash content, lysine, tryptophan, and other amino acids in pearl millet (Rani et al. 2018).

Dehulling boosted *in-vitro* protein digestibility by up to 0.5 percent in a study (Rani et al. 2018). Millet's outer layers have been stripped of polyphenols and anti-nutrients such as phytic acid, which impair protein digestibility by precipitating proteins, which may have contributed to the improvement in protein digestibility (Urooj 2017). Millets improved in terms of protein digestibility (IVPD) and estimated glycemic index after being destitched (eGI). However, kodo millet had a 42 percent rise in eGI, whereas other millets only had a 6 percent increase. The rise in the eGI of decorticated millet may be due to decreases in insoluble dietary fiber, phenolics, and fat content (Annor et al. 2017).

8.5.2 Soaking

By decreasing anti-nutrients, soaking increases mineral bioavailability (Sarita and Singh 2016). On the contrary, Shigihalli et al. (2018) found that soaking finger millet in deionized water for 12 to 48 h reduced the phytic acid level by an average of 250 mg per 100 g of finger millet. To reduce phytic acid, after leaching either seeds or water were hydrolyzed by phytase, according to the study. This could be related to the leaching of minerals from millet grains when they are soaked in water (Bindra 2019). After soaking pearl and finger millet grains for 12 h, another study found that the total polyphenol concentration decreased significantly. From 241.47 to 184.43 GAE mg/100 g in pearl millet, and from 161.64 to 128.97 GAE mg/100 g in finger millet. The polyphenols may be destroyed if the polyphenol oxidase enzyme is active during soaking (Naveena and Bhaskarachary 2013).

8.5.3 Germination/malting

Synonyms for "germination" include "sprouting" and "malting" (Lemmens et al. 2018). Processed in this way, flour becomes lighter and more delicate, and infant/health food formulations become more desirable (Suma 2014). There is a reduction in the antinutrient content of grains, as well as improvements in their nutritional and functional qualities (Ahmed 2019). In a study, pearl millet grain germination boosted the protein content by up to 50 percent (Akinola et al. 2017). When malting pearl millet grains for 24 h, the protein and fiber content increased from 7.52 to 78 percent and 0.77 to 87 percent, respectively. While the fat percentage dropped from 6.34 to 5.55 percent, the carbohydrate content increased (Obadina et al. 2017). During germination, enzymes and fat consumption activities may increase, resulting in a decrease in fat content (Hejazi et al. 2016). At 38.75 °C for 36 h, the protein, fiber, and mineral content of kodo millets increased from 6.7 percent to 7.9 percent, 35.30 g per 100 g, and 231.82 g per 100 g respectively. During germination, carbohydrates are lost by respiration, resulting in an increase in protein content in the seedling. Hemicellulose and cellulose were synthesized in large quantities, which resulted in an increase in the amount of dietary fiber. This was attributable to phytase activation during germination, which caused the mineral content to increase at the same time. Phytate hydrolyzed millet seeds liberate minerals by converting the phytate to inositol and orthophosphate. The glutamate decarboxylase enzyme decarboxylates L-glutamic acid during germination, resulting in an increase in GABA concentration from 9.36 to 47.43 mg/100 g. (Sharma, Saxena, and Riar 2017). It is conceivable that higher protein levels are attributable to protein reserves/synthesis, as well as the formation of certain amino acids as a result of malting (Adebiyi et al. 2017). Several studies show that when millet is germinated, antinutrients such as tannins and phytic acid are reduced (Chauhan 2018). Phytates and tannins in kodo millet decreased from 1.344 to 0.990 mol kg⁻¹ and from 1.603 to 0.232 mg/100 g in just 13 hours after germination. The major reason of this decrease was a high degree of phytase activity during germination (Sharma and Gujral 2020).

8.5.4 Popping/Puffing

One of the methods for preparing expanded cereals using the procedure known as HTST (high-temperature short time) involves popping (Saleh et al. 2013). As the food is being popped or puffed, it imparts the desired scent and taste (Kapoor 2013). When mixed with water or milk or when mixed with other food products, puffed millet grains can be used in the creation of other meals (Rajput et al. 2019). According to Pilat et al. (2016), they also contain high levels of phenolic compounds and minerals that make them suitable as a food component.

Puffing has a significant impact on the nutritional composition of grains. In one study, puffing was used to assess the puffing stability and nutritional composition of kodo millet varieties. After puffing at 230° C for three min with continuous stirring until the puffing sound stopped, the carbohydrate and protein content of millet increased from 68.35 to 74.38 percent and 7.92 to 8.12 percent, respectively. Because the endosperm feeds starch to the kernel, carbohydrate and protein levels increased after puffing, with protein levels increasing due to increased hydrolysis of protein into low molecular-weight protein. Another study revealed that puffing reduced crude fiber and fat content by 1.71 and 0.06 percent, respectively. Calcium levels dropped from 27 mg/100 gm to 18 mg/100 gm.

When grains are puffed, their functional qualities, such as digesting capabilities, are significantly altered and improved. According to one study, during *in vitro* digestion, the starch hydrolysis rate rose during the puffing process (Huang et al. 2018). When kodo millet was popped, the protein digestibility improved (Konapur et al. 2014). Trypsin inhibitors interfere with proteolytic enzyme activity during protein digestion and lower protein digestibility, while heat exposure destroys protease inhibitors (**Table 8.2**). As a result, popping can increase protein digestibility due to the inhibition of trypsin inhibitor activity (Jalgaonkar et al. 2016).

Table 8.2 Comprehensive Nutritional Profiling of Millets

Millet	Free lipid (mg/g)	Soluble dietary fiber (g/100 g)	Insoluble dietary fiber (g/100 g)	Free phenolic content (µg/g)	Bound phenolic content (µg/g)
Foxtail millet	47.0	2.8	23.8	123.0	2775.0
Finger millet	9.4	3.8	20.7	2235	2780
Pearl millet	56.9	2.7	11.6	230	1952
Proso millet	37.3	2.5	16.5	88	1416
Kodo millet	26.4	2.3	32.1	2520	7572
Barnyard millet	47.8	2.2	26.6	129	2326
Little millet	49.5	3.6	22.4	189	1108

Source: Bora et al. 2019; Kaur et al. 2019.

8.6 Uses of kodo millets

Kodo millets can be used in both traditional and innovative recipes. Grain, whether unprocessed or processed, can be cooked whole, decorticated, and milled into flour using traditional or modern processes. Kodo millet flour is used to make pudding in India. It is also cooked as rice in tribal regions, and the flour is used to make a variety of meals. It is cooked similarly to rice in Africa and also used as a cattle feed. *P. scrobiculatum* thrives on steep slopes in Hawaii, where other grasses do not. On hillside farms, it has the potential to be produced as a food source. It may also be used as grass ties on hillside plots to prevent soil erosion while simultaneously providing famine food as a bonus and is a great cover crop. It is becoming more popular as a gluten-free meal and a component in multigrain gluten-free dishes in Europe, despite the fact that it is not a common carbohydrate source.

8.7 Health benefits of kodo millet

8.7.1 Millets as a diabetic treatment

A significant decrease in diabetes rates has been noticed in people who eat millet. Millets are rich in phenolics, such as alpha glucosidase and pancreatic amylase, which decreases postprandial hyperglycemia by preventing the enzymatic breakdown of complex carbs (Shobana et al. 2007). Cataract disease induced by diabetes is decreased by the use of inhibiting molecules such aldose reductase, which helps keep sorbital at low levels (Chethan and Malleshi 2007).

8.7.2 Millets and the ageing process

A chemical reaction between amino group proteins and aldehydes, caused mostly by diabetes and ageing, is thought to be the root cause of a variety of other issues. A reduction group of sugars, termed non-enzymatic glycosylation, is seen when glycosylation takes place without enzymes. Millets, notably kodo millet, are well known for their antioxidant properties due to the presence of phenolics such as - phytates, phenols, and tannins, which could contribute to antioxidant activity in various human body processes, such as health, ageing, and unhealthy metabolic profile (Hegde et al. 2005).

8.7.3 Millets are anti-cancer and anti-celiac

Millets contain "anti-nutrients" like phenolic acids, tannins, and phytates, which can make them less healthy. However, these compounds have also been found to protect the colon, and studies have linked eating animal products to breast cancer. In fact, clinical tests have shown that millets contain phenolics that can prevent cancer from starting and kill cancer cells once they do (Chandrasekara and Shahidi 2010). The growing number of celiac disease diagnoses has led to a greater demand for healthy food options, resulting in a new market for breakfast cereals made from gluten-free grains like corn. Millets and other gluten-free grains are an appealing option for people with the genetic predisposition to celiac disease because it is an immune-mediated enteropathy that results from consuming gluten. Consequently, food items that cater to the needs of those with celiac disease are becoming increasingly popular (Taylor 2006; Ezeogu et al. 2008). Therefore, even though millet may contain "anti-nutrients" that lower their overall nutritional value, these compounds have been found to protect the colon and possibly prevent cancer. Also, the growing number of people with celiac disease has led to a higher demand for healthy, gluten-free food. This has created a new market for breakfast cereals made from corn and millet.

8.7.4 Millets for heart disease

A number of factors, including obesity, smoking, bad eating habits, and lack of exercise, increase one's risk of heart attack and stroke. More and more countries are contending with increasing incidences of cardiovascular disease. Millets have an above-average quantity of free radical scavenging activity, and this keeps the heart healthy (Hegde and Chandra 2005).

8.8 Usage of kodo millet

Millets include amino acids, fatty acids, nutrients, phytochemicals, and sulphur, which have the effect of making them nutritionally enriched. Furthermore, millets are highly suggested for people with celiac disease, as they do not contain gluten (Chandrasekara and Shahidi 2010). The popular consumption of millet foods and beverages, especially in Africa, is evidence of the importance of such products in the diets of locals (Obilana and Manyasa 2002; Amadou et al. 2011). Millet flour is used in the food industry as a core ingredient in cereal. The world's poor and hungry depend on millet in places where it is widely cultivated. Millets are consumed in a variety of ways, including breads, infant formula, and snacks, among others (Chandrasekara and Shahidi 2011b). Flat, thin "rotis" (so called because they are rolled out thin and flat) are used as the bread-like foundation for a meal in Maharashtra state and are often produced from sorghum/millet flour. Flatbreads such as rotis, idlies, dosas, and chakli may be made with 50–75 percent barnyard millet flour (Veena et al. 2010). The large tribal population in Central India relies on kodo millet as a primary food crop. A staple of the Himalayan foothills, people there utilize millet for hearty bread and soups and as a cereal. Finger millet is widely used in the Darjeeling hills and Sikkim in India to make "Kodo ko jaanr," a popular fermented beverage. Chhang is an alcoholic beverage created by fermenting finger millet, which is used by people in the Ladakh area of India. Another fermented beverage that is popular among the Tamil Nadu's ethnic communities is koozh, which is produced from rice and pearl or finger millet flour (Ilango and Antony 2014).

Even though millet can be used to make traditional food at home, farmers do not seem to want to grow it because there is not much demand for it from restaurants and grocery stores. Farmers cannot afford to put time and money into millet production because there is not enough demand (Subramanian and Viswanathan 2003).

Because of this, more and more companies around the world are turning to millet production as a way to make industrial goods. As more companies use millet as a raw material for their products, the demand for millet is likely to rise, which could give farmers new ways to make money from this change in the market. (Subramanian and Viswanathan 2003).

8.9 Potential utilization in value added food products

Due to its impressive nutrient profile, kodo millet is becoming more popular as a healthy food source. Not only is it an effective source of proteins, carbohydrates, and minerals, but it is also low in fat, which makes it a great addition to a healthy diet. In addition to having a lot of macronutrients, kodo millet has a lot of bioactive compounds like total phenolics, total flavonoids, and radical scavenging activity, which have been shown to be very good at killing bacteria. Kodo millet extract has a lot of polyphenols, which makes it a good choice for making nutraceuticals and functional foods that are safe to eat and good for health as a whole. With its wide range of bioactive components, kodo millet has the potential to make a big difference in people's health by giving them a safe and effective way to improve their nutrition and prevent chronic diseases (Sharma et al. 2017). Because they do not contain gluten, kodo millets are a good choice for those who have celiac disease, other wheat allergies, or intolerances. Furthermore, due to its high fiber, polyphenol, and protein content, kodo can significantly contribute to a large portion of the population's nutritional security (Desai et al. 2010). The antioxidant capabilities of plant based polyphenolic compounds and their health-promoting and disease-preventing benefits have been linked in research (Ferguson 2001). As a result, there is a huge opportunity to investigate the technological potential of using kodo millets in the food sector to make a variety of food items.

8.10 Value addition of kodo millet

Millet is an important part of many peoples' diets around the world. Millet grains can be processed domestically and commercially (Obilana and Manyasa 2002). Food technologists and nutritionists are both more interested in millet grains because of their projected health benefits and potential to fight multiple illnesses (Shahidi and Chandrasekara 2013). Millets are frequently under-utilized in countries that have advanced economically. The potential for processing millet grains into value-added foods and drinks is enormous in poorer countries. Many countries around the world have hastened the research to investigate and invent advance processing technologies in the industrial sector to increase the nutritive value and functionality in order to satisfy the needs of their customers.

8.10.1 Multi-ingredient flour

Although the use of millets is unacceptably low in comparison to their nutritive value, greater utilization could be achieved by blending millets with wheat flour (Singh and Raghuvanshi 2012). When millet flour is added to other products, the physical, nutritional, and functional properties of the blended products will be modified (Jaybhaye et al. 2014). Widely consumed in developed countries are food items like extruded products, which are used to ensure convenience. In contrast to quality weaning foods, extruded millet products have superior nutritive value (Almeida-Dominguez et al. 1993). The authors have developed a economic way to prepare snacks out of pearl millet (Balasubramanian et al. 2012). Many scientists have attempted to produce composite millet flours by using substituting cereal flours, which are then used to create various food products like ready-to-eat meals, pastas, and more. This easy way to get nutritious with increased functionality is extremely convenient.

8.10.2 Baked goods

Bakery goods have risen in popularity in the world because of the various tastes, cheap prices, varying textures, long shelf lives, and interesting packaging options (Bunkar et al. 2014). There are a number of benefits to incorporating millets into bakery products: the millets will be packed with important nutrients and fiber, and will be able to have a presence in the bakery industry, leading to the production of a range of tasty products (Verma and Patel 2013). To provide additional alternatives, lessen reliance on wheat, and make gluten-free bread more accessible, certain attempts are being undertaken to substitute some wheat flour with millet flour. The flours of finger millet and foxtail millet are used in bakery goods, including cakes, muffins, and biscuits. According to studies, finger millet flour can replace 40 percent of wheat flour in baked products like cakes and cookies without affecting flavor (Yenagi et al. 2013). Chocolate cupcakes, gel-filled cakes, masala cakes, carrot cakes, soup sticks, biscuits and muffins are among the new items, all of which are produced with finger millet and have received high marks for look, texture, flavor, and general acceptance. Malted Finger Millet Flour was used to improve the nutritional value of a cake by boosting the mineral and fiber content (Desai et al. 2010).

8.10.3 Fermented items

Because of the importance of cereal-based fermented items as a food source, humans have an affinity for them (Mugocha et al. 2000; Gotcheva et al. 2001). Millet fermentation is practiced due to benefits such as in vitro protein digestibility and a decrease in anti-nutrients (Verma and Patel 2013).

When pearl millet grains are ground, they undergo a process of fermentation and chemical transformation, resulting in a significant change in their composition. As a result of this process, protein digestibility of pearl millet significantly increases, with the human body being able to easily digest up to 90 percent of the germinated pearl millet (Ahmed et al. 2009).

Pearl millet was inoculated with *S. diacetylactis*, *S. cerevisiae*, *L. brevis*, and *L. fermentum* to increase starch digestibility; this was done by fermenting the grain with these microbes (Khetarpaul et al. 1990). For their acceptable quality recipes, the scientists mixed and naturally fermented pearl millet flour to create biscuits, cutlets, weaning mixtures, and vermicelli. It is thought that a number of the fermented products can be prepared through enzymatic hydrolysis processes to improve nutritive value and anti-nutritive factors. In South India, people commonly eat fermented foods like Dosa and Idli.

8.10.4 Puffed/popped and flaked millets

Kodo millet has been used for flaking and puffing for centuries (Jaybhaye et al. 2014). After the cereal processing technologies were implemented, ready-to-eat (RTE) products were made by using foxtail millet and treating it with high temperature short time treatment, which resulted in flaked, extruded, and roller-dried, decorticated, and popped grains with a 12 percent moisture content (Ilango and Antony 2014). According to Jaybhyae and Srivastav (2010), steam cooked extrudate samples of millet-based RTE snack food are produced by creating thin, rectangular shaped pieces and then puffing them with HTST puffing.

8.10.5 Extruded items

When products that have been cooked at high temperatures and had little time to sit, this can lead to starch gelatinization and protein denaturation. Improved digestibility and inactivation of anti-nutrient factors happens during this process (Awika and Rooney 2004). Extrusion is beneficial for many reasons such as improving and standardizing quality, increasing production, expanding product offerings, and better in vitro digestibility of proteins (Dahlin and Lorenz 1993). Better iron availability is a benefit of this process, which is used to enrich extruded foods (Cisse et al. 1998). Some common wheat and millet flour blends that are processed using a twin-screw extruder to make various kinds of millet-based extrusions include kodo millet chickpea flour blend (70:30) (Geetha et al. 2014); pearl millet, finger millet and soybean flour blend (Balasubramanian et al. 2012); ragi, sorghum, soy and rice. Devi and Narayanasamy (2013) looked into whether it would be possible to prepare composite millet milk powder using a combination of finger millet and pearl millet, for use in making an RTC extruded product with nutrients in an acceptable range, along with acceptable texture, color, taste, and appearance.

8.10.6 Nutrient-dense foods

Due to their rich nutrient composition, millets have gained popularity as functional foods. Small millets are useful and nutritional. People who eat millet-based diets experience fewer diseases such as heart problems, diabetes, and hypertension. This is supported by clinical trials (Jaybhaye et al. 2014). Patients with celiac disease and gluten allergy are in good hands, as they can be assured of consuming safe food due to the positive characteristics of millets (Saleh et al. 2013). To help diabetics maintaining their blood sugar levels, different food items including roti, upma, and idli were created by blending millets, including kodo millet, with other cereals (Thakkar and Kapoor 2007). Still, a completely millet-based product has not yet made it to the market.

8.11 Toxicity in kodo millet

It is usually regarded as toxic and harmful to both humans and livestock. The fungus is linked to the poisoning of grain. Animals can suffer from spasms, depression, and death if infected with Paspalum ergot, which causes them to display nervousness, uncoordination, a staggering gait, and occasionally more severe symptoms. Humans are vulnerable to kodo poisoning, which results in nausea, vomiting, delirium, depression, drunkenness, and stupor. Mycotoxins generate the poisonous cyclopiazonic acid that causes "kodua poisoning" in kodo millet and is produced by *Aspergillus flavus*, *Aspergillus tamarii*, and *Phomopsis paspalli* (Patwardhan et al. 1974; Rao and Husain 1985; Antony et al. 2003).

8.12 Conclusion

We are certain that everyone on the earth experiences health problems as a result of eating inadequate amounts of fiber. Patients are aware that people can prevent obesity, diabetes, and other illnesses by eating millet for breakfast, lunch, and dinner and staying away from processed foods like rice, wheat, refined flours, meats, oils, and milk. Including these beneficial grains, especially their seeds, in our daily diets will have a positive impact on our health in a number of ways. Many people in the general public are not aware of the advantages of millet or are unaware of its nutritional value. Despite this, millet is one of the older food crops that is better known today. Millets originated in ancient China and have been grown in various countries all over the world for a very long period. Millets, sometimes called "precious" plants in the Bible, are described there as a grain. By providing a nutritious food alternative, this chapter hopes to increase people understanding of the significance of kodo millet in their life. This will help communities all across the world satisfy their dietary demands and reduce malnutrition and other health issues. With its wealth of nutrients, including fiber, kodo millet offers a number of health advantages and helps with problems like diabetes, obesity, and cardiovascular disease. Kodo millet's protein aids in both the growth and development of young children as well as the health maintenance of the elderly. The absence of gluten in kodo millet flour is advantageous for people who have celiac disease or gluten intolerance. The wax layers of kodo millet contain both phytosterols and policosanols, which safeguard the heart. If the kodo millets are processed into flour without removing the hulls, one can benefit from the antioxidants, which may protect the cells from the effects of free radicals.

References

Adebiyi, J.A., Obadina, A.O., Adebo, O.A. and Kayitesi, E. 2017. Comparison of nutritional quality and sensory acceptability of biscuits obtained from native, fermented, and malted pearl millet (Pennisetum glaucum) flour. *Food Chemistry*, 232: 210–217.

Ahmed, A.I., Abdalla, A.A. and El-Tinay, A.H. 2009. Effect of traditional processing on chemical composition and mineral content of two cultivars of pearl millet (Pennisetum glaucum). *Journal of Applied Sciences Research*, 5(12):2271–2276.

Ahmed, O.E.O. 2019. Effect of using simple technique methods to reduce sorghum phytate on performance of broilers and layers (Doctoral dissertation, Sudan University of Science and Technology).

Akanbi, T.O., Timilsena, Y. and Dhital, S. 2019. Bioactives from Millet: Properties and Effects of Processing on Bioavailability. In *Bioactive Factors and Processing Technology for Cereal Foods*, 171–183. Springer, Singapore.

Akinola, S.A., Badejo, A.A., Osundahunsi, O.F. and Edema, M.O. 2017. Effect of preprocessing techniques on pearl millet flour and changes in technological properties. *International Journal of Food Science and Technology*, 52(4): 992–999.

Amadou, I., Gbadamosi, O.S. and Le, G.W. 2011. Millet-based traditional processed foods and beverages – A review. *Cereal Foods World*, 56(3): 115.

Anderson, J.W., Baird, P., Davis, R.H., Ferreri, S., Knudtson, M., Koraym, A., Waters, V. and Williams, C.L. 2009. Health benefits of dietary fiber. *Nutrition Reviews*, 67(4): 188–205.

Annor, G.A. 2013. *Millet Starches: Structural Characteristics and Glycemic Attributes* (Doctoral dissertation).

Annor, G.A., Marcone, M., Bertoft, E. and Seetharaman, K. 2013. *In vitro* starch digestibility and expected glycemic index of kodo millet (*Paspalum scrobiculatum*) as affected by starch-protein-lipid interactions. *Cereal Chemistry*, 90(3): 211–217.

Annor, G.A., Marcone, M., Corredig, M., Bertoft, E. and Seetharaman, K. 2015. Effects of the amount and type of fatty acids present in millets on their in vitro starch digestibility and expected glycemic index (eGI). *Journal of Cereal Science*, 64:76–81.

Annor, G.A., Tyl, C., Marcone, M., Ragaee, S. and Marti, A. 2017. Why do millets have slower starch and protein digestibility than other cereals?. *Trends in Food Science and Technology*, 66: 73–83.

Antony, M., Shukla, Y. and Janardhanan, K.K. 2003. Potential risk of acute hepatotoxicity of kodo poisoning due to exposure to cyclopiazonic acid. *Journal of Ethnopharmacology*, 87(2–3): 211–214.

Awika, J.M. and Rooney, L.W. 2004. Sorghum phytochemicals and their potential impact on human health. *Phytochemistry*, 65(9):1199–1221.

Babiker, E., Abdelseed, B., Hassan, H. and Adiamo, O. 2018. Effect of decortication methods on the chemical composition, antinutrients, Ca, P and Fe contents of two pearl millet cultivars during storage. *World Journal of Science, Technology and Sustainable Development*, 15(3): 278–286.

Balasubramanian, S. 2013. September. Processing of millets. In Paper presented National Seminar on Recent Advances in processing, utilization and nutritional impact of small millets. Madurai Symposium, Thamukkam Grounds, Madurai.

Balasubramanian, S., Singh, K.K., Patil, R.T. and Onkar, K.K. 2012. Quality evaluation of millet-soy blended extrudates formulated through linear programming. *Journal of Food Science and Technology*, 49(4): 450–458.

Bangoura, M.L., Nsor-Atindana, J., Zhu, K., Tolno, M.B., Zhou, H. and Wei, P. 2013. Potential hypoglycaemic effects of insoluble fibres isolated from foxtail millets [*Setaria italica* (L.) P. B eauvois]. *International Journal of Food Science and Technology*, 48(3):496–502.

Bellato, S., Ciccoritti, R., Del Frate, V., Sgrulletta, D. and Carbone, K. 2013. Influence of genotype and environment on the content of 5-n alkylresorcinols, total phenols and on the antiradical activity of whole durum wheat grains. *Journal of Cereal Science*, 57(2):162–169.

Bindra, D. 2019. Whole grains and millets as raw ingredients for recipes in diabetes mellitus. *International Journal of Home Science*, 5(3):214–219.

Bora, P. 2014. Anti-nutritional factors in foods and their effects. *Journal of Academia and Industrial Research*, 3(6):285–290.

Bora, P., Ragaee, S. and Marcone, M. 2019. Characterisation of several types of millets as functional food ingredients. *International Journal of Food Sciences and Nutrition*, 70(6):714–724.

Bunkar, D.S., Jha, A., Mahajan, A. and Unnikrishnan, V.S. 2014. Kinetics of changes in shelf-life parameters during storage of pearl millet based kheer mix and development of a shelf life prediction model. *Journal of Food Science and Technology*, 51(12):3740–3748.

Chandrasekara, A., Naczk, M., & Shahidi, F. 2012. Effect of processing on the antioxidant activity of millet grains. *Food Chemistry*, 133(1):1–9.

Chandrasekara, A. and Shahidi, F. 2010. Content of insoluble bound phenolics in millets and their contribution to antioxidant capacity. *Journal of Agricultural and Food Chemistry*, 58(11):6706–6714.

Chandrasekara, A. and Shahidi, F. 2011a. Antiproliferative potential and DNA scission inhibitory activity of phenolics from whole millet grains. *Journal of Functional Foods*, 3(3):159–170.

Chandrasekara, A. and Shahidi, F. 2011b. Inhibitory activities of soluble and bound millet seed phenolics on free radicals and reactive oxygen species. *Journal of Agricultural and Food Chemistry*, 59(1):428–436.

Chethan, S. and Malleshi, N.G. 2007. Finger millet polyphenols: Characterization and their nutraceutical potential. *American Journal of Food Technology*, 2(7):582–592.

Cisse, D., Guiro, A.T., Diaham, B., Souane, M., Doumbouya, N.T.S. and Wade, S. 1998. Effect of food processing on iron availability of African pearl millet weaning foods. *International Journal of Food Sciences and Nutrition*, 49(5):375–381.

Dahlin, K. and Lorenz, K. 1993. Protein digestibility of extruded cereal grains. *Food Chemistry*, 48(1):13–18.

Dayakar Rao, B., Bhaskarachary, K., Arlene Christina, G.D., Sudha Devi, G., Vilas, A.T. and Tonapi, A. 2017. Nutritional and health benefits of millets. *ICAR_Indian Institute of Millets Research (IIMR): Hyderabad, Indian*: 112.

de Wet, J.M., Brink, D.E., Rao, K.P. and Mengesha, M.H. 1983. Diversity in kodo millet, Paspalum scrobiculatum. *Economic Botany*, 37(2):159–163.

Desai, A.D., Kulkarni, S.S., Sahoo, A.K., Ranveer, R.C. and Dandge, P.B. 2010. Effect of supplementation of malted ragi flour on the nutritional and sensorial quality characteristics of cake. *Advance Journal of Food Science and Technology*, 2(1):67–71.

Deshpande, S.S., Mohapatra, D., Tripathi, M.K. and Sadvatha, R.H. 2015. Kodo millet nutritional value and utilization in Indian foods. *Journal of Grain Processing and Storage*, 2(2):16–23.

Devi, M.P. and Sangeetha, N. 2013. Extraction and dehydration of millet milk powder for formulation of extruded product. *IOSR Journal of Environmental Science, Toxicology and Food Technology*, 7(1):63–70.

Devisetti, R., Yadahally, S.N. and Bhattacharya, S. 2014. Nutrients and antinutrients in foxtail and proso millet milled fractions: Evaluation of their flour functionality. *LWT – Food Science and Technology*, 59(2):889–895.

Duodu, K.G. and Awika, J.M. 2019. Phytochemical–related health-promoting attributes of sorghum and millets. In *Sorghum and Millets*:225–258. AACC International Press.

Dykes, L. and Rooney, L.W. 2006. Sorghum and millet phenols and antioxidants. *Journal of Cereal Science*, 44(3):236–251.

Englyst, K., Goux, A., Meynier, A., Quigley, M., Englyst, H., Brack, O. and Vinoy, S. 2018. Inter-laboratory validation of the starch digestibility method for determination of rapidly digestible and slowly digestible starch. *Food Chemistry*, 245:1183–1189.

Ezeogu, L.I., Duodu, K.G., Emmambux, M.N. and Taylor, J.R. 2008. Influence of cooking conditions on the protein matrix of sorghum and maize endosperm flours. *Cereal Chemistry*, 85(3):397–402.

Fuentes-Zaragoza, E., Sánchez-Zapata, E., Sendra, E., Sayas, E., Navarro, C., Fernández-López, J. and Pérez-Alvarez, J.A. 2011. Resistant starch as prebiotic: A review. *Starch-Stärke*, 63(7):406–415.

Gao, Q., Li, S., Jian, H. and Liang, S. 2011. Preparation and properties of resistant starch from corn starch with enzymes. *African Journal of Biotechnology*, 10(7):1186–1193.

Geetha, R., Mishra, H.N. and Srivastav, P.P. 2014. Twin screw extrusion of kodo millet-chickpea blend: process parameter optimization, physico-chemical and functional properties. *Journal of Food Science and Technology*, 51(11):3144–3153.

Gotcheva, V., Pandiella, S.S., Angelov, A., Roshkova, Z. and Webb, C. 2001. Monitoring the fermentation of the traditional Bulgarian beverage boza. *International Journal of Food Science and Technology*, 36(2):129–134.

Guo, X., Sha, X., Rahman, E., Wang, Y., Ji, B., Wu, W. and Zhou, F. 2018. Antioxidant capacity and amino acid profile of millet bran wine and the synergistic interaction between major polyphenols. *Journal of Food Science and Technology*, 55(3):1010–1020.

Gupta, A. K., Jha, A. K., and Singhal, S. (2021). Optimisation of modification parameters for amaranth starch for the development of pudding and study of the quality traits of developed pudding. *Acta Alimentaria*, 50(1), 22–32.

Hegde, P.S. and Chandra, T.S. 2005. ESR spectroscopic study reveals higher free radical quenching potential in kodo millet (*Paspalum scrobiculatum*) compared to other millets. *Food Chemistry*, 92(1):177–182.

Hegde, P.S., Rajasekaran, N.S. and Chandra, T.S. 2005. Effects of the antioxidant properties of millet species on oxidative stress and glycemic status in alloxan-induced rats. *Nutrition Research*, 25(12):1109–1120.

Hejazi, S.N., Orsat, V., Azadi, B. and Kubow, S. 2016. Improvement of the in vitro protein digestibility of amaranth grain through optimization of the malting process. *Journal of Cereal Science*, 68:59–65.

Huang, R., Pan, X., Lv, J., Zhong, W., Yan, F., Duan, F. and Jia, L. 2018. Effects of explosion puffing on the nutritional composition and digestibility of grains. *International Journal of Food Properties*, 21(1):2193–2204.

Ilango, S. and Antony, U. 2014. Assessment of the microbiological quality of koozh, a fermented millet beverage. *African Journal of Microbiology Research*, 8(3):308–312.

Jalgaonkar, K., Jha, S.K. and Sharma, D.K. 2016. Effect of thermal treatments on the storage life of pearl millet (*Pennisetum glaucum*) flour. *Indian Journal of Agricultural Sciences*, 86(6):762–769.

Jaybhaye, R.V. and PP, S. 2010. Optimization of process parameters for development of millet based puffed snack food. In *International Conference on Food and Health*.

Jaybhaye, R.V., Pardeshi, I.L., Vengaiah, P.C. and Srivastav, P.P. 2014. Processing and technology for millet based food products: a review. *Journal of Ready to Eat Food*, 1(2):32–48.

Jha, A. K., Kumari, S., Gupta, A. K., and Shashank, A. (2021). Improvement in pasting, thermal properties, and in vitro digestibility of isolated Amaranth starch (Amaranthus cruentus L.) by addition of almond gum and gum ghatti powder. *Journal of Food Processing and Preservation*, 45(10), e15829.

Johnson, M., Deshpande, S., Vetriventhan, M., Upadhyaya, H.D. and Wallace, J.G. 2019. Genome-Wide population structure analyses of three minor millets: Kodo Millet, Little Millet, and Proso Millet. *The Plant Genome*, 12(3):190021.

Kapoor, P. 2013. Nutritional and functional properties of popped little millet (*Panicum sumatrense*). A thesis submitted to McGill University.

Kaur, P., Purewal, S.S., Sandhu, K.S., Kaur, M. and Salar, R.K. 2019. Millets: A cereal grain with potent antioxidants and health benefits. *Journal of Food Measurement and Characterization*, 13(1):793–806.

Khetarpaul, N. and Chauhan, B.M. 1990. Fermentation of pearl millet flour with yeasts and lactobacilli: in vitro digestibility and utilisation of fermented flour for weaning mixtures. *Plant Foods for Human Nutrition*, 40(3):167–173.

Konapur, A., Gavaravarapu, S.R.M., Gupta, S. and Nair, K.M. 2014. Millets in meeting nutrition security: issues and way forward for India. *Indian Journal of Nutrition and Dietetics*, 51(3):306–321.

Kulp, K. ed. 2000. *Handbook of Cereal Science and Technology, Revised and Expanded*. CRC Press.

Kumari, S.K. and Thayumanavan, B. 1998. Characterization of starches of proso, foxtail, barnyard, kodo, and little millets. *Plant Foods for Human Nutrition*, 53(1):47–56.

Lestienne, I., Buisson, M., Lullien-Pellerin, V., Picq, C. and Trèche, S. 2007. Losses of nutrients and anti-nutritional factors during abrasive decortication of two pearl millet cultivars (Pennisetum glaucum). *Food Chemistry*, 100(4):1316–1323.

Li, L., Jiang, H., Campbell, M., Blanco, M. and Jane, J.L. 2008. Characterization of maize amylose-extender (ae) mutant starches. Part I: Relationship between resistant starch contents and molecular structures. *Carbohydrate Polymers*, 74(3):396–404.

Liyana-Pathirana, C.M. and Shahidi, F. 2006. Importance of insoluble-bound phenolics to antioxidant properties of wheat. *Journal of Agricultural and Food Chemistry*, 54(4):1256–1264.

Madhujith, T. and Shahidi, F. 2009. Antioxidant potential of barley as affected by alkaline hydrolysis and release of insoluble-bound phenolics. *Food Chemistry*, 117(4):615–620.

Mahajan, P., Bera, M.B., Panesar, P.S. and Chauhan, A. 2021. Millet starch: A review. *International Journal of Biological Macromolecules*, 180:61–79.

Malleshi, N.G. and Hadimani, N.A. 1993. Nutritional and technological characteristics of small millets and preparation of value-added products from them In: *Advances in Small millet Proceeding of Second International Small Millet workshop*. Bulawayo, Zimbabwe.

Mangala, S.L., Malleshi, N.G. and Tharanathan, R.N. 1999. Resistant starch from differently processed rice and ragi (finger millet). *European Food Research and Technology*, 209(1):32–37.

Mugocha, P.T., Taylor, J.R.N. and Bester, B.H. 2000. Fermentation of a composite finger millet-dairy beverage. *World Journal of Microbiology and Biotechnology*, 16(4):341–344.

Naveena, N. and Bhaskarachary, K. 2013. Effects of soaking and germination of total and individual polyphenols content in the commonly consumed millets and legumes in India. *International Journal of Food and Nutritional Sciences*, 2(3):12.

Nithiyanantham, S., Kalaiselvi, P., Mahomoodally, M.F., Zengin, G., Abirami, A. and Srinivasan, G. 2019. Nutritional and functional roles of millets-A review. *Journal of Food Biochemistry*, 43(7): 12859.

Obadina, A.O., Arogbokun, C.A., Soares, A.O., de Carvalho, C.W.P., Barboza, H.T. and Adekoya, I.O. 2017. Changes in nutritional and physico-chemical properties of pearl millet (*Pennisetum glaucum*) Ex-Borno variety flour as a result of malting. *Journal of Food Science and Technology*, 54(13): 4442–4451.

Obilana, A.B. and Manyasa, E. 2002. Millets. In *Pseudocereals and Less Common Cereals*: 177–217. Springer, Berlin, Heidelberg.

Okwudili, U.H., Gyebi, D.K. and Obiefuna, J.A.I. 2017. Finger millet bioactive compounds, bio accessibility, and potential health effects–a review. *Czech Journal of Food Sciences*, 35(1): 7–17.

Patwardhan, S.A., Pandey, R.C., Dev, S. and Pendse, G.S. 1974. Toxic cytochalasins of Phomopsis paspalli, a pathogen of kodo millet. *Phytochemistry*, 13(9):1985–1988.

Piłat, B.E.A.T.A., Ogrodowska, D. and Zadernowski, R. 2016. Nutrient content of puffed proso millet (*Panicum miliaceum* L.) and amaranth (*Amaranthus cruentus* L.) grains. *Czech Journal of Food Sciences*, 34(4):362–369.

Pradeep, P.M. and Sreerama, Y.N. 2015. Impact of processing on the phenolic profiles of small millets: Evaluation of their antioxidant and enzyme inhibitory properties associated with hyperglycemia. *Food Chemistry*, 169:455–463.

Rajput, L.P.S., Parihar, P., Dhumketi, K., Naberia, S. and Tsuji, K. 2019. Development and acceptability of novel food products from millets for school children. *International Journal of Current Microbiology and Applied Sciences*, 8(4):2631–2638.

Ram, S., Narwal, S., Gupta, O.P., Pandey, V. and Singh, G.P. 2020. Anti-nutritional factors and bioavailability: approaches, challenges, and opportunities. In *Wheat and Barley Grain Biofortification*: 101–128. Woodhead Publishing.

Rani, S., Singh, R., Sehrawat, R., Kaur, B.P. and Upadhyay, A. 2018. Pearl millet processing: a review. *Nutrition and Food Science*.

Rao, B.L. and Husain, A. 1985. Presence of cyclopiazonic acid in kodo millet (*Paspalum scrobiculatum*) causing 'kodua poisoning' in man and its production by associated fungi. *Mycopathologia*, 89(3):177–180.

Roopa, S. and Premavalli, K.S. 2008. Effect of processing on starch fractions in different varieties of finger millet. *Food Chemistry*, 106(3):875–882.

Saleh, A.S., Zhang, Q., Chen, J. and Shen, Q. 2013. Millet grains: nutritional quality, processing, and potential health benefits. *Comprehensive Reviews in Food Science and Food Safety*, 12(3):281–295.

Sarita, E.S. and Singh, E. 2016. Potential of millets: nutrients composition and health benefits. *Journal of Scientific and Innovative Research*, 5(2):46–50.

Shahidi, F. and Chandrasekara, A. 2013. Millet grain phenolics and their role in disease risk reduction and health promotion: A review. *Journal of Functional Foods*, 5(2):570–581.

Sharma, B. and Gujral, H.S. 2019. Characterization of thermo-mechanical behavior of dough and starch digestibility profile of minor millet flat breads. *Journal of Cereal Science*, 90:102842.

Sharma, B. and Gujral, H.S. 2020. Modifying the dough mixing behavior, protein and starch digestibility and antinutritional profile of minor millets by sprouting. *International Journal of Biological Macromolecules*, 153:962–970.

Sharma, N. and Niranjan, K. 2018. Foxtail millet: Properties, processing, health benefits, and uses. *Food Reviews International*, 34(4):329–363.

Sharma, R., Sharma, S., Dar, B.N. and Singh, B. 2021. Millets as potential nutri-cereals: a review of nutrient composition, phytochemical profile and techno-functionality. *International Journal of Food Science and Technology*, 56(8):3703–3718

Sharma, S., Saxena, D.C. and Riar, C.S. 2017. Using combined optimization, GC–MS and analytical technique to analyze the germination effect on phenolics, dietary fibers, minerals and GABA contents of Kodo millet (*Paspalum scrobiculatum*). *Food Chemistry*, 233:20–28.

Shigihalli, S., Ravindra, U. and Ravishankar, P. 2018. Effect of processing methods on phytic acid content in selected white finger millet varieties. *International Journal of Current Microbiology and Applied Sciences,* 7(2):1829–1835.

Shobana, S., Usha Kumari, S.R., Malleshi, N.G. and Ali, S.Z. 2007. Glycemic response of rice, wheat and finger millet based diabetic food formulations in normoglycemic subjects. *International Journal of Food Sciences and Nutrition*, 58(5):363–372.

Singh, K.P. 2010. Development of a dehuller for barnyard millet (*Echinochloa frumentacea*) and formulation of millet-wheat composite flour (Doctoral dissertation, IIT Kharagpur).

Singh, P. and Raghuvanshi, R.S. 2012. Finger millet for food and nutritional security. *African Journal of Food Science*, 6(4):77–84.

Sorghum, F.A.O. 1995. Millets in Human nutrition. *FAO Food and Nutrition Series*, 27:16–19.

Subba Rao, M.V.S.S.T. and Muralikrishna, G. 2002. Evaluation of the antioxidant properties of free and bound phenolic acids from native and malted finger millet (Ragi, Eleusine coracana Indaf-15). *Journal of Agricultural and Food Chemistry*, 50(4):889–892.

Sudharshana, L., Monteiro, P.V. and Ramachandra, G. 1988. Studies on the proteins of kodo millet (*Paspalum scrobiculatum*). *Journal of the Science of Food and Agriculture*, 42(4):315–323.

Suma, F.P. and Asna, U. 2014. Influence of processing on physical properties of pearl millet (*Pennisetum typhoideum*) flour. *Advances in Food Sciences*, 36(3):100–108.

Taylor, J.R. and Duodu, K.G. 2015. Effects of processing sorghum and millets on their phenolic phytochemicals and the implications of this to the health-enhancing properties of sorghum and millet food and beverage products. *Journal of the Science of Food and Agriculture*, 95(2):225–237.

Taylor, S.E. 2006. Tend and befriend: Biobehavioral bases of affiliation under stress. *Current Directions in Psychological Science*, 15(6):273–277.

Thakkar, R. and Kapoor, R. 2007. Enrichment of rice and finger millet based preparations with gum acacia and their effects on glycemic response in non-insulin dependent diabetic subjects. *Journal of Food Science and Technology*, 44(2):183–185.

Urooj, A. 2017. Impact of household processing methods on the nutritional characteristics of pearl millet (*Pennisetum typhoideum*): A review. *MOJ Food Processing and Technology*, 4(1):28–32.

Veena, B., Chimmad, B.V., Naik, R.K. and Malagi, U. 2010. Development of barnyard millet based traditional foods. *Karnataka Journal of Agricultural Sciences,* 17(3):522–527.

Verma, V. and Patel, S. 2013. Value added products from nutri-cereals: Finger millet (*Eleusine coracana*). *Emirates Journal of Food and Agriculture*, 25(3):169–176.

Yenagi, N., Joshi, R., Byadgi, S. and Josna, B. 2013. Hand book for school children: importance of millets in daily diets for food and nutrition security; Appendix 11 of the joint technical final report (October 2010–March 2013).

Bioactive and Techno-Functional Properties of Barnyard Millet (*Echinochloa frumentacea*)

Anuradha Dutta, Deepa Joshi, Shweta Suri, Nalini Trivedi, Ranjana Acharya, Rushda Malik, Shayani Bose, and Ankita Dobhal

Contents

DOI: 10.1201/9781003251279-9

9.1 Introduction

The world's fourth most produced minor millet or nutri-cereal, "barnyard millet" belongs to the family *poaceae*, the sub family *panicoideae* and is a species of *Echinochola sp*. It is one of the oldest crops domesticated in Africa and Asia. The other names used for the millet are *barn grass* in North America and Vara, *varai, sanwa, shama, samu, shamula, kudiraivali, samwa* or *swang* in various parts of India. Two major species of the 35 identified species are *Echinochloa esculenta* (Japanese barnyard millet) and *Echoinochloa frumentacea* (Indian barnyard millet).

9.1.1 Origin

Barnyard millet *(Echinochloa frumentacea)* appears to have simultaneous evolution in Africa and India. The *E. colona (L.)* (tropical grass) known as "Jungle Rice," is the wild primogenitor of this millet. It is an annual crop and is mostly grown in the temperate regions of Korea, Germany, Russia, Japan and China (De Wet et al., 1983). Barnyard was domesticated in Japan around 4,000 years ago from its wild antecedent *E. crus-galli (L.) Beauv* (Doggett, 1989). Archaeological studies intimate that the Japanese barnyard millet was cultivated very early, during the Yayoi period (Watanabe, 1999). Nesbitt (2005) talked about domestication of barnyard in Japan as early as the Jomon period in 2000 BCE, whereas, according to Nozawa et al. (2004) barnyard millet was domesticated from *E. crus-galli* population.

Both *E. frumentacea* and *E. esculenta* are different to a large extent from their respective wild ancestors regarding a larger inflorescence, more compact growth habit, reduced vegetative branching, larger seed size and reduced shattering. Archaeological data suggests an increase in seed size with domestication of Japanese barnyard millet. The mean size of *barnyard millet* from the Early Jomon period (5000 BCE) is 20 percent smaller than that of the Middle Jomon period (3470 BCE) in Northern Japan (Takase, 2009).

Indian barnyard millet *(Echinochloa frumentacea)*, also known as billion-dollar grass and *sawa* millet, has its origin in Pakistan, India and Nepal. Today, it is widely distributed in

Australia, Central Africa, tropical Asia, temperate Asia and South America. It grows well in warmer regions at an altitude below 1,900 m.

Japanese barnyard millet (*Echinochloa esculenta*) or Siberian millet, marsh millet, or white millet, originated from Japan, Korea, and China. Nowadays, it is widely distributed in Pacific countries, Australia, India and temperate Asia. It is well suited to warm climates and altitudes of 2500 m.

9.1.2 Botanical description

The herbaceous barnyard millet plant grows annually up to a height of 60 to 120 cm. Its parts are roots, stem/culm, leaves, panicle and seed. The Indian barnyard millet has smaller, awnless spikelets with membranous glumes, whereas, the Japanese barnyard millet has lower lemma, awned spikelets and papery upper glumes. There are substantial variations between the Indian and Japanese barnyard millet related to lower lemma texture glumes and panicle type (Gomashe, 2017).

Barnyard has two types of roots: seminal root and secondary root. The seminal root arises from the germ and helps in the establishment of seedlings. The secondary root emerges from the basal node and becomes the permanent root of the millet, which is fibrous and shallow. Stems or culms are slender or hollow and sparingly branched. Leaves are wide, flat, glabrous or slightly hairy without ligule. The inflorescence is made up of a panicle, which is densely crowded with three to five rows of densely packed spikelets. The color of the glume varies from white to red. The seeds are round at the base, smooth and pointed at the apex. These seeds are 1 to 2 mm wide and 2 to 3 mm long. The caryopsis remains enclosed in the lemma and palea. The color of the grain is usually white or shiny yellow.

9.1.3 Production

Barnyard millet is a short duration crop having the ability to grow in adverse environmental conditions. It is widely cultivated in Asian countries like India, Japan, north eastern parts of China, Korea, and African countries like Burkina Faso, Nigeria, Niger, Sudan,Mali, Chad, Uganda and Ethiopia. In India, it is the second most important millet after finger millet. India is the largest producer of the millet in terms of both area and production. It has shown an average productivity of 1034 kg/ha during last 3 years (IIMR, 2018). In India, barnyard is mainly grown in two different agro climatic zones i.e. the Deccan plateau of Tamil Nadu in the south and mid-hills of the Himalayas in the north. Its cultivation is mainly concentrated in the hills of Uttarakhand and in the tribal belts of Madhya Pradesh, Bihar, Odisha, Maharashtra, Tamil Nadu, Gujrat and Punjab. Some of the major varieties of barnyard millet produced in India are K-1 Kudiraivali, K2 Anurag, Rau-3, VL Madira-8, 21 and 29. Chandan, Kanchan, Sushrutha, Pratap, Sanwa-1, PRJ-1, etc.

General properties: Barnyard millet has a medium amylose content (17.60 g/100 g (db)). Being starchy, it comprises 50–70 g/100 g carbohydrate on dry weight basis (db). Its grain can ripe within 45 days of sowing in affirmative temperature and moisture conditions (Hulse et al., 1980). Usually the moisture content at the time of harvesting (October-November) is 24–26 percent. The grain is then kept under the sun until moisture content reaches to 12–14 percent.

Barnyard millet contains 5–8.5 percent protein, 3.5–4.6 percent fat, 57–66 percent carbohydrate, 2.5–4.0 percent ash, 6.4–12.2 percent fiber at a moisture content of 0.5–0.25 percent. The value of sphericity, mean diameter and surface area of barnyard kernel and grain increases with increase in moisture content, whereas, bulk density of barnyard millet kernel and grain decreases with increase in moisture content. The decrease in bulk density may be attributed to the rise in inter-granular space with increase of moisture content. However, the decrease in true density values with increase of moisture content may be due to increased fraction of water in the seed. While calculating the porosity of kernel and barnyard millet grain, their true density and bulk density are considered. According to researches, the porosity of both kernel and grain increase with increase in moisture content. The moisture content of barnyard millet is a critical attribute in the development of processing machinery and storage. So, it highly affects the polishing time of milling and the quality of the end product. For example, the grain with 8 percent moisture content is better for polishing than the grain with 14 percent moisture.

9.2 Nutraceutical properties and consequent health benefits of barnyard millet (*Echinochloa frumentacea*)

It is well known that millets in general are incredibly beneficial to human health. Millets minimize the risk of cardiovascular disease, cancer and diabetes, improve metabolism and immunity, detoxify the body, strengthen digestive, neural and muscular systems and prevent various degenerative diseases such as Parkinson's and metabolic syndrome.

Indian barnyard millet is abundant in minerals and antioxidants. It is also high in crude fiber and crude protein. Barnyard millet is gluten-free, which makes it a great choice for treating celiac disease as well as constipation, obesity, cancer, cardiovascular diseases and diabetes. Compared to cereal grains like rice, wheat, and semolina (rawa), barnyard millet is a nutritious grain that should not be limited to fasting days. The tiny miracle grain may be used to make a variety of culinary dishes in a flash. Barnyard millet is nature's gift to the contemporary diet and sedentary habits that might contribute to lifestyle ailments.

9.2.1 Celiac sprue

Celiac sprue is also known as celiac disease, non-tropical sprue, gluten sensitivity and gluten-delicate enteropathy. Although the etiology of gluten sensitivity ailment is obscure, researches point towards its genetic component. In the people suffering from this sensitivity of gluten (a wheat protein) when ingested, the immune cells attack normal tissue (especially the small intestine's inner lining tissue) and cause irritation. Prolamins (like gliadin in wheat, horedin in barley, ecalin in rye, and venin in oats) cause the specific response that results in irritation.

Symptoms: In celiac disease, gluten or prolamins cause the body to respond as though they are harmful. The reaction in the small intestine damages the inner lining (mucosal surface). Thus, it is unable to absorb nutrients efficiently due to injured mucosal surface, leaving patients vulnerable to nutritional deficits such as lipids, protein, fat soluble vitamins, iron and calcium. Abdominal cramps, vomiting, diarrhea, failure to thrive, anemia, delayed development and osteoporosis are manifestations of the disease.

Role of barnyard millet in management of celiac disease: As a result of the increased demand for innovative, flavorful, and "healthy" foods, as well as the growing number of persons suffering from celiac sprue, there is an emerging market for food products manufactured from gluten free grains. Millets have taken a unique position in this challenging environment. The consumption of grain is affected to a great extent by a gluten-free diet. These gluten-free diets require grains such as barnyard millet, sorghum, rice, wild rice, amaranth, corn, buckwheat and oats to replace wheat, barley, and rye-based meals. These gluten-free foods are safe for celiac sufferers.

9.2.2 Arthritis: gluten – a trigger for joint pain

Individuals with celiac disease (gluten sensitivity) may have joint discomfort and inflammation. The immune system erroneously targets healthy tissue instead of bacteria, viruses and other harmful pathogens, as it does in many kinds of arthritis. Outside the stomach, inflammation is more likely to damage the joints, but the intensity of joint pain is low in gluten-sensitive people who do not eat it. When partly digested, gluten particles escape from the gut into the circulation. Gluten-sensitive people might experience symptoms all across the body.

Symptoms: The joints have the most frequent indications and symptoms of arthritis. Symptoms may include stiffness, pain, redness and edema in parts of joints, as well as a reduced range of motion, depending on the type of arthritis.

Role of barnyard millet: Patients suffering with arthritis often want to control the condition without the use of medications. Barnyard millet is a lifesaver for such people because, owing to the healing properties of gluten-free proteins, it may contain and prevent inflammation.

9.2.3 Diabetes

Diabetes has become a common serious condition across the world. In 2019, it was responsible for 1.5 million fatalities. Apart from medicine, the major management approaches for affected persons are modifications in lifestyle and dietary control. Diabetics may find relief by conforming to an individual-specific diet. Dietary management includes lowering post-prandial hyperglycemia and maintaining good glycemic control. However, it should be understood that diabetes can be caused by a number of factors, including obesity, heredity, lack of physical exercise and stress.

Symptoms: The most common symptoms are higher or abnormal blood glucose levels, as a result of either insufficient synthesis of insulin or improper response to insulin by the body cells, or both. Polydipsia, polyurea, impaired vision, weight loss and fatigue are all common diabetes symptoms. Diabetes affects vital organs such as the eye, kidney, heart, and nerves over time due to damage to tiny blood arteries.

Role of barnyard millet in management of diabetes: Despite the fact that barnyard millet has a significant amount of starch, it has a low glycemic index. This makes the millet a good choice for diabetics. By swapping barnyard millet for white rice, diabetics can avoid a spike in blood glucose. Populations who consume barnyard millet have been shown to have a lower incidence of diabetes. Millet phenolics, like alpha glucosidase, inhibit the enzymatic breakdown of complex carbohydrates, lowering post-prandial hyperglycemia. Aldose reductase inhibitors prevent sorbitol from aggregating and lower the risk of diabetes-related diseases.

A dietary intervention study on pre-diabetic subjects, who replaced their daily rice based meal with barnyard bhaat/barnyard dalia/rice bean-barnyard khichdi for a period of 4 months revealed significant decrease in HbA1c levels (5.74 v/s 5.1%) and body weight (58.87 v/s 57.77 Kg) with an increase in Hb levels (10.41 v/s 11.57 mg/dL)(AICRP H.Sc.,2019).

9.2.4 Cardiovascular diseases

In cardiovascular disease (CVD) arteries are either obstructed or constricted, which may cause chest discomfort (angina), heart attack, or stroke. Other diseases such as those affecting the heart's valves, muscle or rhythm are also called cardiovascular disease. Chronic illnesses are on the rise in industrialized countries, according to global morbidity and death data. Cardiovascular disease is the most common of them.

Symptoms: Cardiovascular disease symptoms includes chest discomfort, as well as shortness of breath, nausea, and severe tiredness. If the blood arteries in either legs or arms are restricted, people may have numbness, weakness, or coldness in those areas, as well as discomfort in jaw, neck, mouth, back or upper abdomen.

Role of barnyard millet:

Barnyard millet is high in magnesium, which has been shown to reduce the symptoms of heart attacks and migraines. It is also high in phytochemicals, including phytic acid, that are known to reduce cholesterol levels. The high tannin content of the nutri- cereal helps to reduce cholesterol. Extract of barnyard millet protects LDL (bad cholesterol) against oxidation.

After being taken up by macrophages, oxidized LDL is thought to form fatty streaks, which leads to the development of atherosclerosis. Hence, LDL oxidation inhibition is important for prevention of hypertension and CVD. Barnyard has potential to inhibit LDL oxidation that is related to phytochemicals found mostly in the seed coat.

Low physical activity levels, obesity, smoking, and poor eating habits all contribute to an increased risk of strokes and heart attack. The majority of countries around the world have high and increasing rates of CVD. When compared to rice and other minor millets, rats fed a diet of processed starch from barnyard millet had the lowest serum cholesterol, triglyceride and blood glucose levels ascertaining its therapeutic potential.

9.2.5 Cancer

The word cancer is derived from the Greek word carcinoma, which signifies crab. This is attributed to cell growth. The disease is taking a tremendous toll on human life due to lack of a particular therapy. However, if the problem is discovered early on, it may be possible to cure it. According to Cancer Research UK, a poor diet has been related to six malignancies, that is, mouth, upper neck, larynx, lung, stomach, and bowel cancer. Cancer of pancreas, esophagus, upper stomach, liver, gallbladder, endometrium, colon, breast (after menopause), ovaries, uterus, and kidneys have been associated with obesity.

Symptoms: Symptoms may vary depending on the region of cancer. They include emesis and hematemesis, bleeding from nose, sore throat, hemoptysis, change in urination and bowel movement, hoarseness of voice, immediate weight loss, unhealed ulcers (internal or external), difficulty in swallowing, breast enlargement, chronic diarrhea, loss of appetite, sleeplessness, ascitis, irregular menstruation and headache.

Role of barnyard millet: Millets are rich source of fiber because they have bran, germ, and endosperm, whereas processed grains just have endosperm. Numerous nutrients are lost from the germ and bran during the refining process. Furthermore, dehulled millets or whole grains are a good source of anti-cancer minerals, phyto chemicals and vitamins, which might potentially lower the risk of colorectal cancer through a variety of mechanisms. Millets are high in fermentable carbohydrates (resistant starch). These carbohydrates are turned into short-chain fatty acids by intestinal bacteria that may prevent colon cancer. Millets contain valuable lignins, which have been found to help prevent breast and hormone-related cancers.

The impact of bioactive compounds (vanillin) from barnyard millet on cell proliferation was investigated in a recent study. According to the findings, the vanillin component isolated from this millet efficiently triggered apoptotic cell death in a breast cancer cell line. So, barnyard may be beneficial in inhibiting breast cancer cells from proliferating.

9.2.6 Constipation

Constipation is a symptom of illness rather than a sickness itself. It refers to a condition that affects bowel motions or makes it difficult to pass feces. Constipation can cause diverticular illness, hemorrhoids, and a range of other problems. The idea that one might be constipated while having regular bowel motions has recently gained attention. This hypothesis is based on the fact that even when bowel motions are frequent, the colon is not thoroughly evacuated. This indicates that the residue has accumulated on the colon walls and is not adequately expelled by muscular action.

Symptoms: Hard and lumpy stool, decreased frequency of passing stools (< thrice a week), straining to have bowel movements, feeling of unkemptness and blockage in the rectum during defecation and so forth.

Role of barnyard millets: Millets are rich in dietary fiber, and consumption of 40-grams of dietary fiber daily helps relieve constipation. This dietary fiber requirement can be met through consumption of 2–3 servings of millet grains daily.

9.3 Characterization of phytochemical compounds responsible for bioactive properties of barnyard millet

9.3.1 Polyphenols – total phenols and flavonoids

Barnyard millet contains several bioactive phytochemicals, including polyphenolic compounds (Formononetin, Luteolin, Apigenin, Catechin, Kaempferol, 3,7-Dimethylquercetin and Isorhamnetin) (Ofosu et al., 2020). Barnyard is known for its superior antioxidant properties, which are due to the presence of a good amount of polyphenols. Both insoluble and soluble bound phenol extracts, from various varieties of the millet grain, display good antioxidant, metal chelating as well as reducing properties (Chandrasekara and Shahidi, 2010). The flavonoids found in the millet have been shown to have multiple bioactive properties such as antioxidant, anti-inflammatory activities (Shahidi and Chandrasekara, 2013). Flavonoids with subclasses comprising flavanol, isoflavonoid, flavonol, and flavones are commonly observed in barnyard millet (Ofosu et al., 2020). Besides, flavonoids such as Luteolin, Flavones, and Tricin found in barnyard millet have been observed to exhibit high antioxidant properties. Both the whole grain and pearled millet reveal the presence of high flavonoid content, therefore, have potential to be utilized as a functional food (Watanabe, 1999).

Numerous researchers have analyzed and reported on the polyphenol content of diverse varieties of barnyard millet using different solvents for extraction. The extract obtained from barnyard millet displayed total phenolic content of 10.125 mg GAE/g and total flavonoid content of 47.84 mg QE/g (Singh and Naithani, 2014). Likewise, researchers have investigated the DPPH and ABTS radical scavenging activity of the millet and found that the IC_{50} values of the 80 percent ethanol extracts were 141.70 µg/ml for the DPPH radical scavenging activity and

29.59 µg/ml for ABTS radical scavenging activity. They also examined the IC_{50} values of barnyard millet's ethyl acetate extract and reported the radical scavenging activity of DPPH and the radical scavenging activity of ABTS at 31.41 µg/ml and 11.10 µg /ml (Park et al., 2014). Dutta et al. (2019) compared the bioactive compounds and antioxidant capacity of raw, pan cooked and pressure cooked barnyard millet. They found that the total phenolic content was lowest in pressure cooked barnyard (21.32 mg GAE/100 g) and highest in raw barnyard (33.08 mg GAE/100 g). However total flavonoid content was highest in raw barnyard (157.14 mg RE/100 g) and lowest in covered pan cooked barnyard millet (31.42 mg RE/100 g). Results further revealed total antioxidant activity or DPPH radical scavenging activity of 59.89± 0.002 mg TE/100g in raw barnyard millet. A more recent study reported that, among other millet grains, barnyard millet contains the highest DPPH (IC_{50} = 359.6 µg/mL) and ABTS radical scavenging activity (IC_{50} = 362.40 µg/mL)(Ofosuet al., 2020). Similarly, the antioxidant profile of several varieties of barnyard millet was examined in one study and it was found that the 80 percent methanolic extract of millet contained total phenols (38.45 ± 0.45 mg GAE/g) and total flavonoids (28.71 ± 0.27 mg quercetin / G). Also, the DPPH IC_{50} value of the ethyl acetate extract of the millet was 86.7 ± 2.25 mg/ml and ABTS free radical scavenging activity was 74.8 ± 1.85 mg/ml. Of all the extracts used in the study against DPPH and ABTS, the ethyl acetate extract showed the strongest free radical scavenging activity. The study also established that barnyard millet not only has good nutritional content, but also exhibits extremely good antioxidant activity (Kim et al., 2011).

Usually, bioactive compounds like phenols, flavonoids, dietary fiber, and so forth are located in the outer portions of barnyard millet (Daniel, 2005), therefore, processing through a variety of thermal and hydrothermal treatments may be done to improve recovery of bioactive polyphenols to make it fit for human consumption. Modification in the antioxidant nature of millet occurs depending on the severity of the heat treatment, the duration of exposure, and the variety of millet (Liang and Liang, 2019). The bioactive polyphenols present in barnyard millet are significantly affected by the processing method used.

Germination of barnyard millet leads to an increase in total phenols from 29.01 to 77.68 mg GAE/100g. Similarly, the total flavonoids are seen to increase from 29.02 to 71.92 mg/g plant extract in rutin equivalents. This occurs because of the fact that the phenolic compounds such as hydroxyl-cinnamates (ferulic and p-coumaric acids) in the cell walls of the millet bind to polysaccharides other than starch through associations such as esters and ether bonds. The bound phenolic compounds are released by the action of cell wall degrading enzymes (mainly esterase) on these bonds (Duodu, 2019). Germination leads to activation of the enzymes that degrade the cell wall and modifies its structure thereby increasing the concentration of polyphenol compounds(Sharma et al., 2016).Similarly, Murugan et al. (2016) examined the antioxidative action of barnyard millet and found that the methanolic extract of soaked millet had the highest total phenol content (54.5 ± 0.00 mg GAE/g). Another study showed that microwave and steam treatments lowered the phenolic content of barnyard millet. However, roasting of barnyard millet leads to less loss of antioxidants, in contrast to steaming and other processing methods used in the study. According to Pradeep and Sreerama (2015), besides irradiation a non-thermal processing technology enhances antioxidant enzyme and so increases antioxidant properties of the millet. In addition, fermentation of barnyard millet also leads to enhancement of soluble and individual phenols while there is a decrease in bound phenols (Liang and Liang, 2019).Barnyard millet has been shown to contain several compounds with antioxidant properties, that have been seen to increase through techniques like fermentation, malting and germination.

9.3.2 Dietary fiber

Barnyard millet has high soluble and insoluble fiber content (Hadimani et al., 1995), therefore, it can be used as a functional food. Barnyard millet is considered the richest source of dietary fiber when compared to other millet grains. It contains 12.60 percent total fiber with 4.24 percent soluble and 8.36 percent insoluble fiber. They also stated that the fiber content of the millet is higher than that of wheat and rice, which generally have 8.0–9.0 percent fiber (Ugare et al., 2014). Whereas, according to Dutta et al. (2019) barnyard millet contains 2.98 percent of crude fiber. A number of researchers compared the dietary fiber content of barnyard millet with other millet grains and reported that it contains the highest proportion of soluble fiber (6.0 to 6.5%) in contrast to little millet (5.7%) and kodo millet (5.2%)(Veena et al., 2005); Hadimani et al., 1995). A study by Thathola and Srivastava (2002) described the total fiber content of barnyard millet as 31.73 percent. The high dietary fiber content of barnyard millet makes it suitable to be included in therapeutic formulations.

Studies have shown that germination leads to an increase in dietary fiber content from 21.65g/100g to 23.74 g/100 g)(Sharma et al., 2016). Also, the fiber content increases significantly with the increase in germination time due to the modification of the seed's cell wall polysaccharides structure, possibly by influencing the integrity of the tissue histology and altering the protein–carbohydrate interaction. This includes extensive cell wall biosynthesis and leading to the production of new dietary fiber (Martin-Cabrejas et al., 2003).

9.3.3 Resistant starch

Resistant starch is that starch fraction which is resistant to enzymatic hydrolysis in the stomach and small intestine after two hours of ingestion. This part then reaches the colon and is fermented by the gut microflora. Therefore, resistant starch is considered a dietary fiber and has several health benefits (Meenu and Xu, 2019; Nugent, 2005). Barnyard millet has good starch content (48.20–60.20g/100g), amylose fraction (8.90–18.50g/100g), amylopectin (81.50–91.10g/100g), rapid digestible starch (24.8–36.56 g/100g), slowly digestible starch (21.99–30.7 g/100g) and resistant starch (41.45g/100g–44.6 g/100g)(Vali et al., 2018; Bora et al., 2019; Sharma and Gujral, 2020).

Barnyard millet-resistant starch has been shown in rats to reduce blood sugar levels, triglyceride levels and serum cholesterol (Kumari and Thayumanavan, 1998). In addition, a low glycemic index was observed in type-II diabetic patients with regular use of barnyard millet in their meals (Ugare et al., 2014). Besides its use in the food sector as a hypoglycemic raw material, high amylose barnyard millet starch has now attracted attention as an antioxidant packaging material in the biodegradable film industry (Cao et al., 2017).

9.3.4 Bioactive peptides

Some distinctively expressed proteins have also been observed in the *Echinochloa* species. For example, maximal expression of viral nucleoprotein, proteins resistant to quinclorac, defense properties in antimicrobial peptides, cadmium tolerant and Cu/Zn superoxide dismutase have been identified in *Echinochloacrus-galli*. Also, multiple herbicide-resistant proteins were seen in *Echinochloaphyllopogan* variety (Odintsova et al., 2008); (Iwakami et al., 2014). Correspondingly, a new antifungal peptide (EcAMP1) has been recognized in the *Echinochloacrus-galli*seeds. This distinctive antimicrobial peptide is known to have a broad range of antifungal properties against phyto-pathogens, for example; *Alternaria*, *Trichoderma*, *Botrytis* and *Fusarium*. In addition, this peptide acquires a distinct disulfide-stabilized α-helical hairpin structure that strongly binds to the surface of fungal conidia, collects in the cytoplasm, and ultimately hinders hyphal elongation without lysis of the cytoplasmic membrane (Nolde et al.,2011). These therapeutic properties of the barnyard millet variety could be used for the production of new antimicrobial agents in the pharmaceutical and agricultural sectors.

9.4 Techno-functional properties of barnyard millet

9.4.1 Grain structural properties and color profile

Barnyard millet is categorized under small millets like foxtail millet, finger millet, porso millet, kodo millet and little millet (Gupta and Gupta, 2014). As compared to other small millets, the grain of barnyard millet is less hard. The pericarp of mature barnyard millet grain has two epidermal layers (Singh et al., 2010). The aleurone layer of the millet is cutinized cell wall (Zee and O'brien, 1971). In the developing seeds the outer part is multi-layered. The color of the grain differs from white to light gray and dark gray according to the genotype (Renganathan et al., 2020).

The size of the grain also differs according to the genotype. The length of the grain varies from 1.59- 2.84 mm long and the width varies from 1.73–2.25 mm. (Yubuno, 1987). The starch granules shapes are spherical to polygonal, 1.2–10 µm in diameter (Kumari and Thayumanavan, 1998).

9.4.2 Physical properties of grain

The physical properties of barnyard millet grains that have been studied include thousand grain volume and weight, cooking time, swelling capacity and index and hydration capacity and index. The physical properties of grain are given in **Table 9.1**.

9.4.3 Moisture dependent physical properties

There are some physical properties of barnyard millet that depend upon moisture or change in moisture content, such as sphericity, geometric mean diameter, surface area and so forth. Singh et al. (2010) studied the moisture dependent physical properties of barnyard. For the study the samples were kept in tightly sealed polyethylene bags after adding calculated amount of

Table 9.1 Physical Properties of Raw Unprocessed Barnyard Millet

Parameters	Barnyard millet
Thousand grain weight (g)	3.69 ± 0.01
Thousand grain Volume (ml)	4.01 ± 0.01
Hydration Capacity (g/1000 seeds)	2.93 ± 0.05
Hydration Index (%)	78.39 ± 0.1
Swelling Capacity (ml/1000 seeds)	0.23 ± 0.05
Swelling Index (%)	5.83 ± 1.44
Gain in weight after cooking (%)	175.6 ± 10.6
Gain in volume after cooking (%)	41.2 ± 3.76
Cooking time (minutes)	12 minute

Source: Nazni and Devi, 2016.

water. The samples were refrigerated at 5 degrees centigrade for a week. The study revealed the following facts:

9.4.3.1 Geometric mean diameter

The mean diameter of both grain and kernel increases as moisture increases. An increase in mean diameter of 11.91 percent and 12.21 percent was observed for kernel and grain respectively. The increase in moisture was found to be 0.065 to 0.265 kg/kg dry matter, which was lower than other millets and sunflower seed.

9.4.3.2 Sphericity

With increase in moisture content from 0.065 to 0.265 kg kg^{-1} dry matter, the sphericity varied between 0.89 to 0.96, and 0.91 to 0.97 in barnyard millet grain and kernel respectively.

9.4.3.3 Surface area

As the moisture content increased, the surface area also increased. With increase of moisture content from 0.065 to 0.265 percent dry matter, the surface area of grain and kernel increased from 5.58 to 7.13 mm^2 and 9.32 to 12.16 mm^2 respectively. This change occurs due to the changes of dimension with the increasing moisture content.

9.4.3.4 1000 grains and kernels mass

The mass of 1000 kernels and grain was found to increase linearly with increase in moisture content.

9.4.4 Physical properties of starch isolated from barnyard millet

Granules of barnyard millet starch are seen to range between 1.2–10 μm in diameter. The crystallinity degree of barnyard millet starch is 25.91 percent which is lower than porso millet starch but higher than other millet starch.

The water soluble index of barnyard millet is nearly 7 percent which is higher than pearl millet starch but lower than other millet starch. The swelling power, pasting quality, sedimentation volume is about 24 percent, 3 percent and 30 ml/1000 ml respectively, which is lowest among millet starches. The freeze thaw stability, though, is higher than other millet starches (more than 50%) (Wu et al., 2014).

9.4.5 Functional properties of barnyard millet flour

Functional properties of any food component show the interaction between the molecular conformation, nature of environment, the composition, structure, and physico-chemical properties. Functional characteristics are necessary to evaluate how new fat, proteins, carbohydrates and fiber and behave in specific systems. The water absorption capacity is a very important property for food preparation. Water absorption capacity depends on various factors like degree of milling, particle size, percentage of damaged starch and protein content and presence of large proportion of husk in whole flours. High water absorption capacity of barnyard flour shows a good cohesiveness (Nazni and Devi, 2016).

Table 9.2 Functional Properties before and after Processing of Barnyard Millet Flour

Processing Techniques	Swelling power (g/g)	Solubility behavior (%)	Solid loss (%)	Bulk density (g/ml)
Boiling	4.73 ± 0.3	6.3 ± 0.41	27.9 ± 1.45	0.62 ± 0.005
Germination	4.53 ± 0.35	6.8 ± 0.41	26.5 ± 1.30	0.44 ± 0.005
Pressure Cooking	4.7 ± 0.17	6.7 ± 0.60	29.1 ± 1.55	0.52 ± 0.03
Roasting	4.96 ± 0.15	6.3 ± 0.55	30.1 ± 1.34	0.55 ± 0.005
Raw	5.2 ± 0.2	6.2 ± 0.26	30.7 ± 0.90	0.50 ± 0.005

Source: Nazni and Devi, 2016.

Oil absorption capacity helps to induce flavor and mouth feel effect, which depends on surface polarity, amino acid composition and protein conformation of the flour (Chandra and Samsher, 2013).

Sedimentation rate is highly dependent on gluten quantity. Emulsion activity increases the dough strength due to hydrophobic interaction on the surface of the protein that results in a strong oil–water interface and finally in a stable emulsion. Gluten gives texture, appearance and volume to the product (Shrestha and Srivastava, 2017).

9.4.6 Processing techniques and functional properties of barnyard millet

Many functional properties of barnyard millet flour tend to show changes after processing. The processing methods such as germination, boiling, roasting and pressure cooking were experimented by Nazni and Devi (2016). The results of the study are given below (**Table 9.2**).

Functional properties of barnyard millet flour show some changes when it is mixed with other flours. Shrestha and Srivastava (2017) in their study found changes in functional properties of barnyard flour when mixed with refined wheat flour.

9.4.7 Rheological properties of barnyard millet flour
9.4.7.1 Viscosity

Data shows that viscosity of barnyard millet flour increases during heating, decreases just before holding, and increases again during the cooling cycle. However, starch granules are insoluble in water, when a starch suspension is heated above critical temperature, the hydrogen bond weakens and allow the penetration of water (Shinoj et al. 2006).

9.4.7.2 Dynamic moduli

Dynamic moduli consist of two parts: loss modulus and storage modulus. Both of which followed the same trend for barnyard millet flour as viscosity. The trend shows initial increase in heating cycle, then decrease during holding, and finally rise during cooling that remains near constant. Storage module of barnyard millet range from 10.6 to 11.2 and loss modules range from 1.56 to 2.63. This trend is because the elastic nature of the material exceeds the viscous nature. After starch granules swell to a maximum extent, they become very soft, deformable and compressible causing a decrease in storage and loss modules. Storage and loss modules become parallel at the final stages of holding, indicating that the paste has become a stronger. (Shinoj et al., 2006).

9.4.7.3 Phase angle

The phase angle for barnyard millet flour values showed a very broad range. Phase angle ranges from 11 to 9 for barnyard millet. The low values of the phase angle indicates very low tendency of starch paste to undergo deformation by placing the oscillatory strain (Shinoj et al., 2006).

9.4.8 Thermal properties of barnyard millet and barnyard millet flour
9.4.8.1 Specific heat

Along with specific heat, moisture content is increased from 10–30 percent in both grain and flour. 1.48–2.27 kJ kg^{-1} K^{-1} specific heat range were shown in barnyard millet grain, for flour specific heat range was 1.43–2.20 kJ kg^{-1} K^{-1} (Shinoj and Vishwanathan, 2003).

9.4.8.2 Thermal conductivity

As the moisture content increases from 10–30 percent, the thermal conductivity also increases from 0.119 to 0.155 W m^{-1} K^{-1} in barnyard millet grain. Barnyard millet has shown less thermal conductivity among other millets. In the case of other millet flours the thermal conductivity also increases from 0.026 to 0.086 W m^{-1} K^{-1} as the moisture content increases (Shinoj and Vishwanathan, 2003).

9.4.8.3 Thermal diffusivity

The thermal diffusivity decreases in millet grain when moisture content increases. Subramanian and Vishwanathan (2003) reported that thermal diffusivity decreased from 0.770 $\times 10^{-3}$ – 0.600 $\times 10^{-3}$ m^2 h^{-1} in millet grains and 0.723$\times 10^{-3}$ to 0.592$\times 10^{-3} m^2$ h^{-1} for millet flours when moisture increased from 10–30 percent.

9.5 Potential utilization in value added products

Barnyard millet is commonly consumed as boiled rice and as kheer (pudding) like preparations. Consumption of the millet is most common among farmers, who sometimes brew it for beer or use it as cattle feed. Various novel products can be developed with barnyard millet that may help the consumers to diversify their diet and boost their health. Researches indicated that barnyard can be utilized for the formulation of designer and value added food products like bakery products such as cookies (Surekha et al., 2013) and biscuits (Thathola and Srivastava, 2010).

9.5.1 Bakery products

Bakery food items are popular throughout the world, and are in high demand due to variation in texture and taste, low cost, longer shelf life and attractive packaging (Patel and Rao, 1996). The utilization of millets in bakery food is not just for their fiber content and micronutrients but they also have a potential to enter in the bakery industry due to their unique properties (Verma and Patel, 2013). Biscuit prepared by replacing half of refined wheat flour with barnyard flour was reported to have lower glycemic index, GI (50.17) in comparison to the refined flour biscuit (73.58). Surekha et al. (2013) developed value added barnyard millet cookies with the combination of sago flour, pulse flour and vegetable flour. The plain cookies contained

100 percent barnyard flour, vegetable cookies contained 10 percent of dehydrated carrot, and pulse cookies contained 10 percent of soy bean and green gram dhal flour. Among all the three, pulse cookies were more acceptable to the consumers. The formulated cookies had a shelf life of 45–60 days.

Millets can also be used for the formulation of therapeutic food products like low GI biscuits besides their use in making porridge and chapatti. Thathola and Srivastava (2010) prepared low GI biscuits using barnyard millet. The biscuits were prepared with millet flour (45%) and refined wheat flour (55%). These biscuits were found to have higher content of dietary fiber, crude fiber and total ash than the biscuits prepared with refined wheat flour. Biscuits made from barnyard millet flour had a glycemic index of 68. Another study was carried out to explore the possibility of utilization of underutilized but highly nutrient rich barnyard millet in cookies. The quality cookies were prepared from 50 percent *maida* and 50 percent wheat flour, 30 percent *maida* and 70 percent barnyard millet and 20 percent wheat flour and 80 percent barnyard millet flour. Fresh cookies prepared from 50 percent *maida* and 50 percent wheat flour contained 4.0 percent moisture, 11.55 percent protein, 26.30 percent crude fat, 1.10 percent crude fiber, 71.65 percent carbohydrates, 35.50 mg/100 g calcium and 3.80 mg/ 100 g iron. Whereas, fresh cookies prepared from 20 percent wheat flour and 80 percent barnyard millet flour containing 3.0 percent moisture, 7.38 percent protein, 27.10 percent crude fat,8.22 percent crude fiber, 66.28 percent carbohydrates, 4.98 mg/100 g iron and 30.60 mg/100 g calcium. The sensory evaluation of cookies during storage (3 months) was carried out regularly at an interval of one month. Storage study of cookies showed that the cookies prepared by incorporation of *maida*, wheat and barnyard millet flour can be stored up to three months with minimum losses in sensory, nutritional and textural characteristics (Salunke et al., 2019).

Muffins can also be prepared with barnyard millet flour, and these can be well accepted. Nazni and Karuna (2016) developed muffins and rusks from the barnyard millet bran. The wheat flour was combined with barnyard millet bran at different ratios: 100:0, 95:5, 90:10, 85:15, 80:20, 75:25 and 70:30 for both muffin and rusk. The sensory score of rusks showed that combination with proportion 85:15 was more acceptable by the panel of judges and for muffin 70:30 got the highest score among all the samples.

9.5.2 Puffed, popped and extruded products

Puffing of grains is an old conventional process of cooking for snacks or breakfast cereals, either plain or with a few flavors/salt/sugars. Puffed and toasted snack food product from barnyard millet was developed. A study was undertaken to develop an extruded product using three composite formulation of barnyard millet (BM) and maize (M) grits (180 micron). The formulations (BM: M viz; 70:30, 60:40 and 50:50) were extruded using a twin screw extruder. Among the three formulations, the 50:50 was reported to be best based on the sensory evaluation (Devi and Palanimuthu, 2020). In view of the increasing preference of people for oil free products and health conscious demography, a study was conducted to formulate microwave puffed barnyard based fasting foods. The experiment included cold extrudate, then microwave puffing and oven toasting to prepare the product. Sensory evaluation showed that the product was well comparable with other products available in the market. The ash and fat content were least affected while processing, whereas protein content decreased during oven toasting (Dhumal et al., 2014). Srivastava et al. (2003) developed popped grains from barnyard millet, foxtail and little millet. They utilized normal salt as a warming medium in an open iron saucepan. They processed samples and salt in the proportion 1:20 at 240–260 °C for 15–25 s. Further the popped grains were used to make Ladoo's (sphere shaped Indian Sweet). The sensory analysis scores on nine-point hedonic scale for first and second sort Ladoo's were 5.0–6.9 and 7.2–8.1.

9.5.3 Noodles and vermicelli

Noodles are produced using the flours of grains or legumes as the major ingredients, and the dried food items are utilized as ready-to-eat foods. Noodles are the pasta items otherwise called *convenient* food products prepared through the cold expulsion process and which become hard and stiff after drying. Cooking requires only a few minutes and is very convenient (Devaraju et al., 2006). Noodles are one of the favorite food products for all age groups, and they have a long shelf life and profitable value. Barnyard millet has moderately low sugar content (58.56%) having slow edibility of 25.88 percent. This good property was utilized to formulate low GI noodles from barnyard millet in combination with pulse flour, sago flour and Bengal gram leaf powder at different levels to prepare pulse, plain, and vegetable noodles separately (Surekha et al., 2013).

Rice and barnyard flour along with malt from whole green gram, fenugreek powder, carrot powder and xanthan gum were used to formulate nutritionally rich vermicelli. The protein, moisture, fat, ash, and carbohydrate content in the product was found to be 10.07 percent, 6.40 percent, 1.72 percent, 1.44 percent, and 80.31 percent respectively. The optimized product is rich in iron (3.81 mg/100 g) and beta-carotene (1039 μg/100 g)(Goel et al., 2021).

9.5.4 Other traditional recipes

Barnyard, like other minor millets, is nutritionally superior to cereals. By cooking it the way rice is cooked, it can be utilized for formulation of instant *pulav*. Studies revealed that the sensory quality of instant *pulav* (reconstituted) was not significantly different from traditional cooked *pulav*. The glycemic index (GI) of instant *pulav* was 34.12, which classifies it under low glycemic index foods defining its suitability for diabetic subjects. The instant pulav was found to be nutritionally rich. Instant *pulav* can be stored safely at room temperature for a period of one month in summer and two months in winter (Joshi and Srivastava, 2018a). *Laddu, halwa* and *biryani* from barnyard millet was prepared. When compared to control an insignificant difference was observed with regard to the appearance, flavor, color, texture and overall acceptability of barnyard millet laddu and halwa. The fiber, protein and fat content of the product was higher than the rice products (Verma et al., 2014).

Other products like barnyard millet *bhaat*, barnyard *dalia* and ricebean-barnyard *khichdi)* were also developed. The study revealed that barnyard *bhaat* had the highest concentration of protein (11.19 g/100 g) and barnyard *dalia* had the highest concentration of crude fiber (4.25 g/100 g). The glycemic index (GI) of all the products was found to be low (barnyard *bhaat*: 38.49, barnyard *dalia*: 24.14 and ricebean-barnyard *khichdi*: 31.42) (AICRP; *Home Science*, 2019).

Vanithasri and Kanchana (2013) developed barnyard *idlis*. Incorporation of barnyard to the standard recipe at 30, 40 and 50 percent yield acceptable idlis with low fat percent (0.88 g/100 g). Barnyard-rice idlis were found to be nutritionally superior to the control idlis. The protein (6.82 g), fiber (4.64 g), phosphorus (122.01 mg) and iron (4.05 mg) were found to be better than control idli. The idlis were found to be organoleptically good, when analyzed by panelists on a 9-point hedonic scale.

9.5.5 Milk

Barnyard millet can also be used to extract milk. About 75 to 85 ml of milk can be extracted from barnyard after 8 hours of soaking and 18 hours of germination. Processing reduces fat, phytic acid, crude fiber and ash to a great extent, slightly reduces carbohydrate and protein, whereas it increases total sugar and moisture content. Extracted milk can also be used to formulate value added products (Sheela et al., 2018).

9.5.6 Nutri functional snacks

Nutri-Functional Snack (NFS) was prepared using barnyard millet (50–70%), quality protein maize (20–50%) and flax seed (0–20%) with constant level of skim milk powder (5%). There was a significant effect of different percentages of ingredients level on the physical and functional properties. The optimum conditions for maximum acceptability of NFS food was using 70 percent barnyard millet, 28 percent quality protein maize and 2 percent flaxseed (Kumar et al., 2017).

Suri et al. (2020) developed ready to eat iron rich extruded snack using barnyard millet and defatted soy flour in different proportion along with *amla* and rice flour. The optimized product was found to be rich in iron (15.71 mg/100 g) and protein (18.91 g/100 g).Barnyard millet flour and pearl millet flour incorporated wafers were also developed. These wafers were found to be nutritionally and organoleptically superior than the control (Sruthi et al., 2018).

9.5.7 Weaning foods

Complementary foods are the chief source of energy requirement for children during the age of six to twelve months. Four different formulations of weaning foods were developed by Bala and Nazni (2019) to meet the exceeded energy and other nutrient requirements of growing infants. All four formulations displayed adequate amounts of carbohydrates, proteins, fats and micro nutrients. Acceptability scores for weaning food mix ranged from 6.3±0.87 to 8.52± 0.84.

9.6 Conclusion

Barnyard millet is nutritionally rich and has agronomic benefits like potential for bio-fortification of micronutrients, but has received very little attention in terms of research and remains underutilized. Farmers are not aware of its nutritional value and production potential. Therefore, support from government and NGOs (Non-Government Organization) may help to create awareness among the consumers, farmers and nutritionists to cultivate barnyard and promote consumption. There is also an urgent need for advancements in processing techniques and value-addition of barnyard millet. Development of new food products from millets, increase in their market price and a change in consumer choices can fetch better returns to farmers and would lead to a healthier option for consumers. It is hoped that when these problems are overcome, barnyard millet (being nutritionally rich and environmentally hardy) will prove to be the most suitable as sustainable food for achieving nutritional security.

References

AICRP. All India Coordinated Research Project (H.Sc). 2019. *Development of Region Specific Therapeutic Foods for Prevention of Diabetes*. G.B.P.U.A and T, Pantnagar (Project Report).

Bora, P., S. Ragaee and M. Marcone. 2019. Characterization of several types of millets as functional food ingredients. *International Journal of Food Sciences and Nutrition*. 70(6): 714–724.

Cao, T.L., S.Y. Yang and K.B. Song. 2017. Characterization of barnyard millet starch films containing borage seed oil. *Coatings*. 7(11): 183.

Chandra, S. and Samsher. 2013. Assessment of functional properties of different flours. *African Journal of Agricultural Research*. 8(38): 4849–4852.

Chandrasekara, A. and F. Shahidi. 2010. Content of insoluble bound phenolics in millets and their contribution to antioxidant capacity. *Journal of Agricultural and Food Chemistry*. 58(11): 6706–6714.

Daniel, B. 2005. Matering fiber in gastrointestinal foods. In: *Proceedings of Healthy Foods European Summit*.

Devaraju, B., M.J. Begum, S. Begum, and K. Vidhya. 2006. Effect of temperature on physical properties of pasta from finger millet composite flour. *Journal of Food Science and Technology*. 43(4):341–343.

Devi, G.S. and V. Palanimuthu. 2020. Study and Development of Barnyard Millet Based Ready to Eat Product. *Int.J.Curr.Microbiol.App.Sci*. 9(10): 738–746.

De Wet, J., Prasada Rao, K., Mengesha, M., and Brink, D. 1983. Domestication of mawa millet (Echinochloacolona). *Econ. Bot*. 37, 283–291. doi: 10.1007/ BF02858883

Dhumal, C.V., I.L. Pardeshi, P.P. Sutar, and R.V. Jayabhaye. 2014. Development of Potato and Barnyard Millet Based Ready to Eat (RTE) Fasting Food. *Journal of Ready to Eat Foods*. 1(1):11–17.

Doggett, H. (1989). "Small millets-a selective overview," in *Small Millets in Global Agriculture*, eds A. Seetharam, K.W. Riley, and G. Harinarayana (Oxford: Oxford), 3–18.

Duodu, K.G., and Awika, J.M. (2019). Phytochemical-related health-promoting attributes of sorghum and millets. In *Sorghum and millets* (pp. 225–258). AACC International Press.

Dutta, A., P. Shukla, S. Tilara, N. Prasad, R. Khan, S. Suri and S.B. Bharadwaj. 2019. Comparative Evaluation of Antioxidant Potential in Thermally Processed, Underutilized Food Grains of the Himalayan Region. *European Journal of Nutrition and Food Safety*. 9(3): 277–286.

Goel, K., S. Goomer and D. Aggarwal. 2021. Formulation and Optimization of Value-added Barnyard Millet Vermicelli Using Response Surface Methodology. *Asian Journal of Dairy and Food Research*. 1–7.

Gomashe, S.S. 2017. Barnyard millet: present status and future thrust areas. *Millets and Sorghum: Biology and Genetic Improvement*. 134, 184–198. doi: 10.1002/9781119130765.ch7

Gupta, A. and H.S. Gupta. 2014. Textbook of Field Crops Production., R. Prasad (Eds), *Indian Council of Agriculture Research*, New Delhi, November 2014, p. 227.

Hadimani, N.A., S.Z. Ali and N.G. Malleshi. 1995. Physico-chemical composition and processing characteristics of pearl millet varieties. *Journal of Food Science and Technology*. 32(3): 193–198. www.who.int/news-room/fact-sheets/detail/diabetes.

Hulse, J. H., Laing, E. M., Peason, O. E. 1980. Sorghum and millets: Their composition and nutritive Value. Academic Press, London

Iwakami, S., A. Uchino, Y. Kataoka, H. Shibaike, H. Watanabe and T. Inamura. 2014. Cytochrome P450 genes induced by bispyribac-sodium treatment in a multiple-herbicide-resistant biotype of Echinochloaphyllopogon. *Pest Management Science*. 70(4): 549–558.

Joshi, S. and S. Srivastava. 2018a. Development and quality evaluation of low glycemic index barnyard millet instant *Pulav.Agricultural Research Journal*. 55(3):528–535.

Kim, J.Y., K.C. Jang, B.R. Park, S.I. Han, K.J. Choi, S.Y. Kim, S.H. Oh, J.E. Ra, T.J. Ha, J.H. Lee, J.H. Wang, H.W. Kang, and W.D. Seo, 2011. Physicochemical and antioxidative properties of selected barnyard millet (*Echinochloautilis*) species in Korea. *Food Science and Biotechnology*. 20(2): 461–469.

Kumar, K.P., Mani Indra, R.K. Gupta, A. Kumar, S.K. Jha, B. Singh, S. Sarkar. and S.D. Lande. 2017. Development of barnyard millet based nutri-functional snack food: a response surface methodology approach. *Indian Journal of Agricultural Sciences*. 87(11):1430–1436.

Kumari, S.K. and B. Thayumanavan, 1998. Characterization of starches of proso, foxtail, barnyard, kodo, and little millets. *Plant Foods for Human Nutrition*. 53(1) 47–56.

Liang, S., and K. Liang, 2019. Millet grain as a candidate antioxidant food resource: a review. *International Journal of Food Properties*. 22(1): 1652–1661.

Martin-Cabrejas, M.A., N. Ariza, R. Esteban, E. Molla, K. Waldron and F.J. Lopez-Andreu. 2003. Effect of germination on the carbohydrate composition of the dietary fiber of peas (Pisumsativum L.). *Journal of Agricultural and Food Chemistry*. 51(5): 1254–1259.

Meenu, M. and B. Xu. 2019. A critical review on anti-diabetic and anti-obesity effects of dietary resistant starch. *Critical Reviews in Food Science and Nutrition*. 59(18): 3019–3031.

Murugan, S., A. Shanmugam, L. Manoharan, S. Sundaramoorthy, S. Gunasekaran, S. Arunachalam and Sathiavelu, M. 2016. Antioxidant Activity of Aqueous and Methanol Extract of Barnyard Millet. *Research Journal of Pharmacy and Technology*. 9(3):262–266.

Nazni, P. and R. Devi. 2016. Effect of Processing on the Characteristics Changes in Barnyard and Foxtail Millet. *Journal of Food Process Engineering*, 7:3. doi:10.4172/2157-7110.1000566

Nazni, P. and T.D. Karuna. 2016. Development and quality evaluation of Barnyard Millet Bran incorporated Rusk and Muffin. *Journal of Food Industrial and Microbiology*. 2(2):1–6.

Nozawa, S., H. Nakai and Y.I. Sato. 2004. Characterization of micro-satellite and ISSR polymorphism among Echinochloa (L.) *Beauv.* spp. in Japan. *Breeding Research* 6: 187–193.

Nolde, S.B., A.A. Vassilevski, E.A. Rogozhin, N.A. Barinov, T.A. Balashova, O.V. Samsonova, Y.V. Baranov, A.V. Feofanov, T.A. Egorov, A.S. Arsenievand and E.V Grishin. 2011. Disulfide-stabilized helical hairpin structure and activity of a novel antifungal peptide EcAMP1 from seeds of barnyard grass (*Echinochloacrusgalli*). *Journal of Biological Chemistry*, 286(28): 25145–25153.

Nugent, A.P. 2005. Health properties of resistant starch. *Nutrition Bulletin*, 30(1): 27–54.

Odintsova, T.I., E.A. Rogozhin, Y. Baranov, A.K. Musolyamov, N. Yalpani, T.A. Egorov, and E.V. Grishin, 2008. Seed defense of barnyard grass *Echinochloacrusgalli* (L.) Beauv. *Biochimie*, 90(11–12): 1667–1673.

Ofosu, F.K., F. Elahi, E.B.M. Daliri, R. Chelliah, H.J. Ham, J.H. Kim, S.I. Han, J.H. Hur, and D. H. Oh. 2020. Phenolic profile, antioxidant, and antidiabetic potential exerted by millet grain varieties. *Antioxidants*, 9(3): 254.

Park, D.H., Lee, S.T., Jun, D.Y., Lee, J.Y., Woo, M.H., Kim, K.Y., Seo, M.C., Ko, J.Y., Woo, K.S., Jung, T.W., Kwak, D.Y., Nam, M.H. and Kim, Y. H. 2014. Comparative evaluation of antioxidant activities of ethanol extracts and their solvent fractions obtained from selected miscellaneous cereal grains. *Journal of Life Sciences*. 24(1):26–38.

Patel, M.M. and Venkateswara, R.G. 1996. Influence of untreated, heat-treated and germinated blackgram flours on biscuit making quality of wheat flour. *Journal of Food Science and Technology*. 33(1):53–56.

Pradeep, P.M., and Y.N. Sreerama. 2015. Impact of processing on the phenolic profiles of small millets: Evaluation of their antioxidant and enzyme inhibitory properties associated with hyperglycemia. *Food Chemistry*, 169, 455–463.

Renganathan, V.G., C. Vanniarajan, A. Karthikeyan and J. Ramalingam. 2020. Barnyard Millet for Food and Nutritional Security: Current Status and Future Research Direction. *Frontiers in Genetics*. 11:500. doi:10.3389/fgene.2020.00500

Salunke, P.P., U.D. Chavan, P.M. Kotecha and S.B. Lande. 2019. Studies on nutritional quality of barnyard millet cookies. *International Journal of Chemical Studies*. 7(4): 651–657.

Shahidi, F. and A. Chandrasekara. 2013. Millet grain phenolics and their role in disease risk reduction and health promotion: A review. *Journal of Functional Foods*. 5(2): 570–581.

Sharma, B. and H.S. Gujral. 2020. Modifying the dough mixing behavior, protein and starch digestibility and antinutritional profile of minor millets by sprouting. *International Journal of Biological Macromolecules*. 153, 962–970.

Sharma, S., D.C. Saxena and C.S. Riar. 2016. Isolation of functional components β-glucan and γ-amino butyric acid from raw and germinated barnyard millet (*Echinochloafrumentacea*) and their characterization. *Plant Foods for Human Nutrition*. 71(3): 231–238.

Sheela, P., S. Kanchana, T.U. Maheswari and G. Hemalatha. 2018. Optimization of Parameters for the Extraction of Millet Milk for Product Development. *Research Journal of Agricultural Sciences*. 9(6): 1345–1349.

Shinoj, S. and R. Vishwannathan. 2003. Thermal Properties of Minor Millet Grains and Flours. *Biosystems Engineering*. 84(3): 289–296. doi:10.1016/S1537-5110(02)00222-2

Shinoj, S., R. Vishwannathan, M.S. Sajeev and S.N. Moorthy. 2006. Gelatinisation and Rheological Characteristics of Minor Millet Flours. *Biosystems Engineering*. 95(1); 51–59. doi:10.1016/j.biosystemseng.2006.05.012

Shrestha, R. and S. Srivastava. 2017. Functional Properties of Finger Millet and Barnyard Millet Flours and Flour Blends. *International Journal of Science and Research*. 6(6): 775–780.

Singh, K.P., H.N. Mishra and S. Saha. 2010. Moisture-dependent properties of barnyard millet grain and kernel. *Journal of Food Engineering*. 96, 598–606. doi: 10.1016/j. jfoodeng.2009.09.007

Singh, M. and M. Naithani. 2014. Phytochemical estimation and antioxidant activity of seed extract of millets traditionally consumed by common people of Uttarakhand, India. *International Journal of Biology, Pharmacy and Allied Sciences*. 3(10): 2389–2400.

Srivastava, S., M. Dhyani and G. Singh. 2003. Popping characteristics of Barnyard and foxtail millet and their use in preparation of sweets. *Recent Trends in Millet Processing and Utilization*. 38–40.

Sruthi, V., K. Waghray and A.K. Rathod. 2018. Development of Wafers Incorporated with Pearl Millet Flour and Barnyard Millet Flour. *International Journal of Scientific & Technology Research*. 2018. 4(5): 44–51.

Suri, S., A. Dutta, N.C. Shahi, R.S. Raghuvanshi, A. Singh, and C.S. Chopra. 2020. Numerical optimization of process parameters of ready-to-eat (RTE) iron rich extruded snacks for anemic population. *LWT – Food Science and Technology*. 134(110164). https://doi.org/10.1016/j.lwt.2020.110164.

Surekha, N., R.S. Naik, S. Mythri and R. Devi. 2013. Barnyard millet (*Echinochloafrumentacea*) Cookies: Development, value addition, consumer acceptability, nutritional, and shelf life evaluation. *IOSR Journal of Environmental Science, Toxicology and Food Technology*. 7(3):1–10.

Takase, K. 2009. Prehistoric and Protohistoric Plant Use in the Japanese Archipelago. Meiji University Premodern Japan Research Exchange, 0708 DEC 2009, USC, Tokyo, Japan.

Thathola, A. and S. Srivastava. 2002. Physicochemical properties and nutritional traits of millet-based weaning food suitable for infants of the Kumaon Hills, Northern India. *Asia Pacific Journal of Clinical Nutrition*. 11(1): 28–32.

Thathola, A. and S. Srivastava. 2010. Suitability of Foxtail millet (*Setariaitalica)* and Barnyard millet (*Echinochloafrumentacea*) for development of low glycaemic index biscuits. *Malaysian Journal of Nutrition,* 16: 361–368.

Ugare, R., B. Chimmad, R. Naik, P. Bharati and S. Itagi. 2014. Glycemic index and significance of barnyard millet (*Echinochloafrumentacae*) in type II diabetics. *Journal of Food Science and Technology*. 51(2): 392–395.

Vali Pasha, K., C.V. Ratnavathi, J. Ajani, D. Raju,, M.S. Kumarand, S.R. Beedu. 2018. Proximate, mineral composition and antioxidant activity of traditional small millets cultivated and consumed in Rayalaseema region of south India. *Journal of the Science of Food and Agriculture*. 98(2): 652–660.

Vanithasri, J. and S. Kanchana, **2013**. Studies on the quality evaluation of idli prepared from barnyard millet (*Echinochloafrumentacaea*). *Asian Journal of Home Science*. 8(2): 373–378.

Veena, B., B.V. Chimmad, R.K. Naik, and G. Shantakumar. 2005. Physico-chemical and nutritional studies in barnyard millet. *Karnataka Journal of Agricultural Science*. 18:101–105.

Verma, S., S. Srivastava, and N. Tiwari. 2014. Comparative study on nutritional and sensory quality of barnyard and foxtail millet food products with traditional rice products 2014. *Journal of Food Science and Technology*. doi 10.1007/s13197-014-1617-y

Verma, V. and S. Patel, 2013. Value added products from nutri-cereals: Finger millet (*Eleusine coracana*). *Emirates Journal of Food and Agriculture*. 169–176.

Watanabe, M. 1999. Antioxidative phenolic compounds from Japanese barnyard millet (Echinochloautilis) grains. *Journal of Agricultural and Food Chemistry*, 47(11): 4500–4505.

Wu, Y., Q. Lin, T. Cui,, and H. Xiao. 2014. Structural and Physical Properties of Starches Isolated from Six Varieties of Millet Grown in China. *International Journal of Food Properties*, 17:2344–2360. doi: 10.1080/10942912.2013.803119

Yabuno, T. 1987. Japanese barnyard millet (Echinochloa utilis, Poaceae) in Japan. *Econ. Bot.* 41(4): 484–493.

Zee, S.Y., and T.P. O'brien. 1971. Aleurone Transfer Cells and other Structural Features of the Spikelet of Millet. *Australian Journal of Biological Sciences..* 24: 391–396.

10

Little Millet (*Panicum sumatrense*)
Nutraceutical Potential and Techno-Functionality

Hardeep Singh Gujral

Contents

10.1 Introduction

Little millet (*Panicum sumatrense*), commonly known as *kutki,* has its origin in Southeast Asia and is presently grown in India, Nepal, Burma and some African countries. In India, it is mainly grown in the states of Madhya Pradesh, Uttar Pradesh, Karnataka, Andhra Pradesh, Orissa, Tamil Nadu, Bihar, Maharashtra and Jharkhand (Seetharam et al., 1986; Rao et al., 2017). It is considered a relative of Proso millet and can grow up to an elevation of 2,100 meters in the tropics and subtropics (Kumar et al., 2018). Little millet seeds are smaller than proso millet, contain sufficient levels of protein and have the highest fat content of all the millets (Tonapiet al., 2015). Shahidi and Chandrashekhar (2013) reported the nutrient composition of little millet, and they also reported that India is the major producer of little millet. It is grown primarily in the drought-prone areas and used mostly for food purposes, especially by people of economically weaker sections. It is a good source of nutraceuticals such as phenolics, gamma-amino butyric acid (GABA), lignans, resistant starch, sterols, and phytates (Guha et al., 2015).

Millets are known as "miracle grains" due to their remarkable adaptability under diverse agro-climatic conditions of drought, rain-fed areas, barren and acidic land, water scarcity and salinity stress (Chandrasekara and Shahidi, 2010; Hadimani and Malleshi, 1993); also they do not require any chemical fertilizer or pesticide for growth (Behera, 2017). They require less agricultural input, have short growing seasons and can also be stored for longer duration without damage by insects and pests. Millets are popularly known as C4 carbon sequestrating crops that contribute to lowering of carbon dioxide in the environment (Tonapiet al., 2015). They form a part of a multi-cropping system and can be grown along with legumes and oilseeds (Chopra and Neelam, 2004; Pradhan et al., 2010). They are potentially a rich source of protein (6–19%), dietary fiber (12–20%), carbohydrates (60–70%), minerals (2–4%), fat (1.5–6%) (Annoret al., 2017). The major mineral composition of little millet consists of Ca (0.17), P (2.20), K (1.26), Na (0.07) and Mg (2.33 g/kg) and trace minerals like Fe, Cu, Zn and manganese. Little millet contains 3.7 mg/100g of Zn (Kaur et al., 2019). Consumption of millets, which are rich in calcium, such as finger and little can improve the health of women and children (Dayakar, 2017). Little millet has been reported to lower blood glucose and cholesterol levels and helps in the prevention of diabetes mellitus (Guha et al., 2015 and Nithiyanantham et al., 2019). Additionally, they are considered superior to other cereals such as rice and wheat grain in terms of various nutritional attributes such as phytochemicals, fiber, minerals, vitamins especially B1 and B2, energy, lower glycemic index and gluten free behavior (Saleh et al., 2013; Sharma and Gujral, 2019; Manay and Shadaksharaswamy, 2005) and are therefore termed "nutri-cereals." The carbohydrates present in millets are slowly digested, which causes slow absorption of glucose and ultimately modulates the blood glucose level (Chethan and Malleshi, 2008). They are known to provide various health benefits such as preventing cancer, degenerative and cardiovascular diseases, lowering blood pressure and cholesterol, delaying gastric emptying and preventing the incidence of tumors (Chandrasekara and Shahidi, 2010). Therefore, considering the above facts, millets must be viewed as climate change compliant crops that make them the future of the world's food and farming.

Although being nutritionally rich and promising, these grains remain neglected, and their current utilization in food products at an industrial level is still limited due to the presence of antinutritional factors like tannins, phytic acid and insufficient research efforts related to product development, lower preference driven by affluence, their coarse nature, bitter after taste and mouthfeel, dark color, longer time and effort involved in processing and lower cooking quality (Srichuwonget al., 2017; Chandelet al., 2014). However, there are several barriers and loopholes on the part of government policies that have led to a decline in the trend of promotion, production and consumption of millets; including lack of production support and improved methods and technologies in contrast to major cereal crops; lack of appropriate post-harvest processing technologies for small millets; competition from other market friendly remunerative crops; changes in consumption patterns mainly due to inclusion of only major crops like rice and wheat into the Public Distribution System (PDS); lack of public procurement and marketing support and absence of public or private funded promotion of millets as a nutritious food category (NAAS, 2013).

Apart from the higher nutritional value, millets are also known to possess a considerable amount of antinutritional components, such as phytic acid and tannins (Sharma and Gujral, 2019). Food processing methods such as soaking, sprouting, cooking, roasting and decortication can improve the nutritional value. Recently, with growing awareness regarding health, there are increasing concerns with respect to the impact of foods on human health that have led to the improved demands for various forms of healthy foods.

10.1.1 Post-harvesting processing

Little millet is a small hulled grain with the hull forming 16.86 percent of the total kernel (Sahoo et al., 2020). It has a grain size of 3.26 mm length, 1.83 mm width and1.15 mm thickness and a 1,000 kernel weight of 3.18 gm (Rao et al., 2020). It is dehulled to remove the hull, and then it is polished to remove the branny layers. The polished or debranned or milled little millet exhibits improved cooking, sensory, digestive, assimilation, and absorption behavior. There is no exclusive milling machinery for little millets, but existing cereal milling machinery has been adapted to mill the millets. Dehusking is best carried out in some kind of impact dehuller, and debranning is carried out in abrasive dehullers. Hadimani and Malleshi (1993) have reported a yield of 83.1 percent after dehulling and 77 percent after debranning for little millet. Upon debranning, the protein, fat, ash, calcium and phosphorous contents decreased by 10.5, 42.5, 43.7, 66.4 and 36.3 percent respectively whereas the carbohydrate content increased by 5.2 percent.

Little millet can be milled into flour; but it will have poor stability if it has been ground straight-away into a meal as the lipolytic enzymes will bring about hydrolytic changes. If storage stability is to be improved, then the degerming is essential (Abdelrahamet al., 1983) or the millet can be subjected to some kind of thermal treatment to inactivate the hydrolytic and oxidative enzymes.

10.2 Characterization of phytochemical compounds responsible for bioactive properties and effects of processing

Although phytate, tannins, and phenolic acids are considered as anti-nutritional factors, they also have anti-hypertensive properties, which may prevent certain type of cancers. Chandrasekara et al. (2011a) reported that millet phenolics are effective in the prevention of cancer initiation and progression in vitro. The phytocyanin's, phytosterols and polyphenols present in the millet boost the human immune system and delay age-onset diseases (Bravo, 1998). These therapeutic compounds also act as antioxidants and detoxifying systems to reduce the risk of degenerative ailments such as Parkinson's disease and certain other metabolic syndromes. The consumption of little millet is also a boon for diabetic patients as it has low glycemic index, which is a good source of slow digesting carbohydrates and dietary fibers. Since its digestibility is slower than glucose, it takes a longer time to enter into the blood, and a stable blood sugar level is maintained. Also, inhibitors like aldose reductase prevent the accumulation of sorbitol and reduce the risk of diabetes induced cataract diseases (Chethan et al., 2008).

10.2.1 Effects of sprouting on proximate composition of little millet

Sprouting or germination is a simple process employed for improving the nutritional quality of cereals and legumes. During this process significant changes in the biochemical, nutritional

Table 10.1 Impact of Sprouting on Proximate Composition and Protein Solubility of Little Millet

Little millet	Fat (% dry basis)	Protein (% dry basis)	Ash (% dry basis)	Crude fibre (% dry basis)	Total starch (% dry basis)	Protein solubility (% dry basis)
C	6.15[e] ± 0.240	10.33[e] ± 0.163	2.92[e] ± 0.240	3.55[d] ± 0.354	65.39[f] ± 0.297	33.93[a] ± 0.61
S	6.05[e] ± 0.042	10.23[e] ± 0.028	2.73[d] ± 0.078	3.37[d] ± 0.035	64.15[e] ± 0.240	34.72[a] ± 0.73
S12	5.72[d] ± 0.057	9.96[d] ± 0.071	2.53[c] ± 0.021	3.18[d] ± 0.085	61.39[d] ± 0.078	37.69[b] ± 0.61
S24	5.21[c] ± 0.028	8.39[c] ± 0.099	1.99[b] ± 0.170	2.94[c] ± 0.099	59.66[c] ± 0.318	39.49[c] ± 0.51
S36	5.02[b] ± 0.035	7.87[b] ± 0.177	1.60[a] ± 0.106	2.33[b] ± 0.141	58.89[b] ± 0.177	42.03[d] ± 1.17
S48	4.69[a] ± 0.113	7.54[a] ± 0.035	1.45[a] ± 0.057	1.96[a] ± 0.071	58.44[a] ± 0.453	43.73[e] ± 0.62

Notes:

Data presented is mean ± standard deviation (n=3); Values followed by the different superscripts in the same column of same grain are significantly different (p b 0.05); C is control; S is soaked; S12 is 12 hr. sprouting; S24 is 24 hr. sprouting; S36 is 36 hr. sprouting; S48 is 48 hr. sprouting. (Sharma and Gujral, 2020).

and sensory characteristics of cereals occur due to degradation of reserve materials that is used for respiration and synthesis of new cell constituents by the developing embryo in the seed (Danisova et al., 1995). Sprouted millets are considered to be much healthier than the native millets due to the generation of various health promoting bioactive components and reduction of antinutritional factors such as phytic acid, tannins and protease inhibitors that results in better mineral and vitamin bioavailability and nutrient digestibility (Affify et al., 2012). As compared to unsprouted seed, sprouted seeds contain low unsaturated fatty acids and low carbohydrate content. Mineral content such as phosphorous, calcium, zinc and copper were reported to increase in sprouts due to hydrolysis of phytic acid by the phytase enzyme during the sprouting process (Sinha and Kawatra, 2003).

Sharma and Gujral (2020) reported the fat, protein, ash, crude fiber and total starch of little millets as affected by steeping, soaking and sprouting for different durations (**Table 10.1**). The fat, protein, ash and crude fiber all increased upon sprouting duration. Sharma and Gujral (2019b) studied the impact of sprouting on little millet and observed that the protease and amylase activity increased significantly on soaking and with increase in duration of sprouting. Maximum enzymatic activity (proteolytic and amylase) was observed after 48 hr. of sprouting. Upon sprouting, the enzymes get activated and cause the degradation of the nutritional components such as carbohydrates, lipids and proteins.

10.2.2 Dietary fiber

Dietary fiber (DF) is a carbohydrate polymer, found in cereals, legumes, vegetables, and fruits, that contains indigestible components such as lignin, polysaccharides, and oligosaccharides that are transferred to the large intestine for total or partial fermentation (Zhu et al., 2018). The dietary fiber content in millets is 9 to 15 percent approximately, and can be used as a functional food (Shobana et al., 2009). Little millet has the highest content of soluble dietary fibers, around 5.7 percent among all the millets. Debranned little millet has a total dietary fiber content of 15.9 percent (Hadimani and Malleshi, 1993) and 36 percent of the total is soluble dietary fiber, indicating that little millet forms an important source of soluble dietary fiber.

Sharma and Gujral (2019b) reported that the total dietary fiber content of little millet was 14.90 percent while, for whole wheat flour, it was 11.4 percent. Hadimani and Malleshi (1993) reported that millets have a higher total dietary fiber content than rice and wheat. The soluble dietary fiber (SDF) content in little millet was 5.18 percent and in wheat flour it was

3.34 percent while insoluble dietary fiber (IDF) for little millet was found to be 9.72 percent and 8.06 percent for wheat flour. Sharma and Gujral (2019a) studied the chapatti making behavior of little millet flour supplemented with 12 percent exogenous vital gluten. The total dietary fiber content of wheat flour was 11.40 g/100g (dry basis) while the VGSLM (vital gluten supplemented little millet flour) demonstrated higher content that is, 13.02 g/100g (dry basis). Sharma and Gujral (2019c) reported the dough characteristics and chapatti making quality of wheat flour and little millet flour blends in the proportion 3:1. The dietary fiber in wheat flour was 11.40 percent while it increased in little millet composite flour (12.87%).The impact of sprouting of little millet on total dietary fiber (TDF) has been studied by Sharma and Gujral (2020) and the total dietary fiber for little millet (control) was 14.90 percent. while for soaked, it was 15.93 percent. For samples that were soaked and sprouted for 12, 24, 36 and 48 hr., total dietary fiber was found to be 17.47, 20.94, 25.67 and 29.23 percent respectively. The rise in TDF with increasing sprouting time could be due to modification of structure of cell wall polysaccharides of seeds. It involves extensive cell wall biosynthesis and therefore the production of new TDF.

10.2.3 Arabinoxylan content

Sharma and Gujral (2019a) reported that the TAX (total arabinoxylan content) content of VGSLM flour was 1.21 while for wheat flour, it was 1.04 g/100g (dry basis). The total, soluble and insoluble arabinoxylan content of little millet was found to be 1.70, 0.29 and 1.41 percent, respectively while wheat flour displayed 2.11, 0.33 and 1.78 percent, respectively (Sharma and Gujral, 2019b). Millets contain higher amounts of arabinoxylan as compared to oats, wheat and rice (Renger and Steinhart, 2000).The water insoluble fraction was also found to be significantly higher than the soluble arabinoxylan fraction. Courtin, Gelders, and Delcour (2001) reported that the water-soluble fiber fraction contributes to the viscosity of dough and bread and, hence, has good impacts on dough and bread quality. The insoluble fraction of arabinoxylan, on the other hand, is deleterious to bread manufacturing due to its high water-binding capacity.

10.2.4 Total phenolic content

Phenolic compounds act as antioxidants, and these compounds also have anti-glycemic, anti-mutagenic, antioxidative properties. They are being used in healthy food formulations and pharmaceutical products. Pradeep and Guha (2011) reported that the main phenolic fraction in little millet is ferulic acid, followed by gallic and gentisic acid. Chandrasekar and Shahidi (2011b) reported that soluble and insoluble extracts of little millet contained hydroxybenzoic acids 38 and 119, hydroxycinnamic acids 173 and 1242 and flavonoids 49 and 0 (µl/g of defatted meal), respectively. Also, they reported that the major phenolic acids in little millet are p-coumaric (1085.2 µl/g of defatted meal), followed by trans-ferulic (355.3), vanillic (162.4), sinapic (55.4), protocatechnic acid (48.8), p-hydroxybenzoic (32.6) and syringic (23.4).

Pradeep and Sreerama (2017) studied the soluble and bound phenolics of foxtail and little millet and their milled fractions (dehulled, hull, bran and pearled) for their phenolic content, antioxidant properties and inhibition of alpha amylase and alpha glucosidase activities. They reported that hull and bran fractions possess higher phenolic content. More than 83 percent of the total phenolic content in soluble and bound extracts was made up of hydroxycinnamic acid derivatives. Caffeic, ferulic and sinapic acids were the major phenolic acids in soluble fractions whereas ferulic and p-coumaric acids were predominant in the bound fractions. They suggested the use of these millets in diabetic diets as functional food ingredient. Pradeep and Sreerama (2018) studied the inhibitory effects of phenolic antioxidants of foxtail and little millet on α-amylase and α-glucosidase. They reported that little millet showed superior inhibition of alpha-amylase and alpha-glucosidase. Little millet contains higher levels of phenolics and the soluble and bound phenolics of millets as functional food ingredients for regulating postprandial hyperglycemia. Sharma and Gujral (2019b) reported that the total extractable phenols (TEP) of little millet and wheat flour was 4446 and 1772 µg ferulic acid equivalent/g sample. The higher phenolics in little millet could be due to its darker seed coat and pigmented testa (highest a* value), whereas grains with light or yellow testa, such as wheat, had lower phenolics (Dykes and Rooney, 2006). Srinivasan et al. (2021) studied the immunomodulating effect of phenolic acid bound arabinoxylans extracted from little and kodo millet. They reported that the phenolic acid bound arabinoxylans acted as potent immunomodulators by down regulation of ERK and NF-KB signaling pathways.

10.2.5 Total flavonoid content

Sharma and Gujral (2019b) reported that the total flavonoid content (TFC) of little millet was 1189 µg catechin equivalent/g sample, while that of wheat flour was 869 µg catechin equivalent/g sample. As seen in TEP, varieties with dark/pigmented testa had higher TFC than those with white or yellow testa, as seen by a positive correlation of TFC with a* (r = 0.79, p <0.05) and a negative correlation with L* (r = 0.79, p<0.05) and b* (r = 0.42, p<0.05). Sripriya et al. (1996) found that grains with a darker color have greater flavonoid content. Millet grains can be used in a variety of functional meals due to their high flavonoid content, as different epidemiological studies have shown that flavonoid-rich diets can reduce the risk of a variety of chronic diseases. Furthermore, effects of germination, steaming, and roasting on the nutraceutical and antioxidant properties of little millet were also investigated. The results showed that the total phenolic, flavonoid, and tannin contents of processed little millet increased by 21.2, 25.5, and 18.9 mg/100g, respectively, compared to the native sample (Pradeep and Guha, 2011). Little millet has been reported to contain total carotenoid content in the range of 51–104 µg/100g, tocopherol content in the range of 1.3mg/100g (Asharani et al., 2010). The GABA content in little millets as affected by germination time was reported by (Pradeep et al., 2011). It increased from 38.2 to 84.6 nm/100 mg over a 24 hr. germination period and upon further germination to 96 hr. it showed progressive decrease.

10.2.6 Antioxidant activity

Sharma and Gujral (2019b) investigated the antioxidant properties of little millet. They reported that little millet flour had an antioxidant activity of 47.21 whereas it was only 12.55 percent as compared to wheat flour. The antioxidant activity of little millet flour was significantly greater than that of wheat flour due to their higher phenolic content and darker seed coat. Antioxidant activity had a positive association with a* value (r = 0.87, p 0.05), showing that the darker seed coat has a significant impact in grain antioxidant capacity. Millets have stronger radical-scavenging activity than wheat and rice, according to Dykes and Rooney (2006) and Sripriya et al. (1996). The processing treatments also significantly increased the 2,2-Diphenyl-1-picrylhydrazyl (DPPH) radical scavenging activity and reducing power (Pradeep and Guha, 2011).

The little millet flour had a metal chelating activity (MCA) of 54.27 whereas wheat flour had MCA of 26.32 percent (Sharma and Gujral, 2019b). Millets had a stronger ferrous ion chelating activity than wheat and rice, according to Chandrasekara and Shahidi (2010). Because they form complexes with metal ions, proanthocyanidins/tannins are also known to be ferrous ion chelators, as evidenced by a positive correlation ($r = 0.68$, $p < 0.05$) of MCA with tannin concentration. While reducing power (RP) of little millet and wheat flour was 42.43 and 30.90 µmol ascorbic acid equivalent/g sample, respectively. The significant RP of grains is linked to their dark brown pigmented testa, according to Chandrasekara and Shahidi (2012a; 2012b). The Ferulic acid content (FAC) in little millet flour and wheat flour was 0.99 absorbance units (AU) and 0.61AU, respectively. Millets contain hydroxycinnamic/ferulic acid, which has antioxidative, anti-inflammatory, and anticarcinogenic properties (Kumar et al., 2016).

10.2.7 In vitro starch digestibility

Sharma and Gujral (2019c) investigated the effect of little millet composite flour (wheat flour: little millet 3:1) chapatti on starch digestibility. They reported that rapidly digestible starch and glycaemic index of fresh chapatti prepared from little millet composite flour was 29.58 and 71.37 percent, respectively while for wheat flour chapatti, it was 33.24 and 75.85 percent respectively.

The effect of vital gluten incorporation in millet chapatti on starch digestibility was studied by Sharma and Gujral (2019a). They reported that the RDS (rapidly digestible starch) content of wheat flour chapatti (fresh) was 33.24 while for VGSLM chapatti, it was 29.46 g/100g. Millets had more fiber components (DF and AX) than other grains, which could have resulted in less enzyme hydrolysis of starch. After 48 hours of storage, RDS content in wheat flour chapatti decreased by 9.45 percent, whereas RDS content in VGSLM chapatti decreased by 4.17 percent. Furthermore, the addition of VG to millet flour may limit starch hydrolysis by enzymes by forming a layer on the starch matrix that acts as a barrier to enzymatic attack. On storage, the RDS content decreased due to formation of the crystalline structure of starch that are less sensitive to hydrolysis. The SDS (slowly digestible starch) content was found to be 31.53 g/100g for wheat flour chapatti (fresh) whereas for VGSLM chapatti, it was 32.28g/ 100g. The increased SDS content in VGSLM chapatti is attributed to the presence of exogenous vital gluten and fiber components that could have reduced starch hydrolysis because of their shielding abilities. The percent increase in SDS content was 8.71 percent for wheat flour chapatti whereas it was 2.41 percent for VGSLM chapatti after 48 hr. of storage. According to Chung et al. (2006) on retrogradation the SDS content increases. Frie et al. (2003) reported that retrogradation formed crystalline structures such as double helices of amylose and/or amylopectin crystallites, which increased starch resistance to enzyme hydrolysis. RS (resistant starch) content for wheat flour chapatti (fresh) was 0.26 g/100g while it was found to be 1.09 g/100g for VGSLM chapatti, which was significantly ($p < 0.05$) higher. RS content of chapatti increased after 48 hr. of storage. The starch granules became less sensitive to enzyme hydrolysis during storage, resulting in lower RDS levels but higher SDS and RS levels.

Sharma and Gujral (2019b) reported that the RDS (rapidly digestible starch) content was 24.16 in wheat flour while for little millet, it was 13.18 g/100g. They reported that among different millets, RDS content of little millet was lowest, which could be attributable to greater phenolic content, which inhibits enzymatic hydrolysis and reduces starch digestibility. Moza and Gujral (2016) found that in the presence of fibers enzyme hydrolysis of starch is reduced and absorption is delayed. The SDS (slowly digestible starch) content was found to be 20.14 for wheat flour whereas for little millet flour, it was 22.28g/100g. Kim and White (2012) reported that the phenolic components are adsorbed on the surface of starch, inhibiting starch hydrolysis. Antinutritional factors such as tannin and phytic acid also contribute to the reduction of starch breakdown by digestive enzymes. The RS (resistant starch) content of little millet was 29.93 g/ 100g while, for wheat flour, it was found to be 19.27 g/100g. The reduced hydrolysis of millet starch, which is highly influenced by the presence of phenolics, tannins, phytic acid, and fiber components, could explain the increase in RS content of little millet.

Kumar et al. (2020) subjected the proso and little millet in two different methods of cooking (open pan and microwave) and compared the physicochemical and functional properties. They reported the nutritional analysis of the millet after the cooking treatments. The starch digestibility was 41.96 percent in raw little millet and it decreased to 29.71 and 30.09 percent in pan cooked and microwaved cooked little millet, respectively. These underutilized millets could be used by the food industry to formulate new products.

Sharma and Gujral (2020) reported that the total starch was 65.39 for control and 64.15 percent for soaked little millet. For samples that were sprouted for 12, 24, 36 and 48 hr., the total starch content was found to be 61.93, 59.66, 58.89 and 58.44 percent respectively. The decrease in starch content is owned to the hydrolytic activity of amylase enzyme degrading the starch polymer chains.

The impact of sprouting on starch digestibility was reported by Sharma and Gujral (2020)who found that slowly digestible starch (SDS) was measured for control, soaked and sprouted samples on percentage dry basis. For control sample, SDS value was 22.28 percent and for soaked sample, it was 21.71 percent. For samples which were soaked and sprouted for 12, 24, 36 and 48 hr., SDS values were 18.73, 17.82, 14.78 and 14.39 percent, respectively. For control sample, rapidly digestible starch (RDS) was 13.18 percent and for soaked sample, it was 14.77 percent. For samples which were soaked and sprouted for 12, 24, 36 and 48 hr., RDS values were 18.21, 20.05, 24.22 and 25.88 percent respectively. For control sample, resistant starch (RS) was 29.93 percent and for soaked sample, it was found to be 27.67 percent. For samples which were soaked and sprouted for 12, 24, 36 and 48 hr., RS values were 24.99, 21.79, 19.89 and 18.17 percent respectively. Soaking and subsequently sprouting gradually lowered RS and SDS while RDS improved for the millet flour. This may be attributed to the sprouting process which induces both metabolic and structural changes and thus residual components after sprouting show increased susceptibility to enzymatic attack.

10.2.8 Predicted glycemic index (pGI)

Subhulakshmi and Melathi, (2017) studied the hypoglycemic effect of multi millet cookies. Cookies were prepared by blending wheat flour, kodo, little, foxtail and finger millet, each at 20 percent level and results showed that the cookies were highly acceptable and led to reduction in the blood glucose levels. Sharma and Gujral (2019b) studied the Glycaemic Index (GI) of little millet flour and they reported that the little millet GI was 44.27 while, for wheat flour, it was 62.59. Foods with a higher glycaemic index release energy quickly and tend to boost blood glucose levels more quickly, which is undesirable for diabetics. Sharma and Gujral (2019a) reported the Glycaemic index (GI),which was estimated from area under curve (AUC) of digested starch for the fresh flat breads (chapatti) and retrograded flat breads prepared from wheat flour and little millet. The GI of wheat flour flat bread (fresh) was 75.85 while it was found to be 72.22 for VGSLM flat bread, which was significantly ($p < 0.05$) lower. The GI of wheat flour flat bread (retrograded) decreased and it was 73.13 while for VGSLM flat bread (retrograded), it was found to be 66.96 after 48 hr. of storage. The RDS was lower in VGSLM flat breads, which could have resulted in less hydrolysis and, as a result, a lower blood glucose level, resulting in a lower GI.

Geetha et al. (2020) studied the effect of a millet-based food mix on diabetic subjects. The food mix consisted of finger millet, little millet, defatted soya flour, whole green gram, fenugreek seeds, flax seeds, curry leaves, bitter gourd and skim milk powder. The glycemic index was found to be 37, 48 and 53 for dosa, mudde and roti, respectively. Sharma and Gujral (2020) reported predicted glycemic Index (pGI) of 44.27 and 46.27 for control and soaked samples of little millet, respectively. For samples which were soaked and sprouted for 12, 24, 36 and 48 hr., pGI values were 48.12, 50.60, 53.04 and 56.05, respectively. The reason for improved GI during sprouting was the result of enhanced enzyme activity that increased the hydrolysis and digestibility of starch. Increase in sprouting time brings about release of more glucose and subsequently increased postprandial glycemic response.

10.2.9 In vitro protein digestibility (IVPD)

Sharma and Gujral (2019b) reported that the IVPD of little millet flour was as low as 72.88 percent, whereas, for wheat flour it was 80.84 percent. This showed that little millet has lower protein digestibility due to presence of higher levels of anti-nutritional factors such as phytic acid and tannin content, which bind and precipitate proteins. Also, fiber components attributed in lowering the protein digestibility (Duodu, Taylor, Belton, and Hamaker, 2003). Sharma and Gujral (2020) investigated the effect of sprouting on in vitro protein digestibility (IVPD) of control, soaked and sprouted little millet. It was observed to be 72.88 for control and 74.19 percent for soaked, while for little millet which were soaked and sprouted for 12, 24, 36 and 48 hr., IVPD values were 76.20, 77.84, 80.96 and 81.75 percent respectively. The reason for the increase upon sprouting is due to the enhancement of intrinsic proteases/proteolytic activity, weaker association with protein and starch, removal of protease inhibitors, improvement in solubility of proteins and hydrolysis of storage proteins that act as a source of amino acids, nitrogen, and carbon for biomolecules synthesis.

10.2.10 Anti-nutritional factors

The phytic acid has ability to bind various divalent and trivalent metal ions such as iron, calcium and zinc resulting in poor mineral bioavailability (Sharma et al., 2017). Sharma and Gujral (2019b) studied the phytic acid content and tannin content of little millet and wheat flour (**Figure 10.1**). Graf and Eaton (1993) reported that high amounts of phytic acid may have

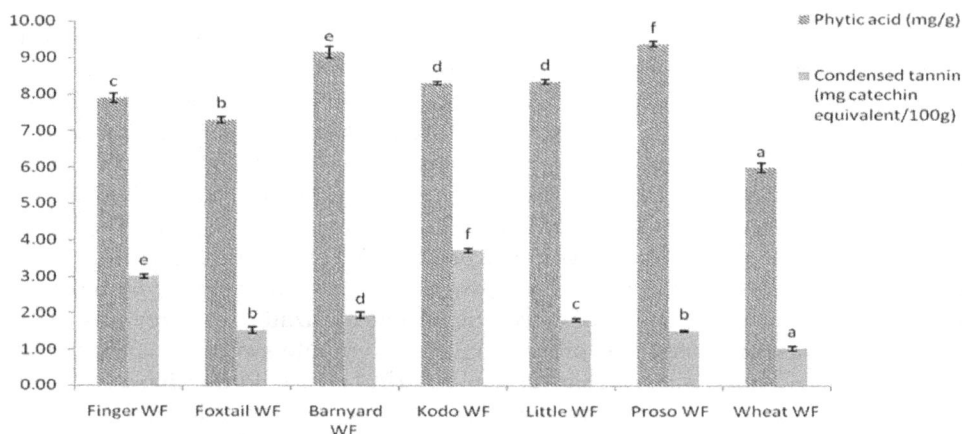

Figure 10.1 *Anti-nutritional factors in various millet flours (Sharma and Gujral 2019b).*

anti-carcinogenic effects by inhibiting the formation of hydroxyl ions in the colon and forming a complex with iron. Dykes and Rooney (2006) claimed that grains with colored testa have higher tannin content than their white counterparts.

The impact of sprouting on phytic acid and condensed tannin were investigated by Sharma and Gujral (2020) and Phytic acid (PA) was measured for control, soaked and sprouted little millet. The PA value was 8.34 and 8.10 mg/g for control and soaked sample respectively. For samples which were soaked and sprouted for 12, 24, 36 and 48 hr., the PA values were 7.53, 5.79, 4.96, 4.56 mg/g respectively. The Reduction in PA content is attributed to increased synthesis of phytase enzyme in sprouted grain.

Condense Tannin (CT) of the little millet control sample was 1.80 percent and for the soaked sample, it was found to be 1.61 percent. For samples which were soaked and sprouted for 12, 24, 36 and 48 hr., the CT values were 1.37, 1.08, 0.83 and 0.60 percent respectively. The decline in CT is due to the formation of hydrophobic association of CT content with seed proteins and enzyme.

10.3 Techno-functional properties and effects of processing

10.3.1 Rheological behavior using Mixolab

Sharma and Gujral (2019a) investigated the effect of exogenous vital gluten on little millet flour using the Chopin Mixolab, thermo-mechanical characteristics showed an improvement in dough formation time, stability, protein weakening, gelatinization temperature, peak viscosity, starch gelatinization rate and retrogradation compared to wheat flour. They reported that in dough development phase, the torque (C1) of wheat flour was 0.981 Nm while VGSLM showed torque of 0.672 Nm. The exogenous VG (vital gluten) is loosely connected, whereas the native gluten protein in wheat flour acted differently since they are tightly bound through covalent bonds (Ortolan and Steel, 2017). The other reason for the decreased consistency of VGSLM could be that little millet had higher amounts of TDF and TAX, which have the ability to bind water molecules, leaving less water accessible for the growth of the gluten network, resulting in lower dough consistency (Moza and Gujral, 2016). The DDT (dough development time) for wheat flour was 0.92 min whereas VGSLM showed higher DDT of 1.26 min. The exogenous VG added to the little millet flour takes longer to hydrate or interact with the intrinsic flour proteins, needing more time to create a dough network than native gluten protein in control wheat flour. Czuchajowska and Paszczynska (1996) reported a similar trend upon introduction of exogenous VG increasing the dough mixing time of bread dough. Increase in DDT is due to the presence of higher amount of TDF and TAX content in little millet as these fiber components interfere in holding the viscoelastic network and give maximum resistance to extension at C1 torque. In protein weakening phase (C2), it was observed that wheat flour displayed lower protein weakening (38.43%) as compared to VGSLM flour (69.37%). In gelatinization phase (C3), the wheat flour showed higher peak viscosity that is, 2.14 Nm while VGSLM flour displayed a torque of 1.45 Nm. The lower peak viscosity of little millet flour may be due to the structure of little millet starch and the hydrophobic protein (kafirins) in millets encircles the starch molecule and prevents it from swelling and gelatinization, as previous research has shown the significance of proteins in preventing starch granule swelling (Hamid et al., 2015) and protein–amylose complex formation (Pomeranz, 1991). In holding phase (C4), the breakdown was 6.43 percent in wheat flour while VGSLM flour displayed higher breakdown of 18.57 percent. This demonstrated a significantly ($p < 0.05$) higher breakdown/shear thinning and thus reduced stability of millet hot starch paste, which could be related to molecular interference produced by exogenous VG, resulting in less resistance to shear thinning of the hot starch paste. In retrogradation phase (C5), the percentage increase in torque from C4 to C5 indicates retrogradation and it was found to be 30.53 percent in wheat flour while VGSLM flour displayed a lower retrogradation percent of 23.43 percent. Lower retrogradation in VGSLM flour could be attributed to starch dilution, as well as hydrogen bonding interactions between starch and exogenous VG, which could inhibit retrogradation. Curti et al. (2014) reported that adding VG to bread reduces starch recrystallization by altering water molecule rearrangement during storage.

Sharma and Gujral (2019b) studied the thermo-mechanical properties of little millet and wheat flour at same hydration of 85 percent and they reported a torque (C1) of 0.13 Nm for little millet and 0.81Nm for wheat flour. The higher torque of wheat flour is due to presence of gluten proteins while little millet is gluten free and displayed the lower C1 value. Torque displayed during protein weakening (C2 in Nm) for little millet was 0.02 Nm which was significantly lower as compared to wheat flour (0.32 Nm). The protein weakening (%) of little millet flour was 45.39 percent and 37.75 percent for wheat flour. The presence of non-starchy polysaccharides and fiber components such as dietary fiber and arabinoxylan in millets may have caused more disruption of the protein network. Peak viscosity (C3 in Nm) was 0.85 for little millet flour and 1.36 Nm for wheat flour. According to Singh and Adedeji (2017), the lower peak viscosity is due to the formation of amylose-lipid complexes which reduce the level of hydration and swelling of starch and inhibit the leaching of amylose. Breakdown (%) that is, disintegration of hot starch paste for little millet was 1.12 percent and higher breakdown was observed for wheat flour (35.91%). This is of significance in foods that are subjected to thermal processing and mechanical shearing. The little millet flour showed significantly lower retrogradation of 30.57 percent as compared to 34.80 percent for wheat flour. Wu et al. (2009) reported that the lower retrogradation observed in millet flour could be related to the type and behavior of starch, as well as the higher phenolic content, which contributes to the lower retrogradation.

Sharma and Gujral (2019c) studied the thermo-mechanical properties of little millet composite flour. The water absorption of the little millet composite flour was 48.52 percent while in the wheat flour, it was observed to be 57.25 percent. This reduction in water absorption in little millet composite flour is due to the dilution of gluten proteins as it lowers the capacity to absorb water. The protein weakening of wheat flour was 43.03 percent while wheat/little composite flour displayed 61.71 percent, this higher weakening is due to its high dietary fiber content. The peak viscosity of wheat flour was found to be 2.02 Nm while in little millet composite flour, it was 2.12 Nm. The starch breakdown percent of wheat flour was 7.36 percent while in wheat/little composite flour, it decreased to 1.32 percent.

The effect of sprouting on thermo-mechanical properties (**Figure 10.2**) was investigated by Sharma and Gujral (2020). They reported that the torque during initial mixing of the dough (C1 in Nm) for control sample was 0.13 and for soaked sample, it was 0.094. For samples which were soaked and sprouted for 12, 24, 36 and 48 hr., C1 values were 0.072, 0.066, 0.57 and 0.053 Nm, respectively. Upon sprouting the increased enzymatic activity and/or the generation of hydrophilic components absorbs large amount of water producing lower C1 torque. Torque required during protein weakening (C2 in Nm) for control sample was 0.02 and for soaked sample, it was 0.019. For samples which were soaked and sprouted for 12, 24, 36 and 48 hr., C1 values were 0.017, 0.015, 0.014 and 0.013 Nm respectively. This could be explained by the degradation of protein as a result of hydrolysis during sprouting process. Peak viscosity (C3 in Nm) for the control sample was 0.847 and for the soaked sample, it was 0.743. For samples which were soaked and sprouted for 12, 24, 36 and 48 hr., the C3 values were 0.651, 0.526,

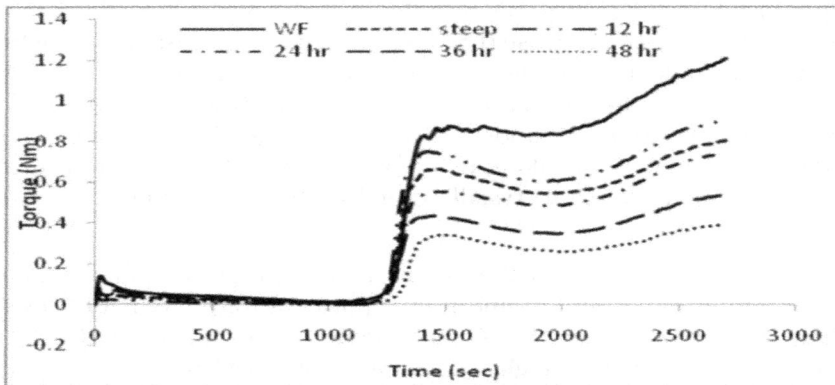

Figure 10.2 *The effect of sprouting on thermo-mechanical properties (Sharma and Gujral 2020).*

0.422 and 0.337 Nm respectively. This decrease in peak viscosity is mainly due to the starch degradation or debranching to smaller units due to amylase action. Proteases also break disulfide linkages and hydrolyze proteins, and thus result in lower viscosity. Breakdown (%), that is, disintegration of hot starch paste for control sample was 1.12 and for the soaked sample, it was 15.43. For samples which were soaked and sprouted for 12, 24, 36 and 48 hr., breakdown values were 16.00, 16.14, 19.14 and 22.17 percent respectively. Final viscosity (C5 in Nm) for control sample was 1.206 and for soaked sample, it was 0.808. For samples which were soaked and sprouted for 12, 24, 36 and 48 hr., C5 values were 0.904, .750, 0.542 and 0.391 Nm respectively. This decrease in C5 is due to lower tendency of aggregation in the degraded starch molecules upon cooling.

10.3.2 Pasting properties of little millet

Figure 10.3 describes the pasting behavior of little millet flour as compared to wheat flour studied using a rapid visco analyzer (RVA) standard protocol (unpublished work). It was observed that the pasting pattern of little millet flour was very different from other flours like wheat and rice flour (Kapoor, 2013). While other flour samples have been reported to show sharp peaks and troughs in their pasting pattern, this was not the case with little millet flour. **Figure 10.3** shows the pasting pattern of little millet flour and wheat flour. The peak viscosity of little millet was 2035 cP while for wheat flour, it was found to be 1652 cP. The higher peak viscosity of little millet indicated that it has higher water binding capacity, causing the starch granules to swell more and also higher fiber content. The final viscosity of little millet was 4819 cP whereas for wheat flour, it was 2647 cP. The breakdown viscosity of little millet was 565 cP whereas for wheat flour, it was 479 cP. The lower breakdown of wheat flour shows the stability of the starch during heating and shearing (Nazni and Shobana, 2016). The setback viscosity of little millet was 3253 cP whereas for wheat flour, it was 1437 cP. Breakdown is a complex phenomenon involving the type and quantity of starch, the temperature gradient, the shear force, and the mixture's composition, such as the presence of enzymes, lipids and proteins, all affect the rate and extent of swelling and breakdown (Debet and Gidley, 2006). The little millet also shows a second breakdown phase during the holding phase at 50° C indicating that it is susceptible to shear thinning both at high and low temperature where this phenomenon is not visible in the pasting behavior of wheat flour only at higher temperature (90°C).

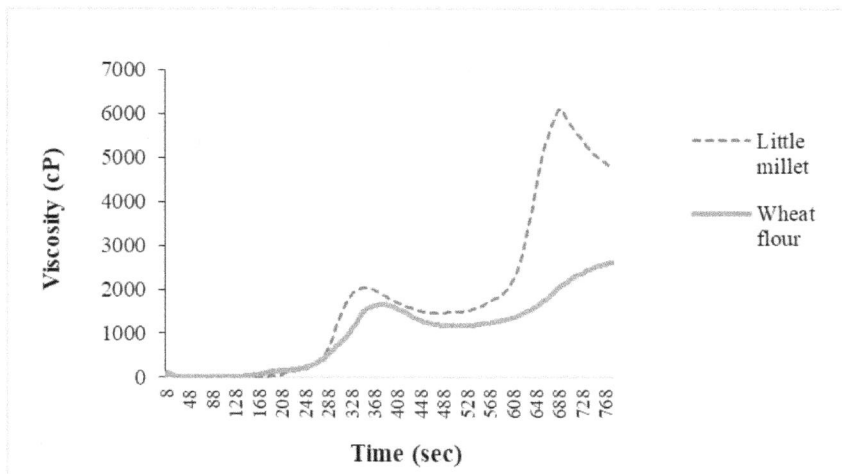

Figure 10.3 *Pasting behavior of little millet flour as compared to wheat flour (Unpublished author's work).*

10.3.3 Flat bread (chapatti) making characteristics

The percent shrinkage in the flat bread (chapatti) prepared from wheat flour was 1.40 percent while for VGSLM flat bread, it was 6.06 percent. This greater shrinkage is attributed to the changes in starch and protein composition in the flours. Also, unlike the native gluten and starch in wheat flour, the exogenous VG added to the little millet flour did not bind to the starch matrix. Baking loss was 18.23 percent in wheat flour while higher baking loss was displayed (23.67%) in VGSLM flat bread. In little millet flat breads, a longer baking time could have resulted in more baking loss (Sharma and Gujral, 2019a). Sharma and Gujral (2019c) reported that the flat breads prepared from wheat/little millet composite flour displayed 1.67 percent shrinkage and 20.06 percent baking loss, while for wheat flour, the shrinkage was 1.40 percent and baking loss was 18.23 percent. The puffing index of wheat flour flat bread was 98.4 percent while the puffing index reduced to 76.60 percent of flat bread prepared from wheat/little millet composite flour.

10.3.4 Color Profile

Sharma and Gujral (2019b) reported the color properties of little millet flour and wheat flour using the Hunter color lab. Little millet flour had a L* value of 69.90 whereas it was 86.35 for wheat flour. The a* was 3.24 for little millet flour and 1.63 for wheat flour. b* was 13.99 for little millet flour and 11.15 for wheat flour. The higher a* and b* values indicated that little millet flour was higher in redness and yellowness as compared to wheat flour.

10.3.5 Sensory acceptability of little millet-based products

Neeharika et al. (2020) reported the organoleptic properties of ready to reconstitute little millet smoothie with fruit juices. The little millet was subjected to washing, germination, drying, roasting, dehulling, pressure cooking, drying, milling into flour to prepare malted and pregelatinized little millet flour. It was mixed with milk and sugar powder in the ratio 45:45:10 to yield a little millet smoothie mix. The mix was reconstituted with fruit juice and had high organoleptic acceptability and potential to commercialize. Khwairakpam et al. (2020) prepared pasta by blending fermented little millet flour with durum wheat semolina at various levels. Fermented little millet flour at 10 and 15 percent showed high acceptability. Saini et al. (2018) developed product based on rice, corn and little millet blends. They suggested that little millet, rice and maize in the ratio 10:45:45 extruded at a feed moisture of 22 percent and at a temperature of 110°C produced extrudates with acceptable sensory properties. Sharma and Gujral (2019c) reported the overall consumer acceptability of the flat bread (chapatti) prepared by incorporating little millet flour with wheat flour in the ratio of 1:3. The overall acceptability of wheat/ little millet composite flour (WLCF) flat bread was lower (6.25) as compared to wheat flour (8.75) flat bread, which was attributed to its coarse nature, flavor and unpleasant mouthfeel.

10.4 Conclusion

Little millet can be considered as a nutraceutical grain due to its rich dietary fiber, mineral, vitamin, macro and micronutrient, and phytochemical content and also because it offers more substantial health advantages and can be used to treat chronic diseases. Minimal processing techniques such as steeping, sprouting and cooking can assist to address the issue of digestibility and antinutrients. More studies need to be carried out on the effects of fermentation, extrusion and roasting to improve digestibility and lower phytic acid and tannins. Finger millet-based products need to be formulated in such a way so as to make them more palatable and organoleptically acceptable without sacrificing the health benefits. Little millet wheat composite flours can find application in a number of food products with increased utilization and improved nutritional quality. Additionally, awareness must be raised about the benefits of little millets' nutritional and nutraceutical advantages at both the commercial and household levels in order to address food insecurity and malnutrition.

References

Abdelrahman, A., Hoseney, R. C., and Varriano-Marston, E. (1983). Milling process to produce low-fat grits from pearl millet. *Cereal Chemistry*, 60(3): 189–191.

Afify, M. R., El-Beltagi, H. S., Abd.El-Salam, S. M., Omran, A. A. (2012). Protein solubility, digestibility and fractionation after germination of sorghum varieties. *PLoS One* 7: 31154.

Annor, G. A., Tyl, C., Marcone, M., Ragaee, S. and Marti, A. (2017). Why do millets have slower starch and protein digestibility than other cereals? *Trends in Food Science & Technology*, 66: 73–83.

Asharani, V. T., Jayadeep, A., and Malleshi, N. G. (2010). Natural antioxidants in edible flours of selected small millets. *International Journal of Food Properties*, 13(1): 41–50.

Behera, M. K. (2017). Assessment of the state of millets farming in India. *MOJ Ecology & Environmental Science*, 2:16–20.

Bravo, L. (1998). Polyphenols: chemistry, dietary sources, metabolism, and nutritional significance. *Nutrition Reviews*, 56(11): 317–333.

Chandel, G., Kumar, M., Dubey, M. and Kumar, M. (2014). Nutritional properties of minor millets: neglected cereals with potentials to combat malnutrition. *Current* Science, 107:1109–1111.

Chandrasekara, A. and Shahidi, F. (2010). Content of insoluble bound phenolics in millets and their contribution to antioxidant capacity. *Journal of Agricultural and Food Chemistry*, 58:6706–6714.

Chandrasekara, A., and Shahidi, F. (2011a). Determination of antioxidant activity in free and hydrolyzed fractions of millet grains and characterization of their phenolic profiles by HPLC-DAD-ESI-MSn. *Journal of Functional Foods*, 3(3), 144–158.

Chandrasekara, A., and Shahidi, F. (2011b). Antiproliferative potential and DNA scission inhibitory activity of phenolics from whole millet grains. *Journal of Functional Foods*, 3: 159–170.

Chandrasekara, A., and Shahidi, F. (2012a). Antioxidant phenolics of millet control lipid peroxidation in human LDL cholesterol and food systems. *Journal of the American Oil Chemists' Society*, 89(2), 275–285.

Chandrasekara, A., and Shahidi, F. (2012b). Bio-accessibility and antioxidant potential of millet grain phenolics as affected by simulated in vitro digestion and microbial fermentation. *Journal of Functional Foods*, 4, 226–237.

Chandrasekara, A., Naczk, M., and Shahidi, F. (2012). Effect of processing on the antioxidant activity of millet grains. *Food Chemistry*, 133(1), 1–9.

Chethan, S., Dharmesh, S. M., and Malleshi, N. G. (2008). Inhibition of aldose reductase from cataracted eye lenses by finger millet (Eleusinecoracana) polyphenols. *Bioorganic & Medicinal Chemistry*, 16(23): 10085–10090.

Chopra, K. and Neelam, M. (2004). Common health problems encountered by the tribal community in Bastar district. *Health and Population–Perspectives and* Issues, 27: 40–48.

Chung, H. J., Lim, H. S., and Lim, S. T. (2006). Effect of partial gelatinization and retrogradation on the enzymatic digestion of waxy rice starch. *Journal of Cereal Science*, 43(3): 353–359.

Courtin, C. M., Gelders, G. G., and Delcour, J. A. (2001). Use of two endoxylanases with different substrate selectivity for understanding arabinoxylan functionality in wheat flour breadmaking. *Cereal Chemistry*, 78(5): 564–571.

Curti, E., Carini, E., Tribuzio, G., and Vittadini, E. (2014). Bread staling: effect of gluten on physico-chemical properties and molecular mobility. *LWT-Food Science and Technology*, 59(1): 418–425.

Czuchajowska, Z., and Paszczynska, B. (1996). Is wet gluten good for baking?*Cereal Chemistry*, 73(4): 483–489.

Danisova, C., Holotnakova, E., Hozova, B., and Buchtova, V. (1995). Effect of germination on a range of nutrients of selected grains and legumes. *ActaAlimentaria*, 3: 287–298.

Debet, M. R., and Gidley, M. J. (2006). Three classes of starch granule swelling: Influence of surface proteins and lipids. *Carbohydrate Polymers*, 64(3): 452–465.

Duodu, K. G., Taylor, J. R. N., Belton, P. S., and Hamaker, B. R. (2003). Factors affecting sorghum protein digestibility. *Journal of Cereal Science*, 38(2): 117–131.

Dykes, L., and Rooney, L. W. (2006). Sorghum and millet phenols and antioxidants. *Journal of Cereal Science*, 44(3): 236–251.

Frei, M., Siddhuraju, P., and Becker, K. (2003). Studies on the in vitro starch digestibility and the glycemic index of six different indigenous rice cultivars from the Philippines. *Food Chemistry*, 83(3): 395–402.

Geetha, K., Yankanchi, G. M., Hulamani, S., and Hiremath, N. (2020). Glycemic index of millet-based food mix and its effect on pre diabetic subjects. *Journal of Food Science and Technology*, 57(7): 2732–2738.

Graf, E., and Eaton, J. W. (1993). Suppression of colonic cancer by dietary phytic acid. *Nutrition and Cancer*, 19(1): 11–19.

Guha, M., Sreerama, Y. N., and Malleshi, N. G. (2015). Influence of Processing on Nutraceuticals of Little Millet (Panicumsumatrense). In *Processing and Impact on Active Components in Food* (pp. 353–360). Academic Press.

Hadimani, N. A. and Malleshi, N. G. (1993). Studies on milling, physico-chemical properties, nutrient composition and dietary fiber content of millets. *Journal of Food Science & Technology* 30: 17–20.

Hamid, S., Muzzafar, S., Wani, I. A., and Masoodi, F. A. (2015). Physicochemical and functional properties of two cowpea cultivars grown in temperate Indian climate. *Cogent Food & Agriculture*, 1(1): 1099418.

Kapoor, P. (2013). Nutritional and functional properties of popped little millet (*Panicumsumatrense*), Montreal: McGill University.

Kaur, P., Purewal, S. S., Sandhu, K. S., Kaur, M., and Salar, R. K. (2019). Millets: A cereal grain with potent antioxidants and health benefits. *Journal of Food Measurement and Characterization*, 13(1): 793–806.

Khwairakpam, M., Kumari, B. A., Suneetha, W. J., and Tejashree, M. (2020). Development and Evaluation of Fermented Little Millet Based Cold Extrudates. *International Research Journal of Pure and Applied Chemistry*, 7–14.

Kim, H. J., and White, P. J. (2012). In vitro digestion rate and estimated glycemic index of oat flours from typical and high β-glucan oat lines. *Journal of Agricultural and Food Chemistry*, 60(20): 5237–5242.

Kumar, A., Tomer, V., Kaur, A., Kumar, V. and Gupta, K. (2018). Millets: a solution to agrarian and nutritional challenges. *Agricultural & Food Security*, 7: 31–45.

Kumar, K. V. P., Dharmaraj, U., Sakhare, S. D., and Inamdar, A. A. (2016). Flour functionality and nutritional characteristics of different roller milled streams of foxtail millet (Setariatalica). *LWT*, 73: 274–279.

Kumar, S. R., Sadiq, M. B., and Anal, A. K. (2020). Comparative study of physicochemical and functional properties of pan and microwave cooked underutilized millets (proso and little). *LWT–Food Science and Technology*, 128: 109465.

Malleshi, N. G., and Hadimani, N. A. (1993). Nutritional and technological characteristics of small millets and preparation of value-added products from them. *Advances in Small Millets*, 270–287.

Manay, S., and Shadaksharaswamy, M. (2005). *Foods: Facts and Principles*. 2nd ed. (pp. 359–375) New Age International. New Delhi.

Moza, J., and Gujral, H. S. (2016). Starch digestibility and bioactivity of high altitude hulless barley. *Food Chemistry*, 194: 561–568.

NAAS (2013). Role of Millets in Nutritional Security of India. Policy Paper No. 66, *National Academy of Agricultural Sciences*, New Delhi, 16.

Nazni, P., and Shobana, D. R. (2016). Effect of processing on the characteristics changes in barnyard and foxtail millet. *Journal of Food Processing and Technology*, 7(3): 1–9.

Neeharika, B., Suneetha, W. J., Kumari, B. A., and Tejashree, M. (2020). Organoleptic Properties of Ready to Reconstitute Little Millet Smoothie with Fruit Juices. *International Journal of Environment and Climate Change*, 78–82.

Nithiyanantham, S., Kalaiselvi, P., Mahomoodally, M.F., Zengin, G., Abirami, A. and Srinivasan, G. (2019). Nutritional and functional roles of millets—A review. *Journal of Food Biochemistry*, 43: e12859

Ortolan, F., and Steel, C.J. (2017). Protein characteristics that affect the quality of vital wheat gluten to be used in baking: A review. *Comprehensive Reviews in Food Science and Food Safety*, 16(3): 369–381.

Pomeranz, Y. (1991). *Functional Properties of Food Components*, 2nd ed.Academic Press, New York.

Pradeep, P. M., and Sreerama, Y. N. (2018). Phenolic antioxidants of foxtail and little millet cultivars and their inhibitory effects on α-amylase and α-glucosidase activities. *Food Chemistry,* 247: 46–55.

Pradeep, P.M., and Sreerama, Y.N. (2017). Soluble and bound phenolics of two different millet genera and their milled fractions: Comparative evaluation of antioxidant properties and inhibitory effects on starch hydrolysing enzyme activities. *Journal of Functional Foods,* 35: 682–693.

Pradeep, S. R., and Guha, M. (2011). Effect of processing methods on the nutraceutical and antioxidant properties of little millet (*PanicumSumatrense*) extracts. *Food Chemistry*, 126(4): 1643–1647.

Pradeep, S.R., Malleshi, N.G., Guha, M., (2011). Germinated millets and legumes as a source of gamma-aminobutyric acid. *World Applied Sciences Journal*, 14: 108–113.

Pradhan, A., Nag, S. K. and Patil, S. K. (2010). Dietary management of finger millet controls diabetes. *Current Science,* 98: 763–765.

Rao, D. B., Bhaskarachary, K., Christina, A. G. D., Devi, S. G., Tonapi, V. A. (2017). Nutritional and health benefits of millets *ICAR-Indian Institute of Millet Research (IIMR)*, Balajiscan. 1–122.

Rao, V. V., Swamy, S. G., Raja, D. S., and Wesley, B. J. (2020). Engineering Properties of Certain Minor Millet Grains. *The Andhra Agricultural Journal*, 67: 89–92.

Renger, A., and Steinhart, H. (2000). Ferulic acid dehydrodimers as structural elements in cereal dietary fibre. *European Food Research and Technology*, 211(6): 422–428.

Sahoo, S., Sarangi, S. S., and Rayaguru, K. (2020). Effect of pretreatment on milling characteristics of little millets. *Journal of Pharmacognosy and Phytochemistry*, 9(5): 554–558.

Saini, R., and Yadav, K. C. (2018). Development and quality evaluation of little millet (*Panicumsumatrense*) based extruded product. *Journal of Pharmacognosy and Phytochemistry*, 7(3): 3457–3463.

Saleh, A. S. M., Zhang, Q., Chen, J. and Shen, Q. (2013). Millet grains: Nutritional quality, processing, and potential health benefits. *Comprehensive Reviews in Food Science and Food Safety,* 12: 281–195.

Shahidi, F., and Chandrasekara, A. (2013). Millet grain phenolics and their role in disease risk reduction and health promotion: A review. *Journal of Functional Foods*, 5(2): 570–581.

Sharma, B. and Gujral, H. S. (2019). Influence of nutritional and antinutritional components on dough rheology and in vitro protein and starch digestibility of minor millets. *Food Chemistry*, 299: 125115.

Sharma, B., and Gujral, H. S. (2019a). Characterization of thermo-mechanical behavior of dough and starch digestibility profile of minor millet flat breads. *Journal of Cereal Science*, 90: 102842.

Sharma, B., and Gujral, H. S. (2019b). Influence of nutritional and antinutritional components on dough rheology and in vitro protein and starch digestibility of minor millets. *Food Chemistry*, 299: 125115.

Sharma, B., and Gujral, H. S. (2019c). Modulation in quality attributes of dough and starch digestibility of unleavened flat bread on replacing wheat flour with different minor millet flours. *International Journal of Biological Macromolecules*, 141: 117–124.

Sharma, B., and Gujral, H. S. (2020). Modifying the dough mixing behavior, protein and starch digestibility and antinutritional profile of minor millets by sprouting. *International Journal of Biological Macromolecules*, 153: 962–970.

Sharma, S., Sharma, N., Handa, S., and Pathania, S. (2017). Evaluation of health potential of nutritionally enriched Kodo millet (*Eleusinecoracana*) grown in Himachal Pradesh, India. *Food Chemistry*, 214: 162–168.

Shobana, S., Sreerama, Y. N., and Malleshi, N. G. (2009). Composition and enzyme inhibitory properties of finger millet (Eleusinecoracana L.) seed coat phenolics: Mode of inhibition of α-glucosidase and pancreatic amylase. *Food Chemistry*, 115(4): 1268–1273.

Singh, M., and Adedeji, A. A. (2017). Characterization of hydrothermal and acid modified proso millet starch. *LWT–Food Science and Technology*, 79: 21–26.

Sinha, R. and Kawatra, A. (2003). Effect of processing on phytic acid and polyphenol contents of cowpeas [*Vignaunguiculata (L) Walp*]. *Plant Foods for Human Nutrition*, 58: 1–8.

Srichuwong, S., Curti, D., Austin, S., King, R., Lamothe, L. and Gloria-Hernandez, H. (2017). Physicochemical properties and starch digestibility of whole grain sorghums, millet, quinoa and amaranth flours, as affected by starch and non-starch constituents. *Food Chemistry*, 233: 1–10.

Srinivasan, A., Aruldhas, J., Perumal, S. S., and Ekambaram, S. P. (2021). Phenolic acid bound arabinoxylans extracted from Little and Kodo millets modulate immune system mediators and pathways in RAW 264.7 cells. *Journal of Food Biochemistry*, 45(1): e13563.

Sripriya, G., Chandrasekharan, K., Murty, V. S., and Chandra, T. S. (1996). ESR spectroscopic studies on free radical quenching action of finger millet (Eleusinecoracana). *Food Chemistry*, 57(4): 537–540.

Subbulakshmi, B., and Malathi, D. (2017). Formulation of multi millet cookies and evaluate its hypoglycaemic effect in albino rats. *Journal of Crop and Weed*, 13(3) 112–116.

Tonapi, V. A., Bhat, V. B., Kannababu, N., Elangovan, M., Kulakarni, A. V. U. R. Tonapi, K. V., Rao, K. V. R. and Rao, T. G. N. (2015). Millet Seed Technology: Seed Production, Quality Control and Legal compliance, *Indian Institute of Millets Research*, Rajendranagar Hyderabad, Telangana, India.

Wu, Y., Chen, Z., Li, X., and Li, M. (2009). Effect of tea polyphenols on the retrogradation of rice starch. *Food Research International*, 42(2): 221–225.

Zhu, Y., Chu, J., Lu, Z., Lv, F., Bie, X., Zhang, C., and Zhao, H. (2018). Physicochemical and functional properties of dietary fibre from foxtail millet (Setaria italic) bran. *Journal of Cereal Science*, 79: 456–461.

Polyphenolic Composition, Nutraceutical Profile, and Techno-Functional Properties of Black Wheat

Arunima S. H., Gopika S. Kumar, Jyoti Singh, Sawinder Kaur,
Prasad Rasane, and Vikas Nanda

Contents

11.1 Introduction

Wheat is a highly consumed crop that provides protein, dietary fiber, B group vitamins and minerals for the human diet (Shewry,P. R. 2018).According to the Food and Agriculture Association (FAO) statistics (2020), world wheat production and utilization in 2019–2020 is estimated to be 761.9 and 750.5 respectively. For 2020–2021 the production and consumption of wheat is forecasted to reach 762.7and 758.0 million tonnes respectively (FAO 2021). Wheat has a carbohydrate content of 78.10 percent, a protein content of 14.70 percent, a fat content of 2.10 percent, and a mineral and vitamin content of 2.10 percent, including thiamine and B group vitamins (Kumar et al. 2011;Tian et al. 2018b). The phenolic acid in whole wheat has antioxidant, anti-carcinogenic properties, anti-inflammatory and also has the potential to modulate critical enzymatic functions in cells (Robbins 2003). Wheat also contains a high amount of soluble dietary fiber (SDF) and insoluble dietary fiber (IDF) (Adom et al. 2003). Insoluble dietary fiber aids in the prevention of colon cancer and other bowel illnesses by reducing intestinal transit time and increasing fecal volume. Soluble fibers reduce glucose absorption and also help to manage diseases such as diabetes and atherosclerosis.

Among wheat varieties the pigmented wheat grains are of great importance because of the presence of phytochemicals in them. Anthocyanin is responsible for the darker colors in wheat grains, such as purple, blue and black (Garg et al. 2016). The amount of anthocyanin in purple, blue and blackwheat is more than that of red and white colored wheat grains. The reddish color of the grain coat is due to catechins and proanthocyanidins that can be produced by the flavonoid biosynthetic pathway and are responsible for the bitter flavor, resistance to sprouting and a lower activity of hydrolytic enzymes (Himi et al. 2011). The colored wheat grains are useful in combating diseases like cancer, cardiovascular diseases, obesity, diabetes, inflammation and aging (Garg et al. 2016); they also possess a therapeutic effect against oxidative liver damage, hyperglycemia and capillary fragility (Ficco et al. 2014b). According to Abdel-Aal et al. (2018), the disease controlling properties of colored wheat are due to high antioxidant activity because of its high phenolic content.

In this chapter genetics and chemistry, health benefits, nutritional composition, antioxidant properties, food product formulation and thermal stability of colored wheat grains are discussed in detail.

11.2 Genetics and chemistry behind pigmented wheat varieties

The variation in color of grain and their product is related to their genetic factor, growing conditions and technological processes. In case blue wheat anthocyanin is present in aleurone layer and blue color of wheat is managed by *Ba1* and *Ba2*genes. According to Qualset et al. (2005)*Ba1* gene is from *Th. Ponticum*, which is responsible for the blue color of the aluerone, and Singh et al. (2007) reported that the *Ba2* is from the *T. boeoticum(syn.Triticummonococcum ssp. aegilopoides)* variety.

The purple colored grain is produced by transferring the purple pericarp (*Pp*) genes from the tetraploid wheat variety named *Triticumturgidum L. subsp. abyssinicumVavilov* into the normal wheat (Eticha et al. 2011). Here, the purple color of pericarp is due to the presence of anthocyanin in the pericarp. *Pp1* on the chromosome 7B and *Pp2*, the name of which was changed to *Pp3a* gene (Dobrovolskaya et al. 2006), on chromosome 7A are the genes present in the tetraploid material that is responsible for the purple color (Musilova et al. 2013).

The red color of the grain is due to one to three dominant alleles on the long arms of the three dominant *R-1* homologous genes located on the long arms of chromosomes 3A, 3B, and 3D. According to McIntosh et al. (1998),the alleles resulting in red color *are R-A1b, R- B1b* and *R-D1b*. A single locus expressing a dominant allele is enough to provide red color of the grain and the intensity of red color is based on the number of R alleles present.

Loci *Psy1* and *Psy2*arepresent in the chromosome 7 and chromosome 5, respectively, which imparts a yellow color to the endosperm (Pozniak et al. 2007). Howitt et al. (2009) reported that *Psy1-A1 (7AL), Psy1-B1 (7BL), Psy1-D1 (7DL),Psy2-A1 (5A)* and *Psy2-B1 (5B)* are studied the most among these two loci.

After a 20-year effort beginning in 1970, black wheat (BW) was initially developed in the lab of the Institute of Crop Genetics, Shanxi Academy of Agricultural Science. Black wheat is created by combining genes from previously existing purple and blue wheat varieties (Li et al. 2005; Li et al. 2006). "Black 76"was the moniker given to the first black wheat. Blue colored hexaploid wheat,"blue 1," was developed in 1970 by scientists (bred from common hexaploid wheat (Triticumaesticum) + Agropyronglaucum, which is linked to the blue wheat cultivar Leymusdasystachys). "Purple 12-1" (a purple-colored hexaploid wheat produced by mating "blue 1" + purple-grained tetraploid wheat) and "blue-purple 114" (a blue-purple colored hexaploid wheat bred by crossing "blue 1" + purple-grained tetraploid wheat) were developed after 20 years. After crossing "purple 12-1" as a male parent and "blue-purple 114" as a female parent, the black colored wheat "black 76" was created (Beta et al. 2019).

The chemistry behind colored wheat grains focuses on the flavonoid synthesis pathway that is responsible for the production of bioactive compounds in them.

Types of anthocyanins and proanthocyanidins responsible for the color and antioxi- dant properties of the pigmented wheat grains are discussed in **Table 11.1**. Garg et al. (2016) identified 26 anthocyanins from black and 23 anthocyanins from purple wheat

Table 11.1 Bioactive Compounds in Colored Wheat Grains

Wheat varieties	Pigment	Compounds	References
Black	Anthocyanin	Cyanidin-3-glucoside	Garg et al. (2016);
		Cyanidin-3-rutinoside	Bartl et al. (2015);
		Cyanidin-3-succinyl-glucoside	Ficco et al. (2014b);
		Delphinidin-3-glucoside	Tyland Bunzel (2012);
		Delphinidin-3- rutinoside	Knievel et al. (2009); Hosseinian et al. (2008);
		Malvidin-3-glucoside	Abdel-Aal and Hucl (2003)
		Malvidin-3-rutinoside	
		Pelargonidin-3-glucoside	
		Peonidin-3-glucoside	
		Peonidin-3-rutinoside	
		Petunidin-3-glucoside	
		Petunidin with hexose and rhamnose	
Blue	Anthocyanin	Cyanidin-3-glucoside	Garg et al. (2016);
		Cyanidin-3-galactoside	Bartl et al. (2015);
		Cyanidin-3-rutinoside	Ficco et al. (2014b);
		Cyanidin-3-succinyl-glucoside	Tyl and Bunzel (2012); Knievel et al. (2009);
		Delphinidin-3-glucoside	Hosseinian et al. (2008); Abdel-Aal and Hucl
		Delphinidin-3-rutinoside	(2003)
		Malvidin-3- glucoside	
		Malvidin-3-rutinoside	
		Pelargonidin-3-glucoside	
		Peonidin-3-arabinoside	
		Peonidin-3-glucoside	
		Peonidin-3-galactoside	

(*continued*)

Table 11.1 *(Continued)*

Wheat varieties	Pigment	Compounds	References
		Peonidin-3-rutinoside	
		Petunidin-3-glucoside	
		Petunidin with hexose and rhamnose	
Purple	Anthocyanin	Cyanidin-3-arabinoside	Garg et al. (2016);
		Cyanidin-3galactoside	Bartl et al. (2015);
		Cyanidin-3-glucoside	Ficco et al. (2014b);
		Cyanidin-3-rutinoside	Tyl and Bunzel (2012); Knievel et al. (2009);
		Cyanidin-3-succinyl-glucoside	Hosseinian et al. (2008); Abdel-Aal and Hucl
		Delphinidin-3-arabinoside	(2003)
		Delphinidin- 3-galactoside	
		Delphinidin-3-glucoside	
		Delphinidin-3-rutinoside	
		Delphinidin with hexose/coumaric acid	
		Malvidin-3-glucoside	
		Malvidin-3-rutinoside	
		Pelargonidin-3-arabinoside	
		Pelargonidin-3-galactoside	
		Pelargonidin-3-glucoside	
		Pelargonidin-3-rutinoside	
		Pelargonidin with hexose and acetic/malonic acid	
		Peonidin-3-arabinoside	
		Peonidin-3-glucoside	
		Peonidin-3-galactoside	
		Peonidin-3-rutinoside	
		Petunidin-3-glucoside	
		Petunidin with hexose and rhamnose	
Red	Proanthoyanidin	Proanthocyanidins(þ)catechin	McCallum and Walker (1990)
		procyanidinB3	
		prodelphinidinB3	
		Phlobaphenes	
	Anthocyanin	Cyanidin-3-glucoside	Garg et al. (2016); Bartl et al. (2015); Ficco
		Cyanidin-3-succinyl-glucoside	et al. (2014a);
		Malvidin-3-glucoside	Tyland Bunzel (2012); Knievel et al. (2009);
		Peonidin-3-galactoside	Hosseinian et al. (2008); Abdel-Aaland Hucl
			(2003)

varieties. Delphinidin-3-rutinoside, delphinidin-3-glucoside, and malvidin-3-glucoside are the compounds present in blue wheat, while purple wheat does notcontain them. The four main anthocyanins identified from blue aleuronic wheat varieties are delphinidin-3-glucoside (45%), cyanidin-3-glucoside (28%), delphinidin-3-rutinoside (22%), and cyanidin-3-rutinoside (2%) (Abdel-Aal et al. 2008). Cyanidin-3-glucoside is the abundant anthocyanin in purple wheat (Abdel-Aal et al. 2003) followed by cyanidin-3-galactoside and then malvidin-3-glucoside (Hosseinian et al. 2008). Phlobaphenes are red colored, water-insoluble phenolic that are co-occurring condensed tannins responsible for the red color of red wheat varieties. The tannins can be easily converted into phlobaphenes compounds (red insoluble) using acids. According to Oomah et al. (1999) most abundant anthocyanins in red wheat were cyanidin, in case of purple wheat petunidin and malvidin and delphinidin in blue wheat.

Due to its excellent nutritional profile, good sensory features and, most significantly, its health-promoting actions, black wheat has received substantial interest among all known colored wheat variants (Beta et al. 2019). It is utilized to make functional foods or related colorants (Abdel-Aal et al. 2006). According to Garg et al. (2016), new products made from unconventional wheat (pigmented/colored) may have improved functional and nutritional qualities. However, few studies have been conducted on black wheat to date, with the majority focusing on its breeding and qualitative analysis (Sun et al. 2014), characteristics and utilization, protein characteristics (Li et al. 2006), total flavonoid content, total phenolic content, antioxidant activity, and free radical scavenging activity (Liu et al. 2018).

11.3 Health benefits of pigmented wheat varieties

Colored wheat is high in dietary fiber, anthocyanin, and proanthocyanidin (condensed tannin) components, which have a variety of health advantages, the majority of which are linked to their antioxidant capabilities (Lachman et al. 2017). Colored wheat contains high amounts of phytochemicals, like phenolics, flavonoids and anthocyanins. The phenolics present in wheat are mostly in bound form and are responsible for numerous health benefits as they are absorbed in the blood stream after digestion by intestinal micro organisms (Van Hung 2016).

Colored wheat cultivars have demonstrated high antioxidant activity, neuroprotection, cholesterol absorption inhibition, cytoprotection, glycaemic regulation, anticancer action, retinal protection, anti-aging, increased immunological response, and anti-hypertension effects in *in-vivo* and *in-vitro* investigations. It also includes body fat reduction, regulation of lipid profile, hepatoprotection, relieved metabolic syndrome (Zhu 2018). Along with all these health benefits, colored pigments offer a safer alternative to the synthetic coloring compounds in food (Somavat et al. 2018).

Anthocyanin is helpful in the preventing cancer, cardio vascular diseases (CVD), obesity, diabetes, aging, inflammation as well as antimicrobial and neuroprotective effect (Chhikara et al. 2020; Khoo et al. 2017; Garg et al. 2016; Tsuda 2012). The amount of anthocyanin is highest in black, blue and purple wheat varieties. Therefore, incorporation of anthocyanin-rich pigmented wheat varieties into a diet may help to boost overall human health (Tian et al. 2018b; Sharma et al. 2018; Giordano et al. 2017; Onipe et al. 2015; Ficco et al. 2014a). A diet high in anthocyanins aids in the correction and enhancement of a variety of visual functions, including preventing myopia, improving dark adaption, lowering eye fatigue, and improving retinal blood flow to reduce glaucoma. This substance also aids normal brain function by preventing age-related neuro-degeneration and cognitive decline (Tsuda 2012).

The exposure to mycotoxins is a critical factor for human health, mainly for malnourished populations, and low-income countries, and this can be prevented by phlobaphene-rich cereal grains (red maize, red wheat, etc.) that will increase the quality of kernel and its healthiness. The positive effects by phlobaphene on humans are because of the high antioxidant activity (Landoni et al. 2020). Phlobaphene is also found to have anti- inflammatory potential and hence can be used for nutraceutical product formulations (Wu et al. 2020).

Black wheat contains high-quality proteins and vital amino acids, with a 30–50 percent higher level of linoleic and linolenic acids than regular wheat. Its consumption has been linked to a variety of health advantages, including endothelial cell protection, heart and cardiovascular disease prevention, and anti-cancer properties (Dykes and Rooney 2007; Liu 2007). It also contains anthocyanins and procyanidins, among other things. These bioactive substances have been linked to a lower chance of chronic diseases like diabetes, obesity, cancer, and cardiovascular disease developing (Sun et al. 2018).

11.4 Comparison of nutritional properties of pigmented wheat grains with common wheat

Colored wheat contains higher nutritional content than common wheat, hence can be used for functional flour production. Common wheat has higher flour yield than colored wheat varieties, because the colored wheat grains have lower volume, weight and plumpness than common wheat. According to Žilić et al. (2012) common wheat contains starch: 61.85 percent, protein: 12.93 percent, oil: 3.33 percent and ash: 2.15 percent and these are considerably lower when compared to colored wheat both qualitatively and quantitatively.

11.4.1 Macronutrients

Comparing nutritional content of common wheat grains with blue wheat, the protein content, total dietary fiber and ash content of blue wheat grain is comparatively more than that of common wheat. However the starch content and fat of blue wheat is comparatively less than common wheat (**Table 11.2**). In case of nutritional values of blue wheat and purple wheat protein, total dietary fiber, ash and fat content of blue wheat is comparatively higher than purple wheat, but starch content is higher in purple wheat. While, compared with the common wheat, approximately 11.74 percent to 18.17 percent higher protein content was found in colored wheat varieties. The Total Amino Acid content (TAA), Essential Amino Acid Content (EAA) and Essential Amino Acid Index (EAAI) of pigmented wheat varieties were higher compared to that of common wheat. For black and blue wheat, TAA content was 13.52g/100g and 12.38g/100g, respectively, which was greater than normal wheat (11.37g/100g). In case of EAA content, also black wheat (4.04g/100g) and blue wheat (3.67g/100g) had higher value than common wheat (3.42g/100g) (Tian et al. 2018b). The fat and dietary fiber content werefound to be almost the same in colored wheat and common wheat varieties, because fat is concentrated in the embryo of wheat and the coating of seed and peel of wheat contain a high amount of dietary fiber (Li et al. 2017).

Table 11.2 Nutritional Composition of Colored Wheat Grains

Wheat varieties	Starch (%)	Protein (%)	Dietary fiber (%)	Ash (%)	Fat (%)	References
Black	–	11–12.9	13–13.3	1.8–2.2	1.7	Tian et al. (2018b); Sharma et al. (2018); Garg et al. (2016)
Blue	12.3–15	12.3–15	12.7–13.5	1.91–2.4	1.2	Tian et al. (2018b); Sharma et al. (2018); Giordano et al. (2017); Garg et al. (2016); Ficco et al. (2014a)
Purple	48.7–59.9	10.3–19.3	9.8–15.1	1.2–2.60	1.21	Gamel et al. (2020); Sharma et al. (2018); Ma et al. (2018) Tian et al. (2018a); Hailu Kassegn (2018); Zanoletti et al. (2017); Giordano et al. (2017); Ficco et al. (2016); Garg et al. (2016)
Red	68.96	9.02	2.76	1.10	–	Duchoňová et al. (2012)
Common wheat	58.9	10.8–14.6	12.5–15.7	1.56–1.9	1.5	Sharma et al. (2018); Tian et al. (2018a); Tian et al. (2018b) Tian et al. (2017); Giordano et al. (2017); Garg et al. (2016); Brandolini et al. (2015); Ficco et al. (2014a)

11.4.2 Carbohydrates and fibers

Wang, Zhang, et al. (2020) found 22 carbohydrates in black wheat. Black wheat has a higher polysaccharide content than conventional/common wheat, according to Liu et al. (2018). Polysaccharides such as cellulose, hemicellulose, pectin, and resistant starch (RS) are found in dietary fiber fractions but are not digested by digestive enzymes. These fibers help to raise the viscosity of the food, which slows the accessibility of starch granules to digestive enzymes in the human digestive system (Angioloniand Collar 2011). Total dietary fibers (TDF) make up about 1.15 percent of the total weight of black wheat grain (Tian et al. 2018). The insoluble type of fiber dominates this TDF, providing volume and body to the stools, preventing constipation, and stimulating intestinal transit (Marlett et al. 2002).

11.4.3 Proteins

Protein quality is linked to bioavailability of amino acids, essential amino acid content, and its capacity to meet the metabolic needs of the human body (Nunes et al. 2016). The nutritional value of each protein is estimated using the amino acid score (AAS), the value of wheat is 43 (WHO 2007). Black wheat is a possible source of protein when compared to regular wheat; 96 amino acids and their derivatives were found in black wheat, according to Wang, Zhang, et al. (2020). The primary storage protein found in black wheat is glutenins and gliadins, which are known as alcohol soluble prolamins.

11.4.4 Micronutrients

Colored wheat cultivars are rich sources of group B vitamins like thiamine, riboflavin, niacin, pyridoxine, and folate and also vitamin E. Theyalso contain minute concentrations of beta carotene, vitamin D and vitamin K (Balyan et al. 2013). The mineral composition of colored wheat varieties and common wheat is given below in **Table 11.3**. From the table it is clear that colored wheat grains are having more micronutrient content than common wheat varieties. While comparing the colored wheat varieties black wheat had higher magnesium and iron content than blue, purple and red wheat varieties. Purple wheat had high calcium content and blue wheat had high copper content than other colored wheat varieties.

11.4.5 Minerals

Calcium is necessary for bone formation, which is necessary for human growth (Liu et al. 2014). Li and Beta (2011) found that black-grained wheat 76 had four times the calcium of Jinchun 9 wheat (conventional wheat variety). In the human body, the mineral phosphorus is

Table 11.3 Mineral Content of Colored Wheat Grains

Wheat grains	Na (mg/kg)	Mg (mg/kg)	Ca (mg/kg)	Fe (mg/kg)	Mn (mg/kg)	Cu (mg/kg)	Zn (mg/kg)	References
Black	21.4	450	184	39–79.3	12.7	3.17	28–30	Bieńkowska et al. (2019)
Blue	30.1	430	310	38.3–46.1	14.8–40.9	6.0–9.6	33.1–40	Tian et al.(2018b)
Purple	–	–	419.6	36.7–46	15–40.1	4.1–6.9	25–41.7	Sharma et al.(2018)
Red	–	–	–	36.71	41.25	6.71	31.54	Tian et al. (2017)
Common wheat	7.7–27.3	320–1160	276–345	26–35.6	10–28.2	3.2	16.4–40.8	Ficco et al. (2014a) Guo et al. (2013)

crucial for the formation of bones and teeth. The black-grained wheat 76 had 4.10 g/kg phosphorus compared to 2.41 g/kg in Jinchun 9. Selenium (Se) is vital for the immune function, oxidative stress reduction, and inflammation reduction. According to Li et al. (2006), black wheat has 1.04 mg/kg Se compared to 0.26 mg/kg in regular wheat. According to Sharma et al. (2018), colorful wheats are higher in zinc and iron minerals than white wheat.

11.4.6 Vitamins

Vitamin B3 (Niacin) aids in the production of energy from carbs, proteins, and lipids, as well as DNA synthesis and skin health. Vitamin B5 aids in haemoglobin synthesis, carbohydrate, protein, and fat metabolism. Women's infertility is exacerbated by a lack of vitamin E (tocopherol). It protects red blood corpuscles from hemolysis. Pantothenic acid (Vitamin B5) is involved in the production of blood cells, hormone secretion, gland function, and the conversion of food to energy. Tian et al. (2018) found significant amounts of vitamin B3, B5, and E in black wheat. BGW 76 has 11.47 mg/kg vitamin K (responsible for blood coagulation and bone metabolism), which was 1.6 times greater than Jinchun 9 according to Li and Beta (2011). (Conventional wheat). Wang, Zhang et al. (2020) reported the presence of 19 vitamins in black carrot.

11.5 Comparison of antioxidant properties of pigmented wheat grains with common wheat

Colored wheat varieties possess higher antioxidant activity due to polyphenol, which can be measured by different antioxidant assays such as 2, 2-diphenyl picrylhydrazyl (DPPH) assay, ferric reducing ability of plasma (FRAP) assay, 2, 2-azino-bis (3-ehtylbenzothiazoline-6-sulfonic acid) (ABTS) assay and oxygen radical absorbance capacity (ORAC) assay. According to Zong et al. (2006) purple, blue and black wheat contain high antioxidant content comparedwith white and red colored wheat grains. The color and antioxidant property of colored wheat grains are related, but according to Mpofu et al. (2006) chromaticity of grains isnot considered as an element of antioxidant level in different colored wheat grains.

11.5.1 Antioxidant content

Sun et al. (2014) observed that the total phenolic content (TPC) in black, red and white colored varieties ranges from 492 to 1313 µmol/100g and inter genotypic variability was 18.5 percent. The inter-genotypic variability was about 24.5 percent in black colored wheat, 17.8 percent in white colored wheat and 10.0 percent in red colored wheat grains. The mean value of the amount of flavonoid present in black colored wheat grains isabout 36.1mg/100g, red colored wheat grains are about 24.1 mg/100g and white colored wheat grains are about 29.0 mg/100g respectively. The total antioxidant capacity (TAC) among black, red and white wheat grains ranged between 3.1–8.3 mmol/100g, where the highest inter genotypic variability was in white colored wheat grains (16.0%), and the lowest inter genotypic variability was in black colored wheat grain (4.4%), which shows a wider antioxidant capacity in white colored wheat grains, even though it has alow antioxidant capacity (5.1 mmol/100g). The antioxidant content of colored wheat grains are mentioned in **Table 11.4**.

TPC value of whole purple wheat is about 197 mg/100g, total ferulic acid content is about 85.15 mg/100g, total vanillic acid content is about 3.50 mg/100gm and total coumaric acid acid is 2.43 mg/100g. The anthocyanin content of blue aleuronic wheat is 24.55 mg/100g, while in

Table 11.4 Comparison of Antioxidant Profile of Colored Wheat Grains with White Wheat

Wheat varieties	TPC	Flavonoid content	Anthocyanin content	TAC	References
Black	660–1313 μmol/100g	32.1–39.7 mg/100g	–	6.7–7.5 mmol/100g	Sun et al.(2014)
Blue	113–134 mg GAE/L	0.08–2.80 Mg RE/L	211.9 μg/g	0.187–0.269 mmolTE/L	Duchoňová et al. (2012); Abdel-Aalet al. (2006)
Purple	100–121 mg GAE/L	0–1.33 mg RE/L	38–95.8 μg/g	0.083–0.151 mmolTE/L	Duchoňová et al. (2012); Abdel-Aal et al. (2006)
Red	650–1088 μmol/100g	21.8–38.9 mg/100g	6.7–7.9 μg/g	4.9–8.3 mmol/100g	Sun et al. (2014); Abdel-Aal et al. (2006)
White Wheat	492–1034 μmol/100g	14.7–35.1 mg/100g	7.1–7.3 μg/g	3.1–6.8 mmol/100g	Sun et al. (2014); Abdel-Aal,et al. (2006)

Notes:
RE: rutin equivalent; TE: trolox equivalent; GAE: gallic acid equivalent

purple colored wheat it ranges from 96–235 μg/g (Gamel et al. 2019). Approximately, during milling all fractions of purple wheat contains similar carotenoid content. According to the yield of purple wheat grains after milling, only 43 percent of total wheat is retained. In case of blue aleurone wheat, it contains double amount of bran and high concentration of wheat endosperm. Approximately 63 percent of total carotenoid content is extracted from blue aleurone wheat flour. Milling procedure, rate of extraction, solvent are the different factors affecting the carotenoid content of wheat. Blue aleurone wheat has an antioxidant content of 0.8 mmol/100g (Siebenhandl et al. 2007). Liu et al. (2010) used methanol and acetone solvents for the extraction of phenolics from the four types of colored wheat varieties (purple, red, white and yellow), and the result obtained showed that TPC of different wheat ranged from 146 to 226mg/100 g in methanol and from 77 to 177 mg/100 g in acetone. Highest phenolic content was seen in purple colored wheat varieties and lower in yellow colored wheat varieties. A purple colored wheat grain contains lipid derived carboxylic acids, high level of alcohols and low level of oxidative degrading lipids. The phytochemicals of purple wheat are present in the bran which contains about 83 percent of total phenolic content and 79 percent of flavonoid. Except colored grains, the anthocyanin content is comparatively low in others. The purple wheat mainly contain acylatedanthocyanidins and includes cyanidins, delphinidin and pelargonidon, which are commonly bound with saccharide residues like glucose, galactose, arabinose as 3–glycosides (Gamel et al. 2019).

Black wheat contains a lot of phenolic acids and has a lot of antioxidant activity. Among all the phenolic acids, ferulic acid is the most prevalent. Ferulic acid is responsible for the kernel's astringent flavor and, as a result, it defends it from insects and animals. Phenolic acid is available in three forms: unbound, soluble conjugated, and insoluble bonded. Only a handful are accessible in a free and soluble conjugated form that may be absorbed through the mucosa of the intestine. As a result, the gastrointestinal system uses enzymes (esterase) found in the intestinal mucosa and bacteria to assist liberate natural phenolic acids (ferulic acid, di-ferulic acid) from the dietary matrix. As a result of this (phenolic acid release in the intestine), the risk of colon cancer is reduced (Gong et al. 2019; Luthria et al. 2015). Wheat phenolic acid alters nuclear factor kappa B (NF-kB), a transcription factor that controls innate immunity processes, cell proliferation, apoptosis, and cell survival, and so regulates pro-inflammatory genes (Luthria et al. 2015). Low-density lipoprotein (LDL) cholesterol lipid peroxidation is also inhibited by phenolic acid (Min et al. 2012).

11.5.2 Antioxidant activity

The antioxidant and anti-inflammatory properties of phenolic acids and anthocyanin depends on the concentration and molecular structure of the food and their fate during digestion,that is, bio-accessibility, bioavailability, absorption and metabolism. A human study conducted by Gamel et al. (2019) suggested that products from purple colored wheat have an ability to positively affect the health because of extensive digestion, metabolism and utilization of these products in the human body.

According to the study by Liu et al. (2010),purple wheat had highest ORAC value, that is up to 6898 µmol/100 g in the methanolic extract and 1920 µmol/100 g in the acetone extract. For the three varieties of purple grains the DPPH radical scavenging activity was in between 715 to 857 µmol/100 g, while for yellow colored wheat grains (669 µmol/100 g) it was slightly higher than red colored wheat grains (648 µmol/100 g) that can be a result of higher total flavonoids of yellow wheat variety (13.44 mg/100 g) when compared to reddish wheat variety (10.72 mg/ 100 g).The DPPH radical scavenging activity was least in the acetone extract of white colored wheat.

In vitro antioxidant capacity of purple wheat whole flour, whole flakes, bran, bars and crackers were examined using ABTS, DPPH and ORAC assays by Gamel et al. (2020), and it was reported that bran showed the highest capacity to scavenge the radicals in order of 74, 100 and 94 percent respectively, while wheat flour and flakes showed 54 and 52 percent of ABTS activity, 38 and 45 percent DPPH activity and 71 and 74 percentORAC activity respectively. The purple wheat bars and crackers exhibited greater ability to restrict radical activity with 76 and 80 percentORAC activity respectively making them suitable to impact metabolic markers.

Sytar et al. (2018) studied the antioxidant potential in blue, purple and yellow colored wheat varieties and their sprouts like citrus yellow, KM 53-14 Blue, KM 178-14 purple, PS Karkulka purple and scorpion blue aluerone. PS Karkulka purple showed the highest antioxidant activity in terms of ABTS and DPPH among all grains. 18 percent higher activity was reported in the seedlings of PS Karkulka purple compared to the grain while other colored grains seedlings just showed 3- to 4-fold increases.

Yu and Beta (2015) used Oelandhvede, Indigo and Konini varieties of purple wheat preparing bread and measured antioxidant activity. The results of DPPH scavenging activities were 139,127 and 130 µmolTE/100 g, respectively. After the mixing, the DPPH scavenging activity of Oelandhvede, Indigo and Konini decreased to 17, 21, and 22 percent. After the first 30-minute level of fermentation, DPPH Scavenging activity increased to 85, 91, and 90percent.The anthocyanin was detected using spectra of absorbance at different wavelength. In Indigo and Konini there was a large change in absorbance of 520 nm and 535 nm and, in the case of Oelandhvede, there was no change in wavelength because of low anthocyanin content.

11.6 Development of food products using pigmented wheat grains

Currently colored wheat grain related research isfocused on products like bread, biscuit, noodles, pasta, and so forth. The stability of the bioactive compounds present in colored wheat depends on the processing technology implied, therefore, selection of appropriate process parameters may help in retaining majority of the bioactive components (Saini et al. 2021). A whole meal of pigmented grains has been used to make anthocyanin-enriched cookies

(Pasqualone et al. 2015), colored food products (Syed et al. 2013), bread and noodles (Li et al. 2015). Whole grain flour can be used to manufacture anthocyanin-rich bioactive baked items, according to Li et al. (2015).

Yu and Beta. (2015) studied the fate of phenolics and antioxidant properties during different stages of bread making process. They reported an increase in the free phenolic content from 65–68 percent during mixing, fermentation and baking process. While bound phenolic showed an initial decrease in 30 minutes of fermentation in comparison to the mixing step but showed a significant increase from 16 to 27 percentduring a 65-minute fermentation and baking step. The total anthocyanin content showed a significant decrease of 21 percent after mixing, but a steep hike to 90 percent of the original value was observed after fermentation. Baking resulted in a drastic decrease in anthocyanin by 55 percent showing the least values in the bread crust. The bound phenolic found in bread crust is protocatechuic, caffeic, syringic and sinapic and free phenolic extract are p- Hydroxybenzoic, p-coumaric, and ferulic acids. Itcan be concluded that the bread-making process resulted in increases in total phneolics and antioxidant activity at the expense of anthocyanin content.

Ficco et al. (2016) discussed the effect of milling and processing on nutritional, textural quality, anthocyanin, carotenoids content and antioxidant properties of purple durum and yellow durum wheat. It was reported that a decrease in dietary fiber content, protein and gluten index was observed from whole meal to semolina, with purple wheat showing a higher value of dietary fiber than yellow durum at every stage of milling, but protein content and gluten index was higher in yellow durum. As anthocyanin are mainly located in the pericarp of the grain so a decrease in anthocyanin content from 73.41 µg/g (whole meal) to 66.49 µg/g (semi-whole meal) and 24.40 µg/g (semolina) was observed in purple durum wheat. The same trend was exhibited for carotenoid content and antioxidant activity in purple durum wheat semi whole meal, and semolina shows an average decrease of 26 percent in anthocyanin content for extruded pasta. The fresh pasta obtained from the extruded pasta retains about 88.5 percent of the anthocyanin while a consistent decrease during drying was observed. After cooking of dried and semi whole meal fresh pasta the anthocyanin content decreased about 20.3 percent. The decrease in antioxidant components during cooking of fresh pasta is because of degradation of soluble antioxidant components. An increase in trolox equivalent antioxidant capacity was observed during cooking of dried pasta, which is because of the release of bound phenolics and maillard reaction products.

Anthocyanin enriched biscuits from purple colored wheat grains shows an anthocyanin content of 13.86 mg/kg. Although, processing resulted in a significant decrease in anthocyanin content, which is affected by a number of factors, such as action of oxidative enzymes endogenous in whole meal, merged effect of light, air and heat treatment and the diluting affect because of presence of other ingredients (Pasqualone et al. 2015).

Duchoňova et al. (2012) prepared a whole-grain loaf of bread from colored purple, blue and red wheat flour. Two of the breads were baked without added beta- glucan and the other two with the addition of beta-glucan. In finished products he evaluated camber (the ratio of height and width) and organoleptic properties like the appearance of the product, sticky, crumb porosity and elasticity, crumb and crust color, smell, taste. During the analysis, it was observed that the red wheat grain had nutritional composition dry matter: 90.4 percent, ash: 1.10 percent. Crude fiber: 2.70 percent, Protein: 9.02 percent, starch: 68.96 percent and beta glucan: 0.46 percent. In case of purple wheat, dry matter, ash, crude fiber, protein, starch, beta glucan was 88.4–89.4, 1.09–1.11 percent, 2.56–3.18 percent, 11.10–13 percent, 66.56–71.15 percent and 0.43–0.67 percent respectively. For blue wheat dry matter, ash, crude fiber, protein, starch, beta glucan was 88.9–90.7 percent, 1.09–1.10 percent, 2.73–2.90 percent, 10.76–13.02 percent, 67.76–70.28

percent and 0.32–0.64 percent, respectively. It was found that most pleasant color, taste, and flexibility bread crumb was present in bread prepared from purple variety konini without β-glucan addition.

11.7 Thermal stability of bioactive compounds in pigmented wheat grains

Thermal processing has a greater influence on anthocyanin degradation. Anthocyanin degradation is higher at long-time less temperatures than low time high temperatures. Co-pigmentation is the major color stabilizing mechanism of anthocyanin. It is usually seen as colorless but, when it is infused into an anthocyanin, it intensifies the color of the solution. Structure and concentration of anthocyanin, pH, temperature, presence of complexing agents like phenols and metals greatly affect the color stability of anthocyanin (Bąkowska et al. 2003).

Abdel-Aal et al. (2003) studied anthocyanin stability at different temperatures and pH in isolated and whole meal of blue colored wheat grains. Anthocyanin will be less stable at pH 5 and more stable at pH 1. From temperatures of 65 to 95°C the degradation of the anthocyanin gradually increases. The average activation energy of anthocyanin present in sunflower hulls in first order kinetics was determined as 23.1kcal/mol. After the addition of sulfur dioxide into blue colored wheat grains, the stabilizing effect of anthocyanin increased compared with the non-adding sulfur dioxide compounds. Also they found that the anthocyanin degradation will mainly depend on temperature and pH where the color degradation highly takes place at pH 1 than pH 5 at 95°C. By the addition of 500–1000 ppm of sulfur dioxide the degradation of the anthocyanin was diminished between 65–95°C, but by the addition of 2000 ppm sulfur dioxide the stabilizing effect is reduced, and there was much pigment in stabilization of common wheat. The anthocyanin degradation is comparatively high at 95°C and low at 65°C, and the addition of sulfur dioxide reduced this degradation. At 60 and 80°C there was slight difference in the pigment stability even after the addition of sulfur dioxide of 2000 ppm. From this study they concluded that, as the addition of sulfur dioxide increases, the anthocyanin stability at 95°C also increases.

Li et al. (2007) proposed that it is not only the heat treatment but also the solvents used for the extraction of antioxidant components that affect the measurement of antioxidant capacity. In the study outcome of thermal treatments on the antioxidant activity of purple wheat bran, thermally treated purple bran and purple colored muffins were measured using ORAC and DPPH assay. Three solvent systems used for extraction were methanol (100%), methanol:HCl (99:1) and ethanol:HCl (99:1). The results of total phenolic content show a notable effect of extracting solvents on the estimation. The results of TPC expressed in terms of mg of ferulic acid equivalent/g of sample of purple wheat bran with methanol (100%), methanol:HCl (99:1) and ethanol:HCl (99:1) were 3.34, 5.98 and 7.70 mg/g respectively. A notable decrease in total phenolics was seen when thermal treatment was given in the preparation of muffins using purple wheat bran, which might be due to the dilution effect of other ingredients as well as degradation due to thermal treatment. The extraction efficiency of methanol: HCl and ethanol: HCl is due to the polarity difference and this might result in the slight hydrolysis of polyphenols under acidic conditions. No anthocyanin was observed in the muffins probably because of anthocyanin degradation at a baking temperature of 177° C for 7–12 min. The ORAC values were significantly lower in both muffins, which indicates the dilution of antioxidant components and interaction with other components during baking.

The effect of different baking temperatures and time combinations on the anthocyanin level of blue colored and purple colored wheat bread showed a significant decrease. Anthocyanin content in blue bread baked at 240° C for 21 minutes showed a 7.1 percent decrease as compared with blue colored whole wheat flour, while the one baked at 180° C for 31 minutes showed a 40.8 percent decrease. The purple wheat bread showed a decrease of 61 percent at 240° C and 72.8 percent at 180° C. The higher loss of anthocyanin in purple wheat bread is because of presence of acetylated anthocyanin, which is sensitive to thermal degradation during baking (Bartl et al. 2015).

According to Li et al. (2015) black and purple wheat varieties were used to produce noodles and steamed bun to check the effect of variety, milling and processing on antioxidant activity. According to the study, black colored wheat variety hasabout 659.8 µggallic acid equivalent (GAE) g^{-1} as total phenolic and 319.3 µg Rutin equivalent (RE) g^{-1} as flavonoid concentration. The flavonoid concentration and total phenolic of refined flour was less than partially debranned flour and whole wheat grain flour of colored wheat grains. Compared with the flour, the noodles and steamed bread have high total phenolic and flavonoid concentration. But the noodles have higher flavonoid concentration and total phenolic compared with steamed bread. Most of the antioxidant compounds in the flour are lost during thermal treatments. After the analysis of antioxidant properties by FRAP assay black wheat flour, the steamed bun and noodles had higher values compared to white wheat. The result shows that the partially debranned and whole black wheat flour are highly beneficial to health.

In the study conducted by Abdel-Aal et al. (2013), baking process affects the phenolic acids in whole grain products depending on the type of the phenolic compounds present in the wheat grain, the recipe used for baking and conditions of heating used for baking. According to them, baking significantly increases the free phenolic acid in muffins, bread and cookies made from colored wheat and decrease in bound phenolic acid taking place in bread, and it is only moderately affected in muffins and cookies. Proper selection of optimized processing technique will help to preserve the antioxidant compounds present in colored wheat.

11.8 Consumer's perception

Compared to other food grains like as rye, oats, and conventional wheat, the use of BW in the food sector has been quite limited, according to Liu et al. (2018). Colored maize genotypes, for example, fetched premium rates, whereas purple wheat cultivars and bakery products generated from them have had limited commercial success in Canada, Europe, and New Zealand. The low crop yield has been a key impediment to the widespread use of BW (Garg et al. 2016). In the same way, the commercial cultivar Skorpion (blue wheat) released in Austria yields 25 percent less than the control varieties (Garg et al. 2016). According to Prashant et al. (2012), the kernel size and shape influence the market price, which is also a component that customers consider. It is necessary to investigate the biosynthetic pathways in order to minimize the bitter compounds through targeted and allelic variation or mutagenesis. Furthermore, flavor characterization using human sensory panels, highly reproducible, robust, and sensitive model systems, or other "consumption" trials [as performed by Kiszonas et al. (2015)] may aid in determining relative consumption preferences for different wheat types (conventional vs. pigmented). However, even those incorporating phenotypes in genetic mapping contribute in a better understanding of flavor/food preferences linked genes/loci, round robin and single elimination tournament designs have failed to establish a unique consumption phenotype for multiple wheat genotypes.

11.9 Conclusion

The colored wheat varieties have gained more interest because of health promoting and disease preventing bioactive compounds. Colored varieties are good sources of micronutrients such as Sodium, Magnesium, Calcium, Iron, Zinc and copper. Black colored wheat contains high nutritional, flavonoid, total phenolic content and antioxidant properties compared with other colored wheat grains. The blue colored wheat grain contains higher amount of protein compared with other colored wheat grains. Colored wheat grains contain seven times more anthocyanin than common wheat grains. The processing of these grains resulted in a decrease in the antioxidant components as well as nutritional components. As anthocyanin is present in the pericarp of the grain the conversion of grain to semolina to flour resulted in reduction of anthocyanin content. Higher loss of anthocyanin is observed during thermal treatments specifically in purple wheat because of presence of acetylated anthocyanin, which are more prone to thermal degradation. Still, the development of products from these colored varieties is less, and detailed studies are required in areas to minimize the degradation of the bioactive compounds so as to get maximum health benefits. The stability of the bioactive compounds present in colored wheat depends on the processing technology implied. Therefore, selection of appropriate process parameters may help in retaining the majority of the bioactive components.

References

Abdel-Aal, El-Sayed M., and Iwona Rabalski. "Effect of baking on free and bound phenolic acids in wholegrain bakery products." *Journal of Cereal Science* 57, no. 3 (2013): 312–318.

Abdel-Aal, El-Sayed M., and Pierre Hucl. "Composition and stability of anthocyanins in blue-grained wheat." *Journal of Agricultural and Food Chemistry* 51, no. 8 (2003): 2174–2180

Abdel-Aal, El-Sayed M., Atef A. Abou-Arab, Tamer H. Gamel, Pierre Hucl, J. Christopher Young, and Iwona Rabalski. "Fractionation of blue wheat anthocyanin compounds and their contribution to antioxidant properties." *Journal of Agricultural and Food Chemistry* 56, no. 23 (2008): 11171–11177.

Abdel-Aal, El-Sayed M., J. Christopher Young, and Iwona Rabalski. "Anthocyanin composition in black, blue, pink, purple, and red cereal grains." *Journal of Agricultural and Food Chemistry* 54, no. 13 (2006): 4696–4704.

Abdel-Aal, El-Sayed M., Pierre Hucl, and Iwona Rabalski. "Compositional and antioxidant properties of anthocyanin-rich products prepared from purple wheat." *Food Chemistry* 254 (2018): 13–19.

Adom, Kafui Kwami, Mark E. Sorrells, and Rui Hai Liu. "Phytochemical profiles and antioxidant activity of wheat varieties." *Journal of Agricultural and Food Chemistry* 51, no. 26 (2003): 7825–7834.

Angioloni, Alessandro, and Concha Collar. "Physicochemical and nutritional properties of reduced-caloric density high-fibre breads." *LWT–Food Science and Technology* 44, no. 3 (2011): 747–758.

Bąkowska, Anna, Alicja Z. Kucharska, and Jan Oszmiański. "The effects of heating, UV irradiation, and storage on stability of the anthocyanin–polyphenol copigment complex." *Food Chemistry* 81, no. 3 (2003): 349–355.

Balyan, Harindra S., Pushpendra K. Gupta, Sachin Kumar, Raman Dhariwal, Vandana Jaiswal, Sandhya Tyagi, Priyanka Agarwal, Vijay Gahlaut, and Supriya Kumari. "Genetic improvement of grain protein content and other health-related constituents of wheat grain." *Plant Breeding* 132, no. 5 (2013): 446–457.

Bartl, Pavel, Alen Albreht, Mihaela Skrt, Bohuslava Tremlová, Martina Ošťádalová, Karel Šmejkal, Irena Vovk, and Nataša Poklar Ulrih. "Anthocyanins in purple and blue wheat grains and in resulting bread: Quantity, composition, and thermal stability." *International Journal of Food Sciences and Nutrition* 66, no. 5 (2015): 514–519.

Beta, Trust, WendeLi, and Franklin B. Apea-Bah. "Flour and Bread From Black, Purple, and Blue-Colored Wheats." In *Flour and Breads and their Fortification in Health and Disease Prevention*, pp. 75–88. Academic Press, 2019.

Bieńkowska, Teresa, Elżbieta Suchowilska, Wolfgang Kandler, Rudolf Krska, and Marian Wiwart. "Triticumpolonicum L. as potential source material for the biofortification of wheat with essential micronutrients." *Plant Genetic Resources* 17, no. 3 (2019): 213–220.

Brandolini, Andrea, Alyssa Hidalgo, Simona Gabriele, and Manfred Heun. "Chemical composition of wild and feral diploid wheats and their bearing on domesticated wheats." *Journal of Cereal Science* 63 (2015): 122–127.

Chhikara, Navnidhi, Amolakdeep Kaur, Sandeep Mann, M. K. Garg, Sajad Ahmad Sofi, and Anil Panghal. "Bioactive compounds, associated health benefits and safety considerations of Moringaoleifera L.: An updated review." *Nutrition and Food Science* 51, no. 2 (2020): 255–277.

Dobrovolskaya, O., V. S. Arbuzova, U. Lohwasser, M. S. Röder, and A. Börner. 'Microsatellite mapping of complementary genes for purple grain colour in bread wheat (Triticumaestivum) L." *Euphytica* 150, no. 3 (2006): 355–364.

Duchonova, Lenka, Maria Vargovicova, and Ernest Sturdik. "Nutritional profile of untradional colored wheat grains and their bread making utilisation." *JLS* 6 (2012): 1008–1015.

Dykes, L., and L. W. Rooney. "Phenolic compounds in cereal grains and their health benefits." *Cereal Foods World* 52, no. 3 (2007): 105–111.

Eticha, F., H. Grausgruber, S. Siebenhandl-Ehn, and E. Berghofer. "Some agronomic and chemical traits of blue aleurone and purple pericarp wheat (Triticum L.)." *Journal of Agricultural Science and Technology B* 1 (2011): 48–58.

FAO. "FAO Food Price Index." FAO Cereal Supply and Demand Brief [accessed: 2021 Mar 15]. www.fao.org/worldfoodsituation/csdb/en/.

Ficco, Donatella B. M., Vanessa De Simone, Salvatore A. Colecchia, Ivano Pecorella, Cristiano Platani, Franca Nigro, Franca Finocchiaro, Roberto Papa, and Pasquale De Vita. "Genetic variability in anthocyanin composition and nutritional properties of blue, purple, and red bread (Triticumaestivum L.) and durum (Triticumturgidum L. ssp. turgidumconvar. durum) wheats." *Journal of Agricultural and Food Chemistry* 62, no. 34 (2014a): 8686–8695.

Ficco, Donatella B. M., Anna M. Mastrangelo, Daniela Trono, Grazia M. Borrelli, Pasquale De Vita, Clara Fares, Romina Beleggia, Cristiano Platani, and Roberto Papa. "The colours of durum wheat: a review." *Crop and Pasture Science* 65, no. 1 (2014b): 1–15.

Ficco, Donatella Bianca Maria, Vanessa De Simone, Anna Maria De Leonardis, Valentina Giovanniello, Matteo Alessandro Del Nobile, Lucia Padalino, Lucia Lecce, Grazia Maria Borrelli, and Pasquale De Vita. "Use of purple durum wheat to produce naturally functional fresh and dry pasta." *Food Chemistry* 205 (2016): 187–195.

Gamel, Tamer H., Amanda J. Wright, Amy J. Tucker, Mark Pickard, Iwona Rabalski, Margaret Podgorski, Nicholas DiIlio, Charlene O'Brien, and El-Sayed M. Abdel-Aal. "Absorption and metabolites of anthocyanins and phenolic acids after consumption of purple wheat crackers and bars by healthy adults." *Journal of Cereal Science* 86 (2019): 60–68.

Gamel, Tamer H., AmandaJ. Wright, Mark Pickard, and El-Sayed M. Abdel-Aal. "Characterization of anthocyanin-containing purple wheat prototype products as functional foods with potential health benefits." *Cereal Chemistry* 97, no. 1 (2020): 34–38.

Garg, Monika, Meenakshi Chawla, Venkatesh Chunduri, Rohit Kumar, Saloni Sharma, Nand Kishor Sharma, Navneet Kaur et al. "Transfer of grain colors to elite wheat cultivars and their characterization." *Journal of Cereal Science* 71 (2016): 138–144.

Giordano, Debora, Monica Locatelli, Fabiano Travaglia, Matteo Bordiga, Amedeo Reyneri, Jean Daniel Coïsson, and Massimo Blandino. "Bioactive compound and antioxidant activity distribution in roller-milled and pearled fractions of conventional and pigmented wheat varieties." *Food Chemistry* 233 (2017): 483–491.

Gong, Lingxiao, Jingwen Chi, Yingquan Zhang, Jing Wang, and Baoguo Sun. "In vitro evaluation of the bioaccessibility of phenolic acids in different whole wheats as potential prebiotics." *LWT* 100 (2019): 435–443.

Guo, Z., Z. Zhang, P. Xu, and Y. Guo. "Analysis of nutrient composition of purple wheat." *Cereal Research Communications* 41, no. 2 (2013): 293–303.

Hailu Kassegn, Hagos. "Determination of proximate composition and bioactive compounds of the Abyssinian purple wheat." *Cogent Food and Agriculture* 4, no. 1 (2018): 1421415.

Himi, Eiko, Masahiko Maekawa, Hideho Miura, and Kazuhiko Noda. "Development of PCR markers for Tamyb10 related to R-1, red grain color gene in wheat." *Theoretical and Applied Genetics* 122, no. 8 (2011): 1561–1576.

Hosseinian, Farah S., Wende Li, and Trust Beta. "Measurement of anthocyanins and other phytochemicals in purple wheat." *Food Chemistry* 109, no. 4 (2008): 916–924.

Howitt, Crispin A., Colin R. Cavanagh, Andrew F. Bowerman, Christopher Cazzonelli, Lynette Rampling, Joanna L. Mimica, and Barry J. Pogson. "Alternative splicing, activation of cryptic exons and amino acid substitutions in carotenoid biosynthetic genes are associated with lutein accumulation in wheat endosperm." *Functional and Integrative Genomics* 9, no. 3 (2009): 363–376.

Khoo, Hock Eng, Azrina Azlan, Sou Teng Tang, and See Meng Lim. "Anthocyanidins and anthocyanins: Colored pigments as food, pharmaceutical ingredients, and the potential health benefits." *Food & Nutrition Research* 61, no. 1 (2017): 1361779.

Kiszonas, Alecia M., E. Patrick Fuerst, and Craig F. Morris. "Use of Student's t statistic as a phenotype of relative consumption preference of wheat (Triticum aestivum L.) grain." *Journal of Cereal Science* 65 (2015): 285–289.

Knievel, D. C., E. S. M. Abdel-Aal, I. Rabalski, T. Nakamura, and P. Hucl. "Grain color development and the inheritance of high anthocyanin blue aleurone and purple pericarp in spring wheat (Triticumaestivum L.)." *Journal of Cereal Science* 50, no. 1 (2009): 113–120.

Kumar, Pawan, R. K. Yadava, Babita Gollen, Sandeep Kumar, Ravi Kant Verma, and Sanjay Yadav. "Nutritional contents and medicinal properties of wheat: a review." *Life Sciences and Medicine Research* 22, no. 1 (2011): 1–10.

Lachman, Jaromír, Petr Martinek, Zora Kotíková, Matyáš Orsák, and Miloslav Šulc. "Genetics and chemistry of pigments in wheat grain–A review." *Journal of Cereal Science* 74 (2017): 145–154.

Landoni, Michela, Daniel Puglisi, Elena Cassani, Giulia Borlini, Gloria Brunoldi, Camilla Comaschi, and Roberto Pilu. "Phlobaphenes modify pericarp thickness in maize and accumulation of the fumonisin mycotoxins." *Scientific Reports* 1, no. 1 (2020): 1–9.

Li, Qian, Rui Liu, Tao Wu, and Min Zhang. "Aggregation and rheological behavior of soluble dietary fibers from wheat bran." *Food Research International* 102 (2017): 291–302.

Li, Wende, and Trust Beta. "Flour and bread from black-, purple-, and blue-colored wheats." In *Flour and Breads and Their Fortification in Health and Disease Prevention*, 59–67. Academic Press (2011).

Li, Wende, Fang Shan, Shancheng Sun, Harold Corke, and Trust Beta. "Free radical scavenging properties and phenolic content of Chinese black-grained wheat." *Journal of Agricultural and Food Chemistry* 53, no. 22 (2005): 8533–8536.

Li, Wende, Mark D. Pickard, and Trust Beta. "Effect of thermal processing on antioxidant properties of purple wheat bran." *Food Chemistry* 104, no. 3 (2007): 1080–1086.

Li, Wende, Trust Beta, Shancheng Sun, and Harold Corke. "Protein characteristics of Chinese black-grained wheat." *Food Chemistry* 98, no. 3 (2006): 463–472.

Li, Yaoguang, Dongyun Ma, Dexiang Sun, Chenyang Wang, Jian Zhang, Yingxin Xie, and Tiancai Guo. "Total phenolic, flavonoid content, and antioxidant activity of flour, noodles, and steamed bread made from different colored wheat grains by three milling methods." *The Crop Journal* 3, no. 4 (2015): 328–334.

Liu, Feng Ru, Li Wang, Ren Wang, and Zheng Xing Chen. "Calcium-induced disaggregation of wheat germ globulin under acid and heat conditions." *Food Research International* 62 (2014): 27–34.

Liu, Qin, Yang Qiu, and Trust Beta. "Comparison of antioxidant activities of different colored wheat grains and analysis of phenolic compounds." *Journal of Agricultural and Food Chemistry* 58, no. 16 (2010): 9235–9241.

Liu, Rui Hai. "Whole grain phytochemicals and health." *Journal of Cereal Science* 46, no. 3 (2007): 207–219.

Liu, Yanping, Ju Qiu, Yanfen Yue, Kang Li, and Guixing Ren. "Dietary black-grained wheat intake improves glycemic control and inflammatory profile in patients with type 2 diabetes: A randomized controlled trial." *Therapeutics and Clinical Risk Management* 14 (2018): 247.

Luthria, Devanand L., Yingjian Lu, and KM Maria John. "Bioactive phytochemicals in wheat: Extraction, analysis, processing, and functional properties." *Journal of Functional Foods* 18 (2015): 910–925.

Ma, Dongyun, JianZhang, JunfengHou, YaoguangLi, XinHuang, ChenyangWang, HongfangLu, YunjiZhu, and TiancaiGuo. "Evaluation of Yield, Processing Quality, and Nutritional Quality in Different-Colored Wheat Grains under Nitrogen and Phosphorus Fertilizer Application." *Crop Science* 58, no. 1 (2018): 402–415.

Marlett, Judith A., Michael I. McBurney, and Joanne L. Slavin. "Position of the American Dietetic Association: Health implications of dietary fiber." *Journal of the American Dietetic Association* 102, no. 7 (2002): 993–1000.

McCallum, J. A., and J. R. L. Walker. "Proanthocyanidins in wheat bran." *Cereal Chem* 67, no. 3 (1990): 282–285.

McIntosh, R. A., G. E. Hart, K. M. Devos, M. D. Gale, and W. J. Rogers. "Catalogue of gene symbols for wheat." In *Proc. 9th Int. Wheat Genet. Symp*, vol. 5, pp.1–235. Saskatoon: Print Crafters (1998).

Min, Byungrok, Liwei Gu, Anna M. McClung, Christine J. Bergman, and Ming-Hsuan Chen. "Free and bound total phenolic concentrations, antioxidant capacities, and profiles of proanthocyanidins and anthocyanins in whole grain rice (Oryza sativa L.) of different bran colours." *Food Chemistry* 133, no. 3 (2012): 715–722.

Mpofu, Archie, Harry D. Sapirstein, and Trust Beta. "Genotype and environmental variation in phenolic content, phenolic acid composition, and antioxidant activity of hard spring wheat." *Journal of Agricultural and Food Chemistry* 54, no. 4 (2006): 1265–1270.

Musilova, Milena, Vaclav Trojan, Tomáš Vyhnánek, and Ladislav Havel. "Genetic variability for coloured caryopses in common wheat varieties determined by microsatellite markers." *Czech Journal of Genetics and Plant Breeding* 49, no. 3 (2013): 116–122.

Nunes, Flávia Aparecida, Marcus Seferin, Vinícius Gonçalves Maciel, Simone Hickmann Flôres, and Marco Antônio Záchia Ayub. "Life cycle greenhouse gas emissions from rice production systems in Brazil: A comparison between minimal tillage and organic farming." *Journal of Cleaner Production* 139 (2016): 799–809.

Onipe, Oluwatoyin O., Afam, I. O., Jideani, and Daniso Beswa. "Composition and functionality of wheat bran and its application in some cereal food products." *International Journal of Food Science and Technology* 50, no. 12 (2015): 2509–2518.

Oomah, B. Dave, and G. Mazza. "Health benefits of phytochemicals from selected Canadian crops." *Trends in Food Science and Technology* 10, no. 6–7 (1999): 193–198.

Pasqualone, Antonella, Anna Maria Bianco, Vito Michele Paradiso, Carmine Summo, Giuseppe Gambacorta, Francesco Caponio, and Antonio Blanco. "Production and characterization of functional biscuits obtained from purple wheat." *Food Chemistry* 180 (2015): 64–70.

Pozniak, C. J., R. E.Knox, F. R. Clarke, and J. M.Clarke. "Identification of QTL and association of a phytoene synthase gene with endosperm colour in durum wheat." *Theoretical and Applied Genetics* 114, no. 3 (2007): 525–537.

Prashant, Ramya, Narendra Kadoo, Charushila Desale, Prajakta Kore, Harcharan Singh Dhaliwal, Parveen Chhuneja, and Vidya Gupta. "Kernel morphometric traits in hexaploid wheat (Triticumaestivum L.) are modulated by intricate QTL× QTL and genotype× environment interactions." *Journal of Cereal Science* 56, no. 2 (2012): 432–439.

Qualset, C. O., K. M. Soliman, C-C. Jan, J. Dvorak, P. E. McGuire, and H. E Vogt. "Registration of UC66049 Triticumaestivum blue aleurone genetic stock." *Crop Science* 45, no. 1 (2005): 432–433.

Robbins, Rebecca J. "Phenolic acids in foods: an overview of analytical methodology." *Journal of Agricultural and Food Chemistry*, 51, no. 10 (2003): 2866–2887.

Saini, Praveen, Nitin Kumar, Sunil Kumar, Peter Waboi Mwaurah, Anil Panghal, Arun Kumar Attkan, Vijay Kumar Singh, Mukesh Kumar Garg, and Vijay Singh. "Bioactive compounds, nutritional benefits and food applications of colored wheat: A comprehensive review." *Critical Reviews in Food Science and Nutrition* 61, no. 19 (2021): 3197–3210.

Sharma, Saloni, Venkatesh Chunduri, Aman Kumar, Rohit Kumar, Pragyanshu Khare, Kanthi Kiran Kondepudi, Mahendra Bishnoi, and Monika Garg. "Anthocyanin bio-fortified colored wheat: Nutritional and functional characterization." *PloS one* 13, no. 4 (2018): e0194367.

Shewry, Peter R. "Do ancient types of wheat have health benefits compared with modern bread wheat?" *Journal of Cereal Science* 79 (2018): 469–476.

Siebenhandl, Susanne, Heinrich Grausgruber, Nicoletta Pellegrini, Daniele Del Rio, Vincenzo Fogliano, Rita Pernice, and Emmerich Berghofer. "Phytochemical profile of main antioxidants in different fractions of purple and blue wheat, and black barley." *Journal of Agricultural and Food Chemistry* 55, no. 21 (2007): 8541–8547.

Singh, Kuldeep, Meenu Ghai, Monica Garg, Parveen Chhuneja, Parminder Kaur, Thorsten Schnurbusch, Beat Keller, and H. S. Dhaliwal. "An integrated molecular linkage map of diploid wheat based on a Triticumboeoticum× T. monococcumRIL population." *Theoretical and Applied Genetics* 115, no. 3 (2007): 301–312.

Somavat, Pavel, Deepak Kumar, and Vijay Singh. "Techno-economic feasibility analysis of blue and purple corn processing for anthocyanin extraction and ethanol production using modified dry grind process." *Industrial Crops and Products* 115 (2018): 78–87.

Sun, De-xiang, Z. U. O. Yi, Chen-yang Wang, Yun-ji Zhu, and Tian-cai Guo. "Diversity of antioxidant content and its relationship to grain color and morphological characteristics in winter wheat grains." *Journal of Integrative Agriculture* 13, no. 6 (2014): 1258–1267.

Sun, Qian, Aiqin Zhang, Zheng Feei Ma, Hongxia Zhang, Fang Li, Yang Yang, and Lingming Kong. "Optimal formulation of a product containing black wheat granules." *International Journal of Food Properties* 21, no. 1 (2018): 2062–2074.

Syed Jaafar, Sharifah N., Johanna Baron, Susanne Siebenhandl-Ehn, Thomas Rosenau, Stefan Böhmdorfer, and Heinrich Grausgruber. "Increased anthocyanin content in purple pericarp× blue aleurone wheat crosses." Plant Breeding 132, no. 6 (2013): 546–552.

Sytar, Oksana, Paulina Bośko, Marek Živčák, Marian Brestic, and Iryna Smetanska. "Bioactive phytochemicals and antioxidant properties of the grains and sprouts of colored wheat genotypes." *Molecules* 23, no. 9 (2018): 2282.

Tian, Shuang-Qi, Yong-HengLi, Zhi-Cheng Chen, and Yong-Feng Qiao. "Effects of layering milling technology on distribution of green wheat main physicochemical parameters." *Journal of Food Quality* (2017).

Tian, Shuang-qi, Zhi-cheng Chen, and Yi-chun Wei. "Measurement of colour-grained wheat nutrient compounds and the application of combination technology in dough." *Journal of Cereal Science* 83 (2018b): 63–67.

Tian, Shuang-Qi, Zhi-Cheng Chen, and Yong-Feng Qiao. "Analysis of main physicochemical parameters in purple wheat with different milling technology." *Journal of Food Processing and Preservation* 42, no. 1 (2018a): e13382.

Tsuda, Takanori. "Dietary anthocyanin-rich plants: biochemical basis and recent progress in health benefits studies." *Molecular Nutrition and Food Research* 56, no. 1 (2012): 159–170.

Tyl, Catrin E., and Mirko Bunzel. "Antioxidant activity-guided fractionation of blue wheat (UC66049 Triticum aestivum L.)." *Journal of Agricultural and Food Chemistry* 60, no. 3 (2012): 731–739.

Van Hung, Pham. "Phenolic compounds of cereals and their antioxidant capacity." *Critical Reviews in Food Science and Nutrition* 56, no. 1 (2016): 25–35.

Wang, Xin, Xiaocun Zhang, Hanxue Hou, Xin Ma, Silong Sun, Hongwei Wang, and Lingrang Kong. "Metabolomics and gene expression analysis reveal the accumulation patterns of phenylpropanoids and flavonoids in different colored-grain wheats (Triticum aestivum L.)." *Food Research International* 138 (2020): 109711.

World Health Organization, and United Nations University. *Protein and Amino Acid Requirements in Human Nutrition*. Vol. 935. World Health Organization, 2007.

Wu, Binning, Rohil Bhatnagar, Vijaya V. Indukuri, Shara Chopra, Kylie March, Nina Cordero, Surinder Chopra, and Lavanya Reddivari. "Intestinal mucosal barrier function restoration in mice by maize diet containing enriched flavan-4-Ols." *Nutrients* 12, no. 4 (2020): 896.

Xuefeng, Zong, Zhang Jiankui, and Li Bangxiu. "Relationship between antioxidation and grain colors of wheat (Triticum aestivum L.)." *Acta Agronomica Sinica* 32 (2006): 237–242.

Yu, Lilei. "Identification and antioxidant properties of phenolic compounds during production of bread from purple wheat grains." *Molecules* 20, no. 9 (2015): 15525–15549.

Zanoletti, Miriam, Parisa Abbasi Parizad, Vera Lavelli, Cristina Cecchini, Paolo Menesatti, Alessandra Marti, and M. Ambrogina Pagani. "Debranning of purple wheat: recovery of anthocyanin-rich fractions and their use in pasta production." *LWT* 75 (2017): 663–669.

Zhu, Fan. "Anthocyanins in cereals: Composition and health effects." *Food Research International* 109 (2018): 232–249.

Žilić, Slađana, Arda Serpen, Gül Akıllıoğlu, Marijana Janković, and Vural Gökmen. "Distributions of phenolic compounds, yellow pigments and oxidative enzymes in wheat grains and their relation to antioxidant capacity of bran and debranned flour." *Journal of Cereal Science* 56, no. 3 (2012): 652–658.

Phytochemical Composition, Nutraceutical, and Techno-Functional Attributes of Amaranth (*Amaranthus cruentus*)

Bababode Adesegun Kehinde, Shafiya Rafiq, and B. N. Dar

Contents

12.1 Taxonomy, distribution, and morphology

Amaranths, or Pigweeds, are eudicots in the Caryophyllales order of the Amaranthaceae family (Montoya-Rodriguez, 2015). This genus of plants has 60 species, 40 indigenous to the Americas, and the remainder spread over Africa, the Asian continents, and Europe. There are seventeen kinds of edible leaves and three varieties of grain amaranths (Zhigila et al., 2014). The taxonomic key list of Sauer 1995 reported three principal species of *Amaranthus*, for grain production: *Amaranthus caudatus* L.; *Amaranthus hypochondriacus* L.; *Amaranthus cruentus* L. The USDA presents the following taxonomic classification of *A. cruentus*:

Kingdom	*Plantae*
Subkingdom	*Tracheobionta*
Superdivision	*Spermatophyta*
Division	*Magnoliophyta*
Class	*Magnoliopsida*
Subclass	*Caryophyllidae*
Order	*Caryophyllales*

Family	*Amaranthaceae*
Genus	*Amaranthus* L.
Species	*Amaranthus cruentus* L.

Amaranthus cruentus is an annual invasive, drought-resistant, herbaceous plant, which reproduces millions of seeds and has a limited growth period of 4–6 weeks (Makinde et al., 2010). Amaranth is well adapted for temperate to tropical climates owing to its C4-photosynthetic pathway, which makes it capable of utilizing carbon dioxide very efficiently by fixing it in the chloroplasts of specialized cells surrounding the leaf vascular bundles. It is not only CO_2 assimilation efficiency that demonstrates the ecological benefit of species using this photosynthetic pathway. Compared to C3 plants, these plants need 60 percent less water to produce biomass. This characteristic of C4 plants is also responsible for maximum photo-synthetic efficiency in high light intensity conditions and high temperatures. Through osmotic adjustment, they can tolerate some lack of water without water wilting (Sage, 2007, Costea, 2003). Furthermore, because of this C4-photosynthetic pathway, amaranth can thrive even with an annual rainfall of 200 mm (Putnam, 1990). On the other side, temperatures below 4 °C can injure such plants (National Research Council, 1984). The Amaranthus plant develops a big, central taproot. Stems are usually thick, straight-ribbed, and red. Furthermore, because of this C4-photosynthetic pathway, amaranth can thrive even with an annual rainfall of 200 mm (Putnam, 1990). On the other side, temperatures below 4 °C can injure such plants (National Research Council, 1984). The Amaranthus leaves are spirally arranged, simple, without stipules, and vary in shape from oval to rhombic-ovate. Leaf and stem surfaces are covered in small, fine hairs. Numerous green unisexual blooms produce finger-like spikes with a long, dense terminal panicle and axillary spikes below. In general, the terminal spike is loose. The inflorescence is prickly due to the presence of five tepal segments that are lanceolate, acute, 2–3 mm long, and have a sharp and long tip. The entire plant turns reddish as it matures. The massive and complicated inflorescence is made up of several axillary concentrated cymes that end in racemes and spikes. The top one is perpendicular, with slim branches approximately 45 cm long (Grubben et al., 2004). The inflorescence is over 50 cm long and has a wide range of colors due to the presence of Betacyanin pigments. Each produces approximately 50,000 thin, smooth seeds in a round or lenticular shape with a diameter of 1–1.5 mm in a variety of colors ranging from white to yellow, golden, red, brown, and black (Robertson and Clemants, 2003). In terms of chromosomes, it is diploid. There is no endosperm in the seed's micropylar region (Valcárcel-Yamani, 2012; Achremowicz, 2015). Either light only or together with high temperatures stimulates germination. Temperatures of 20/35°C and light give the greatest rise (Weaver, 1984).

12.2 Nutritional and phytochemical composition

Amaranths cruentus is a highly nutritional and functional crop (Pasko et al., 2008). Because of its high protein content, beneficial amino acid composition, interesting fatty acid profile, and valuable antioxidants, A. cruentus is gaining popularity among pseudo-cereals as an unconventional grain crop for replacing traditional cereals in the health food market (Rezaei et al., 2014; Seguin et al., 2013; Karamac et al., 2019). Starch is the most abundant polysaccharide in amaranth grains, accounting for 50 to 60 percent of total carbohydrates and 4 to 5 percent dietary fiber (Venskutonis and Kraujalis, 2013; Burisová, 2001). Sucrose, the major sugar in the amaranth seed's carbohydrates profile, is estimated to be 2 to 3 times higher than wheat grain and non-starch polysaccharides. Inositol, maltose, raffinose, and a trace amount of glucose, fructose, and Stachyose are also found in Amaranth grains (Januszewska-Jówiak et al., 2008; Silva-Sánchez et al., 2004). Amaranth grains have a fiber content ranging from 2.2 to 8.1 percent. The soluble dietary fiber (SDF) component accounts for 14 percent of the total fiber content. Pectin, uronic acid, and undigested biopolymers made up of glucose, arabinose, xylose, mannose, and galactose make up the SDF (Escudero et al., 2004). Although the granules of amaranth starch are small in comparison to those of other grains, they have good gelatinization and freeze/thaw stability. Another feature of amaranth starch is that it contains 88.9–99.9 percent amylopectin, which makes it waxy and gives it unique properties like high viscosity and gelatinization at higher temperatures (Venskutonis and Kraujalis, 2013).

On a dry basis, the protein content of *A. cruentus* ranges from 13.2 to 18.2 percent (Computation, 1998). This plant is unique in that it has a balanced amino acid content, with 65 percent of proteins in the embryo and only 35 percent in the perisperm. Albumins (40 percent), globulins (20 percent), prolamins (2–3 percent), and glutelins (25–30 percent) make up the total protein content of *A. cruentus* (Segura-Nieto, 1994). Furthermore, because amaranth does not contain gluten, it can be included in the diet of people who have celiac disease (Wolosik and Markowska, 2019). In general, two primary groups of globulins can be distinguished based on their sedimentation coefficient: 7S and 11S globulins. These globulins are storage proteins in legumes. Storage globulins 7S (conamaranthin) and 12S (amaranthin) have been found in amaranth (Marcone, 1999) and Marcone et al., 1994). Amaranth protein has high biological quality, mainly due to the balance of essential amino acids, with high levels of amino acids considered limiting in most grains, such as lysine (5 percent) and sulphur-containing amino acids (4.4 percent) – methionine and cysteine. Amaranth is also good for baby food because it contains a lot of arginine and histidine, which are important amino acids for newborns. Other amino acids, such as threonine, leucine, valine and isoleucine, are also present in amaranth in smaller amounts (Schmidt et al., 2021). Amaranth has a biological value of 75 (due to the high amount of lysine), which makes it comparable to milk protein (Bressani et al., 1992). According to some studies, amaranth has a biological value of around 85 (Pedersen, et al. 1987). In animal feeding tests conducted by Irving et al. in 1981, amaranth was found to have a PER-value of 1.5. According to Bressani et al. (1988), *A. caudatus* has a protein digestibility of 76 percent. Pedersen, et al. (1987) claimed that *A. caudatus* protein digestibility was as high as 87 percent.

Amaranth grains contain 5.7–9 percent of lipids. The wide range of fat content is very noticeable; different species, as well as geographic locations, cause significant variation. The fat content of the germ and seed coat is higher than that of the perisperm (Irving, 1981; Becker, 1994) Unsaturated fatty acids make up the majority of the lipid fraction, which ranges from 67 to 80 percent. The lipid appears to be attractive not only because of its quantity, but also because of its fatty acid profile, which is dominated by three fatty acids: palmitic (23.4 percent), oleic (24 percent), and linoleic acids (47 percent) (Nasirpour-Tabriz et al., 2020). The degree of unsaturation is greater than 75 percent. (Berghofer, 2002). Phospholipids are found in small amounts in the oil fraction of amaranth, around 5 percent. In amaranth oil, Opute 1979 found 3.6 percent phospholipids, with the cephalin fraction at 13.3 percent, the lecithin fraction at 16.3 percent, and the phosphoinositol fraction at 18.2 percent. Squalene makes up the majority of the unsaponifiable in amaranth oil. It is a highly unsaturated open-chain triterpene that is used as an oxidation-resistant industrial lubricant and in the cosmetic industry.

Regarding mineral composition, amaranth has a valuable level of minerals. Amaranth grains are rich in phosphorus, calcium, manganese, iron, nickel, chromium, zinc, copper and selenium; 66 percent of total minerals are found in the bran and germ fractions. The concentrations of phosphorus, calcium and iron are expressive and higher than cereals and milk (Ferreira et al., 2007; Amaya-Farfan et al., 2005).

The presence and levels of vitamins in amaranth grain vary depending on the species but, generally, beta-carotene (provitamin A) and ascorbic acid (C) are present in nutritionally adequate levels. In addition, Thiamine (B_1), riboflavin (B_2), niacin (B_3), tocopherols (E) and tocotrienols can also be found (Berghofer et al., 2002).

12.3 Phytochemical composition

The outer coats of amaranth seeds contain phytochemicals that serve as chemical protection against insects and bacteria. These substances can be either hydrophilic or lipophilic in nature. The seed coat of amaranth seeds is the primary source of phenolic chemicals, notably phenolic acids. Phenolic acids, flavonoids, and tannins are all examples of hydrophilic phenolics, and they make up the bulk of plant secondary metabolites that have a wide range of physiological effects (Tang and Tsao, 2017). Studies conducted on Amaranthus species have identified a variety of phenolic acids, flavonoids, and glycosides. Overall, the quantity of phenolic acids found in amaranth seeds ranged from 168 to 309mg/kg, with 7 to 14 percent of those phenolic acids being readily extractable. It was reported that amaranth seeds and sprouts had high levels of rutin and phenols, including gallic acid and p-hydroxybenzoic acid. According to Pasko et al., 2008 Gallic, protocatechuic, and p-hydroxybenzoic acid are the three most common free phenolics in amaranth grains. Amaranth seeds contain bound ferulic acids, primarily trans-ferulic acid and cis-ferulic acid that can be extracted with alkali, and enzyme hydrolyses (Bunzel et al., 2005). The list of phenolics identified in amaranth grain with structure and concentration is shown in **Table 12.1**. Betalains are generally categorized as red-violet betacyanins and yellow-orange betaxanthins (Esatbeyoglu et al., 2015). The primary betacyanins in Amaranthus are amaranthine and isoamaranthine. Vitamin E homologs tocopherols and tocotrienols are found in a variety of foods including amaranth grains. All four tocopherol isoforms (α, β, γ, *and* δ) have been reported in amaranth grains, with γ-tocopherol being the most prevalent, followed by α-tocopherol (Alvarez-Jubete et al., 2010).

12.4 Nutraceutical and techno-functional attributes of amaranth

Processed derivatives from amaranth grains and vegetative parts have been used for their various techno-functional potentials in several studies (**Table 12.2**). It has been adopted for food preservation, as a material resource for the preparation of additives, and for the enhancement of certain quality parameters in foods (Sharma et al., 2019). Amaranth is a pseudocereal with a substantial load of nutraceuticals such as bioactive peptides, polyphenols, flavonoids, and phytosterols amongst several others (House et al., 2020). These biologically functional agents are responsible for their health benefits on consumption (Kehinde et al., 2020a; Sharma et al., 2021). Several studies have been conducted using in vivo approaches involving patients or laboratory animals, and in vitro techniques involving aliquots of biochemical assays to investigate these potentials (Kim et al., 2006; Sharma et al., 2019). Such studies usually commence with the extraction of the bioactive component of interest, usually with suitable solvents and processing methodologies, the further purification of such compounds, their identification, and functional characterization (Kehinde et al., 2020b; Kehinde, Sharma and Kaur, 2020). Regarding amaranth, studies related to health benefits such as antidiabetic, antioxidant, antihypertensive, antimicrobial, and anticancer amongst several others have been examined.

Table 12.1 Chemical Structure and Concentration of Phenolic Compounds Present in the Amaranth Grains

Name of Phenolic compound	Chemical Structure	Concentration in amaranth grains (mg/kg)	References
Rutin	(PubChem CID 5280805)	7–592	Hirose et al., 2010 Hemalatha et al., 2016
Gallic acid	(PubChem CID 370)	11–440	Repo-Carrasco-Valencia et al., 2010
p-hydroxybenzoic acid	(PubChem CID 370)	8.5–20.9	Tang et al., 2015

(*Continued*)

Table 12.1 *(Continued)*

Name of Phenolic compound	Chemical Structure	Concentration in amaranth grains (mg/kg)	References
Vanillic acid	(PubChem CID 8468)	15.5–69.5	Tang et al., 2015 Repo-Carrasco-Valencia et al., 2010
Protocatechuic acid	(PubChem CID 72)		
Ferulic acid	(PubChem CID 445858)	120–620.0	Bunzel et al., 2005 Repo-Carrasco-Valencia et al., 2010 Pasko et al., 2008 Miranda et al.,2014

Table 12.1 *(Continued)*

Name of Phenolic compound	Chemical Structure	Concentration in amaranth grains (mg/kg)	References
Quercetin	\n(PubChem CID 5280343)	7–592	Brady et al., 2017
Kaempferol	\n(PubChem CID 5280863)	22.4-59.7	Tang et al., 2015
Kaempferitrin	\n(PubChem CID 5486199)	22.4-55.97	Tang and Tsao, 2017
Amaranthine		151.3	Khanam and Oba, 2013
Isoamaranthine		58.7	Khanam and Oba, 2013

(Continued)

Table **12.1** *(Continued)*

Name of Phenolic compound	Chemical Structure	Concentration in amaranth grains (mg/kg)	References
Betanin	(Pubchem CID 12300103)	17.7	Abderrahim et al., 2015, Khanam and Oba, 2013
Isobetanin	(Pubchem CID 6325438)	5.0	Khanam and Oba, 2013 Steffensen et al., 2011
Ascorbic acid	(PubChem CID 54670067)	41.3–70.5	Steffensen et al., 2011 Miranda, et al., 2014

Table 12.1 *(Continued)*

Name of Phenolic compound	Chemical Structure	Concentration in amaranth grains (mg/kg)	References
Thiamine	(PubChem CID 1130)	0.72–2.4	Steffensen et al., 2011 Vega-Gálvez et al., 2010 Saravacos et al., 2011
Riboflavin	(PubChem CID 493570)	1.8–2.7	Steffensen et al., 2011 Vega-Galvez et al., 2010
Niacin	(PubChem CID 938)	8.9–10	Steffensen et al., 2011 Vega-Gálvez et al., 2010

(Continued)

Table 12.1 *(Continued)*

Name of Phenolic compound	Chemical Structure	Concentration in amaranth grains (mg/kg)	References
Tocopherol	(Pubchem CID14986)	2.97-57.07	
Tocotrienol	(Pubchem CID 9929901)	2.0–48.5	Tang and Tsao, 2017
Lutein	Pubchem CID 5281243	3.55–4.29	Tang et al., 2015 Tang et al., 2016

Table 12.1 *(Continued)*

Name of Phenolic compound	Chemical Structure	Concentration in amaranth grains (mg/kg)	References
Zeaxanthin	Pubchem CID 5280899	0.14–0.32	Tang et al., 2015 Tang et al., 2016

Table 12.2 Evaluated Techno-Functional Attributes of Amaranth and Its Derivatives

Amaranth part	Application type	Study Mechanism	Results obtained	Ref
Vegetable	Multifunctional food packaging	Antioxidant and antimicrobial effect	Effective against *Listeria monocytogenes, Salmonella typhimurium, Staphylococcus aureus,* and *Escherichia coli* and with 6.8 percent to 95.8 percent (at 5 mg/mL) radical scavenging activity	Hu et al., 2020
Ethyl acetate extracts	Antimicrobial and antimycotic effects	Effect against yeast strains *Candida albicans, Candida maltosa* and *Saccharomyces cerevisiae,* and fungal strains *Aspergillus flavus, Fusarium culmorum* and *Alternaria alternata*	Extract was found effective at a concentration of 5 mg/mL	Mosovska and Birosova, 2012
Seed flour	Effect of incorporation on the quality of cookies	Effect on dough and pasting properties of amaranth–wheat flour cookies	Amaranth flour found to improve the quality of cookies even at 25–30 percent	Sindhuja, Sudha, and Rahim, 2005
Milled dried leaves	Effect on nutritional profile of extruded provitamin A-biofortified maize snacks	Evaluation of ash, protein, and provitamin A content	Increase in protein from 9.12 - 10.94 g/100g, increase in ash content from 0.53 - 0.58 g/100g and enhancement of provitamon A composition	Beswa et al., 2016
Flour	Effect of on quality of gluten-free bread	Estimation of how water, albumen and fat content affect amaranth-bread quality	Albumen addition (from 0 to 0.04 g/g of flour) increased crumb elasticity by 20 percent, and 0.6 to 0.8 g/g of flour caused by a 33 percent water content increase	Schoenlechner et al., 2010

(Continued)

Table 12.2 *(Continued)*

Amaranth part	Application type	Study Mechanism	Results obtained	Ref
Flour	Preparation of crackers and tortilla	Effect on chemical composition, physiochemical properties, essential amino acids, essential unsaturated fatty acids and minerals of different bakery products	amaranth flour increased protein, fat, ash and fiber, decreased yellowness and lightness and exhibited higher antioxidant activity than control formulas	Gebreil, Ali, and Mousa, 2020
Leaves	Preparation of canned leaves product	Evaluation of sensory acceptance	Amaranth leaves canned in brine were described as sweet by panelists	Onyeoziri, Kinnear, and de Kock, 2018
Flor from leaves	Preparation of pasta	Evaluation of chemical composition, cooking quality, textural, and sensory/consumer acceptance of amaranth pasta samples	Pasta samples reported to have cooking quality (2.15 pasta weight increase, 4.47 percent residue loss), Iron content: 9.1 mg/100 g, and protein content: 14.18 g/100 g,	Borneo and Aguirre, 2001
Flour	Rice dough and bread preparation	Effect on dough strength	improved crumb porosity of composite bread	Burešová et al., 2017
Expanded seeds	Preparation of grit	Examination of grit incorporation on quality of canned meat	Amarant grit was observed to have a positive effect on the water-holding capacity of stuffing and allowed to reduce the cooking losses in cans	Ostoja et al., 2002
Flour	Preparation of corn-based extruded snacks	Examination of Physical and sensory features	Flour addition was observed to lessen the dominance of roughness and enhance the dominance of crunchiness	Ramos Diaz et al., 2015

12.4.1 Antidiabetic effects

Diabetes circumvents a collection of bodily disorders affecting the utilization of glucose. It is a metabolic syndrome disease characterized by heightened sugar levels in the body over long time durations with symptoms such as increased thirst and appetite and frequent urination as the most prominent ones (Kehinde and Sharma, 2018). Other symptoms that may subtly or rapidly develop over time include hyperphagia (or Polyphagia), which connotes an extreme and unusual desire for food with a concurrent occurrence of overeating; Polydipsia, which involves excessive drinking arising from excessive thirst; Polyuria, which consists of passage and excessive production of urine in volumes well above two and a half liters; weight loss in terms of bodily fluids. The disease can also initiate other symptoms in different parts of the body such as the eyes (blurry vision and vision changes), skin (rashes and itchy skin), prolonged healing of bodily injuries, fatigue and headache.

Extracts and phytochemicals extracted from amaranth and its vegetative parts have been shown to possess anti-diabetic potentials through one or more biochemical pathways. Research focus has been implemented on them based on the perceived advantages they are understood to possess over conventional pharmaceutical agents such as metformin. The most probable advantage is the reduced risks of side effects relative to synthetic drugs. Some potential mechanisms for the management of diabetes and associated disorders by bioactive components derived from amaranth include:

1. Inhibition of digestive enzymes such as a-glucosidase and a-amylase.
2. Inhibition of CD26 (cluster of differentiation 26) or adenosine deaminase complexing protein 2 or Dipeptidyl peptidase-4 (DPPIV) protein which plays a role in glucose metabolism.
3. Regulation of the metabolism of 31-amino-acid-long peptide hormone known as Glucagon-like peptide-1 (GLP-1) and other incretins.
4. Reduction of serum glucose levels.
5. Enhancement of insulin secretion.

Studies involving in vivo mechanisms usually adopt the induction of laboratory rats with streptozotocin through intravenous administration, although other studies simply conduct oral administration of high-calorie diets. Balasubramanian et al. (2017) examined the antidiabetic potentials of amaranth through the oral administration of *Amaranthus hybridus* ethanol leaf extract at doses of 400 and 200 mg/kg to streptozotocin-induced diabetic rats. After 14 days, blood glucose levels of 181.2 and 152.2 mg/dL were reported respectively for treated rats relative to 287.0 mg/dL for control.

Girija et al. (2011) conducted a similar study but with methanol extracts of leaves in three different varieties: *Amaranthus caudatus*, *Amaranthus spinosus* and *Amaranthus viridis*. *After 20 days of oral administration* at doses of 400 and 200 mg/kg, the blood glucose level was reported to decrease from an average of above 90 mg/kg BW, to 61.56±1.88 and 69.69±1.81 mg/kg BW for the *Amaranthus viridis* variety respectively. 63.46±2.2 and 69.92±1.9 mg/kg BW were reported for *Amaranthus caudatus* while the *Amaranthus spinosus* variety was found to be 60.69±1.62 and 68.28±1.72, respectively.

Kim et al. (2006) performed an analogous study with amaranth grain at an oral dosage of (500 g/kg diet) and amaranth oil at an oral dosage of (100 g/kg diet) on Male Sprague-Dawley induced with diabetes through administration of streptozotocin injections at a concentration of 50 mg/kg BW. The study reported that the amaranth oil and grains significantly increased insulin levels and decreased serum glucose in diabetic rats.

12.4.2 Antihypertension

Pressure is fundamentally understood to be the force per unit area, and blood pressure goes by the same trend. It is the amount of force by which blood flows through the arteries. Blood pressure is established by the resistance to blood flow in the arteries and also by the quantity of blood pumped by the heart. Accordingly, narrower arteries having higher pump quantities from the heart will result in higher blood pressures. The blood pressure is expressed in millimeters of mercury (mm Hg) having two categories: the diastolic pressure (bottom number) which is the lower and the second number that indicates the pressures in the arteries between heartbeats; the systolic pressure (top number) which is the upper or first number that offers an estimate of the pressure in the arteries when the heart beats. On another note, hypertension or high blood pressure is a chronic metabolic syndrome disease in which there is a persistent elevation of the blood pressure in the arteries (Sosalagere, Kehinde, and Sharma, 2021). It is usually on a long-term basis and is a potential risk factor for complications with different medical conditions and loss of functionality of different body parts such as vision loss, coronary artery disease, stroke, kidney disease, and heart failure. For normal blood pressure, the systolic and diastolic readings range from 100–130 and 60–80 mmHg, though these values can vary with age, for high blood pressure patients, readings can be around 130–139 and 80–89 mmHg or even higher, depending on the stage of hypertension. The symptoms of this disorder are not its best indicators; medical screening remains the recommended diagnosis. However, reports of headaches, fainting, unpleasant dizziness, recurring anxiety, and vertigo have been observed. Secondary hypertension, on the other hand, is initiated by an underlying and identifiable primary cause, and is less common. It could be instigated by endocrine disorders, adrenal issues, pregnancy, cancers, or even drug intake, amongst a host of others.

Lifestyle changes such as physical exercise for weight loss, decreased alcohol and salt intake, regular sleep, and healthy dieting are encouraged for the management of hypertension, though when insufficient, medications are prescribed for its therapeutic management. In recent years amaranth has been extensively investigated for its antihypertensive properties. Research has been more focused on the potentialities of its storage proteins, hydrolysates, hydrolysate fractions, and peptides, as anti-hypertensive agents. Some studied mechanisms for the working of these peptides as antihypertensive nutraceuticals include:

1. Inhibition of the Angiotensin-converting enzyme (ACE) inhibitors for the relaxation of the arteries and veins to the lowering of blood pressure.
2. Stimulation of endothelial nitric oxide (eNOs) for nitric oxide (NO) syntheses thereby causing retardation of platelet aggregation, vasorelaxation, and enhancing the survival of endothelial cells.
3. Direct renin inhibition for the blockage of the renin enzyme from initiating a course of action for blood pressure regulation, thereby causing a widening and relaxation of blood vessels, enhancing the easier flow of blood through its channels, and consequently lowering blood pressure.
4. Reduction of systolic blood pressure and reduction of peripheral resistance of heart muscles.

The purification of such peptides begins with the extraction of proteins from the parent material seeds or other amaranth parts of interest (Kaur et al., 2020). Hydrolysis follows subsequently with the introduction of enzymes for their specificity advantage, or by other indirect proteases, such as from microbial fermentation. The hydrolysates can be examined for their antihypertensive tendencies as a whole or fractionated based on their molecular weights (in kilodaltons), and each fraction examined separately. The hydrolysates can be further fractionated into peptides using chromatographic techniques or by centrifugation with filtration membranes (Kaur et al., 2019). The peptides can be further purified and then sequenced to determine their amino acid profiles, after which they are being characterized based on one or more antihypertensive mechanisms. Fritz et al. (2011) isolated proteins from amaranth, hydrolyzed them with different enzymes including pronase (0.1 mg/ml), Alcalase (0.8 mg/ml), Trypsin (0.1 -0.3 mg/ml), chymotrypsin (0.1 and 0.3 mg/ml), and papain (0.3 mg/ml) for four hours, and examined their Angiotensin-converting enzyme inhibition *invtro* and *in vivo* using male spontaneously hypertensive rats. The study reported a half-maximal inhibitory concentration (IC50) of 0.12 mg/ml having an equivalence of 300–600 µM. Antihypertensive effects were also reported in the rats. Ramírez-Torres et al. (2017) conducted a study on the use of alcalase enzyme for the optimal hydrolysis of antihypertensive peptides from amaranth protein isolate. The study reported optimal conditions of hydrolysis to be enzyme concentration of 0.04 mU/mg, the temperature of 52 °C, and pH of 7.01 for 6.16 hours. These conditions achieved a hydrolysis degree of 74.77 percent and an ACE-inhibition of 93.5 percent.

12.4.3 Anticancer

Cancer is a collection of a sizeable sum of disorders bearing the common characteristic of involving the development of aberrant cells that divide hysterically and bear the capability to disrupt healthy body tissues by infiltration. It is the generic term for those diseases that entail the rapid creation of abnormal cells that develop beyond their natural profiles into neoplasms and malignant tumors and spread to other organs. Such spreads are termed metastasis and are the fundamental causes of cancer-related fatalities. Unlike cancer cells, benign tumors do not invade or spread to other body parts and when they are surgically removed, there are possible chances of their not growing back. In contrast, healthy body cells grow under natural biochemical processes and undergo their naturally programmed cell death: a phenomenon known as apoptosis. Normal cells also do not spread or invade other tissues but are eliminated by the immune system if damaged. Cancer can be initiated by heredity from the parent, cellular division errors, DNA damage arising from toxic materials in the environment such as free radicals, smoke chemicals, and others. Cancer symptoms could be localized or systemic. For the local ones, the symptoms may be due to tumor ulceration or the tumor mass such as blocking or narrowing of the bowels for colorectal cancer, pneumonia or cough for lung cancer, or lump masses in testicles or breasts. Systemic examples result from the body's reaction to cancer, and they include skin changes, muscle wastage, and fatigue. Because of their large collection, cancers are commonly categorized based on the cell type that their tumors take after. Accordingly, lymphoma cancer originates from the lymphocytes, leukaemia from the bone marrow, sarcoma from connective tissues such as nerves, fat, cartilages or bones, blastoma from embryonic cells, and carcinoma from epithelial cells.

Cancer patients are usually admonished to avoid the consumption of alcohol and tobacco, avoid exposure to harmful radiation, smoke, and other air pollutants, maintaining a healthy diet with substantial amounts of fruits and vegetables, and avoiding a sedentary lifestyle (Sharma et al., 2020). The phenolics and overall polyphenolic content, peptides, phytosterols, and several other phytonutrients in amaranth have been reported to bear anticancer properties. Hyperoside, catechin, myricetin, trans-cinnamic acid, gallic acid, gentisic acid, kaempferol, protocatechuic acid, apigenin, Stigmosterol, Campesterol, and β-sitosterol are a few examples of amaranth phenolics, flavonoids and phytosterols with anticancer potentials (Sarker and Oba, 2020). Some potential anticancer mechanisms of amaranth nutraceuticals include:

1. Initiation of cellular death of cancer cells, a process termed as necrosis.
2. Enhancement of the immune system to counteract cancer cells.
3. Distortion of cellular proteins and DNA of cancer cells.
4. Inhibition of formation or growth of cells, such as retarding angiogenesis.
5. Initiation of apoptosis of cancerous cells.
6. Disruption of membrane integrity of cancer cells and inhibiting their cellular migration.

House et al. (2020) examined the anticancer effects of methanolic extracts from different species of amaranth, namely: Amaranthus viridis, Amaranthus tricolor, Amaranthus spinosus, and Amaranthus dubius using MDAMB231 and MCF7 cells. The Amaranthus spinosus variety was reported to be more effective against the MDAMB231 cells, with an observed IC50 value of 82.11± 7.02 mg/mL while the Amaranthus dubius was reportedly more effective against the MCF7 cells with an IC50 of 48.22 ± 3.65 mg/mL. Taniya et al. (2020) used simulated gastrointestinal digestion to proteolyze amaranth proteins into peptides and measured their anticancer activity against human triple negative breast cancer cells. Results reported indicated that the purified peptides showed a concentration for 50 percent of maximal inhibition of cell proliferation (GI50) of 48.3 ± 0.2 μg/ml.

12.5 Conclusion

The grain flour derivatives, leaves and other vegetative parts of amaranth have been found to bear multiple phytochemical accessories that are beneficial for food and health. Amaranth and its derivatives have been studied to bear potentialities for regulation of blood sugar, management of hypertension, and anticancer effects. Beyond these, amaranth products have been studied to be applicable as natural additives to impart desirable techno-functional effects in foods. They have been found workable to suiting requirements specific to various stages of food processing such as packaging, flavor addition, color enhancement, taste improvement, shelf life extension, and conservation of oxidation-prone constituents. Amaranth is a potentially viable solution to numerous modern concerns regarding food processing and health issues. Continuous research efforts and industrial contribution could significantly aid an optimal exploitation of the underutilized benefits of amaranth.

References

Abderrahim, F., Huanatico, E., Segura, R., Arribas, S., Gonzalez, M. C., and Condezo-Hoyos, L. (2015). Physical features, phenolic compounds, betalains and total antioxidant capacity of coloured quinoa seeds (Chenopodium quinoa Willd.) from Peruvian Altiplano. *Food Chemistry, 183*, 83–90.

Achremowicz, B., Ceglińska, A., Haber, T., Hołownia, J., Just, K., and Obiedziński, M. 2015. General characteristics and technological applicability of Amaranthus seeds. *Part I. Techniki Przetwórstwa Spożywczego, 1*: 118–125.

Alvarez-Jubete, L., Arendt, E. K., and Gallagher, E. 2010. Nutritive value of pseudocereals and their increasing use as functional gluten-free ingredients. *Trends in Food Science and Technology, 21*(2): 106–113.

Amaya-Farfan, J., Marcílio, R., Spehar, C. R., and Deveria, O. 2005. Brasil Investir Em Novos Grãos Para A Sua Alimentação a Proposta Do Amaranto (*Amaranthus* SP). *Seguranca. Alimentar Nutricional-Portal de Periodicos Eletronicos, 12*(1): 47–56. doi: 10.20396/san. v12i1.1838

Balasubramanian, T., Karthikeyan, M., Muhammed Anees, K. P., Kadeeja, C. P., and Jaseela, K. 2017. Antidiabetic and Antioxidant Potentials of Amaranthus hybridus in Streptozotocin-Induced Diabetic Rats. *Journal of Dietary Supplements 14*(4): 395–410. doi:10.1080/19390211.2016.1265037

Becker, R. 1994. Amaranth oil: composition, processing, and nutritional qualities. In: Paredes Lopez (ed.) *Amaranth Biology, Chemistry, and Technology*. CRC Press, London, pp. 133–142.

Berghofer, E., and Schoenlechner, R. 2002. Grain amaranth. In *Pseudocereals and Less Common Cereals*, Springer, Berlin, Heidelberg, pp. 219–260.

Beswa, D., Dlamini, N. R., Siwela, M., Amonsou, E. O., and Kolanisi, U. 2016. Effect of Amaranth addition on the nutritional composition and consumer acceptability of extruded provitamin A-biofortified maize snacks. *Food Science and Technology, 36*(1): 30–39. doi:10.1590/1678-457x.6813

Borneo, R., and Aguirre, A. 2008. Chemical composition, cooking quality, and consumer acceptance of pasta made with dried amaranth leaves flour. *LWT – Food Science and Technology 41*(10): 1748–1751. doi:10.1016/j.lwt.2008.02.011

Bressani, R. 1988. Amaranth: The Nutritive Value and Potential Uses of the Grain and By-Products. *Food and Nutrition Bulletin, 10*(2): 49–59. DOI: 10.1177/156482658801000219

Bressani, R., Sánchez-Marroquín, A., and Morales, E. 1992. Chemical Composition of Grain Amaranth Cultivars and Effects of Processing on Their Nutritional Quality. *Food Reviews International, 8*(1), 23–49. DOI: 10.1080/ 87559129209540928

Bunzel, M., Ralph, J., and Steinhart, H. 2005. Association of non-starch polysaccharides and ferulic acid in grain amaranth (*Amaranthus caudatus* L.) dietary fiber. *Molecular Nutrition and Food Research, 49*: 551–559.

Burešová, I., Tokár, M., Mareček, J., Hřivna, L., Faměra, O., and Šottníková, V. 2017. The comparison of the effect of added amaranth, buckwheat, chickpea, corn, millet and quinoa flour on rice dough rheological characteristics, textural and sensory quality of bread. *Journal of Cereal Science*, 75:158–164. doi:10.1016/j.jcs.2017.04.004

Burisová, A., Tomášková, B., Sasinková, V., and Ebringerová, A. 2001. Isolation and characterization of the non-starch polysaccharides of amaranth seeds. *Chemical Papers*, 55(4): 254–260.

Computation, G. S. 1998. Analysis of Amino Acid Reside Sequences of Amaranth and Some Other Proteins. *Bioscience, Biotechnology, and Biochemistry, 8*(62): 1845.

Costea, F. J., Tardif, M., Costea, M., and Tardif, F. J. 2003. The biology of Canadian weeds. 126. *Amaranthus albus* L., *A. blitoides* S. Watson and *A. blitum* L. *Canadian Journal of Plant Science, 83*(4):1039–1066.

Esatbeyoglu, T., Wagner, A. E., Schini-Kerth,V. B., and Rimbach, G. 2015. Betanir-A food colorant with biological activity. *Molecular Nutrition and Food Research*, 59: 36–47.

Escudero, N. L., de-Arellano, M. L., Luco, J. M., Giménez, M. S., and Mucciarelli, S. I. 2004. Comparison of the chemical composition and nutritional value of *Amaranthus cruentus* flour and its protein concentrate. *Plant Foods for Human Nutrition, 59*(1): 15–21.

Ferreira, T. A. P. C., Matias, A. C. G., and Arêas, J. A. G. 2007. Nutritional and Functional Characteristics of Amaranth (*Amaranthus* Spp.). *Nutrire, 32*: 91–116.

Fritz, M., Vecchi, B., Rinaldi, G., and Añón, M. C. 2011. Amaranth seed protein hydrolysates have in vivo and in vitro antihypertensive activity. *Food Chemistry 126*(3): 878–884. doi:10.1016/j.foodchem.2010.11.06

Gebreil, S., Ali, M. and Mousa, E. 2020. Utilization of Amaranth Flour in Preparation of High Nutritional Value Bakery Products. *Food and Nutrition Sciences*, 11, 336–354. doi: 10.4236/fns.2020.115025

Girija, K., Lakshman, K., Udaya, C., Sabhya, S. G., and Divya, T. 2011. Anti-diabetic and anti-cholesterolemic activity of methanol extracts of three species of Amaranthus. *Asian Pacific Journal of Tropical Biomedicine* 1(2): 133–138. https://doi.org/10.1016/S2221–1691(11)60011-7.

Grubben, G. J. H. and Amaranthus-cruentus, L. 2004. In: Grubben, G. J. H., Denton, O. A, eds. *PROTA (Plant Resources of Tropical Africa. Vegetables/Resources végétales de l'Afrique tropicale. Légumes* Wageningen, The Netherlands: Fondation PROTA, Backhuys Publishers, 2: 73–79.

Hemalatha, P., Bomzan, D. P., Rao, B. S., and Sreerama, Y. N. (2016). Distribution of phenolic antioxidants in whole and milled fractions of quinoa and their inhibitory effects on α-amylase and α-glucosidase activities. *Food chemistry, 199*, 330–338.

Hirose, Y., Fujita, T., Ishii, T., and Ueno, N. (2010). Antioxidative properties and flavonoid composition of Chenopodium quinoa seeds cultivated in Japan. *Food Chemistry, 119*(4), 1300–1306.

House, N. C., Puthenparampil, D., Malayil, D., and Narayanankutty, A. 2020 Variation in the polyphenol composition, antioxidant, and anticancer activity among different Amaranthus species. *South African Journal of Botany 135*: 408–412. doi:10.1016/j.sajb.2020.09.026

Hu, H., Yao, X., Qin, Y., Yong, H., and Liu, J. 2020. Development of multifunctional food packaging by incorporating betalains from vegetable amaranth (*Amaranthus tricolor* L.) into quaternary ammonium chitosan/fish gelatin blend films. *International Journal of Biological Macromolecules 159*: 675–684 doi:10.1016/j.ijbiomac.2020.05.10

Irving, D.W., Betschart, A. A., and Saunders, R. M. 1981. Morphological studies on *Amaranthus cruentus*. Journal of Food Science, 46(4): 1170–1174.

Januszewska-Jóźwiak, K., and Synowiecki, J. 2008. Characteristics and suitability of amaranth components in food biotechnology. *Biotechnologia, 3*: 89–102.

Karamać, M., Gai, F., Longato, E., Meineri, G., Janiak, M. A., Amarowicz, R., and Peiretti, P. G. 2019. Antioxidant Activity and Phenolic Composition of Amaranth (*Amaranthus caudatus*) during Plant Growth. *Antioxidants, 8*: 173.

Kaur, A., Kehinde, B. A., Sharma, P., Sharma, D., and Kaur, S. 2020. Recently isolated food-derived antihypertensive hydrolysates and peptides: A review. *Food Chemistry* 128719. doi:10.1016/j.foodchem.2020.1287

Kaur, N., Sharma, P., Jaimni, S., Kehinde, B. A., and Kaur, S. 2019. Recent developments in purification techniques and industrial applications for whey valorization: A review. *Chemical Engineering Communications* 1–16. doi:10.1080/00986445.2019.1573169

Kehinde, B. A., and Sharma, P. 2018. Recently isolated bioactive hydrolysates and peptides from multiple food sources: a review. *Critical Reviews in Food Science and Nutrition*. https://doi.org/10.1080/10408 398.2018.1528206

Kehinde, B. A., Majid, I., Hussain, S., and Nanda, V. 2020a. Innovations and future trends in product development and packaging technologies. *Functional and Preservative Properties of Phytochemicals*, 377–409. doi:10.1016/b978-0-12-818593-3.00013-0

Kehinde, B. A., Sharma, P., and Kaur, S. 2020. Recent nano-, micro- and macrotechnological applications of ultrasonication in food-based systems. *Critical Reviews in Food Science and Nutrition*, 1–23. doi:10.1080/10408398.2020.1740646

Khanam, U. K. S., and Oba, S. (2013). Bioactive substances in leaves of two amaranth species, Amaranthus tricolor and A. hypochondriacus. *Canadian Journal of Plant Science*, 93(1), 47–58.

Kim, H. K., Kim, M. J., Cho, H. Y., Kim, E.-K., and Shin, D. H. 2006. Antioxidative and anti-diabetic effects of amaranth (*Amaranthus esculantus*) in streptozotocin-induced diabetic rats. *Cell Biochemistry and Function*, 24(3): 195–199. doi:10.1002/cbf.1210

Makinde, E. A., Ayeni, L. S., and Ojeniyi S. O. 2010. Morphological characteristics of *Amaranthus cruentus* L. as influenced by kola pod husk, organomineral and NPK fertilizers in southwestern Nigeria. *New York Science Journal*, 3(5): 130–134.

Marcone, M. F. 1999. Evidence confirming the existence of a 7S globulin-like storage protein in *A. hypochondriacus* seed. *Food Chemistry*, 65: 533–542.

Marcone, M. F., Beniac, D. R., Harauz, G., and Yada, R. Y. 1994. Quaternary structure and model for the oligomeric seed globulin from *Amaranthus hypochondriacus* K343. *Journal of Agricultural and Food Chemistry*, 42: 2675–2678.

Miranda, M., Delatorre-Herrera, J., Vega-Gálvez, A., Jorquera, E., Quispe-Fuentes, I., and Martínez, E. A. (2014). Antimicrobial potential and phytochemical content of six diverse sources of quinoa seeds (Chenopodium quinoa Willd.). *Agricultural Sciences*, 5(11), 1015.

Montoya-Rodríguez, A., Gómez-Favela, M. A., Reyes-Moreno, C., Milán Carrillo, J., and González de Mejía, E. 2015. Identification of bioactive peptide sequences from amaranth (Amaranthus hypochondriacus) seed proteins and their potential role in the prevention of chronic diseases. *Comprehensive Reviews in Food Science and Food Safety*, 14(2): 139–158.

Mosovska S. and Birosova L. 2012. Antimycotic and antifungal activities of amaranth and buckwheat extracts. *Asian Journal of Plant Sciences* 11: 160–162.

Nasirpour-Tabrizi, P., Azadmard-Damirchi, S., Hesari, J., and Piravi-Vanak, Z. 2020. *Amaranth Seed Oil Composition*. In Nutritional Value of Amaranth; IntechOpen: London.

National Research Council 1984. *Amaranth: Modern Prospects for an Ancient Crop*; The National Academies Press: Washington, DC.

Onyeoziri, I. O., Kinnear, M. and de Kock, H. L. 2018. Relating sensory profiles of canned amaranth (*Amaranthus cruentus*), cleome (*Cleome gynandra*), cowpea (*Vigna unguiculata*) and Swiss chard (*Beta vulgaris*) leaves to consumer acceptance. *Journal of the Science of Food and Agriculture* 98: 2231–2242. https://doi.org/10.1002/jsfa.871

Opute, P. I. 1979. Seed lipids of the grain amaranths. *Journal of Experimental Botany*, 30: 601–606.

Ostoja, H., Cierach, M., Konopko, H. and Majewska, K. 2002. Effect of addition of grit made of crude and expanded amaranth seeds on the quality of canned meat. *Nahrung* 46: 270–275.

Pasko, P., Sajewicz, M., Gorinstein, S., and Zachwieja, Z. 2008. Analysis of the selected phenolic acids and flavonoids in Amaranthus cruentus and Chenopodium quinoa seeds and sprouts by HPLC method. *Acta Chromatographica*, 20: 661–672.

Pedersen, B., Kalinowski, L. S., and Eggum, B. O. (1987). The nutritive value of amaranth grain (*Amaranthus caudatus*) 1. Protein and minerals of raw and processed grain. *Plant foods for human nutrition*, 36, 309–324.

Putnam, D. H. 1990. Agronomic practices for grain amaranth. In: *Proceedings of the 4th National Amaranth Symposium. Perspectives on Production, Processing and Marketing*. Minneapolis: American Amaranth Institute: 127–139.

Ramírez-Torres, G., Ontiveros, N., Lopez-Teros, V., Ibarra-Diarte, J., Reyes-Moreno, C., Cuevas-Rodríguez, E., and Cabrera-Chávez, F. 2017. Amaranth protein hydrolysates efficiently reduce systolic blood pressure in spontaneously hypertensive rats. *Molecules* 22(11): 1905. doi:10.3390/molecules22111905

Ramos Diaz, J. M., Suuronen, J. P., Deegan, K. C., Serimaa, R., Tuorila, H., and Jouppila, K. 2015. Physical and sensory characteristics of corn-based extruded snacks containing amaranth, quinoa and kañiwa flour. *LWT – Food Science and Technology* 64(2): 1047–1056. doi:10.1016/j.lwt.2015.07.011

Repo-Carrasco-Valencia, R., Hellström, J. K., Pihlava, J. M., and Mattila, P. H. (2010). Flavonoids and other phenolic compounds in Andean indigenous grains: Quinoa (Chenopodium quinoa), kañiwa (Chenopodium pallidicaule) and kiwicha (Amaranthus caudatus). *Food chemistry*, 120(1), 128–133.

Rezaei, J., Rouzbehan, Y., Fazaeli, H., and Zahedifar, M. 2014. Effects of substituting amaranth silage for corn silage on intake, growth performance, diet digestibility, microbial protein, nitrogen retention and ruminal fermentation in fattening lambs. *Animal Feed Science Technology*, 192: 29–38.

Robertson, K. R., and Clemants, S. E. 2003. Amaranthaceae. *Flora of North America*. 4: 405–456.

Sage, R. F., Sage, T. L., Pearcy, R. W., and Borsch, T. 2007. The taxonomic distribution of C4 photosynthesis in Amaranthaceae sensu stricto. *American Journal of Botany*, 94(12): 1992–2003.

Sarker, U., and Oba, S. 2020. Phenolic profiles and antioxidant activities in selected drought-tolerant leafy vegetable amaranth. *Scientific Reports* 10: 18287 https://doi.org/10.1038/s41598-020-71727-y

Sauer, J. 1955. Revision of the dioecious amaranths. *Madroño*, 13(1): 5–46.

Schmidt, D., Verruma-Bernardi, M. R., Forti, V. A., and Borges, M. T. M. R. (2021). Quinoa and amaranth as functional foods: A review. *Food Reviews International*, 1–20.

Schoenlechner, R., Mandala, I., Kiskini, A., Kostaropoulos, A., and Berghofer, E. 2010. Effect of water, albumen and fat on the quality of gluten-free bread containing amaranth. *International Journal of Food Science and Technology* 45(4): 661–669. doi:10.1111/j.1365-2621.2009.02154.x

Seguin, P., Mustafa, A. F., Donnelly, D. J., and Gélinas, B. 2013. Chemical composition and ruminal nutrient degradability of fresh and ensiled amaranth forage. *Journal of the Science of Food and Agriculture*, 93: 3730–3736.

Segura-Nieto, M., Barbadela-Rosa, A. P., and Paredes-Lopez. 1994. Biochemistry of amaranth proteins. In: Paredes-Lopez (ed.) *Amaranth – Biology, Chemistry, and Technology*. CRC Press, London, pp: 75–101.

Sharma, P., Kaur, H., Kehinde, B. A., Chhikara, N., Sharma, D., and Panghal, A. 2020. Food-derived anticancer peptides: A review. *International Journal of Peptide Research and Therapeutics*. doi:10.1007/s10989-020-10063-1

Sharma, P., Kehinde, B. A., Kaur, N., Chhikara, N. and Panghal A. 2021. Development of whey and turmeric-based functional Synbiotic product, Environmental Sustainability. http://dx.doi.org/10.1007/s42398-021-00211-8

Sharma, P., Kehinde, B. A., Kaur, S., and Vyas, P. 2019. Application of edible coatings on fresh and minimally processed fruits: a review. *Nutrition and Food Science*. doi:10.1108/nfs-08-2018-0246

Silva-Sánchez, C., González-Castañeda, J., de León-Rodríguez, A., and Barba de la Rosa, A. P. 2004. Functional and rheological properties of amaranth albumins extracted from two Mexican varieties. *Plant Foods for Human Nutrition*, 59(4): 169–174.

Sindhuja, A., Sudha, M. L., and Rahim, A. 2005. Effect of incorporation of amaranth flour on the quality of cookies. *European Food Research and Technology*, 221(5): 597–601. doi:10.1007/s00217-005-0039-5

Sosalagere, C., Kehinde, B. A., and Poorva Sharma. 2021. Isolation and functionalities of bioactive Peptides from Fruits and Vegetables: A Review. *Food Chemistry* https://doi.org/10.1016/j.foodchem.2021.130494.

Steffensen, S. K., Rinnan, Å., Mortensen, A. G., Laursen, B., de Troiani, R. M., Noellemeyer, E. J., ... and Fomsgaard, I. S. (2011). Variations in the polyphenol content of seeds of field grown Amaranthus genotypes. *Food Chemistry*, 129(1), 131–138.

Tang, Y., and Tsao R. 2017. Phytochemicals in quinoa and amaranth grains and their antioxidant, anti-inflammatory and potential health beneficial effects: a review. *Molecular Nutrition and Food Research*, 61(7): 600–767.

Tang, Y., Li, X., Chen, P. X., Zhang, B., Hernandez, M., Zhang, H., ... and Tsao, R. (2015). Characterisation of fatty acid, carotenoid, tocopherol/tocotrienol compositions and antioxidant activities in seeds of three Chenopodium quinoa Willd. genotypes. *Food chemistry*, 174, 502–508.

Tang, Y., Li, X., Chen, P. X., Zhang, B., Liu, R., Hernandez, M., ... and Tsao, R. (2016). Assessing the fatty acid, carotenoid, and tocopherol compositions of amaranth and quinoa seeds grown in Ontario and their overall contribution to nutritional quality. *Journal of Agricultural and Food Chemistry*, 64(5), 1103–1110.

Taniya, M. S., Reshma, M.V., Shanimol P. S., Krishnan G., and Priya, S. 202. Bioactive peptides from amaranth seed protein hydrolysates induced apoptosis and antimigratory effects in breast cancer cells. *Food Bioscience* 35, 100588. doi:10.1016/j.fbio.2020.100588

Universidade de São Paulo; Food Research Center. Tabela Brasileira de Composição de Alimentos. www.fcf.usp.br/tbca (accessed December 29, 2020).

Valcárcel-Yamani, B., and da-Silva-Lannes, S. C. 2012. Applications of quinoa (Chenopodium quinoa Willd and amaranth; Amaranthus spp.) and their influence in the nutritional value of cereal based foods. *Food and Public Health*, 2(6): 267–269.

Vega-Gálvez, A., Miranda, M., Vergara, J., Uribe, E., Puente, L., and Martínez, E. A. (2010). Nutrition facts and functional potential of quinoa (Chenopodium quinoa willd.), an ancient Andean grain: a review. *Journal of the Science of Food and Agriculture*, 90(15), 2541–2547.

Venskutonis, P. R., and Kraujalis, P. 2013. Nutritional components of amaranth seeds and vegetables: A review on composition, properties, and uses. *Comprehensive Reviews in Food Science and Food Safety*, 12(4): 381–412.

Weaver, S. E. 1984. Differential growth and competitive ability of *Amaranthus retroflexus*, *A. powellii* and *A. hybridus*. *Canadian Journal of Plant Science*, 64(3): 715–724.

Wolosik, K. and Markowska, A., 2019. Amaranthus Cruentus taxonomy, botanical description, and review of its seed chemical composition. *Natural Product Communications*, *14*(5), p.1934578X19844141.

Zhigila, D. A., Yuguda U. A., Akawu J. J., and Oladele F. A. 2014. Palynomorphs and floral bloom as taxonomic characters in some species of the genus *Amaranthus* L.(Aamaranthaceae). *Bayero Journal of Pure and Applied Sciences*, *7*(2): 164–168.

Non-Food Novel Applications of Nutri-Cereals

Pratibha Singh and B. K. Yadav

Contents

13.1 Introduction

Nutri-cereals is a term collectively attributed to group of small-seeded annual grasses cultivated as grain and feed crops. These crops are commonly grown in semi-arid conditions. Such areas are subjected to frequent monsoon failure and droughts, poor soil fertility and difficult land terrain. The crops are known as "Millets," but due to their high nutritive value for consumption and trade's point of view, are now called nutri-cereals. Nutri-cereals, a term defined by Government of India in 2018, has been dedicated to ten cereal grains owing to their agrarian and nutritive excellence. These ten cereal grains are Sorghum (Jowar), Finger Millet (Ragi/Mandua), Pearl Millet (Bajra), Minor Millets, that is, Proso Millet (Cheena), Foxtail Millet (Kangani/Kakun), Barnyard Millet (Sanwa/Jhangora), Kodo Millet (Kodo), Little Millet (Kutki) and two Pseudo Millets, that is, Amaranth (Chaulai) and Black Wheat (Kuttu). (MoA and FW, *Gazette of India* through F.No. 4-4/2017-NFSM(E)). They have been appreciated for being substantial crops meeting the challenges of the present day, such as global warming, decreasing ground water levels and food security, and confirming their suitability to drylands and soils having poor fertility. They have been termed as "crops of the future" because they are the only crops that will address the important issues in the future, such as climate change, malnutrition and also food, feed and fuel. Nutri-cereal straw is a valuable livestock feed, besides having other uses as building materials, fuel and so forth in the farming systems. They are the main staple food for farm households in several countries and among the poorest peoples. Further, when compared with traditional cereals like wheat and rice, these grains possess better compositional and functional characteristics. These characteristics help them to become ideal candidates in various value-added food products (Mal et. al., 2010; Singh et. al., 2012).

Across the world, these nutri-cereals are commonly utilized as alternative grains. To emphasize the importance of nutri-cereals, in this chapter we highlight their commercial and industrial applications, including pet food and aquatic feed, bio-ethanol, millet starch for bioplastics, films and coatings, fibers and mats, adhesives and resins and other applications. Due to the increase in demand for alternate grains, these novel applications will help for a gradual positioning of nutri-cereals. The nutri-cereals have not been studied extensively for non-food and industrial applications so far.

13.2 Composition of nutri-cereals pertaining to non-food uses

Nutritionally, nutri-cereals contains about 65 percent carbohydrates, 12 to 20 percent dietary fiber, 6 to 19 percent protein, 2 to 4 percent minerals and 1.5 to 5 percent fat (Annor et. al., 2017). Based on their specific type, they are also significant sources of dietary fiber, lipids, polyphenols, vitamin B, and minerals (Shahidi et. al., 2013). The nutri-cereals are more nutritional if protein, macro nutrient content and the energy value are compared to conventional crops including wheat and rice. Due to their high nutritional quality, nutri-cereals are not only beneficial for human diets but similarly contribute to animal diets also. Probably this may be the reason that, in Europe and the United States, nutri-cereals are grown mostly as forage for poultry and as bird feed (Mal et al., 2010).

Sorghum is an important tropical cereal grain. It is generally used as food, feed and fodder crop. It shows high adaptability even in low precipitation agriculture areas. This may be the possible reason that it is one of the major crops for grain and forage of sub-tropical regions. The utilization of sorghum as feed is the main reason for its production and its participation in international trade at a global level. Sweet Sorghum has its application in the production of ethanol and jaggery and also for making paper.

Feed grains are the main resource of energy and a source of protein for livestock and the poultry industry. In sorghum, kafirin governs the protein fraction. So, sorghum contributes in animal nutrition as a source of both protein and energy. The average contents of starch, crude fat and protein on a dry matter basis were reported as 76.4 percent, 3.7 percent, and 10.1 percent, respectively, in nine sorghum cultivars (Yang and Seib, 1995). In another study, 11 sorghum cultivars also showed starch, crude fat and protein as 73.6 percent, 2.7 percent, and 9.8 percent, respectively (Giuberti et al., 2012). These studies concluded that the gross energy of sorghum as 15.6 MJ/kg approximately. This gross energy was mainly obtained from starch (84%), protein (15%) and fat (1%) Leeson and Summers (2001).

The prolamins in sorghum are known as kafirins. There is extensive similarity in sequence and chemical properties among sorghum prolamins, that is, kafirins and maize prolamins, that is, zeins. Due to this, kafirins became categorized into subgroups similar to zeins (Shull et al., 1991, 1992; Mazhar et al., 1993; DeRose et al., 1989; Belton et al., 2006; Xu and Messing, 2009; Esen, 1986). In sorghum, kafirins contains around 50 to 70 percent of the total protein (Taylor et al., 1984a; Hamaker et al., 1995).

When we compare films prepared from protein, like isolated kafirins, zein and oats. It was found that films formed from isolated kafirins were similar in properties as zein. But these films showed more strength and extensibility but low permeability when compared with the films formed from oat prolamins (Gillgren and Stading, 2008).

Behind the formation of protein microparticles, the agglomeration capability of kafirin played a major role (Taylor et al., 2009a, 2009b). Similarly, the permeable nature of the microspheres helps in encapsulation of catechin and condensed tannins (Taylor et al., 2009b). Isolated kafirin also has application in the formation of tablets. The study had been carried out to observe the release of caffeine from the tablets. The results were used to model the continuous release of medicine (Elkhalifa et al., 2009). The function of microparticles formed from Kafirin were also studied in film formation, which was found acceptable (Anyango et al., 2011). The possibility to use these films and kafirin microparticles, as biocompatible materials was also tested (Taylor et al., 2015).

Sorghum is used as main source of starch by some pet food brands. Through some latest human nutrition studies it was concluded that the properties of sorghum, such as slow digestion of starch and low glycemic response, could be targeted as a nutritional benefit for premium pet food products (Simnadis et al., 2016; Anunciação et al., 2018). These pet foods could be specifically designed for obese, diabetic, and geriatric pets. Out of all aquatic feed, the costliest ingredient is the fish meal. So, options for low cost sources of protein are extensively needed. Adedeji et al. (2017) explored the possibility of utilization of distillers dried grain with solubles from sorghum as an alternate resource for shrimp food.

Pearl millet, an important fodder crop, is mainly cultivated in Africa and the Indian subcontinent and covers 50 percent of the total world production of millets. India ranks first in the world in the production of pearl millet. This crop is well adapted to vagaries of drought, high temperature and low soil fertility. Pearl millet, also known as bajra in India, is used both as grain for human consumption and fodder for livestock feeding.

Pearl millet is a crop with high tillering ability, high dry matter yield and protein content of about 10 to 12 percent. The crop exhibits excellent growth and is has high palatability and nutritive value (Shashikala et al., 2013; Bind et al., 2015). The green fodder obtained from pearl millet is leafy, palatable and a highly nutritious feedstock. Traditionally this crop was grown for dry fodder and cattle grazing. This ensures high milk yield in cattle (Upadhyaya et al., 2018). Pearl millet has extensive potential and can be utilized as feed for swine and different types of birds, such as doves, turkeys, song birds, and ducks.

The high nutritional value of pearl millet plays a vital role in the health sector. These characteristics of pearl millet may help in the prevention of various human diseases like diabetes, cancer, cardiovascular and neurodegenerative diseases (Jukanti, et al., 2016).

After observing the value chain of pearl millet in western India, that is, Gujarat, Rajasthan and Haryana, it shows indeed a lot of scope for pearl millet crop in the near future. Other than the cattle and poultry feed industry it can be used in breweries and in the starch industry also (Reddy et al., 2013).

Another nutri-cereal, barnyard millet has the property of rapid growth, and it also produces voluminous fodder. Since the fodder produced from barnyard millet is highly palatable, it can be used for producing hay or silage. An added advantage of its high nutritive value, about 61 percent total digestible nutrients, fiber and high protein make barnyard millet straw an ideal component for animal feed (National Research Council, 1996). Due to high protein and calcium content, its straw is preferred over rice and oat straw (Yabuno, 1987).

Proso millet is highly nutritious nutri-cereal. Proso millet is gluten free. So, it is a perfect choice for gluten intolerant people. It is good for neural health due to its high lecithin content. It is also rich in essential amino acids such as methionine and cysteine. It is mostly considered as bird feed but is also suitable for bioethanol production. Corn is the most common feed stock in the United States. It is used as fuel for ethanol industry and also for distiller's dried grains with solubles. It was found that proso millet contains a similar starch content as in corn. During fermentation higher amounts of distillers' dried grains with solubles (DDGS) were recovered from proso millet in comparison to highly fermentable corn-based fermentation, which contains higher protein content (26.6 percent to 33.4 percent) than the DDGS from corn (17.2 percent to 23.4 percent) (Rose, et al. 2013). This indicates that. like corn-based DDGS, the DDGS derived from proso millet can be an ideal option for livestock feed. So, complete replacement of corn by proso millet or even a blend of both for ethanol production may be the best idea to improve the local economy. This will also help to compensate the occasional scarcity of corn as common raw material.

Another millet grain being found superior to rice regarding contents of protein and iron is foxtail millet grain which has a 12.3 percent protein value and iron content of 2.8mg/100g in comparison to 6.8 percent and 1.8mg/100g protein and iron values of rice, respectively. Fat content of foxtail millet is 4.3 percent, which is superior both to wheat as well as rice. Foxtail millet also contains higher levels of vitamin A precursor, that is, beta carotene (Murugan and Nirmalakumari, 2006). Fermentation of foxtail millet is also done so as to make vinegar, maltose, yellow wine, beer and many other products. Use of its by-product as a bird feed as well as animal feed is well known. Buckwheat is another nutri-cereal that holds a special place in cultivable crops because of its dietetic and nutritional as well as therapeutic properties. Not only the grains, but other plant parts are also utilized in food, cosmetic, feed and pharmaceutical industries.

Amaranth is considered as important because of its richness in antioxidants, including vanillic acid as well as gallic acid. Antioxidants are especially helpful in fighting against free radicals, which actually act as damaging by-products for normal cellular activity and also are helpful in reducing signs of aging as well as in fighting many heart diseases. Amaranth oil has a significant demand as it helps to cure duodenal peptic ulcer and chronic gastritis being caused by *Heliobacter pylori*, a type of bacteria that infects the stomach.

13.3 Nutri-cereals as a feed grain for animals

Nutri-cereals gained momentum in recent years as animal feed, since they are rich in valuable nutrients. Other than human consumption a considerable percentage of nutri-cereals is also utilized for beer production. The other uses are in livestock and bird feed (Obiana AB. 2003).

13.3.1 Cattle feed industry

Previous researches also indicate that as a feed ingredient, pearl millet plays an efficient role in the poultry industry in any form, either whole grain or milled grain, when used in feed for chickens (Cisse et al., 2017).

Pearl millet is also used for cattle and poultry feed. Traditionally, maize was used for this purpose but pearl millet is preferred over maize due to its high protein and low calories. In the year 2020 the maximum consumption of pearl millet was in the cattle feed industry. And this was accompanied by food, alcohol and the poultry industries. One of possible reason may be the decrease in perennial pastures and grazing land while, on the other hand, the downswing in cereals cultivation area, which were considered as main sources of fodder.

About 5 to 10 percent of pearl millet is used by feed manufacturers as an ingredient in feed concentrate. If in the future there would be decline in prices while competing with traditional grains like maize and sorghum, this share may be increased to 15 percent (Reddy, et al., 2013).

13.3.2 Poultry industry

Globally, the predominant feed grains in diets for broiler chickens are maize and wheat. Previous studies show that up to 50 percent inclusion of pearl millet in broiler feed did not affect the performance of chicken. The different varieties of pearl millets indicated commensurable outcomes compared to corn concerning metabolizable energy and digestible amino acids. Likewise, other researchers also concluded that replacing corn in place of sorghum and pearl millet also increases the egg production rate and remarkable improvement in growth and feed efficiency (Baurhoo, et al., 2011) (Issa, et al., 2016). Moreover, by feeding pearl millet to laying hens the eggs will have a high content of omega-3 fatty acids and low content of omega-6 (Jacob, et al., 2015). This definitely states that nutri-cereals may be an alternative for other cereals like maize and wheat.

13.3.3 Ruminant feed

Nutri-cereals like pearl millet and finger millet were markedly utilized to replace typical crops in the feed of small ruminants like sheep and goats. Early studies concluded that partial substitution of corn with pearl millet would not impart any significant effect on the intake of feed and milk production of lactating and growing goats. Whereas pearl millet could be a full replacement of maize in high supplement diets for cattle, nutri-cereals were found instrumental for the animal feed in any form, that is, whole or ground. If nutri-cereals were processed before using for beef cattle, this not only increases the dietary nutrients but also the digestion of dry matter.

13.3.4 Swine feed

Sorghum and pearl millet are considered choices for optimum nutrition for swine. Swine nutritionists prefer sorghum as a primary source of diet for nursery pigs and for gestating and lactating sows. Previous studies concluded that sorghum might be an exemplary substitute for other cereals like maize, wheat, or barley in swine diets if ground to appropriate particle size, since processing methodology results in improvement of nutrient profile. The pearl millet has a higher ratio of protein to amino acid content. This shows that pearl millet can be used for equal-weight substitutions for maize in pig feed formulation for increased supply of total amino acids required for better growth of pigs.

13.4 Application of nutri-cereals as a source of bio-ethanol production

The aspiring target of European Union (EU) commission recorded in 2019 in the name of Green Deal, was to attain climate neutrality by 2050. This may be possible by usage of biofuels.

The substitution of fossil fuels with biodiesel, bioethanol, biogas, and so forth, is imperative. The most propitious alternate for petroleum considered at the moment is bioethanol. Fundamentally, it is ethyl alcohol (C_2H_5OH), which is derived from renewable sources such as plant or algae. The second generation bioethanol production depends on different feedstock sources like sugar cane bagasse, corn stover, various forest residues, sawdust, sweet sorghum and so forth (Niphadkar, et al., 2017).

The undervalued nutri-cereals may be a promising source of feedstock for bioethanol production by virtue of its nutritional characteristics, and also improved tolerance for abiotic and biotic stresses. The main constituent of finger millet is carbohydrate (68.6 to 72.6%), mainly characterized by starch (60 to 69.5%). Previous research findings have shown that compared to maize, wheat and pearl millet, finger millet showed the highest α-amylase activity during malting (Kubo, 2016). Another research reported different ratios while mixing sorghum and finger millet on w/w basis (90/10, 80/20, 70/30). The results shown substantial rise in free sugars levels and fermentability but viscosity and specific gravity values remain unaffected. About 50 percent high output in ethanol was obtained with the ratio 70/30 w/w of mixed feedstock (Lyumugabe, et al., 2015). This shows that to increase the efficiency of alcohol, finger millet malts may be added and it can be categorized as a promising alternate for bioethanol production.

Sorghum has a great capability to be utilized in the production of bio-industrial products, including bioethanol. It is high starch grain with alike composition to maize and as such suitable for the production of bioethanol like maize. Dry-grind ethanol processing typically used for production of bioethanol from sorghum and also considered as one of the best method to convert starch in to bioethanol.

Seven cultivars of proso millet and six breeding lines having waxy starch were evaluated for the production of bioethanol. The results shown similar trend like maize w.r.t. yield of ethanol and fermentation efficiency. It was also mentioned in the results that fermentation efficiency was found higher in waxy proso millet compared to non-waxy. The optimum efficiency was 97 percent (Rose and Santra, 2013).

The husks from three millets (foxtail, barnyard and little millet) also produced bioethanol but the highest concentration was obtained in barnyard millet husk. Fermentation of millet husk can be carried out using the microorganisms isolated from the spoiled fruits and millet husks (Ashwini et al., 2020).

The derived component lignocellulose from amaranth shows prospective feedstock for production of bioethanol. Lignocellulose when subjected to microwave assisted alkali pretreatment results in high concentrations of fermentable sugars and is a feasible feedstock for bioethanol production. (Marx et. al., 2014).

13.5 Value added products from nutri-cereals for non-food applications

13.5.1 Bioplastics

Bioplastics are new and preferred choice of biodegradable polymers (Reddy et al., 2013; European Bioplastic Association, 2018b). Traditional plastics like petroleum-derived and non-biodegradable are dominating the society in everyday life. Even though the bioplastic market share is nearly 1 percent of global plastic production, that is, 370 million tonnes (Coppola et al., 2021). Due to their non-degradable characteristic plastics are dangerous for environment and so the increase in research on bioplastics is the need of the hour. One of the main application of bioplastics is food packaging. Due to resource sustainability, nowadays starch biodegradable films are in demand as a substitute for synthetic packaging material (Paz et al., 2005). Dai et al. found that starch has many beneficial properties, especially good film-forming characteristics, plentiful availability and its low cost, which can reduce the utilization of synthetic polymers. (Dai et al., 2019)

The conventional plastics, that is, petroleum-based plastics may be replaced by polyhydroxyalkanoate (PHA) – produced in plants, which is not only natural thermoplastic but shows the similar properties to petroleum-based plastics and moreover it is biodegradable. This makes PHAs a perfect value-added by-product for plants/crops.

Nearly all the nutri-cereals contains kafirin, a prolamin storage protein which is suitable for the production of bioplastics owing to its extremely hydrophobic nature (Belton et al., 2006; Duodu et al., 2003). Kafirin when used in combination with glycerol, polyethylene glycol 400 (PEG 400), and lactic acid as a plasticizer yields a stable bioplastic with lower strain, higher tensile strength, and high water vapor permeability as compared to commercial zein plastics from maize (Buffo et al., 1997). Kafirin based bioplastics are commercially used in Africa for coating the pears to increase the shelf life, reduce the stem-end shriveling and delay ripening to about 13 days at 20° C storage (Taylor et al., 2006).

13.5.2 Nutri-cereals films and coatings

Starch is an inept biomaterial due to its various applications in foods, textiles, pharmaceuticals, and engineering sectors. Since it is the most abundant polysaccharide in nature; it is cost-effective and is widely utilized for different food and non-food purposes (Singh et al., 2010,). Starch is the main constituent of nutri-cereals. Nutri-cereal starch shows 0.38–28.40 percent amylose content, 7.64–11.48 percent moisture, 0.39–1.60 percent ash, and 0.31–0.58 percent protein content (Wu et al., 2014). Many researchers have shown their interest in biodegradable films to replace petroleum-based plastic materials (Valencia-Sullca et. al., 2018). Starch biodegradable films are gaining popularity. Being a sustainable resource, it may be substitute for the synthetic packaging materials. (Paz; Guillard et al., 2005). This provides a new direction to the starch industry since consumers are aware and sensitive for the safety of food products and eco-friendly packaging.

The prospective application of millet starches for non-food purposes is finite, which may be the possible reason for researchers evaluating the possibility of using these millet starches in the formulation of films. In this context, possible use of proso millet for the production of films was explored. This can be alone or in combination with k-carrageenan gum. It was observed that proso millet starch can be a best option for the formation of edible and biodegradable films. Also, for food safety and shelf life enhancement, it can be used for coatings with bioactive properties in different industrial applications.

The linear nature of amylose content is mainly important for the formation of films (Kramer, 2009). Pearl millet has some outstanding agronomic properties like high tolerance towards unfavorable climatic changes over other traditional sources. It is also cost effective and easily available. But its starch, either native or modified, has never been explored for formation of edible films. Marium Shaikh et al. (2018) conducted a study to illustrate the use of starch from pearl millet crops for the formation of edible films for packaging purposes as a new and non-conventional source. This type of films may be used for packaging of fresh ready-to-eat vegetables like wrapping. But starch modification is needed to avoid the typical drawbacks before usage. This study proposed acetylation and hydroxypropylation as a solution for the modification of starch. This modification also helps to improve moisture-proof property and so there was significant reduction in Water Vapor Permeability and Water Solubility values was observed. Since pearl millet starch films were formed by natural polymer, that is, starch that was environmental friendly. So, pearl millet could be a possible option for biodegradable food packaging (Marium Shaikh et al., 2018).

Another promising nutri-cereal for films and coatings is sorghum. The films prepared from sorghum protein – that is, kafirin – was compared to the films prepared from maize protein, that is, zein (commercial zein). During wet milling of sorghum, the fractions, which were rich in protein, were accumulated for the extraction of kafirin. From this extracted kafirin films were prepared (Buffo et al., 1997). Kafirin films showed similar results for properties like tensile strength and water vapor barrier as obtained from films from zein. The strength and extensibility of kafirin films were found higher compared to zein films (commercial), when bran was included in dry milling fractions of sorghum (Da Silva and Taylor, 2005). These films also showed higher water vapor permeability (WVP) in comparison to zein films.

If we use similar amounts of plasticizer, the strength and extensibility, and water vapor permeability of pearl millet prolamin films were similar to kafirin films. But low WVP and less extensibility with either equal or stronger strength when compared to zein films (Gillgren et al., 2011). The research done up to now studied casting method to produce free-standing kafirin and millet prolamin films.

To improve the shelf life and for extension of quality of maturity enough to be eaten, kafirin coatings are used (Buchner et al., 2011; Taylor et al., 2016). Kafirin coatings are also utilized as the coatings for drug release. Kafirin films have application in pharmaceuticals for coating gelatin and hydroxypropylmethyl cellulose capsules containing paracetamol, when produced using aqueous ethanol which contains sodium hydroxide and polyethylene glycol as plasticizer (Lal et al., 2016).

13.5.3 Microparticles and nanoparticles – application in pharmaceuticals and biomedical

In pharmaceutical and biomedical sector, kafirin, a dominant protein fraction in sorghum showed prospective application in the form of kafirin microparticles and films. Kafirin, a homologous protein to zein, leads over zein when biomedical field is considered for different applications. When compared to Zein, kafirin showed more hydrophobic nature and higher glass transition temperature with less digestibility. Kafirin can be formed into highly vacuolated microparticles, that is, 1 to 10 mm diameters, which provides a large surface area for binding bioactive compounds like polyphenolic nutraceuticals and bone morphogenetic proteins (BMPs).

Kafirin microparticles has the properties of probable conveyors for other drugs and bioactive compounds. The study has shown successful loading of anti-inflammatory drugs such as Prednisolone into kafirin microparticles when kafirin is extracted from distinct sources (Lau et al., 2015a; Lau et al., 2015b).

Curcumin is a yellow pigment which is lipid-soluble. Curcumin has many health benefits. Due to its polyphenolic content it shows anti-carcinogenic, anti-inflammatory and antioxidant activities (Xiao et al., 2015b). For encapsulation of curcumin the nanoparticles of kafirin and kafirin/ carboxymethyl-chitosan were also utilized (Xiao et al., 2015a).

Kafirin is considered safe source, since it belongs to food commodity (www.fda.gov). This allows for considering the use of kafirin as a novel pharmaceutical excipient. In comparison to food products, the value addition is more in pharmaceutical products. So, sorghum growers may be benefitted w.r.t. increase in income through usage of kafirin within this framework.

Proso millet protein can be extracted by either wet milling or ethanol. For the encapsulation of curcumin content and omega fatty acids, this extracted protein can be used as the wall material with tocopherol homologues. Extraction by ethanol is preferred over wet milling due to better performance. The encapsulation helps in lowering the degradation rate compared to free compounds at 60°C. Also, when assessed by the DPPH and ABTS assays, no negative effect was found on the antioxidant activity of curcumin. The study indicated that millet protein and different tocopherol homologues may be used for the formulation of composite nanocarrier systems. They showed good potential for the nanoencapsulation of lipophilic compounds. They can significantly increase the bioavailability of compounds (Khorasani et al., 2018).

13.5.4 Adhesives and resins

In India, petroleum based synthetic resin adhesives are commonly used in the plywood industries. Petroleum resources are required for this purpose. Phenol formaldehyde and urea formaldehyde resins were used to derive the petroleum resources. Long shelf lives and the water resistance property of these formaldehyde-based adhesives helps them to dominate the adhesive market (Ningbo Li et. al., 2011). However, resource limitation, irregular dissemination and unsteady prices are the challenges. Also the environmental issues and cost of the petroleum product, which is increasing on daily basis badly affects the cost of the adhesives and so, the adhesives industry is eagerly looking for bio-based adhesives.

Many natural resources exist that resemble phenol/formaldehyde in their molecular architecture units and are also capable of undergoing reaction similar to phenol/formaldehyde, such as glues from animals, fish, vegetable protein, starch and blood albumen also used in the wood industry. Since they belong to natural origin and so renewable sources, dependence on petroleum resources may be avoided. So, we can minimize the pollutant level by synthesizing these materials to form resin adhesives.

During much research, tannins, lignins, carbohydrates, unsaturated oils, liquefied wood rice bran, soy protein and so forth were explored to check the adhesive potential of these biomaterials. These includes (Pizzi (2006); Smith et al. (2006); Springer, Heidelberg et al. (2007). Out of these Soy protein isolates (SPI) gained support as bio-based and renewable materials and gained focus for research as adhesives in the past 10 years (Wang, 2006). The priority of this research work was improvement in water resistance. This was achieved through chemical modification which results in the increase in hydrophobicity of soy proteins (Wu et al., 2007).

Nutri-cereals like sorghum may be utilized as another resource in the form of isolated sorghum proteins. These isolated sorghum proteins can be used as extenders. The extenders are required in plywood adhesives and in other low-cost adhesives. It can also be used in wallboard and packaging materials. Sorghum protein, that is, kafirin shows better hydrophobic nature compared to Soy proteins and so it is possible to provide better water resistance also (Icoz DZ, et al., 2005). The use of nutri-cereals in the development of the bio-adhesive system would directly reduce the consumption of petroleum based chemicals. This will definitely reflect in the cost reduction of the resin and also that will help in minimizing the emission of formaldehyde.

13.5.5 Fibers and mats

The prolamine protein from nutri-cereals such as sorghum is an encouraging source for the fabrication of renewable and biodegradable materials. although its potentials are still not fully reaped through research. A rapid growth was found in electrospinning of biodegradable polymers. Although not all polymers can be successfully electrospun, this method is well accepted for biodegradable fibers.

The application of electro-spun fiber mats was found in the production of biomaterials. These biomaterials were appropriate for release in a controlled manner, healing of the wounds, or tissue engineering (Xiao et al., 2016).

Many studies shown that few polysaccharides or proteins may be useful for the fabrication of biopolymer nanofiber mats (Mendes et al., 2017). Dissociation of protein in solvent for protein-to-protein interactions in the form of chain entanglements before the electrospinning process is used to produce protein nanofibers (Turasan et al., 2019).

Keratin, collagen, silk, elastin and zein are some examples of effectively electrospun nanofibers. Kafirin is a prolamin found in sorghum. Its amino acid composition is almost identical to zein. The study shown that polycaprolactone polymer was added to the protein solution. This was helpful in the successful electrospun of nanofibers from kafirin (Higashiyama et al., 2021).

13.6 Application of nutri-cereals as natural colorants

Dyes have been used for dyeing of fabrics. Depending on their source of procurement, they may be synthetic or natural. Due to bright colors and fastness properties, there has been excess concentration on synthetic dyes results in down grading of natural dyes. Studies shown that a reliable solution for this problem is nutri-cereals, which were explored for dyes to colored fabrics. The findings revealed that, the millet dye has good affinity for cotton fabrics and more efforts by future researchers were for the modification in the qualities of dye. (Richard Gbadegbe, et al., 2014).

In another study, the leaf sheaths of dye sorghum were used to extract bright red colorant (Akogou et al., 2018; Kayode´ et al., 2012; Oluwaniyi et al., 2009). This can be used as natural food colorants. The other possible uses are colorants for leather, basketwork, and ornamental plants. These can also be used in traditional medicine (Kayode´ et al., 2012). Monascus purpureus is an edible fungus and also produces pigments as secondary metabolites. For the solid-state fermentation of Monascus purpureus as a substrate, rice was compared with different forms of sorghum (Srianta and Harijono, 2015; Srianta et al., 2016). The results showed that the red, yellow and orange pigments were obtained. These pigments can be used in traditional medicine and also as natural food colorants and supplements (Srianta et al., 2016). When compared with rice it was observed that dehulled sorghum showed the highest growth yield and pigment production.

13.7 Nutri-cereals as eco-friendly building materials

Scientists focus on eco-friendly and sustainable innovations to cope with the significant threat of global warming, which further affects climate change, sea level and so forth. Due to increased concern about awareness for sustainable building material and environmental issues, eco-friendly building materials are the need of the hour. Due to its remarkable qualities, such as sustaining indoor air quality, temperature maintenance, and low environmental consequences, adobe blocks are the suitable option. During fabrication, in the entire composition of adobe blocks, about 2–4 percent of the nutri-cereal fibers, can be incorporated. This is required to increase absorption resistance and to decrease thermal conductivity (Babe et al., 2020). The usage of millet fiber in an earth matrix increased the durability of adobe block. This will help to reduce the amount of waste generated as fibers (Ramakrishnan et al., 2021).

Another nutri-cereal in the category of natural raw material is sorghum, which can be utilized as building materials. The different uses are raw material for thatched roofs and fences. Sorghum is a tall crop, having lower woody stalks. These are wastes from crops or are utilized for producing heat earlier by burning leading to air pollution. For the production of boards and panels, usually known as Kirei, sorghum stalks were woven together and afterwards heat pressed using urea and formaldehyde-free adhesive. To keep the Kirei scratch proof they were polished with natural resins. The different applications of these boards are in furniture and cabinetry. They can also be utilized as wall covering, paneling and flooring. They are resistant to chipping, and they do not warp or deform due to low water absorption. The major advantage of sorghum as a building material is that it is renewable and non-polluting.

13.8 Conclusions and future directions

The versatile uses in different sectors and extensive adjustability to varied agro-climatic conditions, nutri-cereals show enormous potential for nonfood applications. Slowly, this potential is starting to be realized, although some serious technical problems remain. The nutri-cereals remain effectively unexplored and their potential untapped.

Animal feeds need a greater number of cereal crops for providing energy and protein in daily intake. Nutri-cereals are one of the excellent options for poultry and livestock feed industries. Anciently, nutri-cereals like sorghum have been disregarded when compared to maize. This is due to its lower feeding value. Although, the newly added sorghum varieties, specific diet compositions, and appropriate techniques for processing are fulfilling this gap. Nutri-cereals are capable of fitting sustainable feed systems for animals. But more studies are required to strengthen the ability of nutri-cereals to serve in a better way for animals' and pets' food requirements.

The research in bioethanol production from nutri-cereals lags behind even though much potential needs to be explored. These nutri-cereals generate huge numbers of derivatives like DDGS and lignin. If methods for proper utlilization of these derivatives would be explored, they might be a good source of improvement in economic viability. DDGS is rich in kafirin protein, that is, about 30 to 40 percent. The kafirin has distinctive properties like high hydrophobic and less digestible nature in comparison to other cereal prolamins. These properties help to produce biomaterials, which consists exceptional functional properties when compared to products prepared from zein or gluten. For chemicals and biofuels, lignin may be explored as a profuse polymer for depolymerization into phenolic compounds and also for the heat generation.

To prioritize sorghum over maize in the area of bioethanol feedstock, it is necessary to focus on sorghum production yield. The new and high-yielding varieties like multi seeded mutant sorghums are required. These varieties have the capability to produce more than two-fold the total amount of seed weight per panicle of the parent (BTx623) (Zhang et al., 2017). Nutri-cereals possess the capability to become an alternate source of feedstock and bioethanol production in place of maize and other cereals respectively. But the limitations of nutri-cereals such as yield and production should be addressed. When compared to regular sorghum, the genotypes of sorghum such as waxy, which contains high amylopectin, are required due to its expeditious starch conversion. Similarly, stalk and grain in the panicle of sweet (sweet stemmed) sorghum has high concentration of soluble sugars, due to which it has substantial potential for bioethanol production. Such varieties are critical for the feasibility of the biofuel industry and so more productive conversion processes are the need of the hour.

For the value-added products, from nutri-cereals for non-food applications, the documentation of research is highly required. Despite, the recognition received by kafirin bioplastics in the biomedical sector as biomaterials is the most encouraging aspect. These biomaterials can be used to introduce into living tissue (Taylor et al., 2013). Due to their superior functional properties like low hydrophilicity, slow biodegradability and bio and cyto compatibility, these high value products show minimum sensitivity to higher production cost (Reddy and Yang, 2011), although, vindication for safety during usage is required, such as implants in live tissue (Taylor et al., 2013). Above all, a consistent quality kafirin with a steady resource is highly needed. So, the research should focus in the area of cost-effective extraction techniques for the development of advanced products.

Also, there is a requirement for novel and efficacious methods of modification to enhance the performance and also the functional properties of kafirin bioplastic. Nutri-cereals protein, like kafirin, has substantial potential in the area of wood veneer adhesives and resins, although lacuna still persist in converting it into mass production and its commercialization with the same performance.

Finally, with the objective of further improvement in economic feasibility, and to explore alternative options, especially from nutri-cereals, they have huge potential as novel, functional raw ingredients for bio-industrial uses like natural colorants and eco-friendly building materials, which are technically feasible and can be easily commercialized. Still, cost reductions in the production processes, use of co-products as bioplastics and other value added products, and enterprising government policies to promote nutri-cereals as "green economy" are likely prerequisites for better adaptability and commercialization.

References

Adedeji, A. A., Zhou, Y., Fang, X., Davis, D. A., Fahrenholz, A., Alavi, S. 2017. Utilization of sorghum distillers dried grains (sDDGS) in extruded and steam pelleted shrimp diets. *Aquacult. Res.* 48, 883–898.

Agra F. E. Gbadegbe, R. S., M. Quashie. 2014. Extraction of Dyes from Natural Sources: An Implication for Exploring Millet (Bajra Pennisetum Americanum), *International Journal Of Humanities and Social Studies*, Vol. 2 Issue 11(ISSN 2321–9203.

Akogou, F. U. G., Polycarpe Kayode´, A. P., Den Besten, H. M. W., Linnemann, A. R. 2018. Extraction methods and food uses of a natural red colorant from dye sorghum. *J. Sci. Food Agric.* 98, 361–368.

Anunciação, P. C., Cardoso, L., Queiroz, V. A. V., de Menezes, C. B., de Carvalho, C. W. P., Pinheiro-Sant'Ana, H. M., Alfenas, R. C. A. 2018. Consumption of a drink containing extruded sorghum reduces glycaemic response of the subsequent meal. *Eur. J. Nutr.* 57, 251–257.

Anyango, J. O., Taylor, J., Taylor, J. R. N. 2011. Improvement in water stability and other related functional properties of thin cast kafirin protein films. *J. Agric. Food Chem.* 59, 12674–12682.

Ashwini R. N. 2020. Bioethanol Production from Husks of Different Small Millets, Biofuels 2020, *11th Edition of International Conference on Biofuels and Bioenergy*, London, March 23–24,.

Babe, C., Tom, A. 2020. Thermomechanical characterization and durability of adobes reinforced with millet waste fibres (sorghum bicolour), *Case Stud. Constr. Mater.* 13, https://doi.org/10.1016/j.cscm.2020.e00422

Baurhoo, N. B., Baurhoo, A. F., Mustafa, Z. X. 2011. Comparison of corn-based and Canadian pearl millet-based diets on performance, digestibility, villus morphology, and digestive microbial populations in broiler chickens. *Poult Sci.*; 90:579–86.

Belton, P. S., Delgalligo, I., Halford, N. G., and Shewry, P. R. 2006. Kafirin structure and functionality. *J. Cereal. Sci.* 44:272–286.

Bind, H., Bharti, B., Kumar, S., Pandey, M. K., Kumar, D. and Vishwakarma, D. N. 2015. Studies on genetic variability, for fodder yield and its contributing characters in bajra [Pennisetum glaucum (L.) r. Br.]. *Agric Sci Dig– Res J* 35(1), 78–80.

Buchner, S., Kinnear, M., Crouch, I. J., Taylor, J., Minnaar, A., 2011. Effect of kafirin protein coating on sensory quality and shelf-life of 'Packham's Triumph'pears during ripening. *J. Sci. Food Agric.* 91, 2814–2820.

Buffo, R. A., Weller, C. L., and Gennadios, A. (1997). Films from laboratory-extracted kafirin. *Cereal Chem*, 74, 473–475.

Cisse R. S., Hamburg, J. D., Freeman, M. E., Davis, A. J. 2017. Using locally produced millet as a feed ingredient for poultry production in Sub-Saharan Africa. *J Appl Poultry Res.* 26(1):9–22. https://doi.org/10.3382/japr/pfw042.

Coppola, G., Gaudio, M. T., Lopresto, C. G. *et al.* 2021. Bioplastic from Renewable Biomass: A Facile Solution for a Greener Environment. *Earth Syst Environ* 5, 231–251. https://doi.org/10.1007/s41748-021-00208-7

Da Silva, L. S., Taylor, J. R. N. 2005. Physical, mechanical, and barrier properties of kafirin films from red and white sorghum milling fractions. *Cereal Chem.* 82, 9–14.

DeRose, R. T., Ma, D.-P., Kwon, I.-S., Hasnain, S. E., Klassy, R. C., Hall, T. C. 1989. Characterization of the kafirin gene family from sorghum reveals extensive homology with zein from maize. *Plant Mol. Biol.* 12, 245–256.

Duodu, K. G., Taylor, J. R. N., Belton, P. S., and Hamaker, B. R. 2003. Mini review: Factors affecting sorghum protein digestibility. *J. Cereal Sci.* 38, 117–131.

Elkhalifa, A. E. O., Georget, D. M. R., Barker, S. A., Belton, P. S. 2009. Study of the physical properties of kafirin during the fabrication of tablets for pharmaceutical applications. *J. Cereal. Sci.* 50, 159–165.

Esen, A. 1986. Separation of alcohol-soluble proteins (zeins) from maize into three fractions by differential solubility. *Plant Physiol.* 80, 623–627.

European Bioplastic Association. 2022. *What Are Bioplastics? Material Types, Terminology, and Labels – an Introduction.* European Bioplastics fact sheet "Bioplastics in Packaging," www.european-bioplastics.org/news/publications/

Gillgren, T., Faye, M.-V., Stading, M. 2011. Mechanical and barrier properties of films from millet protein pennisetin. *Food Biophys.* 6, 474–480.

Gillgren, T., Stading, M. 2008. Mechanical and barrier properties of avening, kafirin, and zein films. *Food Biophys.* 3, 287–294.

Giuberti, G., Gallo, A., Cerioli, C., Masoero, F. 2012. In vitro starch digestion and predicted glycemic index of cereal grains commonly utilized in pig nutrition. *Anim. Feed Sci. Technol.* 174, 163–173.

Hamaker, B. R., Mohamed, A. A., Habben, J. E., Huang, C. P., Larkins, B. A. 1995. Efficient procedure for extracting maize and sorghum kernel proteins reveals higher prolamin contents than the conventional method. *Cereal Chem.* 72, 583–588.

Higashiyama, Y., Turasan, H., Cakmak, M. et al. 2021. Fabrication of pristine electrospun kafirin nanofiber mats loaded with thymol and carvacrol. *J Mater Sci* 56, 7155–7170. https://doi.org/10.1007/s10853-020-05663-7.

Icoz, D. Z., Dogan, H., Kokini, J. L. 2005. Session 51, food chemistry: proteins and enzymes. *2005 IFT annual meeting,* July 15–20. New Orleans.

Issa, S, Jarial, S, Brah, N, Harouna, L. 2016. Are millet and sorghum good alternatives to maize in layer's feeds in NIGER, *West Africa. Indian J AnimSci.* 86(11):1302–5.

Jacob, J. Feeding pearl millet to poultry. 2015. https://articles.extension.org/pages/68861/feeding-pearl-millet-to-poultry.

Jukanti, A. K..., Gowda, C. L. L., Rai, K. N. *et al.* 2016. Crops that feed the world 11. Pearl Millet (*Pennisetum glaucum* L.): an important source of food security, nutrition and health in the arid and semi-arid tropics. *Food Sec.* 8, 307–329. https://doi.org/10.1007/s12571-016-0557-y

Kayode´, A., Bara, C., Dalode´-Vieira, G., Linnemann, A., Nout, M. 2012. Extraction of antioxidant pigments from dye sorghum leaf sheaths. *LWT Food Sci. Technol.* 46, 49–55.

Khorasani, Sepideh et al. 2018. Formulation of nanocarriers composed of millet protein and tocopherols to control bioavailability of lipophilic food components, *J Pharm Sci Emerg Drugs*, Vol.: 6 doi: 10.4172/2380-9477-C5-017

Kramer, M. E. 2009. Structure and function of starch-based edible films and coatings, Edible films and coatings for food applications, Springer, pp. 113–134.

Kubo, R. 2016. The reason for the preferential use of finger millet (Eleusine coracana) in eastern African brewing. *J Inst Brew.* 122: 175–80. http://dx.doi.org/10.1002/jib.309

L. Dai, J. Zhang and F. Cheng. 2019. Effects of starches from different botanical sources and modification methods on physicochemical properties of starchbased edible films, *International Journal of Biological Macromolecules*, https://doi.org/ 10.1016/j.ijbiomac.2019.03.197

Lal, S. S., Tanna, P., Kale, S., Mhaske, S. T. 2016. Kafirin polymer film for enteric coating on HPMC and gelatin capsules. *J. Mater. Sci.* 52, 3806–3820.

Lau, E. T., Johnson, S. K., Stanley, R. A., Mereddy, R., Mikkelsen, D., Halley, P. J., Steadman, K. J. 2015a. Formulation and characterization of drugloaded microparticles using distillers dried grain kafirin. *Cereal Chem.* 92, 246–252.

Lau, E. T., Johnson, S. K., Stanley, R. A., Mikkelsen, D., Fang, Z., Halley, P. J., Steadman, K. J. 2015b. Preparation and in vitro release of drug-loaded microparticles for oral delivery using wholegrain sorghum Kafirin protein. *Int. J. Polym. Sci.* 2015, 8. https://doi.org/10.1155/2015/343647.ID 343647

Leeson, S., Summers, J.D., 2001. Energy. In: *Nutrition of the Chicken*, fourth ed. University Books, Canada. Guelph, Ontario, pp. 34–99.

Lyumugabe, F., Gros, J., Songa, E. B., Thonart, P. 2015. Sorghum beer brewing using Eleusine coracana "Finger Millet" to improve the saccharification. *Am J Food Technol*; 10(4): 167–175. http://dx.doi.org/10.3923/ajft.2015.167.175

Mal, B., Padulosi, S., Ravi, S. B. 2010. Minor millets in South Asia: learnings from IFAD-NUS Project in India and Nepal. Maccarese, Rome: Bioversity Intl and Chennai: M. S. Swaminathan Research Foundation, 1–185.

Marx, S., et. al. 2014. Amaranth lignocellulose as feedstock for bioethanol production: effect of microwave-assisted alkaline pretreatment on reducing sugar yield. www.etaflorence.it/proceedings/index.asp?conference=2014

Mazhar, H., Chandrashekar, A., Shetty, H.S., 1993. Isolation and immunochemical characterization of the alcohol-extractable proteins (kafirins) of Sorghum bicolor (L.) *Moench. J. Cereal. Sci.* 17, 83–93.

Mendes A. C., Stephansen, K,, Chronakis, I. S. 2017. Electrospinning of food proteins and polysaccharides. *Food Hydrocoll* 68:53–68. https://doi.org/10.1016/j.foodhyd.2016.10.022

Ministry of Agriculture and Farmers Welfare, Department of Agriculture, Cooperation and Farmers Welfare, Government of India. 2018. *Gazette* notification on millets dated 10th April 2018 (F.No. 4-4/2017-NFSM(E), *The Gazette of India*.

Murugan, R., Nirmalakumari, A. 2006. Genetic divergence in foxtail millet (*Setaria italica* L.). *Indian J Genet Plant Breed* 66: 339–340.

National Research Council, 1996: *Lost Crops of Africa vol. I: Grains*. Board on Science and Technology for International Development. National Academy Press, Washington, DC.

Ningbo Li, Ying Wang, Michael Tilley, Scott R. Bean, Xiaorong Wu, Xiuzhi Susan Sun, Donghai Wang. 2011. Adhesive Performance of Sorghum Protein Extracted from *Sorghum DDGS and Flour, J Polym Environ*, 19:755–765. doi: 10.1007/s10924-011-0305-5

Niphadkar S, Bagade P, Ahmed S. 2017. Bioethanol production: Insight into past, present and future perspectives. Biofuels, 9(2): 229–38. http://dx.doi.org/10.1080/17597269.2017.1334338

Obiana AB. Overview: importance of millets in Africa. 2003. Published online at www.afripro.org.uk/papers/Paper02Obil. Accessed 7 Feb 2020.

Oluwaniyi, O., Dosumu, O., Awolola, G., Abdulraheem, A. 2009. Nutritional analysis and stability studies of some natural and synthetic food colourants. *Am. J. Food Technol.* 4, 218–225.

Paz, M. H., Guillard, V., Reynes MGontard, N. 2005. Ethylene permeability of wheat gluten film as a function of temperature and relative humidity. *J. Membr. Sci.*, 256, 108–115.

Pizzi, A. 2006. Recent developments in eco-efficient bio-based adhesives for wood bonding: opportunities and issues. *J Adhes Sci Technol* 20(8):829–846.

Ramakrishnan, S., S. Loganayagan, G. Kowshika et al. 2021 Adobe blocks reinforced with natural fibres: A review, *Materials Today*: Proceedings, https://doi.org/10.1016/j.matpr.2020.11.377

Reddy et al. 2013. Utilization Pattern, Demand and Supply of Pearl Millet Grain and Fodder in Western India. Working Paper Series no. 37. Patancheru 502 324, Andhra Pradesh, India: International Crops Research Institute for the Semi-Arid Tropics. 24 pp.

Reddy, M. M., Vivekanandhan, S., Misra, M., Bhatia, S. K., Mohanty, A. K. 2013. Biobased plastics and bionanocomposites: current status and future opportunities. *Prog. Polym. Sci.* 38 (10), 1653–1689.

Reddy, N., Yang, Y. 2011. Potential of plant proteins for medical applications. *Trends Biotechnol.* 29, 490–498.

Rose, D. J., Santra, D. K. 2013. Proso millet (Panicum miliaceum L.) fermentation for fuel ethanol production. *Ind. Crops Prod.*, 43, 602–605.

Shahidi, F., A. Chandrasekara. 2013. Millets grain phenolics and their role in disease risk reduction and health promotion: a review, *J. Funct. Foods* 5, 570–581.

Shaikh, M., Haider, S., Ali, T. M., and Hasnain, A. 2018. Physical, thermal, mechanical and barrier properties of pearl millet starch films as affected by levels of acetylation and hydroxypropylation. Biomac, https://doi.org/10.1016/j.ijbiomac.2018.11.135.

Shashikala, T. K., Rai, N., Balaji, R., Naik, M., Shanti, V., Chandrika., Reddy, K. L. 2013. Fodder Potential of Multicut Pearlmillet Genotypes during Summer Season. *International Journal of Bio-resource and Stress Management* 4(4), 628–630.

Shull, J. M., Watterson, J. J., Kirleis, A. W. 1991. Proposed nomenclature for the alcohol-soluble proteins (kafirins) of Sorghum bicolor (L. Moench) based on molecular weight, solubility, and structure. *J. Agric. Food Chem.* 39, 83–87.

Shull, J. M., Watterson, J. J., Kirleis, A. W. 1992. Purification and immunocytochemical localization of kafirins in Sorghum bicolor (L. Moench) endosperm. *Protoplasma* 171, 64–74.

Simnadis, T. G., Tapsell, L. C., Beck, E. J. 2016. Effect of sorghum consumption on health outcomes: a systematic review. *Nutr. Rev.* 74, 690–707.

Singh, H., Sodhi, N. S., Singh, N. 2010. Characterisation of starches separated from sorghum cultivars grown in India. *Food Chem.* 119, 95–100.

Singh, P., Raghuvanshi, R. S. 2012. Finger millet for food and nutritional security. *Afr J Food Sci* 6(4):77–84.

Smith, A. M., Callow, J. A. 2006. *Preface in biological adhesives.* Springer, Heidelberg; Wang Y, Sun X. S., Wang, D. (2007) *J Adhesion Sci Technol* 21: 1469.

Srianta, I., Harijono, A. 2015. Monascus-fermented sorghum: pigments and monacolin K produced by Monascus purpureus on whole grain, dehulled grain and bran substrates. *Int. Food Res. J.* 22, 377–382.

Srianta, I., Zubaidah, E., Estiasih, T., Yamada, M. 2016. Comparison of Monascus purpureus growth, pigment production and composition on different cereal substrates with solid state fermentation. *Biocatal. Agric. Biotechnol* 7, 181–186.

Taylor, J. R. N., Schüssler, L., van der Walt, W. H. 1984a. Fractionation of proteins from low-tannin sorghum grain. *J. Agric. Food Chem.* 32, 149–154.

Taylor, J. R., Schober, T. J., Bean, S. R. 2006. Novel food and non-food uses for sorghum and millets. *J Cereal Sci* 44(3):252–271.

Taylor, J., Anyango, J. O., Potgieter, M., Kallmeyer, K., Naidoo, V., Pepper, M. S., Taylor, J. R. N. 2015. Biocompatibility and biodegradation of protein microparticle and film scaffolds made from kafirin (sorghum prolamin protein) subcutaneously implanted in rodent models. *J. Biomed. Mater. Res.* 103, 2582–2590.

Taylor, J., Anyango, J. O., Taylor, J. R. N. 2013. Developments in the science of zein, kafirin, and gluten protein bioplastic materials. *Cereal Chem.* 90, 344–357.

Taylor, J., Muller, M., Minnaar, A. 2016. Improved storage and eat-ripe quality of avocados using a plant protein-based coating formulation. *Qual. Assur. Saf. Crops* 8, 207–214.

Taylor, J., Taylor, J. R. N., Belton, P. S., Minnaar, A. 2009a. Formation of kafirin microparticles by phase separation from an organic acid and their characterization. *J. Cereal. Sci.* 50, 99–105.

Taylor, J., Taylor, J. R. N., Belton, P. S., Minnaar, A. 2009b. Kafirin microparticle encapsulation of catechin and sorghum condensed tannins. *J. Agric. Food Chem.* 57, 7523–7528.

Turasan, H., Cakmak, M., Kokini, J. 2019. Fabrication of zeinbased electrospun nanofiber decorated with gold nanoparticles as a SERS platform. *J Mater Sci* 54:8872–8891. https://d oi.org/10.1007/s10853-019-03504-w

Upadhyaya, H. D., Reddy, K. N., Pattanashetti, S. K., Kumar, V., Ramachandran, S. 2018. Identification of promising sources for fodder traits in the world collection of pearl millet at the ICRISAT genebank. *Plant Genetic Resources* 16(2), 127–136. https://doi.org/10.1017/S147926211700003X

Valencia-Sullca, C.; Vargas, M.; Atarés, L.; Chiralt, A. Thermoplastic cassava starch-chitosan bilayer films containing essential oils. *Food Hydrocoll.* 2018, 75, 107–115.

Wang, Y. (2006) "Adhesive performance of soy protein isolate enhanced by chemical modification and physical treatment." Dissertation, Kansas State University; Qi G, Sun XS (2011) *J Am Oil Chem Soc* 88:271.

Wu, X., Zhao, R., Bean, S. R., Seib, P. A., McLauren, J. S., Madl, R. L., Tuinstra, M. R., Lenz, M.C., Wang, D. 2007. *Cereal Chem* 84:130.

Wu, Y., Lin, Q., Cui, T., Xiao, H. 2014. Structural and physical properties of starches isolated from six varieties of millet grown in China. *Int. J. Food Prop.* 17, 2344–2360. www.fda.gov.

Xiao, J., Li, C., Huang, Q. 2015a. Kafirin nanoparticle-stabilized Pickering emulsions as oral delivery vehicles: physicochemical stability and in vitro digestion profile. *J. Agric. Food Chem.* 63, 10263–10270.

Xiao, J., Nian, S., Huang, Q. 2015b. Assembly of kafirin/carboxymethyl chitosan nanoparticles to enhance the cellular uptake of curcumin. *Food Hydrocoll.* 51, 166–175.

Xiao, J., Shi, C., Zheng, H., Shi, Z., Jiang, D., Li, Y., Huang, Q. 2016. Kafirin protein based electrospun fibers with tunable mechanical property, wettability, and release profile. *J. Agric. Food Chem.* 64, 3226–3233.

Xu, J.-H., Messing, J. 2009. Amplification of prolamins storage protein genes in different subfamilies of the Poaceae. *Theor. Appl. Genet.* 119, 1397–1412.

Yabuno, T. 1987. Japanese barnyard millet (Echinochloa utilis, Poaceae) in Japan. *Econ. Bot.* 41, 484–493.

Yang, P., Seib, P. A. 1995. Low-input wet-milling of grain sorghum for readily accessible starch and animal feed. *Cereal Chem.* 72, 498–503.

Zhang, K., Zheng, G., Saul, K., Jiao, Y., Xin, Z., Wang, D. 2017. Evaluation of the multi-seeded (msd) mutant of sorghum for ethanol production. *Ind. Crops Prod.* 97, 345–353.

Processing Technologies of Nutri-Cereals

Uday S. Annapure, Ranjitha Gracy T. Kalaivendan,
Anusha Mishra, and Gunaseelan Eazhumalai

Contents

14.1 Introduction

Nutri-cereals, or millets, are grains with origins way back to the first millennium BC, with the ability to grow in diverse environments and independent fertilization capability (Fuller, 2003). Before undertaking these crops for domesticated farming, their abundance as wild vegetation proved the possibility of including them in organized agriculture (Weber & Fuller, 2008). Their capability to adapt to the varying ecological conditions, soils, rainfall status, and temperatures made them a feasible crop for cultivation in the early agriculture scenario (Singh, 2008). However, as civilization progresses, crops of relatively higher productivity and better organoleptic attributes were preferred for cultivation over the nutri-cereals. Despite widespread global cultivation, many of the African and Asian countries (96.89%), followed by Europe (1.81%), stand out individually in the global production (Chandra, Chandora, Sood, & Malhotra, 2021) (Meena, Joshi, Bisht, & Kant, 2021). Among the existing nutri-cereals, pearl millet, foxtail millet, proso millet, and finger millets account for the highest production globally in the specified order (Das & Rakshit, 2016).

Despite the lesser agricultural production, millets contribute significantly to ensuring the food and nutritional security of the world since it is consumed as the staple cereal in many of the developing nations on the African continent and because of their rich bioactive profile.

Besides providing energy, millets are the sources of various vitamins, minerals, and fibers because of which they are widely called "nutri-cereals." To give an instance, sorghum, foxtail, and pearl millets provide a maximum of 12 g/100 g of protein, and barnyard, proso, and Kodo millets offer up to 10 g/100 g of fiber. Also, minerals, including calcium, iron, and zinc can be obtained in the quantity of 344 mg, 9.3 mg, 3.7 mg, respectively, from per 100 grams of the millets (Aykroyd, Gopalan, & Balasubramanian, 1963). These micro-nutrients are necessary for the healthy functioning and metabolisms of an individual, thereby accounting for nutritional security. In addition, some millet crops, like sorghum, are used for non-food applications such as cattle feed and fodder (Ronda, Aruna, Visarada, & Bhat, 2019), distilleries (Lodge, Stock, Klopfenstein, Shain, & Herold, 1997), and biofuel productions (Dar, Dar, Kaur, & Phutela, 2018). Apart from this, their trivial growth requirements factors help to establish soil sustainability, as their cultivation with less human inputs preclude abandoning the cultivable lands as fallow lands (Maitra & Shankar, 2019). The millets have higher storability and shelf life, which facilitates their use in instances of crop failure. Thus millets are of high importance, both for agriculture and for human consumption.

The structure and chemical composition vary among the millets despite the presence of basic structural elements such as pericarp, endosperm, and germ, which are similar in all the nutri-cereals. The shape of the kernel or caryopsis, the color of endosperm, and the overall appearance of the grain change among the varieties and species of the millets (Serna-Saldivar & Espinosa-Ramírez, 2019). The outermost layer of the caryopsis, that is the pericarp, consists of three layers, namely the epicarp, mesocarp, and endocarp. The epicarp or the farthest seed coat is a layer of cutin wax, and the testa, which would contain the anti-nutritional compounds of the grain. The mesocarp may contain the starch compounds while the endocarp is a moisture reserve utilized for germination of the grain (Serna-Saldivar S., 1995). The endosperm of the caryopsis comprises the aleurone layer, which is the source of proteins, and the peripheral region contains the starch compounds (Taylor, Novellie, & Liebenberg, 1984). And the germ portion would be having other nutritional elements such as fat, vitamins, and minerals (Hama, Icard-Vernière, Guyot, Picq, Diawara, & Mouquet-Rivier, 2011). Out of all the millets, proso and finger millets are considered to be utricles rather than caryopsis as the seed coat is not firmly attached to the endosperm and this facilitates easy removal. The outer seed coat would be removed in primary processing unit operations to make the millets suitable for consumption and better nutrient absorption.

The processing of millets imparts multiple outcomes, including enhanced nutrient digestibility, increased bioavailability of the minerals, modified functionality, and organoleptic attributes (Bookwalter, Lyle, & Warner, 1987). Also, removing or separating the desired nutritional proportion of the millets from the complex matrix would assure extended stability. The separation of starchy components from the millets in the form of milled grain or comminuted flour provides a relatively longer shelf life as the germ portion containing the lipid fractions are parted away discretely (Sruthi & Rao, 2021). Not only in the form of flour, but other products such as porridge mix, baked goods like bread, cakes, biscuits, and cookies, extruded products like pasta, vermicelli, and noodles, and flakes are also prevalent in the current millets processing market (Jaybhaye, Pardeshi, Vengaiah, & Srivastav, 2014). In recent times millets have been explored for gluten substitute products, particularly for gluten intolerant people, because of their functional characteristics similar to that of wheat-based products (Kumar, et al., 2019)(Romero, Santra, Rose, & Zhang, 2017)(Niro, D'Agostino, Fratianni, Cinquanta, & Panfili, 2019). Some millets are used for fermented beverages because of their exceptional malting attributes (Adebiyi, Obadina, Adebo, & Kayitesi, 2018).

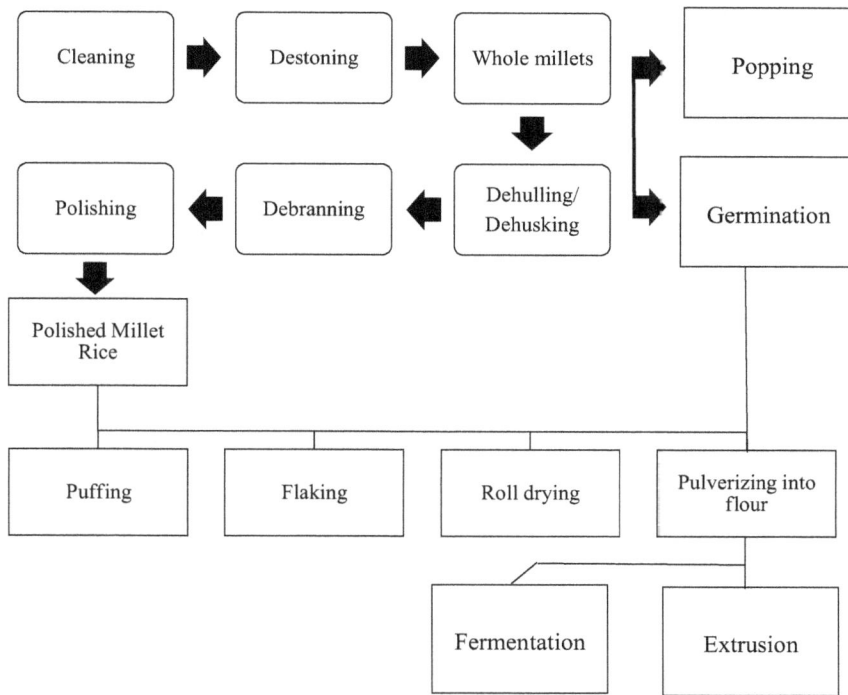

Figure 14.1 *Flow chart of processing of nutri-cereals.*

Thus, it is the need of the hour to have a thorough understanding of the existing conventional and modern processing technologies of such inadequately explored nutri-rich cereals, the millets. The present chapter deals with the primary and secondary processing operations of significant interest in the nutri-cereals such as sorghum, pearl millet, finger millet, barnyard millet, foxtail millet, little millet, Kodo millet, and proso millet with an emphasis on conventional as well as modern technologies. The general overview of processing of millets is shown in **Figure 14.1**.

14.2 Sorghum

A C4 cereal grass, sorghum belongs to the family Gramineae, subfamily Panicoideae. They are diploid (2n = 20 chromosomes), and one of the majorly known species is *Sorghum bicolor Sorghum bicolor L. Moench*, syn S. vulgare. It is widely consumed on the extensive grasslands and savannas of Africa. It requires significantly less rainfall compared to other food crops like rice or maize. It has been popular since 8000 BC.

The color may range from white, red, yellow, or brown with spherical or spindle shapes. For the dry milling of grain, hardness is an essential factor. The caryopsis of the sorghum is naked. The three layers present in it are the outer pericarp (outer layer), endosperm (starchy storage structure), and the germ (embryo) (Dendy, 1994b). The kernel of sorghum is deeply embedded, which leads to difficulty in removal (Rooney and Sullins, 1969). The aleurone layer in sorghum grains contains a thick wall, protein is present in high amounts, with enzymes, ash, and spherosomes in the aleurone layer. Protein and starch are very much present in the endosperm of sorghum grains. Initial dehulling of the grain is required to make it suitable for human consumption. The digestibility of the grain can be improved by steam flaking, micronizing, popping, and reconstitution, leading to disruption of sorghum's structure.

Milling helps develop more value-added products and helps in the utilization of fractions obtained for feed for animals. The dry milling operation begins with the initial cleaning of the grains. Moistening is done to soften the endosperm and then milled by conventional rollers to separate the layers from each other, that is, endosperm, germ, and bran. Decortication or pearling is an important step, performed by wet milling. After cleaning the grains, they are moistened and then pounded in mortar and pestle with water for 2–3 min, leading to the removal of coarse fiber particles and pigments. At the same time, in this entire process minimum cracking of the sorghum needs to be ensured. The detachment of the pericarp at the mesocarp takes place. The grains with round kernels and hard endosperm are known as decortications. In them, maximum polishing of 12 percent can occur, and high protein content of up to 27 percent, high fat, high ash, and low fiber content. To ensure better efficiency and lower wastage, mechanical decortications have been employed. With more advancement, many countries have begun to use commercial milling operations. This technique uses abrasive peelers, followed by determination and further sieving, milling, gravity separation, and sieving to produce low-fat grits, meal, and flour.

The easiest (and a low-cost) process to elevate the quality of raw ingredients in the food processing sector is soaking. Soaking leads to several changes like activation of the enzyme, reduction in the anti-nutritional factor by the digestion of food reserves (Eltayeb, Mohamed, & Fageer, 2017). It also leads to an increment in the total flavonoid content by releasing bound phenolics (Xiong, Zhang, Luo, Johnson, & Fang, 2019). Despite its benefits, however, prolonged soaking is not preferred as it leads to leaching out of soluble minerals and several bioactive substances from the grain (Boniface & Gladys, 2011). Hurdles or a combination of several treatments can be employed to prevent the loss of essential minerals due to leaching.

The development of fermented products through sorghum is now being explored extensively. It is employed to develop porridge, bread, and so forth. The microorganisms in action are yeast and lactic acid bacteria (LAB). The fermented products are preferred for their immense nutritional benefits such as enhanced digestibility and improved amino acid profile. The research conducted to compare the enhancement of nutritional profile illustrated that the grains that have undergone fermentation have better digestibility and a boost in amino acid profile. Inhibitory action of enzymes has reduced, and in-vitro protein digestibility has increased in the fermentation process of sorghum grains (Nkhata, Emmanuel Ayua, & Shingiro, 2018) (Day & Morawicki, 2016).

Sorghum has been exploited to prepare porridge, pancakes, flatbreads, beverages, and other products. Due to the increase in celiac disease and gluten intolerance incidences, the demand for gluten-free products has increased profoundly. Therefore, food industries and researchers have a higher inclination towards gluten-free nutritional storehouses of millets. Various methods, and products that are prepared using sorghum, are:

1. Preparation of sorghum flour: It can be an excellent substitute for conventional wheat flour. Different percentages of sorghum flour can be added based on better consumer acceptability. The flour can be used for various popular products like cookies, porridges, and other snack products.

2. Sorghum flour (SF): In combination with wheat flour can be formulated in varying percentages to develop a nutritionally enhanced sorghum flour. This flour has better anti-oxidant capacity owing to the presence of polyphenols. This can be utilized to prepare flatbreads, porridges, cookies, and snacks (Benhur, Bhargavi, Kalpana, Vishala, Ganapathy, & Patil, 2015) (Cao, Sun, Liu, Yin, & Wu, 2012) (Palavecino, Ribotta, León, & Bustos, 2019) (Sun, Han, Wang, & Xiong, 2014). This flour with the goodness of nutrition can be a preventive measure for numerous lifestyles.

3. Sorghum bread (SB).
Bread is one of the most popular food products consumed by people. It is cheap and readily available; therefore, its consumption has increased consistently, especially in developing nations – the ever-growing population demands healthier and affordable alternatives. Bread production using sorghum is highly preferable. Different kinds of bread, such as kissra bread, sourdough bread (SDB), flatbread, khamir bread, and frybread can be produced (Abdualrahman, Ma, Yagoub, Zhou, Ali, & Yang, 2019) (Marston, Khouryieh, & Aramouni, 2016) (Rose, Williams, Mkandawire, Weller, & Jackson, 2014).

4. Sorghum based extruded products
Extrusion is the most flexible processing technique in the development of the snack food segment. The consumption of snacks among all the sections of society has also led towards healthier innovations and ideas. Millets have proved to be the savior to treat the triple burden of malnutrition globally, that is, under-nutrition, and nutrient deficiencies. (Benhur, Bhargavi, Kalpana, Vishala, Ganapathy, & Patil, 2015) Developed sorghum-based pasta using extrusion technology. The sorghum with 50 percent concentration was acceptable based on sensory characteristics.

14.3 Pearl millet

Pennisetum glaucum, commonly known as pearl millet, has been mainly grown for food, feed and forages (Arora, Sehgal, & Kawatra, 2003), especially in African and Asian countries (Nambiar, Dhaduk, Sareen, Shahu, & Desai, 2011). Pearl millet can survive under extreme conditions like drought and heat, which makes it advantageous to be grown where popular crops like cereals and pulses fail to survive. More than 29 million hectare are used for pearl millet cultivation. It is mostly restricted to the African continent (15 million) and in Asia (around 11 million), which holds the status of largest producer (Rathore, Singh, & Kumar, 2016). India is the largest producer of pearl millet (Basavaraj, 2010). Pearl millet is the storehouse of nutrients. It is a rich source of linolenic (around 74.89%), linoleic acid (around 45.01%) (Jaybhaye, Pardeshi, Vengaiah, & Srivastav, 2014). There are several processing techniques employed to reduce the anti-nutritional content of the pearl millet, like blanching, parboiling and so forth (Singh, Sehgal, Kawatra, & Preeti, 2006) (Legesse, 2013).

Pearl millet is converted into edible forms through different processing procedures, thus enhancing the quality. This not only enhances the nutritional value but also affords better utilization. It can be processed to prepare rice forms millet, flour, roasted, popped, sprouted millet, snacks, porridge and fermented products (Jaybhaye, Pardeshi, Vengaiah, & Srivastav, 2014).

The process of removal of the outer layer of the grain, that is, hull and pericarp is known as dehulling (Taylor J. R., 2015). In case of pearl millet, other small fractions ranged between 1.5–29.3 percent (Jaybhaye, Pardeshi, Vengaiah, & Srivastav, 2014). Research has been conducted on different methods followed for the dehulling of pearl millet on its nutritional composition.

Different decortication methods were employed and compared. There was considerable improvement in the protein content and dry matter digestibility when the decortication was done up to 17.5 percent, as reported by (Serna-Saldivar, Clegg, & Rooney, 1994).

In the decortication of pearl millet, greater decrease in proximate composition of millets, lysine, and other amino acids had been observed, attributed to the removal of the pericarp and the germ layer in the operation. Therefore, in pearl millet techniques to be explored for better retention of nutrients, without the disruption of endosperm (germ embedded in endosperm) post decortication (Hama, Icard-Vernière, Guyot, Picq, Diawara, & Mouquet-Rivier, 2011).

The various advantages of milling include separation of endosperm, bran, and germ, and obtaining fine flour. Pearl millet can be milled using a hammer as well as a roller mill. Large-sized particles are obtained through a hammer mill, which hinders its utilization in the preparation of porridge required with a thin and rough texture, and in baked foods and steamed foods that require a smooth texture. Central Food Technology Research Institute Mysore has commercialized a technique for upgrading the attributes of flour made of pearl millet. In this technique grains were treated, moist heat followed by drying to moisture content of 10–12 percent and thereafter dehulling was carried out up to the stage when the preferred degree of pulverization is obtained. A high content of floury endosperm was obtained after milling. This can be suitable to be stored up to 3–4months as well free fatty acids are below 10 percent during the period of storage (Rai, Gowda, Reddy, & Sehgal, 2008). Roller mills are used to produce low-fat grits. Low-fat grits can be easily produced through roller mills. It was concluded in the study conducted by (Abdelrahman, Hoseney, Varriano & Marston, 1983) that the low-fat grits can be prepared using roller mills from the pearl millet. Decortications and tempering further followed it and milling the grains through corrugated rolls resulted in 61 percent grits from the whole grains with a reduced fat content of around 1.2 percent.

The soaking is a crucial step in processing. It reduces phytic acid content, depending on the variety, pH, soaking conditions, and duration of soaking. A reduction in 15 percent polyphenol content has been observed upon soaking pearl millet for 14 h. (Nithya, Ramachandramurty, & Krishnamoorthy, 2007). Since soaking can lead to loss of essential minerals, Jha, Krishnan, & Meera, 2015, reported that soaking for lesser time in an acidic or alkaline medium resulted in reducing inhibitory factors without any mineral loss.

Blanching is considered to be an essential pre-treatment for preservation and for enhancing the shelf-life of the pearl millet. It leads to a decrease in the activity of enzymes without any significant effect in its nutritive content. In blanching, boiling water is used, followed by dipping the grains in it (ratio being 1:5) for about 30 sec and thereafter drying at 45–50°C for 60 min (Rai, Gowda, Reddy, & Sehgal, 2008). Initial blanching also proved to be beneficial in reduction of inhibitory substances, rancidity, and bitterness in the biscuits manufactured from the pearl millet (Singh, Sehgal, & Kawatra, 2006).

The process is followed by limited sprouting of cereals in a humid atmosphere under the controlled processing conditions referred to as malting. Although this process reduced protein content, the protein quality as well as protein energy ratio was enhanced. (Singh & Saini, 2012). There are certain beneficial phenomena attributed to the process of malting: they are, increased energy density, increased vitamin content, and better digestibility of available nutrients (Preetika, Padmini, & Shobha, 2004). Products that can be prepared through malted flour are milk based beverages and confectionary and cake products (Shobana, et al., 2013). Activation of enzyme occurred due to germination, thus leading to disintegration of carbohydrates, proteins, and lipids into simpler compounds. Proteases led to degradation of proteins, which led to improvement in the bioavailability of the nutrients (Singh, Rehal, Kaur,

& Jyot, 2015). Superior quality of porridge can be prepared by allowing germination for five days followed by fermentation for 24 h and finally roasting of the pearl millet in order to obtain a healthier alternative (Kindiki, Onyango, & Kyalo, 2015). It has also been observed (Bhati, Bhatnagar, & Acharya, 2016) that carrying out malting prior to milling enhances the color, in-vitro iron availability as well reduction in the anti-oxidant activity and free fatty acid proportion.

Fermentation is an essential processing technique that has been successfully incorporated into millet processing. Vitamins B and K, lysine, folate and other micronutrients have found to be increased using lactic acid bacterial fermentation (Tamene, et al., 2019). Increased mineral accessibility is due to the presence of lactic acid bacteria which results in lowering down of the pH. Fermentation is an essential technique for the preservation of food, nutritional quality enhancement and development of flavor (Saleh, Zhang, Chen, & Shen, 2013). Mix cultures of microrgansim – mainly yeast and *Lactobacilli* – are used for the purpose of malting and souring. Malting and souring with mixed cultures of yeast and soluble sugars and starch that is present in millets are broken down into simpler forms by enzymes existing in grain and fermentation media (Rai, Gowda, Reddy, & Sehgal, 2008). Increase in in-vitro protein digestibility and decrease in the anti-nutritional factors of the pearl millet have been a commendable attribute of fermentation processing (Hassan, Ahmed, Osman. Eltayeb, Osman, & Babiker, 2006). This may occur due to formation of simple and soluble end products as a result of partial degradation of complete storage proteins (Chavan, Chavan, & Kadam, 1988). Products prepared using this technology are idli, dosa and ambli.

14.4 Finger millet

One of the oldest millet crops of ancient India is finger millet, referred to as Nrttakondaka in Sanskrit, which means "dancing rain," also called "rajka" or "markata." (Achaya, 2009). It is the fourth most essential millet after pearl miller, sorghum and foxtail millet (Devi, Vijayabharathi, & Sathyabama, 2014). It is cultivated in many states of India, like Maharashtra, where it is called nachni, Bihar (umi), and so forth. The grains can be roasted, sprouted, ground and sieved as per the end product's utilization. Finger millet flour can be eaten as a ball or as gruel, either in the sweet preparation or salty. The main components of the millet kernel are seed coat, embryo and endosperm. The most commonly known variety is the red colored one, the others available are yellow, white, tan, red, brown and violet In the seed coat, the germ and the endosperm cell walls polyphenols have been extensively concentrated (Shobana & Malleshi, 2007).

Processing is an essential step in improving the acceptability of the millet and for enhancing further utilization of the millets.

Reduction of anti-nutrients is one of the crucial steps in processing the grains. Soaking is the widely used technique in order to reduce antinutrients in food. Polyphenols, phytates, saponins, oxalates and trypsin inhibitors have been found to be reduced after soaking ragi for 1–2 days at room temperature in water (1:10 for ragi: water) (Hotz & Gibson, 2007), thus achieving a better nutritional profile of the product.

The metabolic process of converting complex substances into simple forms with the aid of microorganism is called as fermentation. Several scientists have successfully proved the beneficial effects of fermentation. Conventionally, finger millet has been consumed in the form of porridge, chapatti or dosa, or in the form of a beverage. Preparation of most of these involve fermentation (Madhavi & Vaidehi, 1990) (Hadimani & Malleshi, 1993). A significant increase in the concentration of amino acid in fermented finger millet has been found to be observed;

an example, niacin 4.2mg/day, pantothenic acid 1.6mg/100g, riboflavin 0.62mg/100g, that is higher than raw variant (Basappa, Somashekar, Agrawal, Suma, & Bharathi, 1997). Therefore, fermentation holds essential significance in the processing of millets.

Immense efforts have been employed for decortication in the case of finger millet. In finger millet, debranning or decortications is not as effective when compared to other millets. Therefore, for finger millet hydrothermal treatment like hydration, steaming is employed in order to soften the hard and intact structure of the seed coat and to withstand the mechanical influence of decortications.

The debranning or decortication methods followed for most of the cereals were not effective in the case of finger millet owing to the intactness of the seed coat with highly fragile endosperm. Hence, to decorticate, finger millet is hydrothermally processed (hydration, steaming, and drying) to harden the soft endosperm to enable it to withstand the mechanical impact during decortications.

Amylase content of finger millet is higher than other millers, and also it reaches to maximum in 4–5 days (Malleshi, Desikachar, & Tharanathan, 1986). Protection from fungal infection and upgradation in the nutritional value of the malted products are the benefits obtained through germination of the grain. This leads to the generation of alpha and beta amylases resulting in the development of peculiar aroma (during roasting/kilning) that makes it suitable for malt foods (Verma & Patel, 2013). There is a high content if ionisable iron 88.3 percent in the malted finger millet in comparison to the raw one, that is, 7.4 percent (Deosthale, 2002)

One of the important processing techniques commercially utilized to prepare ready-to-thermal technology is *popping*. It is applied to finger millet using sand as the heat transfer media, where gelatinization of finger millet (Ragi, Eleusine coracana L.) starch occurs and the bursting of endosperm occurs. Popped finger millet contains exceptional flavor and aroma. It is used as a snack after seasoning with spices and condiments.

14.5 Barnyard millet

One of the most important millet crop that have experienced rapid upsurge in production worldwide is *Echinocloa* species commonly known as barnyard millet. Among the most cultivated forms are *Echinochloa esculenta* and *Echinochloa frumentacea* for human consumption and livestock feed. They provide food security to millions of people across the world as they constitute the fourth most produced minor millet. India is the largest producer of this millet in terms of area (0.146 m ha−1) as well as production (0.147 mt) and the average productivity in the last three years was 1034 kg/ha (IIMR, 2018).

Consumption of millet as popped snacks has been conventionally used in the form of popped products as snacks with the addition of flavor, sugar and salt, among the major components that form an integral part of various snack products. Popped grains had been prepared from barnyard millet, foxtail millet and little millet by dry heating to specific temperatures. (Srivastava, Dhyani, & Singh, 2003). Normal salt as a warming agent in an open iron saucepan containing samples and salt in the proportion 1:20 at 240–260 °C for 15–25 s were prepared. Bakery products like muffin and rusk have been prepared by (Nazni & Shobana, 2016). Barnyard based muffins have been developed by Goswami, Gupta, Mridula, Sharma, & Tyagi (2015), with increased fiber content and consumer acceptability. Thus it can be concluded that this millet has great potential to be used in the nutritional value of various products.

14.6 Foxtail millet

Foxtail millet is the second most produced millet after pearl millet, and they place among the higher produce yielding crops as well (Ramesh, et al., 2020). Because of their ability to grow even in hard and unfertilized territories, they are widely cultivated in many countries. The potential resistance of the foxtail millet plant against pests and diseases facilitates their organic farming facility. Foxtail millet is one of the proteins and fiber-rich millets containing 12.3 g and 8 g per 100 g, along with the iron and calcium content of 2.8 mg and 31 mg per 100 g (Shankaramurthy & Somannavar, 2019), which makes this millet suitable for diabetics as well as the geriatric population.

The structure of foxtail millet comprises a major portion of the husk, 14 percent, which contains fibrous and anti-nutritional elements. And the lipid-containing germ portion contributes 1.5–2 percent (Ushakumari, Latha, & Malleshi, 2004). The fibrous compounds do not account for the energy production on digestion despite their desirable effects on diseases and disorders like cancer and diabetes. However, the anti-nutritional elements influence the digestibility and bioavailability of the nutritive substances, especially the minor quantity of minerals presented. Also, the lipid molecules in the germ get rancid and affect the quality and shelf life of the millets as a whole. Thus these factions of the grain would be removed generally to fetch better nutrient availability and stability.

Dehusking increases the palatability, nutritional safety, and stability of the millets. Conventionally, attrition mills made of wood or stone were utilized to take away the hulls of the grains. The shearing force on the abrasive surfaces removes the outer husks, which would be separated further using pneumatic operations like winnowing. To make the process less laborious and more efficient, the same mechanism of action is adopted to design types of machinery for the processing of millets. Based on the principle force of action employed, there are two types of dehullers, namely centrifugal and abrasive dehullers (Roy, 2020). In the centrifugal dehullers, the millets are subjected to a centrifugal force driven by the impeller, which throws the millet towards the outer hard envelope that tends to loosen the husk from the grains. The consequent impacts on the passage up to the outlet aids in dehusking of the millets. Whereas, in the case of abrasive dehullers, the millets are allowed to be sheared between two rollers of abrasive surfaces made of stone, steel, or rubber, and dehusked. The dehullers are provided with the aspirating system to obtain the clean dehusked millets free from the removed hulls, husks, and lightweight impurities. Comparatively the abrasive dehullers are more efficient than the centrifugal dehullers, besides the limitations such as pulverization of millets, and contamination of millets with the material of abrasive rollers. To overcome the limitations and to improvise the dehullers, a double chamber centrifugal dehuller was developed, wherein the hulling of millets was efficiently done with a lesser breakage percentage (Durairaj, Gurumurthy, Nachimuthu, Muniappan, & Balasubramanian, 2019). There are different millet huller prototypes developed by the government and non-government research organizations with varying working capacities (Palanimuthu, 2017). Studies have revealed that the process of dehulling the foxtail millet desirably affects the nutrients' digestibility and bioavailability by effectively reducing the anti-nutritional components (Pawar & Machewad, 2006)(Sharma & Niranjan, 2018). However, the polyphenol content have reduced after dehulling despite the incremental ionizable mineral content. Because of the loss in polyphenols, the anti-oxidant activity of the millets was also observed to be reduced (Chandrasekara, Naczk, & Shahidi, 2012).

Along with dehulling, sorting, grading, cleaning, and destoning are considered as primary post-harvest processing operations of millets (Malleshi, 1989). Because of the tiny morphology of millets, separation from the foreign material impurities is carried out with the help of pneumatic and specific gravity separators (Balasubramanian, Kotwaliwale, Kate, & Ambrose, 2021). The lesser dense dust, stalks, and leaves could be blown away with airflow above their terminal velocity, while the high-density stone and metal particles are removed with the help of vibratory specific gravity separators. Also, different sized sieves and sifters are used for further cleaning of the millet grains. These primary processing operations are similar to all the small millets, namely foxtail millet, kodo millet, and little millet.

The structural and molecular attributes of foxtail millet constituents, particularly of the starch, made them relatively tougher for gelatinization among all the millets (Shinoj, Viswanathan, Sajeev, & Moorthy, 2006). The homogeneous nature of crystalline amylopectin molecules contributes to their resistant starch level and also increases cooking time and energy. This necessity made the utilization of pulverized millets into market demand (Amadou, Gounga, Shi, & Le, 2014). Thus the primarily processed foxtail millet is milled into flours for consumption and application in secondary processing operations.

The utilization of foxtail millets was increased via the production of ready-to-eat or ready-to-cook value-added products which enhance the market value as well as the producers' economies. The dehulled foxtail millets are puffed and flaked similarly to rice with the help of high-temperature puffing guns and rollers. The millets are conditioned to a specific moisture content of around 18 percent before subjecting them to high-temperature processing (Ushakumari, Latha, & Malleshi, 2004). These processing operations modify not only the organoleptic attributes of the millet but also affect the nutrient composition and digestibility. For example, the starch digestibility of foxtail millet flakes was less compared to the unprocessed dehulled millet due to resistant starch formation on subsequent heating and cooling processes (Suresha, 2016). This promises the potential of foxtail millet flakes to be used in the diets of the diabetic population. Unlike puffing and flaking, the popping of foxtail millet is done without hulling the grains, because of which, the dietary fiber and the anti-nutritional content of popped millets were found to be higher (Muralikrishna, Malleshi, Desikachar, & Tharanathan, 1986). Also the mineral and protein content was richer in the popped foxtail millets than the unprocessed hulled foxtail millets (Gurupavithra, Jayachitra, & Dilna, 2013).

Extruded foxtail millet whole grain flours also exposed promising responses compared to finger millet and pearl millet. The expanded snack from foxtail millet produced with extruders had a desirable expansion ratio and water/oil absorption capacity (Kharat, et al., 2019). In addition, the foxtail millet along with the composite flour mix consisting of chickpeas, rice, and amaranth was showing the ability of the foxtail millet to blend with other starch-protein complexes to yield products of better palatability (Deshpande & Poshadri, 2011). Similarly, foxtail millet-based extruded pasta and vermicelli products were also available in the market, which shows the feasibility of using foxtail millet in extrusion to produce low-glycemic, protein, and mineral-rich products (Dharmaraj, Rao, Sakhare, & Inamdar, 2016).

To nutritionally enrich conventional baked goods like bread, cake, cookies, and pizza bases, a specific proportion of foxtail millet flour was added improve the nutritional profile without affecting the sensory attributes of such products (Shadang & Jaganathan, 2014). Besides, traditional diet recipes of India, such as instant Upma, Idli Dosa mix, Pongal, Laddu, and Besibila bath mix, were developed from the Indian Institute of Millet Research for commercialization of the foxtail millet products (Dayakar, Sangappa, Vishala, Arlene, & A, 2016).

Further, the undesirable anti-nutritional factors in the millets were reduced with the process of germination and fermentation. The enzymes and microorganisms metabolize the components of the millets in such a way as to improve the nutritional characteristics. To illustrate further, the starch content of foxtail millet is utilized by the endogeneous microflora and converted into reducing sugars via fermentation, however the long-chain fatty acid profile of the millet was unaffected. Also, the quality of protein was increased after fermentation (Antony, Sripriya, & Chandra, 1996). Consumption of fermented millets in the form of porridges and beverages is common among the tribal and rural populations where the production of millets is relatively higher. There is a traditional wine in Korea called "Jeju," fermented from foxtail millet with alcoholic yeast and saccharifying molds (Kim & Koh, 2004). The germinated millet grain in which the starch hydrolyzing enzymes like α- amylase, and proteolytic enzymes break down the complex polysaccharide and protein molecules and enhance their digestibility (Ko, et al., 2011). The polyphenol contents were also increased with germination, which thereby improves the anti-oxidant capacity and radical scavenging activity as well. Thus, germinated foxtail millets are incorporated in composite health mix/malt formulations along with multiple kinds of cereal and nutri-cereals (Malleshi, Desikachar, & Tharanathan, 1986). Therefore, primary and secondary processing technologies of foxtail millets focus on enhancing the sensory, physicochemical, and nutritional attributes of the millets to reach out to more consuming populations in the process of eradicating malnutrition.

14.7 Little millet

The smallest, but one of the most important, crops among small millets is the little millet, which is considered to be an enduring crop as it can be cultivated in fair culturing conditions. However, it is the millet least explored by the researchers as well as consumers due to its relatively lesser production area of about 0.3–0.4 hectares (Ganapathy, 2017). Nutritionally, it is one of the millets with high protein (13.6%) and fiber content (38%), and less fat content (1.9%) (Nazni & Bhuvaneswari, 2015). Apart from the proximate composition, specific bioactive components such as gamma-aminobutyric acid (GABA), phenolics, and sterols are rich in little millets, which makes it a healthy alternative to high carb diets (Guha, Sreerama, & Malleshi, 2015). The nutritional composition of different morphological parts of the little millet varies widely, unlike the other millets. The outer hull or husk of the little millet contributes significantly to the fat content and mineral content, while the dietary fiber content is broadly distributed over the endosperm (Hadimani & Malleshi, 1993). The primary processing of cleaned little millets starts with dehusking, followed by debranning to remove the outermost husk and bran, which imparts a brownish tint to the grains. As discussed in the previous section, the dehullers used for foxtail millet are used for little millet as well as for other millets. However, to preclude the breakage occurring to the small grains, centrifugal dehullers are preferred over the abrasive roller cum shellers with which compression of the grains would bring on a higher fraction of broken grains. In addition, hydro-thermal pre-treatment like parboiling has been observed as a solution to reduce the breakage of grains during milling (Mannuramath & Yenagi, 2015). Removal of the husk and bran would improve the shelf stability of the millet by reducing the likelihood of fat rancidity, yet the loss of mineral and protein content is obvious (Guha, Sreerama, & Malleshi, 2015).

The hulled and the debranned grains would appear polished and called "little millet rice," but with a diminished nutritional profile compared to the native grains in terms of minerals and proteins. Further processing operations of the little millets, such as soaking, roasting, and germination would give a desirable effect over the bioactive profile (Shahidi & Chandrasekara, 2015). These unit operations break down the high molecular weight-bound nutrients and make them more available in comparison with the unprocessed grains, thereby increasing their content prominently (Chandrasekara & Shahidi, 2011). Investigations reveal that the roasted little millets were rich in the total phenolic and flavonoid contents followed by soaked and germinated millets (Pradeep & Guha, 2011).

Secondary processing technologies such as popping and flaking of little millets are also explored for the value addition and to make them part of convenience foods. The optimum moisture content of the grain and the popping temperature significantly affects the popping yield of little millet grains. It was found that grains of 16 percent moisture content popped at 260 °C of the salt medium was giving the highest popping yield (Kapoor, 2013).The popped little millets were observed to have lesser resistant starch content as the high-temperature treatment enhances the amount of digestible starch content. However, the other proximate elements such as protein, fat, and minerals were unaltered, while the total phenolic content was increased (Ravindra, Vijayakumari, Sharan, Raghuprasad, & Kavaloor, 2008). The flakes were prepared in two different ways, with and without prior gelatinization, to yield ready to eat and ready to cook flakes (Kotagi, 2011). The tempered grains gelatinized under pressure cooking followed by flaking with heavy rollers to produce whitish crispy flakes similar to the sensory profile of corn flakes, but with an enhanced nutritional profile comprised of higher protein, lower fat, and higher iron (61.42 mg/100g) and calcium content (29.3 mg/100g) (Patil, Chimmad, & Itagi, 2015). The anti-nutritional phytic acid content in little millet was reduced with the process of fermentation by *Saccharomyces boulardii* and *Lactobacillus thermophilus* (Pampangouda, Munishamanna, & Gurumurthy, 2015), and the fermented little millet slurry is dried and utilized in further value-added products like extrudates because of their improved mineral and protein content (Khwairakpam, Kumari, Suneetha, & Tejashree, 2020). Combined with other millets, the fermented slurry would be developed as malted and alcoholic beverages (Sethi, 2016). The processing technologies of little millet largely concentrate on improvising the nutritional attributes as studies about the effect of different processes shown a promising impact on the bioactive characteristics of the millet. Nevertheless, investigations about the hypocholesterolemic and anti-diabetic effects would reveal further information on the post-digestion behavior of the millet.

14.8 Kodo millet

The nutri-cereal cultivated widely as a cereal only in India is the Kodo millet, while in African and other Asian countries it has been harvested as a wild crop along with the other small cereals (de Wet, Brink, Rao, & Mengesha, 1983). Kodo millet is one of the small millets with a remarkable production of 3 million tonnes and with the cultivation of 9 million hectares of land annually (Neelam, Kanchan, Alka, & Alka, 2013). The protein and fat content of Kodo millet is relatively lesser compared to other nutri-cereals, respectively about 8 percent and 5 percent (Saldivar, 2003) with the iron content of 3.9 mg/100 g of the millets (Amadou, Gounga, & Le, 2013).

As Kodo millet grows well in moist and shady lands, infection of the crop with poisonous spores in the millet spikelet is prominent, but can be overcome with specific processing technologies to make it suitable for safe consumption (Bhide & Aimen, 1959). Thus, decortication of Kodo millet facilitates the removal of the spikelet along with the hull to yield a safe for consumption Kodo millet rice. Kodo millet has organoleptic properties similar to that of rice, due to which, the dehulled millet is cooked and consumed like rice, though the glycemic index of Kodo millet rice is less than that of rice (Annor, Marcone, Bertoft, & Seetharaman, 2013). In this way, traditional Indian delicacies made with rice can be substituted with the low glycemic Kodo millet (Kalpana & Koushikha, 2013) (Padma & Rajendren, 2013).

Apart from increasing the safety and sensory attributes, the removal of husk reduces the anti-nutritional phytate content of the Kodo millet, which in turn enhances the bioavailability of present minerals and proteins (Balasubramanian, Vishwanathan, & Sharma, 2007). However, the anti-oxidant capacity was also reduced in the dehulled millets as the major proportion of ferulic acid is present in the brans of the Kodo millet (Chandrasekara & Shahidi, 2012). Thus it is advisable to employ the cleaned and sterilized whole grains for the application in a value-added product to obtain complete nutritional value. Otherwise, the separated bran and husk should be utilized further for phenolic extraction or cattle feed. Parboiling of Kodo millet has been studied to facilitate the diffusion of nutrients into the endosperm and also for the efficient removal of offal from the grain (Shrestha, 1972).

Puffing technology is also used to develop a convenient food from Kodo millet. The puffed Kodo millet grains are found to be rich in starch and protein digestibility and mineral bioavailability (Patel, Parihar, & Dhumketi, 2018). Also, the structural changes caused by puffing significantly affect the functional characteristics, particularly the solubility of the dietary fiber content (Caprita & Caprita, 2011). Though the processing of Kodo millet into convenient food such as puffs and flakes is feasible, the incorporation of Kodo millet as a pulverized flour with other conventional ingredients for baking and extrusion is quite common, as it enriches the existing product nutritional quality. Correspondingly, the efforts to obtain a conventional product such as cake, cookies, or pasta from a new sole ingredient like whole Kodo millet would be substantial. The addition of 70 percent Kodo millet along with a chickpea blend is studied to give desirable crispy extrudates equivalent to a market product with high protein content (Geetha, Mishra, & Srivastav, 2014). Similarly, ready to cook pasta was prepared with 50 percent incorporation of Kodo millet with wheat and the obtained Kodo millet pasta with its excellent sensory properties is comparable to conventional wheat pasta but with increased mineral content (Sarojani, Suvarna, JS, & Sneha, 2021). Preparation of composite health mix formulations for a low glycemic beverage along with other small millets such as barnyard and foxtail millet is explored for their high fiber content (Arya & Shakya, 2021). Considering the high dietary fiber content and low glycemic attribute of Kodo millet, it should be integrated into the everyday diet of the global population instead of the conventional gluten-rich, high sugar cereals.

14.9 Proso millet

Proso millet (*Panicummiliaceum L.*) is one of the nutri-cereal crop cultivated all through the year for the purpose of feed, fodder and food in India, China. It is stated that the proso millet is widely cultivated in India, Middle Eastern countries, Russia, and North America. In recent times, A few varieties of millets including proso millet is commonly cultivated minor millet in Assam, especially in lower Assamand adjoining northeastern states. Other common names include common millet, hog millet, broomcorn millet, and red millet. In general, the millets are considered to be grown as feed and forage and underutilized for food application and incorporation with other products. However, recent studies are exploring the millets for potential as nutri-cereals and as an alternative to gluten containing flour in food products (Habiyaremye et al., 2017)

The proso millet grains are oval in shape in the size, bigger than finger millet and smaller than pearl millet. it can be in various colors such as cream, black, orange, red, and yellow, depending on the variety. The proso grains are 3 mm in length and 2 mm wide. The kernel weight in proso millet was observed to be 7.1 g. The pericarp in proso is a utricle type, similar to finger millet. The seed coat is 0.2–0.4 pm in thickness. The endosperm to germ ratio of proso is 12:1.

Proso millet is one of the nutritionally rich millet as it contains 65–70 percent starch, 11 percent protein, 2–7 percent dietary fibers, and are rich in minerals (calcium, iron, and magnesium). Proso millet is rich in methionine and cysteine (sulphur-containing essential amino acids) and less in lysine. It is also rich in vitamin B complex like niacin(Sarker, 2015). The primary processing involved in millets and cereals are dehulling, milling, flaking based on the purpose and product application. The milled proso millet flour is been used to replace the wheat flour in products like breads, cookies and pasta. In general, the roller mills are preferred for flour making as in wheat milling, so the proso millet also can be milled in the same way with adjusted moisture content of grain around 10 percent (Kalinová, 2007).

A study reported parboiling of proso millet grains with different conditions. Different parboiling pre-treatments like cold and hot water soaking followed by steaming at different pressures for varying duration were imposed on proso millet grain (Ramya, Palanimuthu, & Ranganna, 2010). Proso millet subjected to 65°C temperature soaking for 60 min and steaming for 10 min took a longer time to cook and had a low swelling ratio. The texture properties of the proso millet soaked for 24 hours in room temperature and soaked for 15 minutes showed better quality than the raw millet grains. The proso millet extrudates are produced and could be utilized as raw materials for instant foods mix as with other cereal and millets (Gulati, Weier, Santra, Subbiah, & Rose, 2016). A study demonstrated that puffed proso millet can be considered as nutritive products like other puffed product made from common cereals as they have relatively high amino acid and dietary fiber content (Piłat, Ogrodowska, & Zadernowski, 2016). McSweeney, Seetharaman, Ramdath, & Duizer (2017) utilised refined proso millet in order to produce four different products, namely muffins, couscous, extruded snacks, and porridge at various concentrations (100%, 75%, 25% proso millet). In almost every product, the glycaemic index decreased with increase in incorporation of flour percentage in the product.

14.10 Conclusion

Millets are storehouse of all macronutrients and micronutrients. They are rich in fiber and loaded with essential vitamins and minerals, making them superb ingredients to overcome diet related and lifestyle disorders. Therefore, adequate knowledge and efficient application of different processing techniques discussed in this chapter will help in healthy product formulation. Utilization and proper combination of processing techniques like soaking, dehulling, milling, germination, malting, fermentation and extrusion technology can be employed to exploit the lesser utilized grains.

References

Abdelrahman, A., Hoseney, R. C., & Varriano-Marston, E. (1983). Milling process to produce low-fat grits from pearl millet. Cereal Chem, 189–191.
Abdualrahman, M. A., Ma, H., Yagoub, A. E., Zhou, C., Ali, A. O., & Yang, W. (2019). Nutritional value, protein quality and anti-oxidant activity of Sudanese sorghum-based kissra bread fortified with bambara groundnut (Voandzeia subterranea) seed flour.
Achaya, K. T. (2009). *The Illustrated Food of India A–Z*. New Delhi: Oxford University.
Adebiyi, J. A., Obadina, A. O., Adebo, O. A., & Kayitesi, E. (2018). Fermented and malted millet products in Africa: Expedition from traditional/ethnic foods to industrial value-added products. *Critical Reviews in Food Science and Nutrition*, 58(3), 463–447.

Amadou, I., Gounga, M. E., & Le, G. W. (2013). Millets: Nutritional composition, some health benefits and processing-A review. *Emirates Journal of Food and Agriculture*, 501–508.

Amadou, I., Gounga, M. E., Shi, Y. H., & Le, G. W. (2014). Fermentation and heat-moisture treatment induced changes on the physicochemical properties of foxtail millet (Setaria italica) flour. *Food and Bioproducts Processing*, 92(1), 38–45.

Annor, G. A., Marcone, M., Bertoft, E., & Seetharaman, K. (2013). In vitro starch digestibility and expected glycemic index of kodo millet (Paspalum scrobiculatum) as affected by starch–protein–lipid interactions. *Cereal Chemistry*, 90(3), 211–217.

Antony, U., Sripriya, G., & Chandra, T. S. (1996). The effect of fermentation on the primary nutrients in foxtail millet (Setaria italica). *Food Chemistry*, 56(4), 381–384.

Arora, P., Sehgal, S., & Kawatra, A. (2003). Content and HCl-extractability of minerals as affected by acid treatment of pearl millet. *Food Chemistry*, 141–144.

Arya, S. S., & Shakya, N. K. (2021). High fiber, low glycaemic index (GI) prebiotic multigrain functional beverage from barnyard, foxtail and kodo millet. *LWT*, 130, 109991.

Aykroyd, W. R., Gopalan, C., & Balasubramanian, S. C. (1963). *The Nutritive Value of Indian Foods and the Planning of Satisfactory Diets*. Indian Council of Medical Research.

Balasubramanian, S., Kotwaliwale, N., Kate, A., & Ambrose, D. C. (2021, Jan). Entrepreneurship Opportunities in Nutri-Cereal Processing Sector. *Indian Food Industry Magazine*, 3(1).

Balasubramanian, S., Vishwanathan, R., & Sharma, R. (2007). Post harvest processing of millets: An appraisal. *Agricultural Engineering Today*, 31(2), 18–23.

Basappa, S. C., Somashekar, D., Agrawal, R., Suma, K., & Bharathi, K. (1997). Nutritional composition of fermented ragi (chhang) by phab and defined starter cultures as compared to unfermented ragi (Eleucine coracana G.). *International Journal of Food Science and Nutrition*, 313–319.

Basavaraj, G. R. (2010). Availability and utilization of pearl millet in India. *SAT e Journal*.

Benhur, D. R., Bhargavi, G., Kalpana, K., Vishala, A. D., Ganapathy, K. N., & Patil, J. V. (2015). Development and standardization of sorghum pasta using extrusion technology. *Journal of Food Science and Technology*, 6828–6833.

Bhati, D., Bhatnagar, V., & Acharya, V. (2016). Effect of pre-milling processing techniques on pearl millet grains with special reference to in-vitro iron availability. *Asian Journal of Dairy and Food Research*, 76–80.

Bhide, N. K., & Aimen, R. A. (1959). Pharmacology of a tranquillizing principle in Paspalum scrobiculatum grain. *Nature*, 183(4677), 1735–1736.

Boniface, O. O., & Gladys, M. E. (2011). Effect of alkaline soaking and cooking on the proximate, functional and some anti-nutritional properties of sorghum flour. *Au JT*, 210–216.

Bookwalter, G. N., Lyle, S. A., & Warner, K. (1987). Millet processing for improved stability and nutritional quality without functionality changes. *Journal of Food Science*, 52(2), 399–402.

Cao, W., Sun, C., Liu, R., Yin, R., & Wu, X. (2012). Comparison of the effects of five pre-treatment methods on enhancing the enzymatic digestibility and ethanol production from sweet sorghum bagasse. *Bioresource Technology*, 215–221.

Caprita, A., & Caprita, R. (2011). The effect of thermal processing on soluble dietary fibre fraction in wheat. *Journal of Food, Agriculture & Environment*, 9, 14–15.

Chandra, A. K., Chandora, R., Sood, S., & Malhotra, N. (2021). Global production, demand, and supply. In *Millets and Pseudo Cereals* (pp. 7–18). Woodhead Publishing.

Chandrasekara, A., & Shahidi, F. (2011). Bioactivities and antiradical properties of millet grains and hulls. *Journal of Agricultural and Food Chemistry*, 59(17), 9563–9571.

Chandrasekara, A., & Shahidi, F. (2012). Bioaccessibility and anti-oxidant potential of millet grain phenolics as affected by simulated in vitro digestion and microbial fermentation. *Journal of Functional Foods*, 4(1), 226–237.

Chandrasekara, A., Naczk, M., & Shahidi, F. (2012). Effect of processing on the anti-oxidant activity of millet grains. *Food Chemistry*, 133(1), 1–9.

Chavan, U. D., Chavan, J. K., & Kadam, S. S. (1988). Effect of fermentation on soluble proteins and in vitro protein digestibility of sorghum, green gram and sorghum-green gram blends. *Journal of Food Science*, 1574–1575.

Dar, R. A., Dar, E. A., Kaur, A., & Phutela, U. G. (2018). Sweet sorghum-a promising alternative feedstock for biofuel production. *Renewable and Sustainable Energy Reviews*, 82, 4070–4090.

Das, I. K., & Rakshit, S. (2016). Millets, their importance, and production constraints. *Biotic Stress Resistance in Millets*, 3–19.

Day, C. N., & Morawicki, R. O. (2016). Effects of fermentation by yeast and amylolytic lactic acid bacteria on grain sorghum protein content and digestibility. *Journal of Food Quality*.

Dayakar, R. B., Sangappa, Vishala, A. D., Arlene, C. G., & A, T. V. (2016). *Technologies of Millet Value Added Products*. Rajendra Nagar, Hyderabad: Centre of Excellence on Sorghum, ICAR-Indian Institute of Millet Research.

de Wet, J. M., Brink, D. E., Rao, K. P., & Mengesha, M. H. (1983). Diversity in kodo millet, Paspalum scrobiculatum. *Economic Botany*, 37(2), 159–163.

Dendy, D. A. V. (1994). Structure and chemistry of sorghum and the millets. In: Dendy, D. A. V. (eds.) *Sorghum and Millets Chemistry and Technology*. St. Paul, MN: AACC International.

Deosthale, Y. G. (2002). *The Nutritive Value of Foods and the Significance of Some Household Processes*. National institute of Nutrition, Hyderabad.

Deshpande, H. W., & Poshadri, A. (2011). Physical and sensory characteristics of extruded snacks prepared from Foxtail millet based composite flours. *International Food Research Journal*, 18(2).

Devi, P. B., Vijayabharathi, R., & Sathyabama, S. M. (2014). Health benefits of finger millet (Eleusine coracana L.) polyphenols and dietary fiber: a review. *Journal of Food Science and Technology*, 1021–1040.

Dharmaraj, U., Rao, B. S., Sakhare, S. D., & Inamdar, A. A. (2016). Preparation of semolina from foxtail millet (Setaria italica) and evaluation of its quality characteristics. *Journal of Cereal Science*, 68, 1–7.

Durairaj, M., Gurumurthy, G., Nachimuthu, V., Muniappan, K., & Balasubramanian, S. (2019). Dehulled small millets: The promising nutricereals for improving the nutrition of children. *Maternal and Child Nutrition*, 15, e12791.

Eltayeb, L. F., Mohamed, M. A., & Fageer, A. S. (2017). Effect of Soaking on Nutritional Value of Sorghum. *International Journal of Science and Research*.

Fuller, D. Q. (2003). African crops in prehistoric South Asia: a critical review. *Food, Fuel and Fields: Progress in African Archaeobotany*, 239–271.

Ganapathy, K. N. (2017). Genetic improvement in little millet. In *Millets and Sorghum: Biology and Genetic Improvement* (pp. 170–183).

Geetha, R., Mishra, H. N., & Srivastav, P. P. (2014). Twin screw extrusion of kodo millet-chickpea blend: process parameter optimization, physico-chemical and functional properties. *Journal of Food Science and Technology*, 51(11), 3144–3153.

Goswami, D., Gupta, R. K., Mridula, D., Sharma, M., & Tyagi, S. K. (2015). Barnyard millet based muffins: Physical, textural and sensory properties., 64(1), *LWT–Food Science and Technology*, 374–380.

Guha, M., Sreerama, Y. N., & Malleshi, N. G. (2015). Influence of Processing on Nutraceuticals of Little Millet (Panicum sumatrense). *Processing and Impact on Active Components in Food*, 353–360.

Gulati, P., Weier, S. A., Santra, D., Subbiah, J., & Rose, D. J. (2016). Effects of feed moisture and extruder screw speed and temperature on physical characteristics and antioxidant activity of extruded proso millet (Panicum miliaceum) flour. *International Journal of Food Science and Technology*, 51(1), 114–122. https://doi.org/10.1111/ijfs.12974.

Gurupavithra, S., Jayachitra, A., & Dilna, K. (2013). Study on biochemical and nutritive value of popped foxtail millet. *International Journal Pharmaceutical and Biological Sciences*, 4(2), 549–558.

Habiyaremye, C., Matanguihan, J. B., D'Alpoim Guedes, J., Ganjyal, G. M., Whiteman, M. R., Kidwell, K. K., & Murphy, K. M. (2017). Proso millet (Panicum miliaceum L.) and its potential for cultivation in the Pacific Northwest, U.S.: A review. *Frontiers in Plant Science*, 7(January), 1–17. https://doi.org/10.3389/fpls.2016.01961.

Hadimani, N. A., & Malleshi, N. G. (1993). Studies on milling, physico-chemical properties, nutrient composition and dietary fibre content of millets. *Journal of Food Science and Technology* (India), 17–20.

Hama, F., Icard-Vernière, C., Guyot, J. P., Picq, C., Diawara, B., & Mouquet-Rivier, C. (2011). Changes in micro- and macronutrient composition of pearl millet and white sorghum during in field versus laboratory decortication. *Journal of Cereal Science*, 54(3), 425–433.

Hassan, A. B., Ahmed, I. A., Osman, N. M., Eltayeb, M. M., Osman, G. A., & Babiker, E. E. (2006). Effect of processing treatments followed by fermentation on protein content and digestibility of pearl millet (Pennisetum typhoideum) cultivars. *Pakistan Journal of Nutrition*, 86–89.

Hotz, C., & Gibson, R. S. (2007). Traditional food-processing and preparation practices to enhance the bioavailability of micronutrients in plant-based diets. *The Journal of Nutrition*, 1097–1100.

IIMR (2018). *Annual Report 2017–18*. Hyderabad: Indian Institute of Millets Research.

Jaybhaye, R. V., Pardeshi, I. L., Vengaiah, P. C., & Srivastav, P. P. (2014). Processing and technology for millet based food products: a review. *Journal of Ready to Eat Food*, 1(2), 32–48.

Jha, N., Krishnan, R., & Meera, M. S. (2015). Effect of different soaking conditions on inhibitory factors and bioaccessibility of iron and zinc in pearl millet. *Journal of Cereal Science*, 46–52.

Kalinová, J. (2007). Nutritionally Important Components of Proso Millet (Panicum miliaceum L). *Foods,* Global Science Books.

Kalpana, C. A., & Koushikha, N. M. (2013). Development and Evaluation of Varagu Incorporated Recipes. National Seminar on Recent Advances in processing, utilization and nutritional impact of small millets. Thamukkam Grounds: Madurai Symposium.

Kapoor, P. (2013). Nutritional and functional properties of popped little millet. Dissertation. Canada: McGill University.

Kharat, S., Medina-Meza, I. G., Kowalski, R. J., Hosamani, A., Ramachandra, C. T., Hiregoudar, S., & Ganjyal, G. M. (2019). Extrusion processing characteristics of whole grain flours of select major millets (foxtail, finger, and pearl). *Food and Bioproducts Processing*, 114, 60–71.

Khwairakpam, M., Kumari, B. A., Suneetha, W. J., & Tejashree, M. (2020). Development and Evaluation of Fermented Little Millet Based Cold Extrudates. *International Research Journal of Pure and Applied Chemistry*, 7–14.

Kim, J. Y., & Koh, J. S. (2004). Fermentation characteristics of Jeju foxtail millet-wine by isolated alcoholic yeast and saccharifying mold. *Applied Biological Chemistry*, 47(1), 85–91.

Kindiki, M. M., Onyango, A., & Kyalo, F. (2015). Effects of processing on nutritional and sensory quality of pearl millet flour. *Food Science and Quality Management*, 13–19.

Ko, J. Y., Song, S. B., Lee, J. S., Kang, J. R., Seo, M. C., Oh, B. G.,…. Woo, K. S. (2011). Changes in chemical components of foxtail millet, proso millet, and sorghum with germination. *Journal of the Korean Society of Food Science and Nutrition*, 40(8), 1128–1135.

Kotagi, K. S. (2011). Little millet (Panicum Miliare) flakes: Development, value addition, quality evaluation, consumer acceptability and commercialization. Doctoral dissertation. UAS Dharwad: Krishikosh.

Kumar, C. M., Sabikhi, L., Singh, A. K., Raju, P. N., Kumar, R., & Sharma, R. (2019). Effect of incorporation of sodium caseinate, whey protein concentrate and transglutaminase on the properties of depigmented pearl millet based gluten free pasta. *LWT*, 103, 19–26.

Legesse, E. (2013). Effect of processing on qualtiy characteristics of pearl millet (pennisetum glaucum) value based products. *Food Chemistry*.

Lodge, S. L., Stock, R. A., Klopfenstein, T. J., Shain, D. H., & Herold, D. W. (1997). Evaluation of corn and sorghum distillers byproducts. *Journal of Animal Science*, 75(1), 37–43.

Madhavi, M., & Vaidehi, M. P. (1990). Fermented beverages from minor millets, their acceptability and chemical composition. *Beverage Food World*, 30–32.

Maitra, S., & Shankar, T. (2019). Agronomic management in little millet (Panicum sumatrense L.) for enhancement of productivity and sustainability. *International Journal of Bioresource Science*, 6(2), 91–96.

Malleshi, N. G. (1989). Processing of small millets for food and industrial uses. In *Small Millets in Global Agriculture* (pp. 325–339).

Malleshi, N. G., Desikachar, H. S., & Tharanathan, R. N. (1986). Physico-chemical properties of native and malted finger millet, pearl millet and foxtail millet starches. *Starch-Stärke*, 38(6), 202–205.

Mannuramath, M., & Yenagi, N. (2015). Optimization of hydrothermal treatment for little millet grains (Panicum miliare). *Journal of Food Science and Technology*, 52(11), 7281–7288.

Marston, K. K. (n.d.). Effect of heat treatment of sorghum flour on the functional properties of gluten-free bread and cake. *LWT–Food Science and Technology*, 637–644.

Marston, K., Khouryieh, H., & Aramouni, F. (2016). Effect of heat treatment of sorghum flour on the functional properties of gluten-free bread and cake. *LWT–Food Science and Technology*, 637–644.

McSweeney, M. B., Seetharaman, K., Ramdath, D. D., & Duizer, L. M. (2017). Chemical and physical characteristics of proso millet (Panicum miliaceum)-based products. *Cereal Chemistry*, 94(2), 357–362. https://doi.org/10.1094/CCHEM-07-16-0185-R.

Meena, R. P., Joshi, D., Bisht, J. K., & Kant, L. (2021). Global Scenario of Millets Cultivation. In *Millets and Millet Technology* (pp. 33–50). Singapore: Springer.

Muralikrishna, G., Malleshi, N. G., Desikachar, H. S., & Tharanathan, R. N. (1986). Effect of popping on the properties of some millet starches. *Starch/Staerke*, 38(2), 48–51.

Nambiar, V. S., Dhaduk, J. J., Sareen, N., Shahu, T., & Desai, R. (2011). Potential functional implications of pearl millet (Pennisetum glaucum) in health and disease. *Journal of Applied Pharmaceutical Science*, 62.

Nazni, P., & Bhuvaneswari, J. (2015). Analysis of physico chemical and functional characteristics of finger millet (Eleusine coracana L) and little millet (P. sumantranse). *International Journal of Food and Nutritional Sciences*, 4(3), 109.

Nazni, P., & Shobana, D. R. (2016). Effect of processing on the characteristics changes in barnyard and foxtail millet. *Journal of Food Process Technology* .

Neelam, Y., Kanchan, C., Alka, S., & Alka, G. (2013). Evaluation of hypoglycemic properties of kodo millet based food products in healthy subjects. *IOSR Journal of Pharmacy*, 3, 14–20.

Niro, S., D'Agostino, A., Fratianni, A., Cinquanta, L., & Panfili, G. (2019). Gluten-free alternative grains: Nutritional evaluation and bioactive compounds. *Foods*, 8(6), 208.

Nithya, K. S., Ramachandramurty, B., & Krishnamoorthy, V. V. (2007). Effect of processing methods on nutritional and anti-nutritional qualities of hybrid (COHCU-8) and traditional (CO7) pearl millet varieties of India. *Journal of Biological Sciences*, 643–647.

Nkhata, S. G., Emmanuel Ayua, E. H., & Shingiro, J.-B. (2018). Fermentation and germination improve nutritional value of cereals and legumes through activation of endogenous enzymes. *Food Science and Nutrition*, 2446–2458.

Padma, A., & Rajendren, R. (2013). Standardization of Spirulina and Kodo Millet Incorporated Cookies. National Seminar on Recent Advances in processing, utilization and nutritional impact of small millets. Thamukkam Grounds: Madurai Symposium.

Palanimuthu, V. (2017, September 22). Small millets. Retrieved from Dhan Foundation: www.dhan.org

Palavecino, P. M., Ribotta, P. D., León, A. E., & Bustos, M. C. (2019). Gluten-free sorghum pasta: starch digestibility and antioxidant capacity compared with commercial products. *Journal of the Science of Food and Agriculture*, 1351–1357.

Pampangouda, P., Munishamanna, K. B., & Gurumurthy, H. (2015). Journal of Applied and Natural Science. Effect of Saccharomyces boulardii and Lactobacillus acidophilus fermentation on little millet (*Panicum sumatrense*), 7(1), 260–264.

Patel, A., Parihar, P., & Dhumketi, K. (2018). Nutritional evaluation of Kodo millet and puffed Kodo. *International Journal of Chemical Studies*, 6(2), 1639–1642.

Patil, K. B., Chimmad, B. V., & Itagi, S. (2015). Glycemic index and quality evaluation of little millet (Panicum miliare) flakes with enhanced shelf life. *Journal of Food Science and Technology*, 52(9), 6078–6082.

Pawar, V. D., & Machewad, G. M. (2006). Processing of foxtail millet for improved nutrient availability. *Journal of Food Processing and Preservation*, 30(3), 269–279.

Piłat, B., Ogrodowska, D., & Zadernowski, R. (2016). Nutrient content of pufed proso millet (Panicum miliaceum L.) and Amaranth (Amaranthus cruentus L.) Grains. *Czech Journal of Food Sciences*, 34(4), 362–369. https://doi.org/10.17221/405/2015-CJFS

Pradeep, S. R., & Guha, M. (2011). Effect of processing methods on the nutraceutical and anti-oxidant properties of little millet (Panicum sumatrense) extracts. *Food Chemistry*, 126(4), 1643–1647.

Preetika, A., Padmini, G., & Shobha, U. (2004). Nutrient dense mixes for enteral feeding in India. *Nutrition & Food Science*, 277–281.

Rai, K. N., Gowda, C. L., Reddy, B. V., & Sehgal, S. (2008). Adaptation and potential uses of sorghum and pearl millet in alternative and health foods. *Comprehensive Reviews in Food Science and Food Safety*, 320–396.

Ramesh, B., Kavitha, G., Gokiladevi, S., Balachandar, R. K., Kavitha, K., Gengadharan, A. C., & Puvanakrishnan, R. (2020). Effect of Extremely Low Power Time-Varying Electromagnetic Field on Germination and Other Characteristics in Foxtail Millet (Setaria italica) Seeds. *Bioelectromagnetics*, 41(7), 526–539.

Ramya, K. G., Palanimuthu, V., & Ranganna, B. (2010). Effect of parboiling pre-treatments on cooking quality of proso millet. *Mysore Journal of Agricultural Sciences*, 44(4), 749–754.

Rathore, S., Singh, K., & Kumar, V. (2016). Millet grain processing, utilization and its role in health promotion: A review. *International Journal of Nutrition and Food Science*, 318–329.

Ravindra, U., Vijayakumari, J., Sharan, S., Raghuprasad, K. P., & Kavaloor, R. (2008). A comparative study of postharvest processing methods for little Millet (Panicum miliare L.). *Tropical Agricultural Research*, 20, 115–122.

Romero, H. M., Santra, D., Rose, D., & Zhang, Y. (2017). Dough rheological properties and texture of gluten-free pasta based on proso millet flour. *Journal of Cereal Science*, 74, 238–243.

Ronda, V., Aruna, C., Visarada, K. B., & Bhat, B. V. (2019). Sorghum for animal feed. In *Breeding Sorghum for Diverse End Uses* (pp. 229–238). Woodhead Publishing.

Rooney, I.W. and Sullins, R.D. (1969). Laboratory method for milling small samples of sorghum grain. *Cereal Sci. Today*. 46: 486–490.

Rose, D. J., Williams, E., Mkandawire, N. L., Weller, C. L., & Jackson, D. S. (2014). Use of whole grain and refined flour from tannin and non-tannin sorghum (Sorghum bicolor (L.) Moench) varieties in frybread. 2. *Food Science and Technology International*.

Roy, T. C. (2020). Millet processing machine. Retrieved from tcrconnectingagriculture.com: https://tcrconnecting agriculture.com/2020/10/millet processing-machine/

Saldivar, S. (2003). Cereals: dietary importance. In B. Caballero, L. Trugo, & P. Finglas, *Encyclopedia of Food Sciences and Nutrition* (pp. 1027–1033). Academic Press, Agosto, London.

Saleh, A. S., Zhang, Q., Chen, J., & Shen, Q. (2013). Millet grains: nutritional quality, processing, and potential health benefits. *Comprehensive Reviews in Food Science and Food Safety*, 281–295.

Sarker, A. (2015). *Effect of Pre-processing on the Nutritive, Physical, and Sensory Properties of Proso Millet*. The University of Guelph.

Sarojani, J. K., Suvarna, C. H., Hilli, J. S., & Sneha, S. (2021). Development of Ready to cook Kodo millet pasta. *The Pharma Innovation Journal*, 10(6), 910–915.

Serna-Saldivar, S. (1995). Structure and chemistry of sorghum and millets. *Sorghum and Millets: Chemistry and Technology*, 69–124.

Serna-Saldivar, S. O., & Espinosa-Ramírez, J. (2019). Grain structure and grain chemical composition. *Sorghum and Millets*, 85–129.

Serna-Saldivar, S. O., Clegg, C., & Rooney, L. W. (1994). Effects of parboiling and decortication on the nutritional value of sorghum (Sorghum bicolor L. Moench) and pearl millet (Pennisetum glaucum L.). *Journal of Cereal Science*, 83–89.

Sethi, N. (2016). Optimization of fermentation parameters for production of malted and alcoholic beverage from Kodo and little millet. *International Journal of Farm Sciences*, 6(1), 191–198.

Shadang, C., & Jaganathan, D. (2014). Development and standardisation of formulated baked products using millets. *International Journal of Research in Applied, Natural and Social Sciences*, 2(9), 75–78.

Shahidi, F., & Chandrasekara, A. (2015). Processing of millet grains and effects on non-nutrient anti-oxidant compounds. *Processing and Impact on Active Components in Food,* 345–352.

Shankaramurthy, K. N., & Somannavar, M. S. (2019). Moisture, carbohydrate, protein, fat, calcium, and zinc content in finger, foxtail, pearl, and proso millets. *Indian Journal of Health Sciences and Biomedical Research* (KLEU), 12(3), 228.

Sharma, N., & Niranjan, K. (2018). Foxtail millet: Properties, processing, health benefits, and uses. *Food Reviews International*, 34(4), 329–363.

Shinoj, S., Viswanathan, R., Sajeev, M. S., & Moorthy, S. N. (2006). Gelatinisation and rheological characteristics of minor millet flours. *Biosystems Engineering*, 95(1), 51–59.

Shobana, S., & Malleshi, N. G. (2007). Preparation and functional properties of decorticated finger millet (Eleusine coracana). *Journal of Food Engineering*, 529–538.

Shobana, S., Krishnaswamy, K., Sudha, V., Malleshi, N. G., Anjana, R. M., Palaniappan, L., et al. (2013). Finger millet (Ragi, Eleusine coracana L.): a review of its nutritional properties, processing, and plausible health benefits. *Advances in Food and Nutrition Research*, 1–69.

Shrestha, K. B. (1972). Dehusking of varagu and its utilization for edible purposes. Unpublished Thèse de maîtrise. Karnataka, India: University of Mysore.

Singh, A. K., Rehal, J., Kaur, A., & Jyot, G. (2015). Enhancement of attributes of cereals by germination and fermentation: A review. *Critical Reviews in Food Science and Nutrition*, 1575–1589.

Singh, G., Sehgal, S., & Kawatra, A. (2006). Mineral profile, anti-nutrients and in vitro digestibility of biscuit prepared from blanched and malted pearl millet flour. *Nutrition & Food Science*, 231–239.

Singh, G., Sehgal, S., Kawatra, A., & Preeti. (2006). Mineral profile, anti-nutrients and in vitro digestibility of biscuit prepared from blanched and malted pearl millet flour. *Nutrition and Food Science*, 231–239.

Singh, N. B., & Saini, R. S. (2012). Products, diversification, marketing and price discovery of pearl millet in India. Documentation, National Rainfed Area Authority, Todapur Village, Pusa.

Singh, P. (2008). History of millet cultivation in India. In *History of Science, Philosophy and Culture in Indian Civilization* (pp. 107–119).

Srivastava, S., Dhyani, M., & Singh, G. (2003). Popping characteristics of Barnyard and Foxtail millet and their use in preparation of sweets. *Recent Trends Millet Process*; Util Hisar, India Chaudhary Charan Singh Hisar Agril Univ. 38–40.

Sruthi, N. U., & Rao, P. S. (2021). Effect of Processing on Storage Stability of Millet Flour: A Review. *Trends in Food Science & Technology*.

Sun, Q., Han, Z., Wang, L., & Xiong, L. (2014). Physicochemical differences between sorghum starch and sorghum flour modified by heat-moisture treatment. *Food Chemistry*, 756–764.

Suresha, K. B. (2016). *Development of Ready to Use Millet Based Complementary Health Food*. Bidar: Karnataka Veterinary, Animal and Fisheries Sciences University.

Tamene, A., Baye, K., Kariluoto, S., Edelmann, M., Bationo, F., Leconte, N., et al. (2019). Lactobacillus plantarum P2R3FA isolated from traditional cereal-based fermented food increase folate status in deficient rats. *Nutrients*, 2819.

Taylor, J. N., Novellie, L., & Liebenberg, N. W. (1984). Sorghum protein body composition and ultrastructure. *Cereal Chemistry*, 61(1), 69–73.

Taylor, J. R. (2015). Effects of processing sorghum and millets on their phenolic phytochemicals and the implications of this to the health-enhancing properties of sorghum and millet food and beverage products. *Journal of the Science of Food and Agriculture*, 25–237.

Ushakumari, S. R., Latha, S., & Malleshi, N. G. (2004). The functional properties of popped, flaked, extruded and roller-dried foxtail millet (Setaria italica). *International Journal of Food Science & Technology, 39*(9), 907–915.

Verma, V., & Patel, S. (2013). Value added products from nutri-cereals: Finger millet (Eleusine coracana). *Emirates Journal of Food and Agriculture,* 169–176.

Weber, S. A., & Fuller, D. Q. (2008). Millets and their role in early agriculture. *Pragdhara,* 18(69).

Xiong, Y., Zhang, P., Luo, J., Johnson, S., & Fang, Z. (2019). Effect of processing on the phenolic contents, antioxidant activity and volatile compounds of sorghum grain tea., 85. *Journal of Cereal Science,* 6–14.

15

Storage Stability and Quality Management of Nutri-Cereals and Associated Products

Mamta Thakur and Vikas Nanda

Contents

15.1 Introduction

Nowadays, the coarse cereals such as millets are more popular among researchers, nutritionists, and food scientists due to their concentrated essential nutrients and functional compounds. The Government of India recognized the value and potential of millets to fulfill the country's food and nutritional security and, therefore, designated millets as "nutri-cereals." The major millets like sorghum, finger millet and pearl millet and minor millets such as barnyard millet, kodo millet, proso millet and little millet are considered as nutri-cereals in addition to two pseudocereals, that is, amaranthus and buckwheat (Ministry of Agriculture and Farmers Welfare 2018). These are recognized to endure current issues like global warming, declining ground water levels, and food security because of their tolerance to drylands and soils with low fertility. Attributed to their nutraceutical value and environmental responsibility, they have been considered as "future cereals."

Human civilization started the consumption of millets as the first grains that can be evidenced by the development of millet flour derived noodles in northern part of China around 4,000 years ago (Lu et al. 2005). Millets contain numerous micronutrients and macronutrients, higher or equivalent to the major cereals. They are rich in quality protein and dietary fiber ranging from 10–11 percent and up to 38 percent, respectively (Kumar et al. 2018). For the development of functional food products, the millets are used nowadays to provide maximum concentration of minerals ranging from 1.70–4.30 percent. The protein content of pearl millet (14.50%), foxtail millet (11.70%), proso millet (11%), kodo millet (8.30%) and little millet (7.70%) is more than that present in rice (7.50%). Millets also contain phytonutrients, biologically active components like phytosterols, flavonoids, dietary fiber, phenolic acids, and so forth, which are comparable to cereals (Chandra et. al. 2016). Regular intake of millets is associated with many benefits and a lower chance of developing a number of degenerative illnesses. They offer additional health advantages like anti-ulcerative, antioxidative, and anti-inflammatory properties besides having cholesterol-lowering and hypoglycemic capabilities (Schoenlechner et al. 2013; Sharma and Gujral 2019). The carbohydrates in millets usually have a reduced starch digestibility thereby causing delayed absorption, which alters the blood glucose levels positively. Millets are continuously explored in the food processing sector as promising bioactive components because of their phytochemical composition and significance towards human wellbeing for the development of gluten-free and low-glycemic products.

Owing to such nutritional, biological, and functional benefits, nutri-cereals have been proposed as a staple diet. Millet grains are processed before consumption and during food preparation using common traditional methods like flaking, decorticating, fermentation, malting, roasting, and crushing for improving their functional, nutritional, and organoleptic attributes (Saleh et al. 2013). Aside from the obvious benefits of adopting millets as a main food crop, many significant obstacles are linked to nutri-cereals' processing and storage. Because of millet's smaller size, the germ having lipid content is separated with severe problems from the endosperm. This germ portion can pose difficulties during storage, particularly for flour because during grinding the fats from pericarp and germ are exposed to the environment, leading to lipid hydrolysis and further oxidation of the produced de-esterified unsaturated fats. Off-flavors are often detected as a result of these chemical changes at some stage during storage, especially when moisture levels are high and oxygen is present (Duodu and Dowell 2019; Liang et al. 2021). The quality of its preservation is highly dependent on the pre-treatments used and the storage circumstances.

Several pre-treatment and processing strategies for treating nutri-cereals have been developed to increase storage stability, reduce rancidity concerns, and decontaminate spoiling microorganisms. Nevertheless, a specific and comprehensive chapter that highlights the storage stability and quality management of nutri-cereals and their associated products while highlighting the effect on essential storage attributes and requirements has not been studied yet. In this scenario, this chapter mainly focuses on and covers the necessary practices for the storage stability of nutri-cereals.

15.2 Nutri-cereals: an overview

Nutri-cereals are much better than the other cereals due to their high nutritional and health benefits. They are used as a food by a significant malnourished population, and suffer huge temperature fluctuations during production. In India, the highest areas under millets is in Rajasthan, Gujarat, and Uttar Pradesh (for pearl millet), trailed by Maharashtra and Karnataka (for sorghum). In parts of Karnataka, Tamil Nadu, Telangana, and Uttarakhand, the finger millet constitutes an essential component of the diet. Other minor millets have maximum area in Madhya Pradesh and Chhattisgarh. The total cultivated area in the country is over 12 million hectares, with an annual production of around 15 million tonnes (Likhi 2021). From the past few decades, the millet production has been shifted to other competing crops like sunflower, sugarcane, soybean, maize, and cotton. However, there is a need to improve and promote the production of nutri-cereals with new and higher productive variety, value addition, the scientific package of practices, price support, bio-fortification, and farm-gate processing.

Millets are concentrated sources of different vitamin and essential nutrients like thiamine 0.15–0.60 mg/100g, iron 2.2–17.7 mg/100g, riboflavin 0.28–1.65 mg/100g, phosphorus 200–339 mg/100g, calcium 10–348 mg/100g, niacin 0.09–1.11 mg/100g and zinc 32.7–60.6 mg/100g, which make them ideal food (Kumar et al. 2018; Sharma et al. 2021). The millets generally contain high amounts of protein, fat, crude fiber, and carbohydrates, ranging from 6.20–12.50 percent, 1.10–5.00 percent, 1.20–9.80 percent, and 60.90–72.60 percent, respectively (Kulthe et al. 2016; Hassan et al. 2021). The higher amounts of calcium, proteins, polyphenols, and dietary fiber in millets makes them distinctive from others. The nutritional characteristics of nutri-cereals are summarized in **Table 15.1**.

Nutri-cereals contain carbohydrates, which range from 50 percent to 88 percent depending on the particular variety, species, crop management, and agro-climatic conditions. Carbohydrates contain starch, non-starchy polysaccharides and free sugars ranging from 60–75 percent, 15–20 percent, and 2–3 percent, respectively (Chauhan et al. 2018). Nevertheless, dietary fiber is also found that includes arabinoxylans, β-glucans, lignins, celluloses, and hemicelluloses. Among millets, the finger millet, pearl millet, and kodo millet are typically higher in starch. Several cultivars of finger millet, foxtail millet, and proso millet are glutinous because of the presence of waxy starches in them (Patil 2016).

Because of genetic diversity and agro-geographical considerations, protein content varies greatly between the species and cultivars. Higher protein content is found in proso and little millet varying from 10–15 percent, and that is greater or equivalent to certain protein rich cereal grain species. Although, several agronomic factors like soil nitrogen concentration and growing atmosphere greatly impact the protein content. Due to their high protein content, millets are excellent targets for formulating the nutrient-dense foods for malnourished and other vulnerable populations. Lysine is usually low in most cereals, but millets like finger millet and kodo millet have 2.20–5.50 g lysine/100 g protein, whereas pearl millet may have 6.50 g lysine/100 g protein.

The lipid content in millets varies from 1–6 percent, and particular kinds may have larger amounts, which can often impact storage stability. Slama et al. (2019) found that pearl millet contained 5.06 percent total lipid content out of which 77.22 percent was present in form of mono- or polyunsaturated fat. Mineral deficiency is a reason for concern since it has a dramatic influence on metabolic processes and tissue structure, potentially leading to severe and chronic illnesses (Vali Pasha et al. 2018). Despite the fact that climatic conditions, soil fertility, agronomic approaches, and geographical constraints have a significant impact on the mineral composition of each crop, the minerals phosphorus and potassium are the most abundant in most nutri-cereals. Millets also contain significant

Table 15.1 Nutritional Composition (g/100g) of Nutri-Cereals

Nutri-cereals	Carbohydrates (g)	Proteins (g)	Total fats (g)	Dietary fibre (g)	Ash content (g)	Ca (mg)	P (mg)	Fe (mg)	Zn (mg)
Sorghum (*Sorghum bicolor*)	72.60	10.40	1.73	10.22	1.60	25	222	5.29	3.00
Pearl Millet (*Pennisetum glaucum*)	67.50	11.60	5.00	11.49	2.30	42	296	10.30	3.10
Finger Millet (*Eleusine coracana*)	72.00	07.0	1.30	11.18	2.70	344	283	4.27	36.60
Foxtail Millet (*Setaria italica*)	60.90	12.30	4.30	4.25	3.30	31	290	3.50	60.60
Proso Millet (*Panicum miliaceum*)	70.40	12.50	1.10	2.20	1.90	14	206	2.20	3.70
Kodo Millet (*Paspalum scrobiculatum*)	65.90	8.30	1.40	6.39	2.60	27	188	3.17	32.70
Barnyard Millet (*Echinochloa esculenta*)	74.30	11.60	5.80	2.98	4.70	14	121	17.47	57.45
Little Millet (*Panicum sumatrense*)	75.70	8.70	5.30	6.39	1.70	17	220	9.30	3.70
Buckwheat (*Fagopyrum esculentum*)	70.60	12.60	3.10	10.00	1.80	48	362	13.35	2.20
Amaranthus (*Amaranthus cruentus*)	61.4	16.50	5.70	20.6	2.8	190	448	14.70	4.45

Sources: D'Amico and Schoenlechner (2017); Longvah et al. (2017); Tömösközi and Langó (2017); Kumar et al. (2018); Ministry of Agriculture and Farmers Welfare (2021).

amounts of calcium, sodium, and magnesium. Except for B12, which is found mostly in yeast and animal products, millets also constitute a major source of the B-complex vitamins. Millets' bran, pericarp, and aleurone layer are similarly high in vitamin content. Millets contain 0.25–0.57 mg/100 g of thiamin and 0.05–0.23 mg/100 g of riboflavin, respectively (Sharma et al. 2021).

Besides basic nutrients, millets also contain a variety of physiologically active chemicals, the most important of which are tannins and polyphenolic compounds, including phenolic acids and flavonoids (Duodu and Dowell 2019). Such chemicals offer several health advantages by acting as antioxidants and thereby reducing chronic and degenerative diseases. These phytochemicals, together with minerals, vitamins, and fibers, are concentrated in the outer bran layers. The phytochemicals including phytic acids make the millets capable of decreasing cholesterol, and leading to minimize the cancer. Flavonoids have anti-human immunodeficiency virus functions (Yao et al. 2004). Millets may be used as raw materials for functional foods and nutraceuticals since they contain enough amounts of proteins, phytochemicals, dietary fiber, minerals, and vitamins. Nutri-cereals are therefore, a healthy alternative for persons with celiac diseases, diabetes, and cardiovascular diseases (El Khoury et al. 2018). Because millets have several nutritional and health advantages, efforts are being made to increase their storage stability as well as nutrient bioavailability Further, the millet storage and processing may pose substantial challenges which must be handled effectively.

So far many antinutritional compounds of cereals grain are reported, such as protease inhibitors, phytates, tannins, lectins, and alkaloids; yet, the most researched antinutritional elements in millet grains are tannins and phytates (Ram et al. 2020). Proso millet and foxtail millet have phytic acid contents of 7.20 and 9.90 mg/g, respectively.

15.3 Shelf life

There is no doubt that millet is more beneficial than conventional cereals such as rice and wheat. The process for the harvesting of millet is cutting the earheads then stalks later. The straw (stalks) harvested after a week then needs to be dry for stacking. For the dried grain, the moisture content needs to be reduced up to 14 percent. But for long-term storage (usually above six months), the moisture content of grain must be less than 13 percent (Kumar and Kalita 2017). Nutri-cereals are typically good storage grains that can be stored for 1–2 years without any particular concerns. However, for long-term storage, airtight containers must be used in addition to oxygen absorbers. Oxygen absorbers are cheap and simple to use. They will absorb any remaining oxygen in the millets, depriving any remaining, live insect eggs of oxygen and therefore killing them.

For the preparation of several food formulations, the millet grains need to be dehulled or crushed into flour. When kernels are no longer protected by chaff, they begin to deteriorate and have a harsh flavor. Rancidity will also develop during storage, resulting in an increased phytic acid and free fatty acid (FFA) levels (Sruthi and Rao 2021). Lipase hydrolysis of triglycerides, which reduces pH and increases grain rancidity, is likely to be causing the increase in FFAs. The polyunsaturated fatty acids having 1-cis, 4-cis-pentadiene structure are oxidized from lipoxygenase (LOX), creating the hydroperoxides with conjugated dienes, which may cause off-flavor and rancidity (Wang et al. 2014). As a result, limiting automatic and enzymatic oxidation of millet fatty acids is a difficult challenge in millet quality control.

Pearl millet grains are treated with roasting, acid, and blanching to change their color, minimize anti-nutritional components, and increase lipase inactivation, which can otherwise cause the rancid odor on milling. Such treatments can extend the product shelf life (Bhati et al. 2016). During roasting, the millet kernels or flours are subjected to dry heat at higher temperatures (120–180° C) for 75–120 minutes. Various acids like tartaric acid, acetic acid, or diluted HCl are used in acid treatment. On the other hand, the blanching involves dipping the grains in water to temperatures below 100 ° C for 10–90 seconds (Dias-Martins et al. 2018). Furthermore, the recent techniques like irradiation can be employed in cases of millets, however, very few investigations have been published in this regard. Pearl millet meal was irradiated up to 2 kGy as a potential substitute to chemical and heat methods and minimize the incidence of bitter taste during flour storage (Mohamed et al. 2010).

15.4 Storage stability properties

In the supply chain, storing the millets is a vital part of their production, which enables numerous producer aims such as preserving millets for the future and eliminating food scarcity, supplying seed for the subsequent cultivation season, and allowing the producer to sell millets when prices would be higher in the market. However, excellent and safe storage is required in order to fulfil these goals. Storage can permit their delayed consumption and the millets can be exported at any time of year. Storage stability of grain is so necessary from an industrial, governmental, and cultural standpoint.

Often millet grains are kept in metal bins, silos, thatch/mud/clay based granaries with or without cowdung plaster. Spikes are quite often suspended from the roofs of buildings of developing nations, and kernels are stored in pottery jars and bamboo or fiber based baskets and bags. Such stored millets can be substantially lost qualitatively and quantitatively due to insect, fungal, and rodent infestations (Kumar and Kalita 2017). Nevertheless, the native quality of millet is lost more without considerable dry matter reductions, such as decreased nutritional value, vitamin or essential fatty acid losses, the development of toxicity, off-flavors or variations in color owing to microbial growth, and insect contamination. To reduce such losses, it is critical in every circumstance to always have a thorough awareness about local climatic, economical, cultural, and ecological variables, as well as a comprehension of potential causes of grain degradation during storage. The most substantial losses are usually caused by insect infestations or the proliferation of bacteria. Biochemical changes in the grain itself, to a lesser extent, can also be significant (Richard-Molard 2003). The details of changes occurring in nutri-cereals during storage and factors responsible for them are explained in **Table 15.2**.

15.4.1 Moisture content

The safe storage of millets and their flour is significantly influenced by moisture content which, in return, is greatly influenced due to storage environments. Millet grains and flour typically have less moisture and are, therefore, regarded as microbiologically safe. Nonetheless, the extended storage may affect the quality entirely. Different studies showed that millet flour typically contain moisture varying from 6 to 13 percent amid variations in processing, milling and handling conditions. The moisture value has been shown to increase considerably with storage (Akinola et al. 2017; Abioye et al. 2018). Increased moisture levels are linked to higher lipolytic as well as proteolytic reactions, thereby resulting in free fatty acids' generation and nutritional loss, which may cause poor organoleptic attributes. The pH and acidity of many products are affected due to variations in moisture content or RH and storage temperatures (Goyal et al. 2017). For instance, in a study, the flour from different pearl millet varieties showed higher peroxide value due to rise in their moisture value. Varsha and Narayanan (2017) and authors ascribed this as moisture-driven hydrolytic rancidity. During storage, the moisture increase was 22.50 percent (*Proagro*-RF), 70.20 percent (*Dhanshakthi*-UP), 53.80 percent (*Dhanshakthi*-RT), 49.20 percent (*Dhanshakthi*-RF), 31.90 percent (*Proagro*-RT), and 44.30 percent (*Proagro*-UP).

15.4.2 Water activity (a_w)

Usually the sorghum and millet kernels have water activity (a_w) of less than 0.65, which makes them quite stable against microbiological proliferation. But several biochemical reactions still exist at this low a_w value because of the starch and protein injury in shelled millets and their flour (Rathore and Singh 2018). Further, at higher water activity levels, fungi growth is common, and several insects can also grow.

Table 15.2 Summary of Different Factors Affecting Storage Stability of Nutri-Cereals and Associated Products

Nutri-cereal	Storage conditions	Factors affecting storage stability	Impact on quality	References
Sorghum	Stored in jute sacks at 30±5°C	Insect, larvae	Increase in number of infected seeds from 0 to 12.89% after 8 months of storage and poor sensory properties of porridge prepared from grains during storage	Rahman et al. (2020)
Sorghum	Earthenware pots, gunny sacks and reed baskets, and Kotlu - earth floors surfaced with cowdung and dried neem (*Azadirachta indica*) leaves	Fungi	Maximum growth of *Aspergillus* spp. and *Fusarium* spp in in Kotlu than other storage structures	Sashidhar et al. (1992)
Sorghum flour	100 g sorghum flour packed in individual polypropylene bags (200 mm), sealed and stored at 28±2°C,65% RH under dark	Lipid content	increase in free fatty acid (FFA) content, higher lipase activity, reduced flour whiteness	Meera et al. (2011)
Pearl millet	Pearl millet grains stored in Polypropylene sacks, underground pit storage system and clay pot	Fungi	Growth of *Cryptolestes ferrugineus, Tribolium castaneum, Liposcelis bostrychophilus* and *Rhyzopertha dominica*; higher damaged pearl millet grains stored in underground pit (80.1%)	Lale and Yusuf (2000)
Pearl millet flour	Aluminium boxes & temperature: 15°C, 82 % RH	Lipid content and enzymes	Increased fat acidity, maximum peroxidase and polyphenoloxidase activity	Goyal et al (2017)
Pearl millet flour (from *Proagro* and *Dhanshakthi* cultivars)	Packed in double HDPE cover and stored under ambient conditions (27.5°C) and refrigerated temperature (4°C)	Moisture content, lipid content and microbial growth	Increased FFA value and moisture content in all samples with highest rise in control *Proagro* flour; decrease in protein and fat content, refrigerated (RF) samples showed minimum microbial growth; loss of iron content in all samples	Varsha and Narayanan (2017)

(Continued)

Table 15.2 *(Continued)*

Nutri-cereal	Storage conditions	Factors affecting storage stability	Impact on quality	References
Proso bran fraction	Stored in textile bags, ambient or refrigerator conditions for 180 day	Lipid content	No significant change in iodine value of 137 2; increased FFA content due to hydrolytic rancidity development; peroxide value increased in sample under ambient condition than refrigerator; rise in anisidine value under ambient conditions; stability of proso bran fraction for 150 days in refrigerator, but highly unstable under ambient conditions	Mustač et al. (2020)
Pearl millet flour (var. PHB-2168)	Stored in polyethylene bags (75μ) at room temperature (15–35 °C)	Lipid content	Increased peroxide vale at 20th day of storage and then decreased sharply due to production of aldehydes and secondary peroxides products; increased FFA content	Yadav et al. (2012)
Pearl millet flour (cv. Pusa Composite 443)	Stored in LDPE bags (10 μm) under ambient condition at 30±1.5°C, 20.7±2.0°C (minimum) and 72±2% RH	Lipid content	Significant increase in fat acidity, FFA and phytic acid content; sensory parameters like color, appearance, taste as well as overall acceptability were not acceptable	Tiwari et al. (2014)
Little millet biscuit	Stored in room temperature for 120 days	Lipid and moisture content	Increased acid value and moisture content of little millet based biscuits than control biscuits	Hemalatha et al (2006)
Foxtail millet pearl millet, barnyard millet, finger millet and proso millet (whole grain, flour and groats)	Stored in room temperature for 6 weeks	Lipid content	Maximum titratable acidity found in whole grains of pearl millet with minimum changes in foxtail millet whole grains	Čepková et al. (2014)

Product	Parameter studied	Storage conditions	Observations	Reference
Proso millet (whole grain and groats)	Lipid content	Stored for 14 weeks under RH 65%, temperature 35°C.	Increased titratable acids in all samples but maximum in ground grain prepared from intact whole grains; bitter flavor detected from stored samples with maximum flavor stability for intact whole and intact scoured grains	Dvoracek et al. (2010)
Finger millet pasta	Moisture content	Stored in 400 gauge LDPE bags, sealed and stored under room conditions for 90 days	Increased moisture content of products; increased cooking time of pasta from 8.32 to 8.55 min; minimum solid loss (1.23–1.36%) recorded in finger millet pasta during storage	Seema et al. (2016)
Buckwheat flour (25% and 50%) containing chips	Moisture content and lipid content	stored in polypropylene (75 μ) and metallized polyester (90 μ) pouches for 6 months under ambient conditions	Increased moisture content, peroxide value, FFA value and thiobarbituric acid value in samples packed in polypropylene films and metallized polyester; acceptability was maintained for 6 months in both samples under room temperature	Goel et al. (2018)
Buckwheat	Moisture content	Stored in gunny bag, polythene bag and metallic bin for 1 year	Increased moisture content in gunny bags at 3 months after that it decreased due to relatively higher temperature and less RH in environment during May-June; significant decrease in protein and total sugar content but increase in free amino acid content at end of storage	Dogra and Awasthi (2012)

(Continued)

Table 15.2 *(Continued)*

Nutri-cereal	Storage conditions	Factors affecting storage stability	Impact on quality	References
Amaranth flour	Stored in air-tight glass container at room temperature for 6 months	-	First increase (up to 2 months) and then decrease in moisture content due to changes in RH of storage environment; decrease in starch, fat and fibre content during storage; significant reduction in minerals like Ca, Fe, and Mg	Nidhi and Indira (2012)
Buckwheat cultivar *Bedija*	Packed in paper bags and stored on wooden shelves in dry storage room without light for 30 days and 1 to 6 years	-	Increasing storage duration decreased the seed mass as well as the value of germination energy significantly	Jevdjović et al. (2003)

15.4.3 Free fatty acid (FFA) content

Normally, the lipids in millets are stable inside the cell's functional spherosomes. Due to microbial enzymes, the lipid portion degrades relatively faster during open-air storage which causes the buildup of fatty acids. However, numerous factors besides enzymes, like physical shocks and other external agents, weaken the lipid barrier during storage, thereby allowing the lipids to readily decompose (Mujahid et al. 2005). Attributed to this, the FFAs amount is increased in millet which, when oxidized creates aldehydes and ketones. Further oxidation of unsaturated acids can cause volatile components with rancid smell and off-flavors. According to Varsha and Narayanan (2017), the level of FFA slowly raised to 0.826±0.010 (*Proagro-RF*) from 0.071±0.002 (*Proagro-RT*) with minimum value in refrigerated samples.

After storing amaranth grains for 1 year at 25° C and relative humidity of 55 percent, they contained 6.0 percent of fat (with 6.5% in the original sample), and after 12 months' storage at 25° C with relative humidity of 75 percent, the amount of fat decreased to 5.0 percent (Stankevych et al. 2021). Millet storage leads to the greatest changes in the composition and properties of the lipid fractions of grain, as they are most susceptible to oxidative and hydrolytic processes. Various enzymatic processes take place in the lipid complex: phospholipids and glycolipids are cleaved, and free fatty acids accumulate. Peroxides, hydroperoxides, and other oxidation products accumulate and can form complexes with proteins and carbohydrates. All this affects the quality of fats during storage. The simplest way of controlling the changes in the quality of fat in grain is monitoring the acid value and acidity of fat (Stankevych et al. 2021).

15.4.4 Microbial count

During storage, mold growth can be prevented by properly drying the moist millet grains obtained from harvesting. The microbes can grow intensively and quickly under humid and hot environments. At 25° C, the millet kernels should have a moisture level of 13 percent or less. For several years, nothing significant can happen in millets below this moisture limit, excluding the probable development of lipid based derived products (Akinola et al. 2017). Many mold species, including *Aspergillus, Eurotium,* and *Monascus*, may develop over this moisture limit, which is really quite low. Those molds have an exceptional physiologic potential to reproduce in incredibly dry environments and will inevitably emerge if further hurdles like refrigeration, altered atmospheres, or use of antifungal chemicals are not used. Molds use fats and starches or sugar from millets to multiply by generating heat, $CO2$ and water and heat. Such conditions further leads to the growth of molds having increased water needs like *Penicillium* spp., *Aspergillus* spp., and even *Fusarium* spp. which damage millets. These molds can create very harmful mycotoxins like immunosuppressive T2 toxin or hepatotoxic aflatoxin B1 (Raghavender et al. 2007). Therefore, the urgent chilling/cooling and drying the millets is essential using forced air circulation in such cases so that the potential difficulties can be avoided throughout the bulk contained inside a silo.

15.4.5 Biological factors

Insects seem to be the most detrimental animal pests to stored nutri-cereals in addition to mites, birds and rats. Even in circumstances of minor infestation, their occurrence diminishes the price of sorghum and millets (Kumar and Kalita 2017). The metabolic products such as quinones and uric acid are released from insects, which can cause unpleasant tastes and odors. Insects might even spread harmful microorganisms to stored millets and/or deposit debris (with excrement) in flours as well as other commodities. Such filth strongly repels customers and, therefore, insects destroy far more grain due to contamination in addition to ingestion. *Coleoptera* and *Lepidoptera* are mainly considered as storage insects. Most are adapted forms of crop destroying species. Most dangerous species, which are exceedingly adaptive and versatile with just low nutritional needs, include the genera *Cryptolestes, Rhyzopertha, Sitophilus, Sitotroga, Trogoderma,* and *Tribolium* (Banga et al. 2020). Further, rats and mice, which are common across the world, are also damage-causing agents, wreaking havoc mostly on fields, where millets are kept in open or unsecured constructions. Traps, baits with rodenticides, and fumigation are all methods of controlling rats and mice.

15.4.6 Other biochemical changes

During storage, the starch content gradually decreases at all temperatures and relative humidity. When amaranth is stored at a low temperature (5° C) and relative humidity (55%), the starch content decreases more slowly (from 61.40 to 59.70%) than at higher temperatures and humidity (from 61.40 to 56.50%) (Stankevych et al. 2021). The rise of humidity and temperature intensifies the grain's aerobic respiration, which accelerates the consumption of carbohydrates. Besides this, longer storage intensifies the natural aging processes in grain. Stankevych et al. (2021) stored the amaranth throughout a year at 5° C, 15° C, and 25° C, at RH 55 percent and 75 percent, and noticed alterations in grain quality parameters as the content of protein, starch, fiber, fat, and ash, the acid value, and the acidity of fat.

A thorough inspection of grains from farms, gunny sacks, transportation vehicles, and other places should be followed at different stages of the supply chain. The most efficient pest management strategy is to store the well-dried millets following the hygienic practices in sanitary facilities. Government and commercial warehouses with sufficient storage infrastructure are required in addition to better handling. Some developing nations are successfully employing the airtight storage as well as aeration methods under low temperatures.

15.5 Quality management

As discussed above, the storage stability of millets can be compromised due to several biological and biochemical reasons. A significant reduction in the consequences of degradation would be achieved by appropriate management of stored grains, which need in-depth understanding of the deterioration principles as well as information about the storage behavior of individual grains. The quality attributes of sorghum and millets may be examined using a variety of methods, starting from basic sight to light microscope based assessments. Such simple as well as sophisticated analytical testing can indicate and confirm the chemical contaminants, existence of foreign matter, and microbiological safety. Besides these, several other tests such as for seed viability, presence of mycotoxins, free fatty acid (FFA) levels, and the prevalence of farm as well as storage mites, molds, and insects. At the farm level, easier tests are frequently performed with more accuracy. As the millets move from farm level through the supply chain, the more expensive and sophisticated tests are utilized. When a country's economy grows, the testing standards frequently become much stringent because of charging greater costs for a product of proven quality by sellers. Subsequently, the higher product quality is demanded by consumers in these nations, which emphasize improved quality management practices (Henry and Kettlewell 2012).

Before storage of nutri-cereals, their cleaning should be done because freshly harvested millets may carry impurities, such as small or large weed seeds, other grain, dust, straw, chaff, small stones, broken kernels or other debris. Nevertheless, the safe storage of millets can be adversely affected due to the occurrence of a substantial amount of bigger impurities such as chaff, husk, ear fragments, stalks, weed parts leaves, and so forth. Such contaminants may lead to a severe risk to safe storage of millets due to their potential to cause or favor the microbiological spoilage by offering suitable environments for insect and mite proliferation and growth. To reduce this problem, cleaning procedures must be performed using a variety of equipment, from basic sieves employed in underdeveloped nations for the removal of minor particles to highly advanced time saving industrial machinery. Aside from cleaning, moisture content, particularly water activity (a_w), of nutri-cereals at the beginning is also affecting storage life. The several water activity levels that permit the emergence and growth of fungus species (which commonly thrive in stored nutri-cereals) are now well understood. To retain the native properties of millets, the mathematical models based on sorption curves must be adopted, which assist in predicting storage conditions such as moisture and temperature. The quality of stored sorghum and millets can be forecasted using these models at any phase, as well as the calculation of time remaining until the nutri-cereals reach to the least acceptable level of safety. During storage, cooling of millets through cold air aeration can be done to lower the millet temperature. Millets that has 14 percent moisture content are stored under temperatures as high as 25–35° C, they cannot be stored for an extended duration unless they are chilled at lower temperatures, which can prevent the reduction in quality (Fleurat-Lessard and D'Ornon 2004). No reproduction of maximum insects occur at temperatures below 12° C or above 34° C, because the ideal temperature for reproduction is around 26° C.

Additionally, the removal of lipid rich anatomical components or inactivation of lipase and LOX enzymes is important, and several mechanical or chemical approaches must be followed for this purpose. The technologies target to create DNA damage in fungal or microbial organisms, hence preventing deterioration. Besides this, the techno-functional properties of millets can be affected, which cause variations in the nutritional profile of nutri-cereals and their associated products. Several traditional methods like mechanical processing, germination, fermentation, and so forth, and novel processing approaches can be employed for the improvement of storage stability and techno-functional as well as nutritional quality of nutri-cereals before their use (Sruthi and Rao 2021), as shown in **Figure 15.1**. Moreover, the natural or chemical preservatives as well as antioxidants are often used to extend the shelf life of millets and associated processed products, which are capable of inhibiting microbial spoilage and degradation. The use of acid treatment can remove the enzyme lipase, found in millets and thus assist in decreasing the amount of FFA generated. For instance, the pearl millet grains were exposed to 0.2 N HCl for different duration which caused a 73.90 percent and 62.50 percent decrease in FFA content of millet flour after 24 h and 18 h of treatment, respectively, at 25–30° C (Bhati et al. 2016). The manmade as well as natural antioxidants offer protection to cells against any kind of injury by the reduction of lipid auto-oxidation. In a study, butylated hydroxytoluene (0.02%) was added in two varieties from pearl millet (*Hreabry* and *Ashana*), which were then stored for three months under ambient conditions without any quality loss (Abdalgader et al. 2019).

Depending on the criteria, such as usefulness and economy, the nutri-cereal based processing firm will choose the above-mentioned strategies to raise or improve the quality of sorghum and millets. After that, appropriate quality requirements would need to be defined based on the individual food end-use. **Figure 15.2** illustrates the quality management of nutri-cereals for their value-added products, such as bread, alcoholic beverages, porridge, and so forth.

Further, the monitoring of quality is essential for minimizing millet damage. Accurate and precise information about grain quality can be procured from sensor technology advancements, especially cellular (mobile) phone-based technologies, which can benefit the entire grain sector

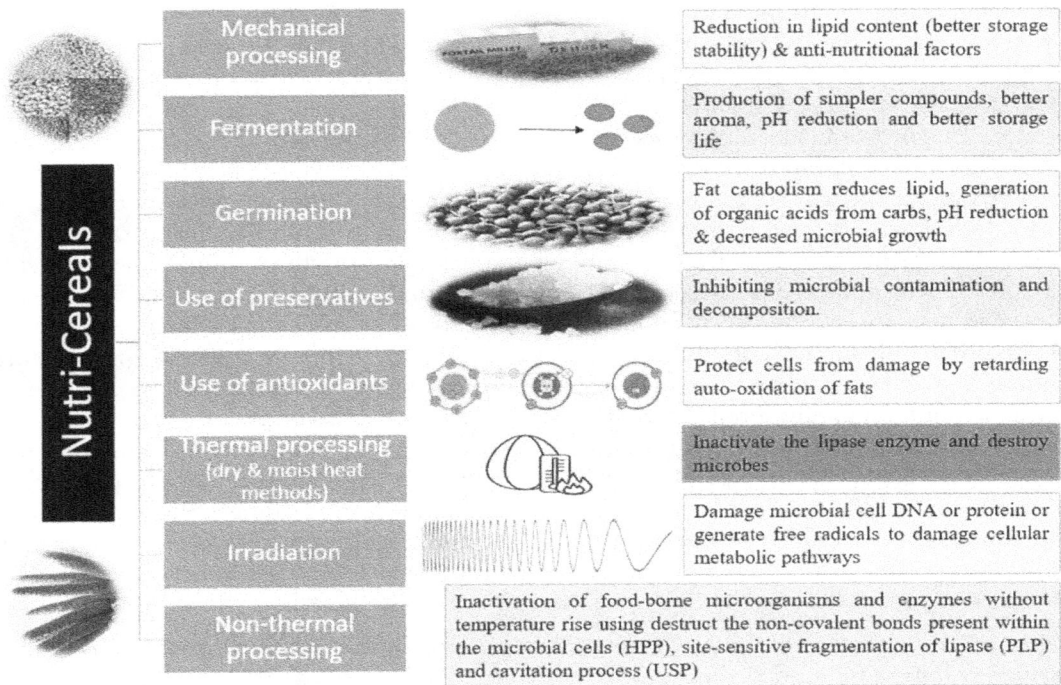

Figure 15.1 *Summary of techniques and their respective mechanisms to improve the storage stability of millets and associated products.*

Note: HPP: High pressure processing, PLP: Pulsed light processing, USP: Ultrasound processing.

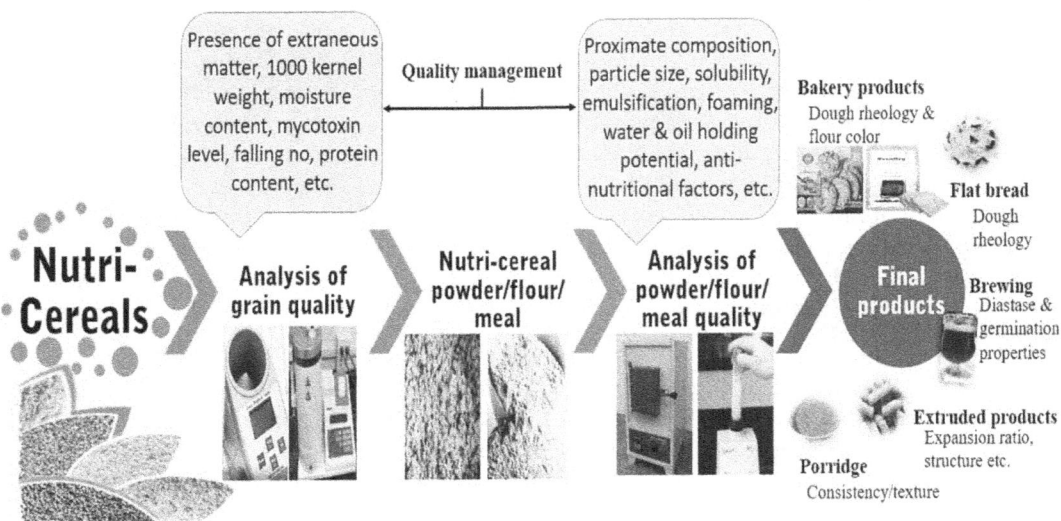

Figure 15.2 *Quality management of nutri-cereals for their value-added products.*

(Kaushik and Singhai 2018). Farmers as well as customers will be assisted by quick information about potential millet yields and grain quality using the aerial drones as well as space satellites. Breeders will be able to choose seeds with particular qualities using rapid sensing and sorting techniques, allowing them to produce lines with precise agronomic plus quality attributes. Fast and simple sensors will allow the determination of grain quality and safety as well as the rapid measurement of toxins. Grain processors can make clear decisions about grain safety and quality using the online sensors. Portable or mobile sensors will further help to get information about quality and safety from any place, including processing, field, transportation, warehousing, or marketing. Consumers can also use such mobile sensors to acquire quality and safety related information while buying or eating the product (Duodu and Dowell 2019). With these economic sensors, the regional, national and global buyers can get access into emerging markets for understanding the precise quality attributes of grains. Nowadays, the development of functional and nutraceutical foods from nutri-cereals is a growing trend, and such products are rich in bioactive compounds. Sorghum, pearl millet, and other millets are exceptionally rich in micronutrients, which can assist the government and other bodies to promote millets on this basis.

15.6 Government initiatives

Although nutri-cereals offer multiple benefits, they are mostly grown and consumed by tribal societies. Nutri-cereals and their associated products are not available to urban populations and middle-class people because of the scarcity of ready-to-eat (RTE) millet-based value added products. However, they have acquired recent popularity among health-conscious customers. Therefore, several processors as well as the governments of various countries are undertaking initiatives for producing accessible, easy to cook or ready to eat and other processed items. Several developed countries have also acknowledged the importance of nutri-cereals in the food system.

In India, there are several collaborative research and development programs between various regional, global and national bodies. Different organizations like International Crops Research Institute for Semi-Arid Tropics (ICRISAT), Indian Institute of Millets Research, Hyderabad, Directorate of Millets Development, Jaipur, as well as the All India Coordinated Pearl Millet Improvement Project are continuously working in India to promote millet production and consumption. Similarly, in Africa there are the Collaborative Research Support Programme on Sorghum and Pearl Millet, Sahel Institute, and Research Institute for Tropical Agronomy and Food Crops. Such national and international organizations must create and support a multidisciplinary perspective to pest management during millet storage and aid in boosting the nutritional value of grains. For promoting the cultivation and intake of nutri-cereals, the Indian Government had declared 2018 as the "National Year of Millets" (The Hindu 2018). With the introduction from the Government of India, the United Nations General Assembly has overwhelmingly accepted a resolution to celebrate 2023 as the "International Year of Millets" in collaboration with Bangladesh, Kenya, Nepal, Nigeria, Russia, and Senegal (The Hindu 2021). For the 2023 ceremony, three major targets have been identified by the Indian Government: (a) promoting awareness of millets' contribution to food security and nutrition; (b) encouraging interested parties, stakeholders, and also national governments for enhancing the production, yield, and quality of millets by joint actions, and (c) strengthening and emphasizing the greater investments in millet research and development as well as extension activities (Invest India 2022). Over the past few years, millets have been included in India's food security benefit plans for assisting and meeting the country's nutritional needs. Some of the existing programs of the

Indian Government for millets production include Integrated Cereals Development Programs in Coarse Cereals-Based Cropping Systems Areas under Macro Management of Agriculture. In order to promote cultivation of nutri-cereals, the Initiative for Nutritional Security through Intensive Millet Promotion (INSIMP) and Rainfed Area Development Program (RADP) are the extensive programs only, which are also a part of Rashtriya Krishi Vikas Yojana (RKVY) (Ministry of Agriculture and Farmers Welfare 2016).

Additionally, there are Codex Alimentarius regulations for grain and flour from pearl millet and sorghum, which were last amended in 1995. These guidelines are primarily concerned with food safety, and they address concerns like presence of heavy metals, hazardous seeds, pesticide/insecticide/fumigant residues and toxins produced by fungi (Taylor and Duodu 2017). In case of sorghum, there are the Grain Inspection, Packers and Stockyards Administration (GIPSA) requirements established by the US Department of Agriculture (GIPSA 2013). These guidelines mainly cover food safety criteria, however they are largely trading norms that correlate to end-use quality in certain ways. Grain Trade Australia in Australia has established regulations for sorghum but not for other millets. This organization regulates the trade norms, grain trading regulations, and grain contracts throughout the Australian grain sector. This regulation has removed the varietal differences among grains. The sorghum grain can be labelled as red, white, or yellow as per the appearance (Grain Trade Australia 2015. The economic organization from East African Community (made up of six nations – Kenya, Uganda, Tanzania, Burundi, Rwanda and South Sudan) has also set standards for sorghum and finger millet grains for public use. Different variables taken into account include physico-chemical properties of grains and food safety (Taylor and Duodu 2017).

15.7 Future perspectives

Conventional crops or grasses like millets are cultivated in dry regions and are considered drought-tolerant. However, millets often decline in quality owing to inherent (e.g., respiration) and/or external factors (e.g. insects, molds, and rodents). The most significant environmental element influencing insect growth and reproduction is temperature. Grain moisture content and water activity are other important considerations affecting storage stability of millets. The interaction of moisture, respiration, and mold development raises the stored millets temperature as well as moisture, which make them increasingly susceptible to additional harms. As a result, it is crucial to improve the storage systems in order to reduce losses and enhance the integrity of millets and associated foods, particularly for individuals with limited economic means. These days, the customers are demanding safe products, which means that items should be free of chemical as well as microbiological residues. In the next years, this approach may replace the application of contact pesticides and maybe even fumigants with a little more reasonable options such as physical strategies like modified or controlled atmospheric storage.

To achieve best outcomes, the storage stability and quality management approaches of stored millets sometimes need to combine the diverse methodologies. Nevertheless, one of the most significant difficulties now is that the majority of the research is concentrated on major millets – sorghum, finger millet and pearl millet. There is a need to recognize the nutritive importance of minor millets, which have been currently underused. Hence, extensive study in this domain is required to create shelf-stable minor millets, flours, and their products exhibiting remarkable sensory and nutritional qualities. Aside from these factors, the decrease in production of millet is also affecting nutri-cereal sector because of increased labor intensity, reduced crop yield, time consuming and labor based post-harvest handling, and an absence of enticing farm level pricing. A few millets, mainly major millets, can be obtained through public distribution networks, and the arduous task of polishing and decorticating small millets restricts the domestic producers. In addition, the pervasive perception of poor social status linked with their intake decreases the consumption as well as utilization of millets. Their attractiveness and growth is further hampered by the high cost of millets and the scarcity of millet based products. Scientist and researchers in this field must work on such limitations to improve the significance and utilization of millets in future.

15.8 Conclusion

Nutri-cereals are nutrient-dense and contain several minerals, vitamins, and dietary fiber also. However, their high fat content and lipase activity in addition to other biological factors mainly reduce the storage stability of millets. Therefore, several treatments and proper management with simple practices before and during storage may ensure biochemical as well as biological stability for longer periods. Such methods have been shown to increase millet safety while also providing excellent sensory and nutritional qualities. However, very little information is available about the minor millets. Specific studies on under-utilized millets need to be considered for assisting in their reduction of qualitative and quantitative losses. Further, researchers must also look towards chemical-free initiatives or treatments to improve storage stability as well as wholesomeness of nutri-cereals in upcoming years.

References

Abdalgader, M. A. A., S. A. Ashraf, A. M. Awadelkareem, M. W. A. Khan, and A. I. Mustafa. 2019. Effect of natural and synthetic antioxidant on shelf life of different Sudanese *Pennisetum glaucum* L. flour. *Bioscience Biotechnology Research Communications* 12:652–657 DOI: 10.21786/bbrc/12.3/15.

Abioye, V. F., G. O. Ogunlakin, and G. Taiwo. 2018. Effect of germination on anti-oxidant activity, total phenols, flavonoids and anti-nutritional content of finger millet flour. *Journal of Food Processing and Technology* 9: 719. DOI:10.4172/2157-7110.1000719.

Akinola, S. A., A. A. Badejo, O. F. Osundahunsi, and M. O. Edema. 2017. Effect of preprocessing techniques on pearl millet flour and changes in technological properties. *International Journal of Food Science and Technology* 52: 992–999.

Banga, K. S., S. Kumar, N. Kotwaliwale, and D. Mohapatra. 2020. Major insects of stored food grains. *International Journal of Chemical Studies* 8: 2380–2384.

Bhati, D., V. Bhatnagar, and V. Acharya. 2016. Effect of pre-milling processing techniques on pearl millet grains with special reference to in-vitro iron availability. *Asian Journal of Dairy and Food Research* 35: 76–80.

Čepková, P. H., Z. Dvořáková, D. Janovská, and I. Viehmannova. 2014. Rancidity development in millet species stored in different storage conditions and evaluation of free fatty acids content in tested samples. *Journal of Food, Agriculture and Environment* 12:101–106.

Chandra, D., S. Chandra, and A. K. Sharma. 2016. Review of Finger millet (*Eleusine coracana* (L.) Gaertn): A power house of health benefiting nutrients. *Food Science and Human Wellness* 5:149–155.

Chauhan, M., S. K. Sonawane, and S. S. Arya. 2018. Nutritional and nutraceutical properties of millets: A review. *Clinical Journal of Nutrition and Dietetics* 1:1–10.

D'Amico, S. and R. Schoenlechner. 2017. Amaranth: Its unique nutritional and health-promoting attributes. In: *Gluten-Free Ancient Grains: Cereals, Pseudocereals and Legumes*. ed. J. R. N. Taylor and J. M. Awika. Cambridge: Woodhead Publishing. 131–159.

Dias-Martins, A. M., K. L. F. Pessanha, S. Pacheco, J. A. S. Rodrigues, and C. W. P. Carvalho. 2018. Potential use of pearl millet (*Pennisetum glaucum* (L.) R. Br.) in Brazil: Food security, processing, health benefits and nutritional products. *Food Research International* 109:175–186.

Dogra, D. and C. P. Awasthi. 2012. Effect of different storage containers and duration on biochemical constituents of buckwheat (*Fagopyrum esculentum Moench*) Grains. *Indian Journal of Agricultural Biochemistry* 25:57–62.

Duodu, K. G. and F. E. Dowell. 2019. Chapter 14 – Sorghum and Millets: Quality Management Systems. In: *Sorghum and Millets: Chemistry, Technology and Nutritional Attributes*. ed. J. R. N. Taylor and K. G. Duodu, 421–442. Duxford, UK: Elsevier.

Dvoracek, V., D. Janovska, L. Papouskova, and E. Bicanova. 2010. Post-harvest content of free titratable acids in the grain of proso millet varieties (*Panicum milliaceum* L.), and changes during grain processing and storage. *Czech Journal of Genetics and Plant Breeding*. 46:S90–S95 DOI: 10.17221/699-CJGPB.

El Khoury, D., S. Balfour-Ducharme, and I. J. Joye. 2018. *A Review on the Gluten-Free Diet: Technological and Nutritional Challenges*. Nutrients 10:1410. DOI: 10.3390/nu10101410.

Fleurat-Lessard, F. and V. D'Ornon. 2004. Stored grain - Physico-Chemical treatment. In: *Encyclopedia of Grain Science*. ed. C. Wrigley, 254–263. Sydney: Academic Press DOI:10.1016/b0-12-765490-9/00167-1.

GIPSA. 2013. Grain Inspection, Packers and Stockyards Administration (GIPSA). *Sorghum, Grain Inspection Handbook II*, United States Department of Agriculture, Washington, DC (Chapter 9). www.gipsa.usda.gov (accessed December 12, 2021).

Goel, C., A. D. Semwal, P. Ananthan, and G. K. Sharma. 2018. Development and storage stability of buckwheat chips using response surface methodology (RSM). *Journal of Food Science and Technology* 55:5064–5074.

Goyal, P., L. K. Chugh, and M. K. Berwal. 2017. Storage effects on flour quality of commonly consumed cereals. *Journal of Applied and Natural Science* 9: 551–555.

Grain Trade Australia. 2015. Grain Trade Australia Section 2 – Sorghum Trading Standards, 2015/16 Season. www/graintrade.org.au/commodity_standards (accessed December 27, 2021).

Hassan, Z. M., N. A. Sebola, and M. Mabelebele. 2021. The nutritional use of millet grain for food and feed: A review. *Agriculture and Food Security* 10: 1–14.

Hemalatha, G., S. Amutha, D. Malathi, P. Vivekanandan, and G. Rajannan. 2006. Development of little millet (*Panicum sumatrense* Roth ex Roem and Schult.) substituted biscuits and characterization of packaging requirement. *Tropical Agricultural Research* 18:1–10.http://192.248.43.153/bitstream/1/1954/2/PGIA TAR-18-143.pdf.

Henry, R. and Kettlewell, P. 2012. *Cereal Grain Quality*. Springer Science and Business Media.

Invest India. 2022. Promoting Millet production for enhanced nutritional and women empowerment outcomes. *Invest India National Investment Promotion and Facilitation Agency*, New Delhi. www.investindia.gov. in/team-india-blogs/promoting-millet-production-enhanced-nutritional-and-women-empowerment-outco mes (accessed January 2, 2022).

Jevdjović, R. and R. Maletić. 2003. Effect of buckwheat seed storage duration on its quality. *Journal of Agricultural Sciences* (Belgrade) 48:135–141.

Kaushik, R. and J. Singhai. 2018. Sensing technologies used for monitoring and detecting insect infestation in stored grain. *International Journal of Engineering and Technology* 7:169–173.

Kulthe, A. A., S. S. Thorat, and S. B. Lande. 2016. Characterization of pearl millet cultivars for proximate composition, minerals and antinutritional contents. *Advances in Life Sciences* 5:4672–4675.

Kumar, A., V. Tomer, A. Kaur, V. Kumar, and K. Gupta. 2018. Millets: A solution to agrarian and nutritional challenges. *Agriculture and Food Security*, 7:31 DOI: 10.1186/s40066-018-0183-3.

Kumar, D. and P. Kalita. 2017. Reducing postharvest losses during storage of grain crops to strengthen food security in developing countries. *Foods* 6:8. DOI: 10.3390/foods6010008.

Lale, N. E. S., and B. A. Yusuf. 2000. Insect pests infesting stored pearl millet *Pennisetum glaucum* (L.) R. Br. in Northeastern Nigeria and their damage potential. *Cereal Research Communications* 28:181–186.

Liang, K., Y. Liu, and S. Liang. 2021. Analysis of the characteristics of foxtail millet during storage under different light environments. *Journal of Cereal Science* 101:103302. DOI:10.1016/j.jcs.2021.103302.

Likhi, A. 2021. Promoting nutri-cereal (millet) farming in India. National Institute of Agricultural Extension Management, Hyderabad. www.manage.gov.in/millets.html (accessed January 10, 2022).

Longvah, T., R. Ananthan, K. Bhaskarachary, and K. Venkaiah. 2017. *Indian Food Composition Tables*. National Institute of Nutrition (Indian Council of Medical Research) Department of Health Research Ministry of Health and Family Welfare, Government of India Hyderabad – 500 007, Telangana State, India.

Lu, H., X. Yang, M. Ye, et al. 2005. Millet noodles in late Neolithic China. *Nature* 437:967–968.

Meera, M. S., M. K. Bhashyam, and S. Z. Ali. 2011. Effect of heat treatment of sorghum grains on storage stability of flour. *LWT–Food Science and Technology* 44:2199–2204.

Ministry of Agriculture and Farmers Welfare. 2016. State of Indian Agriculture 2015–16. *Ministry of Agriculture and Farmers Welfare*, Government of India. https://agricoop.nic.in/sites/default/files/State_of_Ind ian_Agriculture%2C2015-16.pdf (accessed January 12, 2022).

Ministry of Agriculture and Farmers Welfare. 2018. *Ministry of Agriculture and Farmers Welfare*, Government of India. www.aicpmip.res.in/pmnutricereals.pdf (accessed January 12, 2022).

Ministry of Agriculture and Farmers Welfare. 2021. 2023 International Year of Millets. *Ministry of Agriculture and Farmers Welfare*, Govt. of India. https://nutricereals.dac.gov.in/# (accessed January 20, 2022).

Mohamed, E. A., I. A. M. Ahmed, A. E. A. Yagoub, and E. E. Babiker. 2010. Effects of radiation process on total protein and amino acids composition of raw and processed pearl millet flour during storage. *International Journal of Food Science and Technology* 45:906–912.

Mujahid, A., I. Haq, M. Asif, and A. H. Gilani. 2005. Effect of various processing techniques and different levels of antioxidant on stability of rice bran during storage. *Journal of the Science of Food and Agriculture* 85:847–852.

Mustač, N. C., D. Novotni, M. Habuš, et al. 2020. Storage stability, micronisation, and application of nutrient-dense fraction of proso millet bran in gluten-free bread. *Journal of Cereal Science* 91:102864. DOI:10.1016/j.jcs.2019.102864.

Nidhi, B., and Indira, V. 2012. Effect of storage on nutritional and sensory qualities of grain amaranth (*Amaranthus hypochondriacus*) flour. *Asian Journal of Dairying and Foods Research* 31:297–300.

Patil, J. V. 2016. *Millets and Sorghum: Biology and Genetic Improvement*. New York: John Wiley.

Raghavender, C. R., B. N. Reddy, and G. Shobharani. 2007. Aflatoxin contamination of pearl millet during field and storage conditions with reference to stage of grain maturation and insect damage. *Mycotoxin Research* 23:199–209.

Rahman, R. Y. A., M. I. Abdalla, and A. H. R. Ahmed. 2020. Effect of using gamma radiation on storability and sensory acceptability of Sudanese sorghum. *International Journal for Research in Applied Sciences and Biotechnology* 7:5–11.

Ram, S., S. Narwal, O. P. Gupta, V. Pandey, and G. P. Singh. 2020. Anti-nutritional factors and bioavailability: Approaches, challenges, and opportunities. In: *Wheat and Barley Grain Biofortification*. ed. O. P. Gupta, V. Pandey, S. Narwal, P. Sharma, S. Ram, and G. P. Singh. 101–128. Cambridge, USA: Elsevier.

Rathore, S., and K. Singh. 2018. Application of response surface methodology for optimization study of equilibrium moisture sorption content for efficient drying and storage of pearl millet flour. *Journal of Food Measurement and Characterization* 12:2020–2031.

Richard-Molard, D. 2003. Cereals - Bulk storage of grain. In: *Encyclopedia of Food Sciences and Nutrition*. ed. L. Trugo and P. M. Finglas. 1014–1018 Maryland: Academic Press. DOI: 10.1016/B0-12-227055-X/00187-5.

Saleh, A. S., Q. Zhang, J. Chen, and Q. Shen. 2013. Millet grains: Nutritional quality, processing, and potential health benefits. *Comprehensive Reviews in Food Science and Food Safety* 12:281–295.

Sashidhar, R. B., Y. Ramakrishna, and R. V. Bhat. 1992. Moulds and mycotoxins in sorghum stored in traditional containers in India. *Journal of Stored Products Research* 28:257–260.

Schoenlechner, R., M. Szatmari, A. Bagdi, and S. Tomoskozi. 2013. Optimisation of bread quality produced from wheat and proso millet (*Panicum miliaceum* L.) by adding emulsifiers, transglutaminase and xylanase. *LWT–Food Science and Technology* 51:361–366.

Seema, B. R., K. P. Sudheer, N. Ranasalva, T. Vimitha, and K. B. Sankalpa. 2016. *Effect of Storage on Cooking Qualities of Millet Fortified Pasta Products*. 5:6658–6662.

Sharma, B. and H. S. Gujral. 2019. Modulation in quality attributes of dough and starch digestibility of unleavened flat bread on replacing wheat flour with different minor millet flours. *International Journal of Biological Macromolecules* 141:117–124.

Sharma, R., S. Sharma, B. N. Dar, and B. Singh. 2021. Millets as potential nutri-cereals: A review of nutrient composition, phytochemical profile and techno-functionality. *International Journal of Food Science and Technology* 56:3703–3718. DOI: 10.1111/ijfs.15044.

Slama, A., A. Cherif, F. Sakouhi, S. Boukhchina, and L. Radhouane, 2019. Fatty acids, phytochemical composition and antioxidant potential of pearl millet oil. *Journal of Consumer Protection and Food Safety*, 15:145–151.

Sruthi, N. U. and P. S. Rao. 2021. Effect of processing on storage stability of millet flour: A review. *Trends in Food Science and Technology* 112:58–74.

Stankevych, G., N. Valentiuk, L. Ovsiannykova, and D. Zhygunov. 2021. Changes in quality of amaranth grain in the course of postharvest handling and storage. *Food Science and Technology* 15:80–90.

Taylor, J. R. N. and K. G. Duodu. 2017. Chapter 13 – Sorghum and millets: Grain-quality characteristics and management of quality requirements. In: Cereal *Grains: Assessing and Managing Quality*. Ed. C. Wrigley, I. Batey and D. Miskelly. 317–351. Cambridge: Woodhead Publishing. DOI:10.1016/B978-0-08-100719-8.00013-9.

The Hindu. 2018. Centre to declare 2018 as "national year of millets." *The Hindu Group*, Chennai, Tamil Nadu, India. www.thehindu.com/news/cities/bangalore/centre-to-declare-2018-as-national-year-of-millets/article22478125.ece (accessed January 20, 2022).

The Hindu. 2021. UNGA unanimously adopts India-sponsored resolution declaring 2023 as International Year of Millets. *The Hindu Group*, Chennai, Tamil Nadu, India. www.thehindu.com/news/international/unga-unanimously-adopts-india-sponsored-resolution-declaring-2023-as-international-year-of-millets/article33986118.ece (accessed January 20, 2022).

Tiwari, Ajita, S. K. Jha, R. K. Pal, S. Sethi, and L. Krishan. 2014. Effect of pre-milling treatments on storage stability of pearl millet flour. *Journal of Food Processing and Preservation* 38:s1215–1223.

Tömösközi, S. and B. Langó. 2017. Buckwheat: Its Unique Nutritional and Health-Promoting Attributes. In: *Gluten-Free Ancient Grains: Cereals, Pseudocereals and Legumes.* ed. J. R. N. Taylor and J. M. Awika. Cambridge: Woodhead Publishing. 161–177.

Vali Pasha, K., C. V. Ratnavathi, J. Ajani, D. Raju, M. S. Kumar, and S. R. Beedu. 2018. Proximate, mineral composition and antioxidant activity of traditional small millets cultivated and consumed in Rayalaseema region of south India. *Journal of the Science of Food and Agriculture* 98:652–660.

Varsha, R. and A. Narayanan. 2017. Storage stability of bio fortified pearl millet flour. *International Journal of Agriculture Innovations and Research* 5:2319–1473.

Wang, R. C., Y. J. Chen, J. H. Ren, and S. T. Guo. 2014. Aroma stability of millet powder during storage and effects of cooking methods and antioxidant treatment. *Cereal Chemistry* 91:262–269.

Yadav, D. N., J. Kaur, T. Anand, and A. K. Singh. 2012. Storage stability and pasting properties of hydrothermally treated pearl millet flour. *International Journal of Food Science and Technology* 47:2532–2537.

Yao, L. H., J. You-Ming, and F. A. John Shi, et al. 2004. Flavonoids in food and their health benefits. *Plant Foods for Human Nutrition* 59:113–122.

Economic, Social, and Market Feasibility of Nutri-Cereals

Rachna Gupta, Himani Singh, R. Prasanth Kumar, and Murlidhar Meghwal

Contents

16.1 Introduction

In recent years, the public has increasingly become more interested in food not only because of its nutritional value but also due to its functional role to prevent diseases. Currently, the consumption of cereals around the world is increasing, and cereals are used as the main food source. However, there is a high demand for mainstream/tropical grains such as rice, wheat, corn and pigeon peas despite their high trade volume in local and export markets. In addition, economic constraints make it impossible for them to be consumed by local/rural people. In addition, a type of nutri-cereals crop termed "underutilized/neglected" has many hidden potentials in occupying a niche among the local ecology, production and consumption system. Despite their practical uses, they are still under-characterized and ignored by research. Research on these underutilized/neglected species provides ecosystem sustainability, protection and diversification, economic empowerment of rural towns and nutrition and health security, the greater needs of today's world (Gopalan, Rama Sastri and Balasubramanian, 2009; McDonough, Rooney and Zaldivar, 2000). These kinds of cereals, with abundant essential macro and trace elements, and non-nutritive biological activity/secondary metabolites/phytochemicals are important in the food system. Wheat and rice are the main grains in the world. The cultivation of wheat and rice requires specific types of environmental conditions so there are different growing areas. Unlike rice and wheat, millet can grow almost anywhere. The Indian government named millets "nutri-cereals" in 2018. Nutri-cereals once had a place in Indian cuisine and, over time, with the increasing popularity of crops such as wheat and rice they lost their luster. Nutri-cereals have been ignored and relegated to being grown primarily by farmers in tribal areas of India. All the states on the Indian peninsula have a certain amount of nutri-cereals still grown by marginal farmers.

Nutri-cereals, as a whole grain, provides consumers with dietary fiber, minerals, phenols and vitamins. Most of the nutri-cereals contains less cross-linked gliadin – perhaps this is a factor that leads to the higher digestibility of nutri-cereals protein. The cooking and preparation process of nutri-cereals is also similar to that of staple foods. Years of neglect have caused the value-added potential of nutri-cereals to not be fully studied. Its value chain is limited to harvesting, threshing, cleaning, shelling and decortication with nutri-cereals as the final product. Many nutri-cereals grains are very small, their bran layer very thin and the germ is just below. Dehulling causes the grain to be exposed to the air, which leads to challenges in the storage of grains and flour. According to the Food and Agricultural Organization (FAO, 2001), traditional food processing methods – decortications, grinding, sprouting, fermenting, malting and roasting of nutri-cereals – are recommended to inhibit its anti-nutritional properties and improve its quality for consumption. The functional properties of nutri-cereals such as oil and water holding capacity, viscosity, foaming activity, bulk density, expansion force, and so forth, are the basic physical and chemical properties that reveal the intricate relationship between the structure, the molecular composition and the composition of food. ingredients and physical and chemical properties (Ramashia, Gwata, Meddows-Taylor), Anyasi and Jideani, 2017). Nutri-cereals are considered a high-energy nutritious food that helps to solve the problem of malnutrition. They are eaten as flour, rolled into balls, steamed and then eaten as milk porridge (FAO, 2009). Nutri-cereals can be processed and consumed as a traditional local food such as popcorn, porridge, crepes, dosa, pasta, bread and cookies (Adebiyi, Obadiina, Adebo and Kayitesi, 2018 Jalgaonkar and Jha, 2016; Omoba, Taylor and Cork, 2015). In most African countries, nutri-cereals based foods and beverages are the main diet (Amadou, Mahamadou and Le, 2011). Due to the presence of natural biologically active compounds such as those rich in calcium, dietary fiber, polyphenols and proteins in nutri-cereals crops, these compounds have many health benefits associated with whole grains such as lowering plasma cholesterol (Davidson et al., 1991) and decreasing in glycemic index (Amadou, Gounga and Le, 2013; Annor, Tyl, Marcone, Ragaee and Marti, 2017; Izadi, Nasirpour, Izadi and Izadi, 2012; Srikanth and Chen, 2016). Various characteristics of nutria-cereals have been summarized in **Table 16.1**. In the last decade, the agricultural and food industries of India have readjusted their methods to intervene in production technology practices through customized value-added operations for nutri-cereals, thus reviving this traditional Indian crop that has been long neglected. Therefore, these forgotten nutri-cereals are slowly being accepted by the processed food industry. People can find these nutri-cereals in the form of raw materials, semi-finished products and finished products on the shelves of supermarkets and shopping centers. At this time, there are enough opportunities for massive entrepreneurship in the processing and value-added aspects of nutri-cereals.

16.2 Origin and distribution

The word millet is derived from the French word "mille," which means one thousand, and a handful of nutri-cereals contains up to 1,000 grains (Shahidi and Chandrasekara, 2013). Nutri-cereal belongs to the small seed species of annual cereals or cereal crops (Shihii et al., 2011). There are generally seven kinds of nutri-cereals with different colors, shapes, sizes and origins. These grains are the oldest known to mankind and may be the first grains for household use; they are small, round grains that belong to the *Poaceae* family (**Table 16.1**). Due to its nutritional content and climate-friendly agronomy, the time has come to consider nutri-cereals a priority. Millet (finger, foxtail, kodo, little, barnyard and proso) is an ancient crop with a cultivation history of more than five thousand years and is famous for being suitable for arid land, hills and tribes. They hardly need irrigation, mature early, are very suitable for growing under resource-poor conditions and are a good source of food, feed and fodder. As a staple food, nutri-cereal species are widely distributed and cultivated all over the world. Nutri-cereals are likely to grow in Asian and African countries and parts of Europe,. They currently grow in tropical and subtropical regions of the world (Saleh, Zhang, Chen, and Shen, 2013). Nutri-cereals are considered to be the first domesticated grain. They are characterized by being very easy to survive in less fertile soil, be resistant to drought, pests and diseases and exhibit a short growing season ,usually 45 to 60 days (Awika, 2011). Currently, millions of poor people in Africa and Asia eat nutri-cereals as their staple food. In terms of nutri-cereals production, India occupies an important position in world productivity and tradability. For a long time, nutri-cereals crops have been valued for their nutrition and eating quality (Bukhari, Ayub, Ahmad, Mubeen and Waqas, 2011).

16.3 Entrepreneurship opportunities in nutri-cereal processing sector

Nutri-cereals are taking their position on the plate and becoming part of the diet of health-conscious people, not by force but by choice. The demand for nutri-cereals base products is increasing in the urban areas and thus creating ample opportunities for the new market entrant's entrepreneurs. Entrepreneurship development (ED) refers to the process of enhancing entrepreneurial skills and knowledge through structured training and institution-building programs. According to Global Entrepreneurship Monitor (2007), in poorer countries women are key to the development of entrepreneurship in any given society. They also help to eliminate some societal problems through their entrepreneurial activities in informal sectors. Many institutions are involved in entrepreneurial development programs to assist communities to start an enterprise (Ahamed and Julion, 2012). Entrepreneurs face many challenges in establishing, running and managing their enterprises. Besides challenges,

Table 16.1 Different Characteristics of Nutri-Cereals

Nutri-cereals's name	Color, Shape Size, and origin	Annual Production (2016) Tonnes	Major geographical regions	Potential Marketing and entrepreneurship opportunities
Foxtail millet (*Setariaitalica*)	Pale yellow to orange, Ovoid, 2 mm long, China	50.2	It is grown mainly in Andhra Pradesh, Karnataka, Telangana, Rajasthan, Maharashtra, Tamil Nadu, Madhya Pradesh, Uttar Pradesh, and to a small extent in the northeast states of India	Multigrain atta, beverages, instant mix, flaked products, puffed products, cookies, cakes, baked products, extruded products, vermicelli, pasta, noodles, extruded, snacks, extruded
Pearl millet (*Pennisetum glaucum*)	White, grey, pale yellow, brown, or purple, Ovoid 3–4 mm in length, Tropical West Africa	10280	It is extensively grown in Rajasthan, Gujarat and Haryana	Multigrain atta, beverages, instant mix, flaked products, puffed products, cookies, cakes, baked products, extruded products, vermicelli, pasta, noodles, extruded, snacks, extruded
Proso millet (*Panicum miliaceum*)	White cream, yellow, orange Spherical to oval 3 mm long and 2 mm in diameter Central and eastern Asia	20.0	In India it is largely grown in Madhya Pradesh, eastern Uttar Pradesh, Bihar, Tamil Nadu, Maharashtra, Andhra Pradesh and Karnataka.	Multigrain atta, beverages, instant mix, flaked products, puffed products, cookies, cakes, baked products, extruded products, vermicelli, pasta, noodles, extruded, snacks, extruded
Kodo millet (*Paspalum scrobiculatum*)	Blackish brown to dark brown Elliptical to oval 1.2 to 9.5 μm long Mainly in India also in west Africa	84.2	Both kodo and little millet are grown in parts of Madhya Pradesh, Chhattisgarh and Tamil Nadu	Multigrain atta, beverages, instant mix, flaked products, puffed products, cookies, cakes, baked products, extruded products, vermicelli, pasta, noodles, extruded, snacks, extruded
Little millet (*Panicum sumatrense*)	Grey to straw white. Elliptical to oval, 1.8 to 1.9 mm long, Southeast Asia	119.9		Multigrain atta, beverages, instant mix, flaked products, puffed products, cookies, cakes, baked products, extruded products, vermicelli, pasta, noodles, extruded, snacks, extruded
Barnyard millet (*Echinochloacrusgalli*)	White Tiny round 2–3 mm long	151.0	Himalayan hills are the largest producer. Karnataka and Andhra Pradesh	Multigrain atta, beverages, instant mix, flaked products, puffed products, cookies, cakes, baked products, extruded products, vermicelli, pasta, noodles, extruded, snacks, extruded
Finger millet (*Eleusine coracana*)	Light brown to dark brown. Spherical 1–2 mm in diameter East Central Africa (Uganda)	1822.0	More than 60% of finger millet is produced by the state of Karnataka in India	Multigrain atta, beverages, instant mix, flaked products, puffed products, cookies, cakes, baked products, extruded products, vermicelli, pasta, noodles, extruded, snacks, extruded

Crop	Characteristics		Distribution	Products
Sorghum (*Sorghum bicolor*)	Tan, cream and white, rounded and bluntly pointed 4 – 8 mm in diameter, north-eastern quadrant of Africa	4410	It is grown extensively in north-western, western and central India and southern peninsula, with maximum acreage in Maharashtra and Karnataka	Multigrain atta, beverages, instant mix, flaked products, puffed products, cookies, cakes, baked products, extruded products, vermicelli, pasta, noodles, extruded, snacks, extruded
Amaranth (*Amaranthus tricolor*)	Creamish white, red, and black colour, circular or spherical, 1 to 1.5 mm in diameter, Asia, Africa, Australia and Europe	70.7	Kerala, Tamilnadu, Karnataka, Maharashtra, Andhra Pradesh, Telangana	Multigrain atta, beverages, instant mix, flaked products, puffed products, cookies, cakes, baked products, extruded products, vermicelli, pasta, noodles, extruded, snacks, extruded
Wheat (*Triticum aestivum*)	Light black colour, oval shape, 4 to 6 mm in length, China, Western countries, Ethiopia	98.38	Patiala and Jalandhar in Punjab to Vidisha in Madhya Pradesh	Multigrain atta, beverages, instant mix, flaked products, puffed products, cookies, cakes, baked products, extruded products, vermicelli, pasta, noodles, extruded, snacks, extruded

Source: IIMR based on FAO/DES-GOI data.

there are unlimited opportunities for entrepreneurship development. Challenges faced by the entrepreneurs sampled for this chapter are classified as marketing, raw material, manpower, finance, power, procedural formalities and machinery (Pradhan and Nath, 2012). The environments in which entrepreneurial activities are propagated contribute significantly to the development of the private sector. Consequently, favorable conditions constitute the bedrock for the survival, growth and competitiveness of entrepreneurship development (UNESCAP, 2012). Availability and access to adequate and sustainable finance therefore is critical for entrepreneurs and small and medium enterprises (SMEs) since the life cycle of businesses requires varied needs for cash, taking cognizance of the start-up, growth and transition stages of development. The increasing emphasis on the significance of entrepreneurship as a decisive factor for national development has dovetailed into the search, through a wide range of schemes, targeted at hastening the tempo of new business activities in the organized communities (Okpala, 2012). Entrepreneurs identify innovative approaches to seizing an opportunity, mobilizing money and management skills, taking calculated risks to open markets for new products, processes and services. To build and sustain SMEs, the entrepreneur needs access to diverse forms of resources, such as financial capital, human capital and physical capital with each one playing significant but different roles during the life cycle of a new business (Fatoki, 2014).

16.3.1 Nutri-cereal enterprises by category

There are four broad categories of nutri-cereals based enterprises; nutri-cereals processors, food enterprises traders and organic product companies as described below.

16.3.1.1 Nutri-cereals processors

These refer to small millers who de-husk the grain, sort and grade, and polish it. Major nutri-cereals and finger nutri-cereals do not have the husks and so do not need dehulling and only have to be cleaned. Smaller nutri-cereals need specific processing machinery to remove the hulls. These can be processed into fully polished or unpolished grains. The unpolished grains are nutritionally superior to polished grains, as the process of polishing/dehusking bran also removes the nutritional content. The average processing capacity of a miller of unpolished grain is between 2–3 tonnes per day, and investment can be around USD 12,000 or more. There are at least ten millers in South India who mill unpolished rice. The capacity of mills that provide fully polished rice is much higher, at 10–20 tonnes per day; setting up such a mill would require an investment of over USD 244,000. These millers typically sell their products to supermarkets/large distributors who brand the product in their name. There are around 70 such millers in the country.

16.3.1.2 Nutri-cereals food manufacturers and value addition

These enterprises manufacture several products from nutri-cereals rice and grains, including nutri-cereals flour, "ready to cook" and "ready to eat" products. Value addition is the process of taking a raw commodity and changing its form to produce a high-quality end product. The enterprise size is highly variable and can range from home based business to a fully equipped enterprise that has both nutri-cereals milling as well as product manufacturing. These enterprises sell their products in their brand name. Some of them also provide contract manufacturing for other brands.

16.3.2 Why value addition of nutri-cereals is required?

- To meet taste/preferences of the consumers
- Reducing post-harvest losses
- Nutrient enhancement
- Ready to eat (RTE), ready to cook (RTC) – reduces cooking time
- Enhance shelf life and make the product available for a long time
- Diversified nutri-cereals products can solve our food needs as climate changes
- Improve the consumption of nutri-cereals products and help to overcome malnutrition
- Nutri-cereals farmers will have more post-harvest technologies thus enhancing the economic value of nutri-cereals as well as improving the status of farmers

16.3.2.1 Traders

Traders procure grains, products and sell them under their brand name or as a commodity. These include supermarkets and several players across the country. Most of the traders deal in many commodities.

16.3.2.2 Health food/organic product companies

These enterprises sell several organic or health products along with nutri-cereals based products. These include companies such as Sresta Natural Bio products (sells under their brand 24 Mantra) and Southern Health Foods (Manna foods).

16.3.3 Nutri-cereals enterprises can be divided into cottage industry, micro, small and medium enterprises

a. Cottage enterprise

These are typically run by women/single person from their homes. They usually get training via an NGO/government sponsored program and sell their goods in the neighborhood shops/to people known to them. The investment in such enterprises is typically less than USD 1,200.

b. Micro enterprise

Most of the nutri-cereals enterprises fall into this category; these can be a small nutri-cereals mill (under five tonnes per day) or a food enterprise. Average investment is less than USD 24,400. Such enterprises usually have independent premises, and they may also have a retail outlet to sell their goods. These enterprises typically employ up to five people.

c. Small enterprise

These include a nutri-cereals mill or a food enterprise that manufactures several products. Such enterprises have a larger factory (as compared to micro enterprises) with a separate storage area for raw material and grain. These enterprises may also have a vehicle for transporting their goods to market. They typically employ 5–10 people, including one sales person. The investment in such enterprises would be under USD 122,000.

d. Medium enterprises

These refer to health food/organic food companies that sell branded products, including nutri-cereals.

16.3.4 Nutri-cereals business plan for nutri-cereals seed entrepreneurs

The demand for nutri-cereals is increasing, and there is also demand from the farmers' side for good quality nutri-cereals seeds. In India, there is a good presence of quality seeds of sorghum and pearl nutri-cereals. But there is ample scope for the minor nutri-cereals. Good quality minor nutri-cereals seeds will definitely increase the production and productivity of nutri-cereals. Farmers can also join in Government Seed Production Programme for producing quality seeds. There is a huge potential business in the seed sector.

16.3.5 Nutri-cereals aggregators

Aggregating nutri-cereals in the warehouse for supermarkets and the food industry is another nutri-cereals business plan. Large quantities of nutri-cereals can be purchased and kept in the warehouse and can fulfil the needs of the food industry. The specialty of nutri-cereals grains is that they can be stored for years together with husks by following the storage safety principles. There is always a problem at the farmers' level for proper storage of nutri-cereals grains. Farmers prefer to sell their produce just after harvesting. This opportunity can be captured by procuring the nutri-cereals from the farmers for storing in big warehouses.

16.3.6 Nutri-cereals based bio-degradable films for supermarkets and retail industry

In recent years, several investigations focused on the development of eco-friendly, edible, and bio-degradable films using natural resources. Materials that are recognized as safe substances are used, such as lipids, proteins, and polysaccharides, to develop edible film and coatings (Cho et al. 2004). These materials can be consumed and work effectively as a barrier layer between the food and the surrounding environment (Carrillo et al. 2000). The main reason for developing films of nutri-cereals starch is to prevent the changes of taste, color, flavor, and appearance in food products (Chiumarelli, et al.2012). Starch from nutri-cereals may be utilized for various industrial purposes for making food products and starch-based innovative films (Li et al. 2020; Sanderson et al. 2017). Starch is mainly present in pearl nutri-cereals and represents 59 to 80 percent of the endosperm. Barnyard nutri-cereals contain 66 percent starch, in addition to various micronutrients (Kim et al. 2011). Starch containing high amylose content is considered as a raw material for edible films presenting good oxygen barrier properties (Weiwei et al. 2016). Some studies reported that starch-based edible films possess the ability to transfer the water vapor, and for this reason, waxes, lipid additives, and essential oils were used to improve the hydrophobic fraction side of the film (Sanchez et al. 2011). Exploring the potential of nutri-cereals for achieving sustainability in the agricultural and food sector may open a new era for small-scale packaging industries and marginal farmers. It is flexible, transparent, eco-friendly, and has good anti-oxidant and antimicrobial properties. It can be cooked by steaming, baking, and frying.

16.3.7 Nutri-cereals business plan for bakery industry (biscuits and cakes)

Today, we see biscuits in the market that are mostly made of refined cereals (maida), but when made with nutri-cereals, they are a healthy choice. For establishing a nutri-cereals bakery unit, you require three machines: planetary mixer, cookies cutting machines, and a rotary convection oven. People are becoming more health conscious, and when it is time for a birthday celebration, they are preferring cakes made of nutri-cereals. We hear of Mumbai-based chef Natasha Gandhi, founder of the House of Nutri-cereals, who makes beautiful cakes with nutri-cereals that are healthy, gluten-free, vegan, and free from refined sugar. She delivers all over Mumbai.

16.3.8 Nutri-cereals business plan for export industry

India is the largest producer of nutri-cereals, and there is ample demand in the international markets. The potential export markets for Indian nutri-cereals in 2020 were Indonesia, Austria, Netherlands, Philippines, and New Zealand. India's export of nutri-cereals was US $24.5 million in 2017–2018; $25.76 million in 2018–2019; and $21.05 million in 2019–2020 (April to December). According to a report, the North America nutri-cereals market was valued at $841.87 million in 2018 and expected to grow at a compound annual growth rate (CAGR) of 4.0 percent during 2019–2027, to reach $1,12 billion by 2027. Agricultural and Processed Food Products Export Development Authority (APEDA) is taking the initiative for export of nutri-cereals from India. Recently, Ragi and Barnyard nutri-cereals from Uttarakhand were exported to Denmark, and this has created export opportunities in European countries.

16.3.9 Nutri-cereals business plan for the farmer producer organization (FPO)

Creating FPOs will successfully deal with the constraints and challenges faced by the farmers. This will give strength and bargaining power to access financial and non-financial inputs, services, and technologies. This will help farmers in generating more revenue from the collective sale of nutri-cereals grains. Under the Odisha nutri-cereals mission, FPOs are created at the block level and are involved in the procurement of Ragi for PDS and ICDS. For this, one can get lot of support from Government.

16.3.10 Creating e-commerce online platform on nutri-cereals products

Now, people are preferring online platforms to buy their household items, including groceries. Online sales are increasing at a faster rate in India. Establishing an e-Commerce platform for nutri-cereals products will create more business as it will reach major customers. What is needed is to build a good website and add all ranges of nutri-cereals products.

16.4 Precautions of nutri-cereals

Some of the common precautions for use of nutri-cereals:

- **They Are Not Recommended for People with Thyroid Issues**
- Ragi generally contains goitrogenic compounds that are bad for hypothyroidism so it is better to avoid this kind of food interference with thyroid diseases.
- **Over Consumption of Ragi**

- In ragi disadvantages, which is the crucial one because overconsumption of grains like ragi, nutri-cereals, barley, and others may cause small intestinal damage. Which further leads to health problems like bloating, gas, diarrhea, and constipation.
- **Digestive Issues Due to higher consumption**
- For a healthy weight, people avoid ragi in cold seasons, which is the best choice because it reduces body heat and raises digestive problems like suppressing appetite, bloating, indigestion.
- **Constipation in Some People**
- Ragi causes constipation in some babies and others. It happens when you arre not drinking a sufficient amount of water because of its huge fiber content.
- **Not Good for Kidney Problems**
- Higher consumption of ragi increases the oxalic acidity in the body, so try to avoid this kind of food with patients who suffer from kidney stones and unary issues.
- **Do not Eat too Much for Weight Gain**
- People who want to gain weight then need to reduce the ragi-related recipes because it is gluten-free, which is why it is not recommended for weight gaining. Gluten is an essential component in gaining weight and abdominal fat.
- **Less Consumption for Low Appetite Individuals**
- For people who are struggling with appetite levels, it is better to avoid this food, because it acts as a hunger suppressant due to its high fiber.
- **Not Good in Cool Weather**
- Ragi provides a cooling effect in the body, so it becomes a major disadvantage of ragi when it comes to colder weather conditions. When the weather is cool or raining, ragi is not recommended for eating in any form.

16.5 Social feasibility

Nutri-cereals consumption could address the problem of undernourishment. The reason for declining nutri-cereals intake is affordable food grains (rice and wheat), which may provide over 50 percent of calories of an Indian household (Umanath, M., Balasubramaniam, R., and Paramasivam, R., 2018), which often has increased in per capita income, with growing urbanization, and changing tastes and preferences (Chand, 2007). For example, processing of nutri-cereals for household consumption is tedious and time consuming. There seems, recently, to be a notable shift in the dietary pattern of households, away from cereals to high value food commodities such as livestock products, fruits, vegetables and beverages (Kumar et al. 2011; Chatterjee et al. 2006; Radhakrishna, 2005). However, nutri-cereals are among the ancient food grains, and now they are accessible in modern forms of food products, ready to cook (RTC) and ready to eat (RTE) due to healthier food habits for health, nutrition and fitness (Pravallika at. al, 2020).

16.6 Economic feasibility

The economic viability of nutri-cereals production and consumption must be justified using social, economic, and environmental dimensions. There is little literature available on the economic analysis of nutri-cereals consumption and its determinants (Green and Park, 1998; Jones and Akbay, 2000; Hsu and Kao, 2001; Schmit et al. 2002; Hatirli et al. 2004; Pazarlioglu et al. 2007). Owing to easy accessibility to irrigation and markets for high-value crops, the nutri-cereals crops have been slowly and steadily replaced by high-value crops (Chandrakanth and

Akarsha, 2011). The transition in food consumption pattern is influenced by the availability of expected fresh and processed food products in the market, improved logistics and warehouse facilities and the rise of supermarkets (Kumar et al. 2011; Vasileska and Rechkoska, 2012; Chengappa et al. 2007). The household characteristics, such as prices of foods, disposable income of households, age, education level and gender of the head of household, asset position, household size and location of the dwelling (Quah and Tan, 2009; Pazarlioglu et al. 2007) influence the purchase of nutri-cereals. The determinants of nutri-cereals production being treated as inferior food, low to negative income elasticity of demand and positive price elasticity.

16.7 Production and market value chain

Nutri-cereals constitute the sixth most productive grain in the world. Total world nutri-cereals production in 2018 is estimated at 31,019,370 tonnes; however, India is the largest producer of nutri-cereals, followed by Niger, Sudan, and other countries. The production data have been summarized in Table 16.2. It can be estimated that more than 96 percent of nutri-cereals crops are grown in Africa and Asia because they are different from other grains and are suitable for agroclimatic conditions that favor nutri-cereals growth. Global nutri-cereals consumption is declining at a rate of nearly 1 percent, and positive changes are expected during 2019–2024 (Anbukkani et al., 2017). In the past two decades, the importance of nutri-cereals as a staple food in India (Michaelraj and Shanmugam, 2013) and globally has been declining due to supply and demand factors such as income increase, urbanization, and political government (King, 2017). Currently, more than 50 percent of nutri-cereals produced is used for alternative purposes, not just consumed as a staple food (Gowri et al., 2011). More than 98 percent of global nutri-cereals production is contributed by Asia and Africa. The contribution of Asian countries in global nutri-cereals production has observed a general increase from 48.72 percent to 52.25 percent. A drastic increase in world sorghum production was observed in the year 2014, from 60 million tonnes to 68.9 million tonnes.

Table 16.2 Trend in Area Production and Yield of Nutri-Cereals in India (1950–1951 to 2018–2019)

Year	Area (000 ha)	Production (000 ton)	Productivity (kg/ha)
1950-51 to 1954-55	5144	2113	409
1955-56 to 1959-60	5098	1987	389
1960-61 to 1964-65	4755	1960	413
1960-61 to 1969-70	4697	1697	361
1970-71 to 1974-75	4512	1758	389
1975-76 to 1979-80	4465	1813	405
1980-81 to 1984-85	3623	1462	403
1985-86 to 1989-90	2895	1204	417
1990-91 to 1994-95	2040	931	456
1995-96 to 1999-2000	1540	688	447
2001-05 to 2004-05	1246	533	428
2005-06 to 2009-10	970	466	480
2011-12 to 2014-15	725	429	596
2015-16 to 2018-19	623	401	655
CCG	-16.21	-13.58	3.23

Source: Gowri et al., 2020.

Nutrient to nutrient, every grain of nutri-cereals is much better than rice and wheat, so it is a solution to the malnutrition that affects the huge population of India. However, the cultivation of these nutri-cereals is now facing many restrictions and limitations resulting in a reduction in the planting area of these crops, a high yield gap (Gowri et al., 2017), and a low priority on the agenda, therefore, the technological advancement for the growth of these crops is required. In addition, public and private investment is limited to the development and production of nutri-cereals seeds (Pray and Latha, 2009). The international price of nutri-cereals fluctuates greatly, mainly determined by supply, and usually has nothing to do with other major coarse grains such as corn, sorghum or barley. Due to its nutritional content, any improvement or development of nutri-cereals in planting, supply, storage, price and processing technology may make a significant contribution to the food and nutrition security of the Indian population. In addition, these nutri-cereals helps to enrich our food basket, which is currently very limited due to excessive dependence on the main cereals such as rice and wheat. The area, yield and yield trends of nutri-cereals from 1950–1951 to 2018–2019 are shown in Table 16.2. It can be seen that the nutri-cereals planting area in India decreased from 5,144,000 hectares (1950–1955) to 623,000 hectares (2015–2019) and the nutri-cereals area decreased by 16.21 percent; similarly, nutri-cereals production increased from 21.13 lakh tons to 4.01 thousand tons, with an annual loss of 13.58 percent. The productivity of Indian nutri-cereals was declining until 2005 and then showed an upward trend. This is significantly lower than the 37 million hectares reported before the pre-Green Revolution era (1965–1966). This decrease is mainly due to the low production of nutri-cereals and greater dependence on the main cereals as food. To feed and maintain industrial demand, these crops are continuously grown in various countries. Due to low profitability, farmers mainly grow pearl nutri-cereals, sorghum, finger nutri-cereals and other important nutri-cereals, resulting in low yields of other nutri-cereals varieties.

Markets for minor nutri-cereals are mostly unorganized (Hayenga, 1979). The nutri-cereals market is imperfect and shallow in nature, predominantly during surplus production. In the absence of formal trading structures, local markets may presume greater importance as a potential source for both seed (especially for locally adapted, farmer varieties) and grains (Nagarajan et al., 2010). In turn, nutri-cereals producers received much lower prices as compared to the nutri-cereals sold at the retail stores (Nagarajan et al. 2005). Although the human consumption demand is likely to play a major role in nutri-cereals production, the information asymmetry persists. Such information is essential for producers and market players to take market-oriented production decisions (Lapar et al. 2010). The information on households' diet is crucial for policy makers and other stakeholders involved in production, marketing and processing of nutri-cereals. Nutri-cereals cultivation fascinated farmers due to the low investment requirements such as water, fertilizer, pesticide and so forth, accompanied by good income potential and market access. Besides, an increase in MSP (Minimum Support Price) of the nutri-cereals can also be considered to have a positive impact on farmers' cropping pattern decisions towards nutri-cereals (Pravallika at. al, 2020).

16.8 Future prospects

1. **Strengthen breeding programs:** It is necessary to increase research efforts in the collection and improvement of traditional varieties to provide improved seeds with high yield potential and other desirable characteristics (such as drought tolerance). Variety/hybrid development is best carried out with the active participation of farmers

(participatory nutri-cereals breeding) to make full use of local resources and local knowledge. Traditional breeding and market-based bio-enhanced breeding, combined with other characteristics, such as better root structure.

2. **Seed production:** Take effective seed production measures together with public and private associations and non-governmental organizations, and establish seed villages to ensure a sufficient quantity of high-quality seeds.

3. **Improving agricultural technology:** Work needs to be done at the same time to formulate specific management measures to take full advantage of the potential of newly developed varieties. A no-tillage strategy needs to be implemented in fallow paddies. It would be beneficial to verify and improve indigenous technologies in various agricultural situations.

4. **Processing facilities:** Develop small processing devices and packaging facilities suitable for shelling and pulverizing of nutri-cereals.

5. **Added value:** Processing technology for the production of instant or instant products, and secondary and tertiary processing technology for the production of instant value-added products.

6. **Government support:** Given the poor financial situation of nutri-cereals producers, it is necessary to support producers with subsidies and loans for various agricultural operations. Because of the severe decline in production due to climate change, better purchase prices and insurance coverage can alleviate the difficulties for nutri-cereals producers.

7. **Post-harvest technology:** Research on better post-harvest management, especially to improve the shelf life of nutri-cereals and prevent waste.

8. **Verification of health benefits:** Research can be conducted as a consortium to verify the benefits of nutri-cereals as a health and functional food.

9. **Marketing:** Use modern and innovative methods and village-level marketing and warehousing facilities for market and business development.

10. **Public awareness:** Appropriate advertising strategies need to be adopted to raise awareness of the nutritional advantages and health benefits of nutri-cereals.

16.9 Conclusion

Nutri-cereals are a staple food in many communities in India and abroad. Due to low productivity and short post-processing shelf life, their importance in the human diet has decreased, but now they are gaining again due to nutritional advantages and the availability of proper processing technologies. Whole grains, flour, fermenting, and baking are some popular forms of nutri-cereals, whether eaten alone or fortified. Although different studies have demonstrated the use of nutri-cereals in food, the recommendations of these studies have not been accepted by the competent authorities and are rarely adopted, which reduces their general acceptance. As the global demand to feed a growing population increases, it is imperative to explore food crops grown locally and consumed by low-income families in developing countries. Looking ahead, the agricultural productivity of this crop must be increased substantially to reduce costs, making it more advantageous and attractive compared to other grains. The application of industrial processing technology using modern equipment and optimized conditions will increase the commercial scale production of diverse and high-quality foods. This will further increase the productivity of nutri-cereals and make a major contribution and promotion to India's economic development and food.

References

Ahmed Zafer, U. and Julion Carign, C. (2012) International Entrepreneurship in Lebanon, *Global Business Review*, Sage Publications, 13(1), 25–38.

Amadou, I., Gounga, M. E., & Le, G. W. (2013). Millets: Nutritional composition, some health benefits and processing – A review. *Emirates Journal of Food and Agriculture*, 25(7), 501–508.

Amadou, I., Mahamadou, E. G., & Le, G. (2011). Millet-based traditional processed foods and beverages – A review. *Cereal Food World*, 56(3), 115–121. https://doi.org/10.1094/CFW-56-3-0115

Anbukkani, P., Balaji, S. J., & Nithyashree, M. L. (2017). Production and consumption of minor millets in India-A structural break analysis. Ann. Agric. Res. New Series, 38(4), 1–8.

Annor, G. A., Tyl, C., Marcone, M., Ragaee, S., & Marti, A. (2017). Why do millets have slower starch and protein digestibility than other cereals? *Trends in Food Science and Technology*, 66, 73–83. https://doi.org/10.1016/j.tifs.2017.05.012

Awika, J. M. (2011). Major cereal grains production and use around the world. In *Advances in Cereal Science: Implications to food processing and health promotion* (pp. 1–13). Washington, DC: American Chemical Society.

Bukhari, M. A., Ayub, M., Ahmad, R., Mubeen, K., & Waqas, R. (2011). Impact of different harvesting intervals on growth, forage yield and quality of three pearl millet (Pennisetum americanum L.) Cultivars. *International Journal for Agro Veterinary and Medical Sciences*, 5(3), 307–315. https://doi.org/10.5455/ijavms.20110619114620

Carrillo-Garcia, A., Bashan, Y., & Bethlenfalvay, G. J. (2000). Resource-island soils and the survival of the giant cactus, cardon, of Baja California Sur. *Plant and Soil*, 218, 207–214.

Chand, R. (2007). Demand for food grains. *Economic & Political Weekly*, 42(52): 10–13.

Chandrakanth, M. G., and Akarsha, B. M. (2011). Green development for sustainable agriculture. *FKCCI Journal*, Karnataka.

Chatterjee, S., Rae, A. and Ray, R. (2006). Food consumption, trade reforms and trade patterns in contemporary India: How do Australia and NZ fit in?. In: Conference Paper, Massy University, New Zealand: Department of Applied and International Economics.

Chengappa, P. G., Achoth, L., Mukherjee, A., Reddy, B. R., Ravi, P. C. and Dega, V. (2007). *Evolution of Food Retail Chains in India. Agricultural Diversification and Smallholders in South Asia*. New Delhi, Academic Foundation.

Chiumarelli, M.; Hubinger, M. D. (**2012).** Stability, solubility, mechanical and barrier properties of cassava starch–Carnauba wax edible coatings to preserve fresh-cut apples. *Food Hydrocoll*, 28, 59–67.

Cho, S. Y.; Rhee, C. (2004). Mechanical properties and water vapor permeability of edible films made from fractionated soy proteins with ultrafiltration. *L.W.T. Food Science and Technology*, 37, 833–839.

Davidsson, L., Cederblad, A., Lonnerdal, B. O., Sandstrom, B. (1991). The effect of individual dietary components on manganese absorption in humans. *The American Journal of Clinical Nutrition*, 54(6), 1065–1070.

FAO. (2001). Post Harvest Operations. Millet: Kajuna. www.fao.org/fileadmin/user_upload/inpho/docs/Post-Harvest-Compendium

FAO. (2009). FAOSTAT. Food and Agriculture Organisation of the United Nations. FAOSTAT. Retrieved from http://faostat.fao.org/site/339/ default.aspx

Fatoki, O. (2014). The financing options for new small and medium enterprises in South Africa. *Mediterranean Journal of Social Sciences*, 5(20), 748.

Gaertn] based extrudates. *Global Science Research Journals*, 3(6), 239–246. Retrieved from www.globalsciencerresearchjournals.org/gjas/877382015392.pdf

Garcia, M. A.; Martino, M. N.; Zaritzky, N. E. (2000). Lipid addition to improve barrier properties of edible starch-based films and coatings. *Journal of Food Science*, 65, 941–944.

Global Entrepreneurship Monitor. (2007). *2006 Report on Women and Entrepreneurship*. Centre for Women Leadership.

Gopalan, C., Rama Sastri, B. V., & Balasubramanian, S. C. (2009). *Nutritive Value of Indian Foods*. Hyderabad, India: National Institute of Nutrition, Indian Council of Medical Research.

Gowri, M. U., & Prabhu, R. (2017). Millet production and its scope for revival in India with special reference to Tamil Nadu. *International Journal of Farm Sciences*, 7(2), 88–93.

Gowri, M. U., & Shivakumar, K. M. (2020). Millet scenario in India. *Economic Affairs*, 65(3), 363–370.

Gowri, S., Devi, T. U., Sajan, D., Bheeter, S. R., & Lawrence, N. (2011). Spectral, thermal and optical properties of L-tryptophanium picrate: a nonlinear optical single crystal. *Spectrochimica Acta Part A: Molecular and Biomolecular Spectroscopy*, 81(1), 257–260.

Green, G. M. & Park, J. L. (1998). Retail demand for whole vs. low-fat milk: new perspectives on loss leader pricing. In: *Annual Meeting of the American Agricultural Economics Association Held at Salt Lake City, UT.*

Hatirli, S. A., Ozkan, B., Aktas, A. R. (2004). Factors affecting fluid milk purchasing sources in Turkey. *Food Quality and Preference*, 15(6): 509–515.

Hayenga, M. L. (ed.) (1979). Pricing problems in the food industry (with emphasis on thin markets), *Monograph No. 7, N.C. Project 117.*

Hsu J. L. and Kao, J. S. (2001). Factors affecting consumers' fluid milk purchasing patterns in Taiwan: product comparisons and marketing implications. *Journal of Food Products Marketing*, 7(3): 41–51. https://eands.dacnet.nic.in/PDF/At%20a%20Glance%202019%20Eng.pdf

Izadi, Z., Nasirpour, A., Izadi, M., & Izadi, T. (2012). Reducing blood cholesterol by a healthy diet. *International Food Research Journal*, 19(1), 29–37.

Jalgaonkar, K., & Jha, S. (2016). Influence of particle size and blend composition on quality of wheat semolina-pearl millet pasta. *Journal of Cereal Science*, 71, 239–245. https://doi.org/10.1016/j. jcs.2016.09.007

Joint FAO/WHO Expert Committee on Food Additives. Meeting, & World Health Organization. (2001). Safety evaluation of certain mycotoxins in food (No. 74). Food & Agriculture Org.

Jones, E. and Akbay, C. (2000). An analysis of consumers' purchasing behavior for high-and low-fat milk: A focus on healthy drinking. *Journal of Food Distribution Research*, 31(1): 124–131.

King, G., Pan, J., & Roberts, M. E. (2017). How the Chinese government fabricates social media posts for strategic distraction, not engaged argument. *American Political Science Review*, 111(3), 484–501.

Kumar, P., Kumar, A., Parappurathu, S. and Raju, S. S. (2011). Estimation of demand elasticity for food commodities in India. *Agricultural Economics Research Review*, 24(1): 1–14.

Lapar, M..L..A., Choubey, M., Patwari, P., Kumar, A., Baltenweck, I., Jabbar, M. A. and Staal, S. (2010). Consumer preferences for attributes of raw and powdered milk in Assam, Northeast India. *ILRI*, pp. 103.

Li, K., Zhang, T., Narayanamoorthy, S., Jin, C., Sui, Z., Li, Z., Li, S., Wu, K., Liu, G., Corke, H. (2020). Diversity analysis of starch physicochemical properties in 95 proso millet (Panicum miliaceum L.) accessions. *Food Chem.*, 324, 126863.

McDonough, C. M., Rooney, L. W., & Saldivar, S. (2000). The millets. In K. Kulpand& J. G. Ponte, Jr. (Eds.), *Handbook of Cereal Science and Technology* (pp. 177–195), New York: Marcel Dekker.

Michaelraj, P.S.J. and Shanmugam, A.A., Study on millets based cultivation and consumption in India, *International Journal of Marketing, Financial Services and Management Research*, 2013, 2, 49–58.

Nagarajan, L., E. D. I. Oliver King, M. Smale, and T. J. Dalton (2010). Access to minor millet genetic resources in rural market towns of Dharmapuri District, Tamil Nadu, India, in L. Lipper, C.L. Anderson and T.J. Dalton (eds), *Seed Trade in Rural Markets, Rome: FAO and London*: Earthscan, pp. 125–145.

Nagarajan, Latha, Smale, Melinda and Glewwe, Paul (2005). Comparing farm and village-level determinants of millet diversity in marginal environments of India: the context of seed systems. Discussion paper 139; International Food Policy Research Institute. Washington D C.

Okpala, K. E. (2012). Venture capital and the emergence and development of entrepreneurship: A focus on employment generation and poverty alleviation in Lagos state. *International Business and Management*, 5(2), pp: 134–141.

Omoba, O. S., Taylor, J., & Kock, H. L. (2015). Sensory and nutritive profiles of biscuits from whole grain sorghum and pearl millet plus soya flour with and without sourdough fermentation. *International Journal of Food Science and Technology*, 50(12), 2554–2561. https:// doi.org/10.1111/ijfs.12923.

Pazarlioglu, M.V., Miran, B., Üçdoğruk, S. and Abay, C. (2007). Using econometric modelling to predict demand for fluid and farm milk: A case study from Turkey. *Food Quality and Preference*, 18(2): 416–424.

Pradhan, R. K. and Nath, P. (2012), "Perception of Entrepreneurial Orientation and Emotional Intelligence: A study on India's Future Techno-Managers" *Global Business Review,* Sage Publications, Vol. 13(1) pp 89.

Pravallika, D. R., Rao, B. D., Chary, D. S., & Devi, N. S. (2020). Market Strategies for Promotion of Millets: A Critical Analysis on Assessment of Market Potential of Ready to Eat (RTE) and Ready to Cook (RTC) Millet Based Products in Hyderabad. *Asian Journal of Agricultural Extension, Economics & Sociology*, 147–155.

Pray, C. E., & Latha, N. (2009). Improving crops for arid lands: pearl millet and sorghum in India. Millions fed: Proven successes in agricultural development, 83–88.

Quah, S. H. and Tan, A. K. (2009). A Heckman sample selection approach to the demand for organic food products: An exploratory study using Penang data. *Journal of Food Products Marketing*, 15(4): 406–419.

Radhakrishna, R. (2005). Food and nutrition security of the poor: emerging perspectives and policy issues. *Economic and Political Weekly*, 40(18): 1817–21.

Ramashia, S. E., Gwata, E. T., Meddows-Taylor, S., Anyasi, T. A., & Jideani, A. I. O. (2017). Some physical and functional properties of finger millet (Eleusine coracana) obtained in sub-Saharan Africa. *Food Research International*, 104, 110–118. https://doi.org/10.1016/j. foodres.2017.09.065

Rani, S., Singh, R., Sehrawat, R., Kaur, B. P., & Upadhyay, A. (2018). Pearl millet processing: a review. *Nutrition & Food Science*, 48(1), 30–44.

Saleh, S. M., Zhang, Q., Chen, J., & Shen, Q. (2013). Millet grains, nutritional quality, processing and potential health benefits. *Comprehensive Reviews in Food Science and Technology*, 12(3), 281–295. http://dx.doi.org/10.1111/1541-4337.12012

Sanchez-Gonzalez, L.; Vargas, M.; Gonzalez-Martinez, C.; Chiralt, A.; Chafer, M. (2011). Use of essential oils in bioactive edible coatings: A review. *Food Eng Rev*, 3, 1–16.

Sanderson, E.; Duizer, L. M.; McSweeney, M. B. (2017). Descriptive analysis of a new proso millet product. *International Journal of Gastronomy and Food Science*, 8, 14–18.

Schmit, T. M., Dong, D., Chung, C., Kaiser, H. M. and Gould, B. W. (2002). Identifying the effects of generic advertising on the household demand for fluid milk and cheese: a two-step panel data approach. *Journal of Agricultural and Resource Economics*, 27(1): 165–186.

Shahidi, F., & Chandrasekara, A. (2013). Millet grain phenolics and their role in diseases risk reduction and health promotion – review. *Journal of Functional Foods*, 5(2), 570–581. http://dx.doi.org/10.1016/j.jff.2013.02.004.

Shiihii, S. U., Musa, H., Bhati, P. G., & Martins, E. (2011). Evaluation of physicochemical properties of Eleusine coracana starch. *Nigerian Journal of Pharmaceutical Sciences*, 10(1), 91–102.

Umanath, M., Balasubramaniam, R., & Paramasivam, R. (2018). Millets' consumption probability and demand in India: an application of Heckman sample selection model. *Economic Affairs*, 63(4), 1033–1044.

UNESCAP (United Nations Economic and Social Commission for Asia and Pacific). (2012). *Policy Guidebook for SME Development in Asia and the Pacific*. Bangkok. United Nations.

Vasileska, A. and Rechkoska, G. 2012. Global and regional food consumption patterns and trends. *Procedia-Social and Behavioral Sciences*, 44: 363–369.

Weiwei, L.; Juan, X.; Beijiu, C.; Suwen, Z.; Qing, M.; Huan, M. (2016). Anaerobic biodegradation, physical and structural properties of normal and high-amylose maize starch films. *International Journal of Agricultural and Biological Engineering*, 9, 184–193.

Index

For Product Safety Concerns and Information please contact our EU
representative GPSR@taylorandfrancis.com
Taylor & Francis Verlag GmbH, Kaufingerstraße 24, 80331 München, Germany

www.ingramcontent.com/pod-product-compliance
Lightning Source LLC
Chambersburg PA
CBHW080706220326
41598CB00033B/5322

9 781032 169842